Basic Applied Mathematics for the Physical Sciences

Based on the Latest Syllabus of the University of Delhi

Third Updated Edition

R. D. Sarma

Associate Professor, Department of Mathematics
Rajdhani College, University of Delhi

Umesh Kumar

Assistant Professor, Department of Mathematics
Rajdhani College, University of Delhi

To
my parents,
Sh. Manu Dev Sarma and Smt. Savitri Devi

A humble token of gratitude from their 'Naman Bapa'

R. D. Sarma

To
my parents and other family members

Umesh Kumar

Published by Dorling Kindersley (India) Pvt. Ltd., licensees of Pearson Education in South Asia.

ISBN: 978-81-317-8782-3

First Impression, 2013
Thirteenth Impression, 2022
Fourteenth Impression

Published by Pearson India Education Services Pvt.Ltd., CIN: U72200TN2005PTC057128.

Head Office: World Trade Tower, 15th Floor, Plot No: C – 01, Sector – 16, Noida 201301 U.P., India.

Registered Office: 7th Floor, SDB2, ODC 7, 8 & 9, Survey No.01 ELCOT IT/ ITES SEZ,
Sholinganallur,
Chennai – 600119, Tamil Nadu, India. Website: in.pearson.com, Email:
companysecretary.india@pearson.com

Compositor: White Lotus Infotech Pvt. Ltd.

Printed In India : India Binding House, Noida

CONTENTS

Preface to the Third Updated Edition vii

Preface viii

Syllabus xi

PART I

Chapter 1 Matrices **1–40**

1.1 Introduction to Matrices 3
- 1.1.1 Types of Matrices 5
- 1.1.2 Operations on Matrices 6
- 1.1.3 Properties of Matrix Addition 6
- 1.1.4 Properties of Scalar Multiplication 7

1.2 Elementary Row Operations 13
1.3 Inverse of a Matrix 18
1.4 Rank of a Matrix 25
1.5 Systems of Linear Equations 30

Chapter 2 Vectors in R^2 and R^3 **41–76**

2.1 Introduction 41
2.2 Vector Operations 45
- 2.2.1 Standard Basis for R^2 and R^3 50

2.3 Linear Combination of Vectors 52
2.4 Linear Independence 56
2.5 Basis and Dimension 61
2.6 Subspaces 67

Chapter 3 Linear Transformations **77–114**

3.1 Introduction 77
3.2 Some Special Transformations 84
- 3.2.1 Projection 84
- 3.2.2 Dilation and Contraction 84
- 3.2.3 Reflection 86
- 3.2.4 Rotation 87
- 3.2.5 Shear 91
- 3.2.6 Translation 92

3.3 Matrix Representation of a Linear Transformation 97

3.3.1 Composition of Two Transformations 105

3.3.2 Geometric Effect of Multiplication by a Matrix 107

Chapter 4 Eigenvalues and Eigenvectors 115–146

4.1 Eigenvalues and Eigenvectors 115

4.2 Eigen Space 129

4.3 Diagonalization 136

PART II

Chapter 5 Sequences 147–182

5.1 What is a Sequence? 149

5.2 Recursion Formula for Sequences 151

5.2.1 Fibonacci Sequence 152

5.2.2 Tower of Hanoi Game (Tower of Brahma) 156

5.3 Difference Equation 158

5.4 Types of Sequences 161

5.4.1 Subsequences 161

5.4.2 Other Important Types of Sequences 161

5.5 Convergent Sequences 165

5.5.1 Convergence of Sequences 165

5.5.2 Algebra of Convergent Sequences 170

5.5.3 More on Convergence of Sequences 172

5.5.4 Some Important Limits 180

Chapter 6 Functions and Their Graphs 183–218

6.1 Introduction 183

6.1.1 Graph of a Function 184

6.1.2 Some Useful Graphs 194

6.1.3 Logarithmic Functions 205

6.1.4 Trigonometric Functions 207

6.1.5 Inverse Trigonometric Functions 212

6.1.6 Hyperbolic Functions 214

Chapter 7 Differential Equations in Mathematical Modelling 219–240

7.1 Introduction 219

7.2 Exponential Growth Model 219

7.3 Exponential Decay Model 225

7.4 Population Growth with Logistic Growth Model 232

Chapter 8 Successive Differentiation 241–260

8.1 Higher Order Derivatives 241
8.2 The *n*th Derivatives of Some Important Functions 247
8.3 Leibnitz's Theorem 253

Chapter 9 Polynomial Approximation of Functions 261–286

9.1 Taylor Polynomial and Maclaurin Polynomial 261
9.2 Polynomial Approximations of Functions and Error Estimation 265
9.3 Series Expansion of Functions 274
9.3.1 Taylor and Maclaurin Series 275
9.4 Convergence of Taylor Series 278

Chapter 10 Functions of Two Variables 287–314

10.1 Functions of Two or More Variables 287
10.1.1 Graphs and Level Curves of Functions of Two Variables 289
10.2 Partial Derivatives 304
10.3 Applications of Partial Derivatives 309
10.3.1 The Wave Equation 309
10.3.2 Laplace's Equation 310
10.3.3 Diffusion Equation 310
10.3.4 Heat Equation 312

PART III

Chapter 11 Geometry of Complex Numbers 315–352

11.1 Introduction 317
11.1.1 Geometrical Representation of Sum, Difference, Product and Quotient of Complex Numbers 322
11.1.2 Point Dividing a Line Segment in a Given Ratio 328
11.1.3 Regarding Angles in the Argand Plane 330
11.2 Straight Lines in Argand Plane 337
11.2.1 General Equation of a Straight Line 337
11.2.2 Parametric Equation of a Straight Line 339
11.2.3 Half-Planes 342
11.3 Circles in the Argand Plane 344
11.3.1 Equation of a Circle with Given Centre and Radius 344
11.3.2 The General Equation of a Circle 346
11.3.3 Equation of a Circle with Given Endpoints of a Diameter 347
11.3.4 Discs in a Plane 348
11.4 Fundamental Theorem of Algebra 350

Chapter 12	**De Moivre's Theorem**	**353–372**
	12.1 De Moivre's Theorem for a Rational Index	353
	12.2 Roots of Complex Numbers and Solution of Equations	362
	Answers to Exercises	**373–394**
	Appendix	**395–400**
	Examination Paper 2008–09	**401–404**
	Examination Paper 2009–10	**405–406**
	Examination Paper 2010–11	**407–410**
	Solution of Examination Paper 2010–11	**411–412**
	Examination Paper 2011–12 (Sem. I)	**413–416**
	Solution of Examination Paper 2011–12 (Sem. I)	**417–418**
	Examination Paper 2011–12 (Sem. II)	**419–422**
	Solution of Examination Paper 2011–12 (Sem. II)	**423–424**
	Index	**425–428**

PREFACE TO THE THIRD UPDATED EDITION

The warm welcome received for the first and second revised edition of our book is an indication that the book is serving its purpose. However, in a fast growing subject like mathematics, complacency has no place, particularly when our universities are changing and improving their syllabi. Therefore, we have come up with the third edition of the book. In the previous edition, we have removed chapters pertaining to statistics as these chapters are no longer a part of the first year syllabus of the B.Sc. (Physical Sciences) course of the University of Delhi under the semester system.

We are thankful to the dedicated team of Pearson Education, the publisher of the book, for the painstaking job they have undertaken in bringing out this edition.

We hope that this updated edition will also be well received by students as well as by faculty members.

R. D. SARMA
UMESH KUMAR

PREFACE

The beginning of the present millenium is marked by explosion of knowledge, particularly in the field of science and technology. Advancement has never been so rapid before, whether in information technology, biotechnology, medical science, space science or any other discipline of science. Inter-disciplinary research and study is playing a major role in this advancement. The undergraduate science syllabi of our universities are being revamped in order to accommodate this change. The University of Delhi has already taken up this challenge by introducing B.Sc. (Physical Sciences) and B.Sc. (Applied Physical Sciences) courses in the year 2005. The syllabi for these courses are radically new and there is a paradigm shift in their content from those of the existing courses. In mathematics, emphasis is laid on its practical applications in physical and biological sciences.

The Mathematics Department, University of Delhi, has organized several workshops for educating the mathematics teachers of the university for this new course. Based on this, we had prepared 'Lecture Notes in Mathematics for Physical and Applied Physical Sciences' in December 2005 to cater to the immediate needs of students. The overwhelming response from our students as well as from teachers has encouraged us to write a proper textbook for these courses. We hope that this book will be able to fulfil the needs of the students pursuing these courses. At the same time, we also hope that the students of other courses as well as of the other universities offering mathematics will also benefit from this book.

We request our readers to provide their valuable suggestions and point out any possible mistakes by contacting us at *ratna_sarma@yahoo.com* so that the quality of the book can be improved further in the future.

Acknowledgements

We take this opportunity to acknowledge the help, encouragement and guidance offered to us, directly or indirectly, by our friends and colleagues. Thanks are due to all our colleagues in the Mathematics Department, Rajdhani College, and Dr Vijay Laxmi Pandit, Principal, Rajdhani College. Dr C. S. Lalitha, Dr Subhash Prabha and Ms Krishna Threja of Rajdhani College deserve a special mention. We offer our sincere thanks to Professor S. C. Arora and Professor Ajay Kumar of the Department of Mathematics, University of Delhi, Dr Naseem Ajmal of Zakir Hussain College, Dr Sonia Davar of St. Stephen's College, Dr Kanta Garg and Dr Seema Khera of Hindu College, Mr Ravi Srivastava of Swami Shraddhanand College, Ms Preeti Garg of Kirori Mal College, Dr Pratibha Mehrotra of ARSD College, and Dr C. M. Upadhye of Gargi College. Mr Rajeev Arora deserves special thanks for the hard work he put in to prepare the electronic manuscript for this book. We thank our publisher Pearson Education and its team of editors for bringing out this book in a brief span of time. Lastly, and above all, thanks to God, for his infinite grace.

R. D. SARMA
UMESH KUMAR

REVIEWERS

The publishers would like to thank the following reviewers for their encouraging feedback towards development of the book:

Harjeet Arora
Hans Raj College, Delhi

Sonia Davar
St. Stephen's College, Delhi

Kanta Garg
Hindu College, Delhi

Renu Khaneja
Maitreyi College, New Delhi

Dharmendra Kumar
Sri Guru Tegh Bahadur Khalsa College (Morning), Delhi

Sadanand Prasad
Acharya Narendra Dev College, New Delhi

K.V. Shantha
Miranda House College, Delhi

Virender Thukral
Deen Dayal Upadhyaya College, New Delhi

C. M. Upadhye
Gargi College, New Delhi

Syllabus for B.Sc. (Physical Sciences)

First Semester (Mathematics)

(University of Delhi)

MAPT-101: CALCULUS AND MATRICES

Six lecturers per week (including practicals)

Maximum marks: 100 (incuding internal assessment)

Examination hours: 3 hrs

THEORY

Unit I. Matrices (20 L)

$\mathbf{R}$, $\mathbf{R}^2$, $\mathbf{R}^3$ as vector spaces over $\mathbf{R}^n$. Standard basis for each of them. Concept of Linear Independence and examples of different bases. Subspaces of $\mathbf{R}^2$, $\mathbf{R}^3$. **-----Chapter 2**

Translation, Dilation, Rotation, Reflection in a point, line and plane. Matrix form of basic geometric transformations. **-----Chapter 3**

Interpretation of eigenvalues and eigenvectors for such transformations and eigenspaces as invariant subspaces. Matrices in diagonal form. Reduction to diagonal form upto matrices of order 3. **-----Chapter 4**

Computation of matrix inverses using elementary row operations. Rank of matrix. Solutions of a system of linear equations using matrices. **-----Chapter 1**

Illustrative examples of above concepts from Geometry, Physics, Chemistry, Combinatorics and Statistics.

Unit II. Calculus (34 L)

Sequences to be introduced through the examples arising in Science beginning with finite sequences followed by concepts of recursion and difference equations. For instance, the sequence arising from Tower of Hanoi game, the Fibonacci sequence arising from branching habit of trees and breeding habit of rabbits. Convergence of a sequence and algebra or convergent sequences. Illustration of proof of convergence of some simple sequences such as $(-1)^n/n$, $1/n^2$, $(1+1/n)^n$, $\sin n/n$, x^n with $|\bar{x}| < 1$. **-----Chapter 5**

Graphs of simple concrete functions such as polynomial, trigonometric, inverse trigonometric, exponential, logarithmic and hyperbolic functions arising in problems or chemical reaction, simple pendulum, radioactive decay, temperature cooling/heating problem and biological rhythms. **-----Chapter 6**

Successive differentiation, Leibnitz theorem, Recursion formulae for higher derivative. **-----Chapter 8**

Functions of two variables. Graphs and Level Curves of functions of two variables. Partial differentiation upto second order. **-----Chapter 10**

Computation of Taylor's Maclaurin's series of functions such as e^x, $\log(1 + x)$, $\sin(2x)$, $\cos x$. Their use in polynomial approximation and error estimation. **-----Chapter 9**

Formation and solution of Differential equations arising in population growth, radioactive decay, administration of medicine and cell division. **-----Chapter 7**

Unit III. (L14)

Geometrical representation of addition, subtraction, multiplication and division of complex numbers. Lines half planes, circles, discs in terms of complex variables. Statement of the Fundamental Theorem of Algebra and its consequences, De Moivre's theorem for rational indices and its simple applications. **----- Chapters 11 and 12**

PART I

CHAPTER ONE

Matrices

■ 1.1 INTRODUCTION TO MATRICES

Suppose there are 2 families–family **A** and family **B**. Family **A** consists of 7 members–3 males and 4 females. Family **B** consists of 6 members–4 males and 2 females. We can express this information as follows:

	Males	**Females**
Family **A**	3	4
Family **B**	4	2

If we suppress the headings and enclose the numbers in square brackets, we obtain a rectangular array having two rows and two columns.

$$\begin{bmatrix} 3 & 4 \\ 4 & 3 \end{bmatrix} \tag{I}$$

This rectangular array is called a matrix.

> **? DEFINITION:** A 'matrix' is a rectangular array of numbers. Its generalized form is
>
> $$\mathbf{A} = \begin{bmatrix} a_{11} & a_{12} & \cdots & a_{1n} \\ a_{21} & a_{22} & \cdots & a_{2n} \\ \cdots & \cdots & \cdots & \cdots \\ a_{m1} & a_{m2} & \cdots & a_{mn} \end{bmatrix} \tag{II}$$
>
> where a_{11}, a_{12} and so on are numbers, real or complex. These numbers are called the elements of the matrix.

The matrix **A** given above is a matrix having mn elements arranged in m rows and n columns. We say that **A** is a matrix of order $m \times n$.

For example, $\begin{bmatrix} 3 & 4 \\ 4 & 2 \end{bmatrix}$ is a 2×2 matrix.

We use the notation $[a_{ij}]$ to denote matrix **A** where a_{ij} is the $(i, j)^{\text{th}}$ entry in the matrix **A**.

EXAMPLE 1.1 The maximum and minimum temperatures in Delhi were recorded for 3 consecutive days–15 June 2006, 16 June 2006 and 17 June 2006–and put in the following table

	June 15	June 16	June 17
Max. Temp.	41°C	39°C	34.5°C
Min. Temp.	22°C	18°C	18°C

The information can be expressed as a 2 × 3 matrix

$$\begin{bmatrix} 41 & 39 & 34.5 \\ 22 & 18 & 18 \end{bmatrix}.$$

EXAMPLE 1.2 Consider the system of linear equations

$$\begin{aligned} 2x + 3y + z &= 4 \\ x - y + 2z &= 5 \\ 3x + 2y - z &= 7 \end{aligned}$$

Then the coefficients of x, y, z in the above equations can be expressed as the 3 × 3 matrix

$$\mathbf{A} = \begin{bmatrix} 2 & 3 & 1 \\ 1 & -1 & 2 \\ 3 & 2 & -1 \end{bmatrix}.$$

EXAMPLE 1.3 The following table gives the vitamin contents of 3 food items in conveniently chosen units.

Vitamin	A	B	C	D
Food I	0.4	0.5	0.1	0
Food II	0.5	0	0.2	0.1
Food III	0.2	0.1	0.5	0.2

The above information can be expressed as a 3 × 4 matrix

$$\begin{pmatrix} 0.4 & 0.5 & 0.1 & 0 \\ 0.5 & 0 & 0.2 & 0.1 \\ 0.2 & 0.1 & 0.5 & 0.2 \end{pmatrix}$$

Equality of two matrices Two matrices $\mathbf{A} = [a_{ij}]$ and $\mathbf{B} = [b_{ij}]$ are said to be equal if

(i) they are of the same order, and

(ii) corresponding elements of **A** are equal to the corresponding elements of **B**, that is, $a_{ij} = b_{ij}$ for all i, j.

For example, $\mathbf{A} = \begin{bmatrix} 1 & 2 \\ 3 & 4 \end{bmatrix}$ and $\mathbf{B} = \begin{bmatrix} 1 & 2 \\ 4 & 3 \end{bmatrix}$ are not equal because $a_{21} = 3 \neq 4 = b_{21}$.

$\begin{bmatrix} a & b \\ c & d \end{bmatrix} = \begin{bmatrix} 1 & 2 \\ 3 & 4 \end{bmatrix}$ if and only if

$$a = 1, b = 2, c = 3 \text{ and } d = 4.$$

Transpose of a matrix If $\mathbf{A} = [a_{ij}]$ is an $m \times n$ matrix, then the transpose of **A**, denoted as $\mathbf{A}^{\mathrm{T}}$, is the $n \times m$ matrix $\mathbf{A}^{\mathrm{T}} = [a_{ji}]$.

For example, if $\mathbf{A}=\begin{bmatrix}1 & 2\\ 3 & 4\end{bmatrix}$, then $\mathbf{A}^{\mathrm{T}}=\begin{bmatrix}1 & 3\\ 2 & 4\end{bmatrix}$.

We will now discuss some special types of matrices.

1.1.1 Types of Matrices

1. **Zero or null matrix** A matrix $\mathbf{A}=[a_{ij}]$ is said to be a 'zero matrix' or 'null matrix' if all its elements are zero i.e. $a_{ij}=0, \forall i, j$. We denote a zero matrix by $\mathbf{O}$.
2. **Square matrix** A matrix in which the number of rows is equal to the number of columns, say n, is called a 'square matrix' of order $n \times n$ or order n.

   For example, $\begin{bmatrix}1 & 2\\ 3 & 4\end{bmatrix}$ is a square matrix of order 2.
3. **Diagonal matrix** A square matrix $\mathbf{A}=[a_{ij}]$ is said to be a 'diagonal matrix' if $a_{ij}=0$ for $i \neq j$. It is of the form

$$\mathbf{A}=\begin{bmatrix}a_{11} & 0 & \dots & 0\\ 0 & a_{22} & \dots & 0\\ \dots & \dots & \dots & \dots\\ 0 & 0 & \dots & a_{mn}\end{bmatrix}.$$

   If $\mathbf{A}=[a_{ij}]$ is a square matrix of order n, then elements $a_{11}, a_{22}, \dots, a_{nn}$ are called the 'diagonal elements' of $\mathbf{A}$. They are said to constitute the main diagonal or the principal diagonal of $\mathbf{A}$. Thus, a square matrix is a diagonal matrix if all elements except those in the main diagonal are zero.

   A diagonal matrix $\mathbf{A}=[a_{ij}]$ of order n is sometimes written as

$$\mathbf{A} = \operatorname{diag}[a_{11}, a_{22}, \dots, a_{nn}].$$

4. **Scalar matrix** A diagonal matrix in which all the diagonal elements are equal is called a 'scalar matrix'.

   For example, $\begin{bmatrix}1 & 0\\ 0 & 1\end{bmatrix}, \begin{bmatrix}3 & 0 & 0\\ 0 & 3 & 0\\ 0 & 0 & 3\end{bmatrix}$
5. **Unit matrix or identity matrix** A scalar matrix in which all the diagonal elements are '1' is called a 'unit matrix'. Thus, a square matrix $\mathbf{A}=[a_{ij}]$ of order n is a unit matrix if

$$a_{ij} = \begin{cases}1 & \text{if } i=j\\ 0 & \text{if } i \neq j\end{cases}$$

   The unit matrix of order n is denoted by $\mathbf{I}_n$.

   For example, $\mathbf{I}_2=\begin{bmatrix}1 & 0\\ 0 & 1\end{bmatrix}, \mathbf{I}_3=\begin{bmatrix}1 & 0 & 0\\ 0 & 1 & 0\\ 0 & 0 & 1\end{bmatrix}$
6. **Symmetric matrix** A square matrix $\mathbf{A}=[a_{ij}]$ is said to be 'symmetric' if $a_{ij}=a_{ji}$ for all i and j. In other words, $\mathbf{A}$ is symmetric if $\mathbf{A}=\mathbf{A}^{\mathrm{T}}$.

   For example, $\mathbf{A}=\begin{bmatrix}1 & 3\\ 3 & 2\end{bmatrix}$ is symmetric because $\mathbf{A}=\begin{bmatrix}1 & 3\\ 3 & 2\end{bmatrix}=\mathbf{A}^{\mathrm{T}}$.
7. **Skew-symmetric matrix** A square matrix $\mathbf{A}=[a_{ij}]$ is said to be 'skew-symmetric' if $a_{ij}=-a_{ji}$ for all i and j. In other words, $\mathbf{A}$ is skew-symmetric matrix if $\mathbf{A}^{\mathrm{T}}=-\mathbf{A}$.

   For example, $\begin{bmatrix}0 & -1\\ 1 & 0\end{bmatrix}$.

8. **Hermitian matrix** A square matrix $\mathbf{A} = [a_{ij}]$ is said to be 'Hermitian' if $\bar{a}_{ij} = a_{ji}$, for all i and j where $\bar{a}_{ij}$ is complex conjugate of a_{ij}.

   If $\mathbf{A} = [a_{ij}]$, we define $\bar{\mathbf{A}} = [\bar{a}_{ij}]$ and $\mathbf{A}^{\theta} = (\bar{\mathbf{A}})^{\mathrm{T}} = [\bar{a}_{ji}]$. Thus, $\mathbf{A}$ is Hermitian if $\mathbf{A} = \mathbf{A}^{\theta}$.

   For example, $\mathbf{A} = \begin{bmatrix} 1 & 1-i \\ 1+i & 2 \end{bmatrix}$ is Hermitian.

9. **Skew-Hermitian matrix** A square matrix $\mathbf{A}$ is said to be 'skew-Hermitian' if $a_{ij} = -\bar{a}_{ji}$ for all i and j.

   In other words, $\mathbf{A}$ is skew-Hermitian if $\mathbf{A}^{\theta} = -\mathbf{A}$.

   For example, $\mathbf{A} = \begin{bmatrix} i & 1+i \\ -1+i & 2i \end{bmatrix}$ is skew-Hermitian.

10. **Triangular Matrix** A matrix $\mathbf{A} = [a_{ij}]$ is said to be a 'triangular matrix' if $a_{ij} = 0$ whenever $i > j$.

    For example, $\begin{bmatrix} 2 & 3 \\ 0 & 1 \end{bmatrix}, \begin{bmatrix} 2 & 1 & 2 \\ 0 & 1 & 0 \\ 0 & 0 & 3 \end{bmatrix}, \begin{bmatrix} 1 & 1 & 0 & 1 \\ 0 & 2 & 1 & 2 \\ 0 & 0 & 3 & 3 \end{bmatrix}$

1.1.2 Operations on Matrices

If $\mathbf{A} = [a_{ij}]$ and $\mathbf{B} = [b_{ij}]$ are two matrices of order $m \times n$, then their sum is the $m \times n$ matrix $\mathbf{A} + \mathbf{B} = [c_{ij}]$ where $c_{ij} = a_{ij} + b_{ij}$ for all i, j. Similarly we define difference of the matrices $\mathbf{A}$ and $\mathbf{B}$ as the matrix $\mathbf{A} - \mathbf{B} = [a_{ij} - b_{ij}]$ for all i and j.

EXAMPLE 1.4 If $\mathbf{A} = \begin{bmatrix} 1 & 2 & 3 \\ 1 & 0 & 2 \end{bmatrix}$ and $\mathbf{B} = \begin{bmatrix} 3 & 1 & 2 \\ 0 & -1 & 4 \end{bmatrix}$

then
$$\mathbf{A} + \mathbf{B} = \begin{bmatrix} 1 & 2 & 3 \\ 1 & 0 & 2 \end{bmatrix} + \begin{bmatrix} 3 & 1 & 2 \\ 0 & -1 & 4 \end{bmatrix}$$
$$= \begin{bmatrix} 1+3 & 2+1 & 3+2 \\ 1+0 & 0+(-1) & 2+4 \end{bmatrix}$$
$$= \begin{bmatrix} 4 & 3 & 5 \\ 1 & -1 & 6 \end{bmatrix}$$

and
$$\mathbf{A} - \mathbf{B} = \begin{bmatrix} 1-3 & 2-1 & 3-2 \\ 1-0 & 0-(-1) & 2-4 \end{bmatrix} = \begin{bmatrix} -2 & 1 & 1 \\ 1 & 1 & -2 \end{bmatrix}$$

NOTE: Two matrices of different orders cannot be added or subtracted.

1.1.3 Properties of Matrix Addition

1. **Matrix addition is commutative** That is, if $\mathbf{A}$ and $\mathbf{B}$ are two matrices of same order, then $\mathbf{A} + \mathbf{B} = \mathbf{B} + \mathbf{A}$.

   Proof Let $\mathbf{A} = [a_{ij}]$ and $\mathbf{B} = [b_{ij}]$. Then
   $$\begin{aligned} \mathbf{A} + \mathbf{B} &= [a_{ij}] + [b_{ij}] \\ &= [a_{ij} + b_{ij}] \\ &= [b_{ij} + a_{ij}] \quad \text{(because addition of numbers is commutative)} \\ &= \mathbf{B} + \mathbf{A} \end{aligned}$$

2. **Matrix addition is associative** That is, if **A, B** and **C** be three matrices of the same order, then
   $$(\mathbf{A} + \mathbf{B}) + \mathbf{C} = \mathbf{A} + (\mathbf{B} + \mathbf{C})$$

3. **Existence of additive identity** If **A** is any $m \times n$ matrix and **O** be the $m \times n$ zero matrix, then

$$\mathbf{A} + \mathbf{O} = \mathbf{A} = \mathbf{O} + \mathbf{A}$$

4. **Existence of additive inverse** If **A** is any $m \times n$ matrix, then we can find an $m \times n$ matrix **B** such that

$$\mathbf{A} + \mathbf{B} = \mathbf{O} = \mathbf{B} + \mathbf{A}$$

The matrix **B** is called 'additive inverse' or negative of **A** and is denoted by $-\mathbf{A}$.

If $\mathbf{A} = [a_{ij}]$, we have $-\mathbf{A} = [-a_{ij}]$.
Thus, $\mathbf{A} + (-\mathbf{A}) = \mathbf{O} = (-\mathbf{A}) + \mathbf{A}$.
The properties 2, 3 and 4 can be proved similarly as we proved property 1.

If $\mathbf{A} = [a_{ij}]$ is an $m \times n$ matrix and k is a number called a scalar, then the scalar multiple of **A** by k, denoted by $k\mathbf{A}$, is the $m \times n$ matrix whose $(i,j)^{\text{th}}$ element is ka_{ij} for all possible values of i and j. Thus $k\mathbf{A} = [ka_{ij}]$.

EXAMPLE 1.5 Let $\mathbf{A} = \begin{bmatrix} 1 & 2 & 3 \\ -1 & 0 & 4 \end{bmatrix}$. If $k = 2$, then

$$k\mathbf{A} = 2\mathbf{A} = \begin{bmatrix} 2 & 4 & 6 \\ -2 & 0 & 8 \end{bmatrix}$$

and if $k = 3$, then

$$k\mathbf{A} = 3\mathbf{A} = \begin{bmatrix} 3 & 6 & 9 \\ -3 & 0 & 12 \end{bmatrix}.$$

REMARK: Note that $(-1)\mathbf{A} = -\mathbf{A}$ and $\mathbf{A} + (-1)\mathbf{B} = \mathbf{A} - \mathbf{B}$.

1.1.4 Properties of Scalar Multiplication

The following properties of multiplication of a matrix by a scalar can be easily proved.

1. If **A** and **B** are two matrices of the same order and if k is any scalar, then

$$k(\mathbf{A} + \mathbf{B}) = k\mathbf{A} + k\mathbf{B}.$$

2. If k_1 and k_2 are two scalars and **A** is a matrix, then

$$(k_1 + k_2)\mathbf{A} = k_1\mathbf{A} + k_2\mathbf{A}.$$

3. If k_1 and k_2 are two scalars and **A** is any matrix, then

$$k_1(k_2\mathbf{A}) = (k_1k_2)\mathbf{A}.$$

4. If **A** is any matrix, then

$$1\mathbf{A} = \mathbf{A}.$$

Now we define multiplication of two matrices.

Let $\mathbf{A} = [a_{ij}]$ be an $m \times n$ matrix and $\mathbf{B} = [b_{ij}]$ be an $n \times p$ matrix. Then the product **AB** is the $m \times p$ matrix $[c_{ik}]$ where $c_{ik} = \sum_{j=1}^{n} a_{ij}\, b_{jk}$.

EXAMPLE 1.6 Let $\mathbf{A} = \begin{bmatrix} 1 & 2 & 3 \\ 1 & 0 & 1 \end{bmatrix}$ and $\mathbf{B} = \begin{bmatrix} 2 & 1 \\ 3 & 5 \\ 2 & 3 \end{bmatrix}$.

$$\text{Then } \mathbf{AB} = \begin{bmatrix} 1 & 2 & 3 \\ 1 & 0 & 1 \end{bmatrix}\begin{bmatrix} 2 & 1 \\ 3 & 5 \\ 2 & 3 \end{bmatrix}$$

$$= \begin{bmatrix} 1\times 2 + 2\times 3 + 3\times 2 & 1\times 1 + 2\times 5 + 3\times 3 \\ 1\times 2 + 0\times 3 + 1\times 2 & 1\times 1 + 0\times 5 + 1\times 3 \end{bmatrix}$$

$$= \begin{bmatrix} 14 & 20 \\ 4 & 4 \end{bmatrix}$$

NOTE: The product **AB** of two matrices **A** and **B** is possible only if the number of columns in **A** is equal to the number of rows in **B**.

The following properties of matrix multiplication can be easily proved.

1. **Matrix multiplication is associative** That is, if **A**, **B** and **C** are matrices of the order $m \times n$, $n \times p$ and $p \times q$, respectively, then

$$(\mathbf{AB})\mathbf{C} = \mathbf{A}(\mathbf{BC}).$$

2. **Matrix multiplication distributes over matrix addition** That is, if **A** is a matrix of order $m \times n$ and if **B** and **C** are matrices of order $n \times p$, then

$$\mathbf{A}(\mathbf{B} + \mathbf{C}) = \mathbf{AB} + \mathbf{AC}$$

REMARK: Matrix multiplication is not commutative. If **A** is an $m \times n$ matrix and **B** is $p \times q$ matrix, then **AB** exists if $n = p$ and **BA** exists if $q = m$. This means that **AB** and **BA** both may not exist together. For example, if **A** is a 2×3 matrix and **B** is a 3×4 matrix then **AB** exists and is a 2×4 matrix while **BA** is not defined.

It is also clear that if **AB** and **BA** both exist they may not be of the same order. For example, if **A** is a 2×3 matrix and **B** is a 3×2 matrix, then both **AB** and **BA** exist but **AB** is a 2×2 matrix while **BA** is a 3×3 matrix. So, they are not comparable for equality. Even if **AB** and **BA** both exist and are of same order, they may still be not equal, that is, $\mathbf{AB} \neq \mathbf{BA}$.

EXAMPLE 1.7 Let $\mathbf{A} = \begin{bmatrix} 1 & 1 \\ 0 & 0 \end{bmatrix}$ and $\mathbf{B} = \begin{bmatrix} 1 & 0 \\ 1 & 0 \end{bmatrix}$.

$$\text{Then } \mathbf{AB} = \begin{bmatrix} 1 & 1 \\ 0 & 0 \end{bmatrix}\begin{bmatrix} 1 & 0 \\ 1 & 0 \end{bmatrix} = \begin{bmatrix} 2 & 0 \\ 0 & 0 \end{bmatrix}$$

$$\text{and } \mathbf{BA} = \begin{bmatrix} 1 & 0 \\ 1 & 0 \end{bmatrix}\begin{bmatrix} 1 & 1 \\ 0 & 0 \end{bmatrix} = \begin{bmatrix} 1 & 1 \\ 1 & 1 \end{bmatrix}.$$

So $\mathbf{AB} \neq \mathbf{BA}$.

EXAMPLE 1.8 If $\mathbf{A} = \begin{bmatrix} 1 & 0 \\ 0 & 0 \end{bmatrix}$ and $\mathbf{B} = \begin{bmatrix} 0 & 0 \\ 0 & 1 \end{bmatrix}$, then

$$\mathbf{AB} = \begin{bmatrix} 1 & 0 \\ 0 & 0 \end{bmatrix}\begin{bmatrix} 0 & 0 \\ 0 & 1 \end{bmatrix} = \begin{bmatrix} 0 & 0 \\ 0 & 0 \end{bmatrix} = \mathbf{O}.$$

Thus, the product of two matrices can be a zero matrix even though neither of the matrices is itself zero.

Positive integral powers of a square matrix If $\mathbf{A}$ is a square matrix, it can be multiplied by itself, that is, $\mathbf{A} \cdot \mathbf{A}$ is defined. We denote the product $\mathbf{A} \cdot \mathbf{A}$ by $\mathbf{A}^2$. Moreover, using the associative law, we can define

$$\begin{aligned} \mathbf{A}^3 &= \mathbf{A}^2 \cdot \mathbf{A} = \mathbf{A} \cdot \mathbf{A} \cdot \mathbf{A} \\ \mathbf{A}^4 &= \mathbf{A}^3 \cdot \mathbf{A} = \mathbf{A} \cdot \mathbf{A} \cdot \mathbf{A} \cdot \mathbf{A} \\ &\vdots \\ \mathbf{A}^n &= \mathbf{A}^{n-1}\mathbf{A} = \mathbf{A} \cdot \mathbf{A} \cdots \mathbf{A} \ (n \text{ factors}) \end{aligned}$$

These are naturally called the 'powers' of the matrix $\mathbf{A}$. $\mathbf{A}^2$ is called square of $\mathbf{A}$, $\mathbf{A}^3$ is called cube of $\mathbf{A}$ and so on.

EXAMPLE 1.9 If $\mathbf{A} = \begin{bmatrix} 1 & 0 \\ 3 & 4 \end{bmatrix}$, then

$$\mathbf{A}^2 = \begin{bmatrix} 1 & 0 \\ 3 & 4 \end{bmatrix}\begin{bmatrix} 1 & 0 \\ 3 & 4 \end{bmatrix} = \begin{bmatrix} 1 & 0 \\ 15 & 16 \end{bmatrix}$$

$$\mathbf{A}^3 = \begin{bmatrix} 1 & 0 \\ 15 & 16 \end{bmatrix}\begin{bmatrix} 1 & 0 \\ 3 & 4 \end{bmatrix} = \begin{bmatrix} 1 & 0 \\ 63 & 64 \end{bmatrix} \text{ and}$$

$$\mathbf{A}^4 = \begin{bmatrix} 1 & 0 \\ 63 & 64 \end{bmatrix}\begin{bmatrix} 1 & 0 \\ 3 & 4 \end{bmatrix} = \begin{bmatrix} 1 & 0 \\ 255 & 256 \end{bmatrix}.$$

Determinant of a square matrix To every square matrix $\mathbf{A} = [a_{ij}]$, we can associate a number called 'determinant' of $\mathbf{A}$ denoted by $|\mathbf{A}|$ or det $\mathbf{A}$.

The matrices which are not square do not have determinants.

The association of the number det $\mathbf{A}$ to the square matrix $\mathbf{A}$ is done in the following manner.

1. If $\mathbf{A} = [a_{11}]$ is a 1×1 matrix, then $|\mathbf{A}|$ is defined to be a_{11}.
2. If $\mathbf{A} = \begin{bmatrix} a_{11} & a_{12} \\ a_{21} & a_{22} \end{bmatrix}$ is a 2×2 matrix, then we define

   $$|\mathbf{A}| = a_{11}a_{22} - a_{21}a_{12}.$$
3. If $\mathbf{A} = \begin{bmatrix} a_{11} & a_{12} & a_{13} \\ a_{21} & a_{22} & a_{23} \\ a_{31} & a_{32} & a_{33} \end{bmatrix}$ is a 3×3 matrix, then $|\mathbf{A}|$ is defined as

   $$\begin{aligned} |\mathbf{A}| &= a_{11}\begin{vmatrix} a_{22} & a_{23} \\ a_{32} & a_{33} \end{vmatrix} - a_{12}\begin{vmatrix} a_{21} & a_{23} \\ a_{31} & a_{33} \end{vmatrix} + a_{13}\begin{vmatrix} a_{21} & a_{22} \\ a_{31} & a_{32} \end{vmatrix} \\ &= a_{11}(a_{22}a_{33} - a_{32}a_{23}) - a_{12}(a_{21}a_{33} - a_{31}a_{23}) + a_{13}(a_{21}a_{32} - a_{31}a_{22}). \end{aligned}$$
4. If $\mathbf{A} = [a_{ij}]$ is an $n \times n$ matrix, then

   $$|\mathbf{A}| = \begin{vmatrix} a_{11} & a_{12} & \cdots & a_{1n} \\ a_{21} & a_{22} & \cdots & a_{2n} \\ \cdots & \cdots & \cdots & \cdots \\ a_{n1} & a_{n2} & \cdots & a_{nn} \end{vmatrix}. \qquad (1)$$

By deleting the ith row and the jth column in $|\mathbf{A}|$, we obtain a determinant $\mathbf{M}_{ij}$ of order $(n-1)$. This determinant is called minor of the element a_{ij}. The value of determinant of order n given in (1) can now be defined as

$$|\mathbf{A}| = \sum_{j=1}^{n} (-1)^{i+j} a_{ij}\mathbf{M}_{ij}, \quad \text{for } i \text{ fixed, } i = 1, \ldots, n.$$

EXAMPLE 1.10 Let $\mathbf{A} = \begin{bmatrix} 1 & 2 & 3 \\ 4 & 5 & 6 \\ 7 & 8 & 9 \end{bmatrix}$.

Then
$$\begin{aligned} |\mathbf{A}| &= \begin{vmatrix} 1 & 2 & 3 \\ 4 & 5 & 6 \\ 7 & 8 & 9 \end{vmatrix} \\ &= 1\begin{vmatrix} 5 & 6 \\ 8 & 9 \end{vmatrix} - 2\begin{vmatrix} 4 & 6 \\ 7 & 9 \end{vmatrix} + 3\begin{vmatrix} 4 & 5 \\ 7 & 8 \end{vmatrix} \\ &= 1(45-48) - 2(36-42) + 3(32-35) \\ &= -3 + 12 - 9 = 0. \end{aligned}$$

? DEFINITION: A square matrix **A** is called 'singular' if $|\mathbf{A}| = 0$. However, If $|\mathbf{A}| \neq 0$, then the matrix is called 'non-singular'.

The 3×3 matrix **A** given in above example is a singular matrix.

EXAMPLE 1.11 The matrix $\begin{bmatrix} \cos\theta & -\sin\theta \\ \sin\theta & \cos\theta \end{bmatrix}$ is non-singular because $\begin{vmatrix} \cos\theta & -\sin\theta \\ \sin\theta & \cos\theta \end{vmatrix} = \cos^2\theta + \sin^2\theta = 1 \neq 0$.

EXERCISE 1.1

1. If $\mathbf{A} = \begin{bmatrix} 1 & 2 & 3 \\ -1 & 0 & 2 \end{bmatrix}$, $\mathbf{B} = \begin{bmatrix} 4 & 0 & -1 \\ -3 & 2 & 1 \end{bmatrix}$ find

(a) $\mathbf{A} + \mathbf{B}$ **(b)** $\mathbf{A} - \mathbf{B}$ **(c)** $2\mathbf{A}$ **(d)** $-\mathbf{B}$

2. If $\mathbf{A} = \begin{bmatrix} 2 & 4 \\ 3 & 2 \end{bmatrix}$, $\mathbf{B} = \begin{bmatrix} 1 & 3 \\ -2 & 5 \end{bmatrix}$, $\mathbf{C} = \begin{bmatrix} -2 & 5 \\ 3 & 4 \end{bmatrix}$ find

(a) $\mathbf{A} + \mathbf{B}$ **(b)** $\mathbf{A} - \mathbf{B}$ **(c)** $3\mathbf{A} - \mathbf{C}$

(d) $\mathbf{AB}$ **(e)** $\mathbf{BA}$ **(f)** $(\mathbf{BC})^{\mathrm{T}}$

3. If $\mathbf{A} = \begin{bmatrix} 1 & 2 & -4 \\ -2 & 4 & 1 \end{bmatrix}$, $\mathbf{B} = \begin{bmatrix} 5 & 1 \\ 0 & -2 \\ 1 & 3 \end{bmatrix}$, $\mathbf{C} = \begin{bmatrix} 2 & 2 \\ 1 & 3 \\ 1 & -2 \end{bmatrix}$, compute

(a) $\mathbf{BA}$ **(b)** $\mathbf{AC}$ **(c)** $\mathbf{CA}$

(d) $\mathbf{A}(2\mathbf{B} - 3\mathbf{C})$ **(e)** $\mathbf{A}(\mathbf{B} + \mathbf{C})$ **(f)** $\mathbf{AB} + \mathbf{AC}$

4. Let $\mathbf{A} = \begin{bmatrix} 1 & 2 \\ 3 & 2 \end{bmatrix}$ and $\mathbf{B} = \begin{bmatrix} 2 & -1 \\ -3 & 4 \end{bmatrix}$. Show that $\mathbf{AB} \neq \mathbf{BA}$.

5. For $\mathbf{A} = \begin{bmatrix} 1 & 2 & -1 \\ 0 & 1 & 2 \\ 1 & 0 & -2 \end{bmatrix}$, $\mathbf{B} = \begin{bmatrix} 5 & 0 & 1 \\ 1 & 0 & 1 \\ 1 & -2 & 1 \end{bmatrix}$ and $\mathbf{C} = \begin{bmatrix} 3 & 1 & -2 \\ 0 & 1 & 4 \\ 1 & 0 & 5 \end{bmatrix}$ verify that

(a) $(\mathbf{AB})\mathbf{C} = \mathbf{A}(\mathbf{BC})$

(b) $\mathbf{A}(\mathbf{B} + \mathbf{C}) = \mathbf{AB} + \mathbf{AC}$

(c) $(\mathbf{A} + \mathbf{B})\mathbf{C} = \mathbf{AC} + \mathbf{BC}$

6. If $\mathbf{A} = \begin{bmatrix} \cos\alpha & -\sin\alpha \\ \sin\alpha & \cos\alpha \end{bmatrix}$, $\mathbf{B} = \begin{bmatrix} \cos\beta & -\sin\beta \\ \sin\beta & \cos\beta \end{bmatrix}$ show that $\mathbf{AB} = \mathbf{BA}$.

7. If $\mathbf{A} = \begin{bmatrix} \cos\theta & \sin\theta \\ -\sin\theta & \cos\theta \end{bmatrix}$, verify that $\mathbf{AA}^{\mathrm{T}} = \mathbf{A}^{\mathrm{T}}\mathbf{A} = \mathbf{I}_2$.

8. Prove that

(a) $(\mathbf{A}^{\mathrm{T}})^{\mathrm{T}} = \mathbf{A}$ **(b)** $(\mathbf{A} + \mathbf{B})^{\mathrm{T}} = \mathbf{A}^{\mathrm{T}} + \mathbf{B}^{\mathrm{T}}$

(c) $(k\mathbf{A})^{\mathrm{T}} = k\mathbf{A}^{\mathrm{T}}$ **(d)** $(\mathbf{AB})^{\mathrm{T}} = \mathbf{B}^{\mathrm{T}}\mathbf{A}^{\mathrm{T}}$

assuming that both sides of equalities are defined.

9. If **A** be any matrix, prove that the matrices $\mathbf{AA}^{\mathrm{T}}$ and $\mathbf{A}^{\mathrm{T}}\mathbf{A}$ are both symmetric.

10. Let **A** be a square matrix. Show that $\frac{1}{2}(\mathbf{A} + \mathbf{A}^{\mathrm{T}})$ is a symmetric matrix and $\frac{1}{2}(\mathbf{A} - \mathbf{A}^{\mathrm{T}})$ is a skew-symmetric matrix. Conclude that any square matrix can be written as sum of a symmetric matrix and a skew-symmetric matrix.

11. If $\mathbf{A} = \begin{bmatrix} 0 & -\tan\frac{\alpha}{2} \\ \tan\frac{\alpha}{2} & 0 \end{bmatrix}$ and **I** is the unit matrix, show that $\mathbf{I} + \mathbf{A} = (\mathbf{I} - \mathbf{A})\begin{bmatrix} \cos\alpha & \sin\alpha \\ \sin\alpha & \cos\alpha \end{bmatrix}$.

12. If $\mathbf{A} = \begin{bmatrix} 1 & 0 \\ 0 & 1 \end{bmatrix}$ and $\mathbf{B} = \begin{bmatrix} 0 & 1 \\ -1 & 0 \end{bmatrix}$ show that $(a\mathbf{A} + b\mathbf{B})(a\mathbf{A} - b\mathbf{B}) = (a^2 + b^2)\mathbf{A}$.

13. Let **A** be any 3×3 matrix and **I** be the identity matrix of order 3. Show that **AI** = **IA** = **A.**

14. Let **A** be any 3×3 matrix and **O** be the null matrix of order 3. Show that **AO** = **OA** = **O.**

15. Let $\mathbf{A} = \begin{bmatrix} 0 & 1 \\ 0 & 2 \end{bmatrix}$. Find all 2×2 matrices **B** such that **(a)** $\mathbf{AB} = \mathbf{O}$ **(b)** $\mathbf{BA} = \mathbf{O}$.

16. Find all 2×2 matrices **A** such that $\mathbf{A}^2 = \mathbf{O}$.

17. Let $f(x) = x^2 - 5x + 6$. Find $f(\mathbf{A})$ if $\mathbf{A} = \begin{bmatrix} 2 & 0 & 1 \\ 2 & 1 & 3 \\ 1 & -1 & 0 \end{bmatrix}$.

18. If $\mathbf{A} = \begin{bmatrix} 1 & 0 & -2 \\ 2 & 2 & 4 \\ 0 & 0 & 2 \end{bmatrix}$ and $f(x) = x^2 - 3x + 2$ show that $f(\mathbf{A}) = \mathbf{O}$. Use the result to find $\mathbf{A}^5$.

19. The matrix $\mathbf{J} = \begin{bmatrix} 0 & -1 \\ 1 & 0 \end{bmatrix}$ has properties similar to those of the imaginary number $i = \sqrt{-1}$. Show that $\mathbf{J}^2 = -\mathbf{I}$, where **I** is the second order identity matrix, and that $\mathbf{J}^3 = -\mathbf{J}$, $\mathbf{J}^4 = \mathbf{I}$, $\mathbf{J}^5 = \mathbf{J}$ and so on.

20. Let $\mathbf{A} = \begin{bmatrix} \cos\theta & -\sin\theta \\ \sin\theta & \cos\theta \end{bmatrix}$. Verify that $\mathbf{A}^2 = \begin{bmatrix} \cos 2\theta & -\sin 2\theta \\ \sin 2\theta & \cos 2\theta \end{bmatrix}$ and compute $\mathbf{A}^n$.

21. Let $\mathbf{A} = \begin{bmatrix} 1 & 1 \\ 0 & 1 \end{bmatrix}$. Find $\mathbf{A}^2$, $\mathbf{A}^3$, $\mathbf{A}^4$. Hence compute $\mathbf{A}^n$.

22. If $\mathbf{A} = \begin{bmatrix} 1 & 1 \\ 1 & 1 \end{bmatrix}$, find $\mathbf{A}^2$, $\mathbf{A}^3$, $\mathbf{A}^n$.

23. If $\mathbf{A} = \begin{bmatrix} 1 & 1 \\ 1 & 0 \end{bmatrix}$, find $\mathbf{A}^2$, $\mathbf{A}^3$, $\mathbf{A}^4$, $\mathbf{A}^5$. Hence, discuss the nature of $\mathbf{A}^n$. [Hint: Observe that the elements of A^n follow the pattern 1, 1, 2, 3, 5, 8, 13, 21, … where each term is the sum of preceding two terms. Such a sequence is called Fibonacci sequence. We shall learn more about it in a later chapter.]

24. Let $\mathbf{A} = \begin{bmatrix} 1 & 2 \\ -1 & 3 \end{bmatrix}$ and $\mathbf{B} = \begin{bmatrix} 2 & 1 \\ 0 & 1 \end{bmatrix}$. Find whether the following identities are true or not. If not, why?

(a) $(\mathbf{A} + \mathbf{B})^2 = \mathbf{A}^2 + 2\mathbf{AB} + \mathbf{B}^2$

(b) $\mathbf{A}^2 - \mathbf{B}^2 = (\mathbf{A} + \mathbf{B})(\mathbf{A} - \mathbf{B})$.

25. Show that if a diagonal matrix is commutative with every matrix of some order, it is necessarily a scalar matrix.

26. Prove that every diagonal matrix is symmetric.

27. Find which of the following matrices are singular.

(a) $\begin{bmatrix} 2 & 3 \\ 1 & 4 \end{bmatrix}$ **(b)** $\begin{bmatrix} 3 & 1 \\ 6 & 2 \end{bmatrix}$

(c) $\begin{bmatrix} 1 & 2 & -3 \\ 1 & -2 & 1 \\ 5 & -2 & -3 \end{bmatrix}$ **(d)** $\begin{bmatrix} 1 & 1 & 1 \\ 0 & 2 & 4 \\ 5 & 5 & 1 \end{bmatrix}$

28. Given $\mathbf{A} = \begin{bmatrix} 1 & 2 \\ 3 & 4 \end{bmatrix}$ and $\mathbf{B} = \begin{bmatrix} 5 & 6 \\ 7 & 8 \end{bmatrix}$. Show that $\mathbf{AB} \neq \mathbf{BA} \neq \mathbf{A}^{\mathrm{T}}\mathbf{B} \neq \mathbf{AB}^{\mathrm{T}} \neq \mathbf{A}^{\mathrm{T}}\mathbf{B}^{\mathrm{T}} \neq \mathbf{B}^{\mathrm{T}}\mathbf{A}^{\mathrm{T}}$ but that the determinant of each product is 4.

29. A furniture manufacturer makes chairs and tables, each of which must go through an assembly process and a finishing process. The times required for these processes are given (in hours) by the matrix

$$\begin{array}{cccc} & \text{Assembly process} & \text{Finishing process} & \\ \mathbf{A} = & \left[\begin{matrix} 2 \\ 3 \end{matrix}\right. & \left.\begin{matrix} 2 \\ 4 \end{matrix}\right] & \begin{matrix} \text{Chair} \\ \text{Table} \end{matrix} \end{array}$$

The manufacturer has a plant in Delhi and another in Chennai. The hourly rates for each of the processes are given (in rupees) by the matrix

$$\mathbf{B} = \begin{array}{c} \text{Delhi} \quad \text{Chennai} \\ \begin{bmatrix} 9 & 10 \\ 10 & 12 \end{bmatrix} \end{array} \begin{array}{l} \\ \text{Assembly process} \\ \text{Finishing process} \end{array}$$

What do the entries in the matrix product AB tell the manufacturer?

30. The following matrix gives the vitamin contents of three foods in conveniently chosen units.

$$\begin{array}{l} \text{Vitamin} \\ \text{Food I} \\ \text{Food II} \\ \text{Food III} \end{array} \begin{array}{c} \begin{array}{cccc} \text{A} & \text{B} & \text{C} & \text{D} \end{array} \\ \begin{bmatrix} 0.5 & 0.5 & 0 & 0 \\ 0.3 & 0 & 0.2 & 0.1 \\ 0.1 & 0.1 & 0.2 & 0.5 \end{bmatrix} \end{array}$$

If we eat 5 units of Food I, 10 units of Food II and 8 units of Food III, how much of each type of vitamin have we consumed? If we pay only for the vitamin content of each food, paying Rs 10, Rs 20, Rs 25 and Rs 50 respectively for units of the four vitamins, how much does a unit of each type of food cost ? Compute the total cost of the food we ate.

1.2 ELEMENTARY ROW OPERATIONS

An elementary row operation (or an E-row operation) is an operation of any one of the following three types.

Type-I The interchange of two rows.

Type-II The multiplication of a row by a non-zero number.

Type-III The addition of a multiple of one row to another row.

Consider the following matrices

$$\mathbf{A} = \begin{bmatrix} 1 & 2 & 3 \\ 2 & 1 & 1 \end{bmatrix}, \quad \mathbf{B} = \begin{bmatrix} 2 & 1 & 1 \\ 1 & 2 & 3 \end{bmatrix}$$

$$\mathbf{C} = \begin{bmatrix} 1 & 2 & 3 \\ 4 & 2 & 2 \end{bmatrix}, \quad \mathbf{D} = \begin{bmatrix} 7 & 5 & 6 \\ 2 & 1 & 1 \end{bmatrix}$$

Matrix **B** can be obtained from matrix **A** by interchanging the first and the second row. This is a type I elementary row operation. We denote this E-row operation by $\mathbf{R}_1 \leftrightarrow \mathbf{R}_2$.

Matrix **C** is obtained from matrix **A** by multiplying the second row by 2. This is a type II elementary row operation. We denote this E-row operation by $\mathbf{R}_2 \rightarrow 2\mathbf{R}_2$.

Matrix **D** is obtained from **A** by adding thrice the second row to the first row of **A**. This is a type III elementary row operation. We denote this E-row operation by $\mathbf{R}_1 \rightarrow \mathbf{R}_1 + 3\mathbf{R}_2$.

> ? **DEFINITION:** If a matrix **B** is obtained from a matrix **A** by a finite chain of elementary row operations, we say that **A** is 'equivalent' to **B** and write it as $\mathbf{A} \sim \mathbf{B}$.

The following results are obvious:

(i) Every matrix is equivalent to itself i.e. $\mathbf{A} \sim \mathbf{A}$.

(ii) If $\mathbf{A} \sim \mathbf{B}$, then $\mathbf{B} \sim \mathbf{A}$.

(iii) If $\mathbf{A} \sim \mathbf{B}$, $\mathbf{B} \sim \mathbf{C}$, then $\mathbf{A} \sim \mathbf{C}$.

From the results, it is clear that the four matrices **A, B, C** and **D** discussed above are equivalent to each other.

EXAMPLE 1.12 The matrices $\mathbf{A} = \begin{bmatrix} 1 & 4 & 7 \\ 2 & 5 & 8 \\ 3 & 6 & 9 \end{bmatrix}$ and $\mathbf{B} = \begin{bmatrix} 1 & 4 & 7 \\ 0 & -3 & -6 \\ 0 & 0 & 0 \end{bmatrix}$ are equivalent.

This is because we can obtain **B** from **A** by a finite sequence of E-row operations as shown below:

$$\mathbf{A} = \begin{bmatrix} 1 & 4 & 7 \\ 2 & 5 & 8 \\ 3 & 6 & 9 \end{bmatrix}$$

$$\sim \begin{bmatrix} 1 & 4 & 7 \\ 0 & -3 & -6 \\ 3 & 6 & 9 \end{bmatrix} \quad \text{By applying E-row operation } \mathbf{R}_2 \to \mathbf{R}_2 - 2\mathbf{R}_1.$$

$$\sim \begin{bmatrix} 1 & 4 & 7 \\ 0 & -3 & -6 \\ 0 & -6 & -12 \end{bmatrix} \quad \text{By applying E-row operation } \mathbf{R}_3 \to \mathbf{R}_3 - 3\mathbf{R}_1.$$

$$\sim \begin{bmatrix} 1 & 4 & 7 \\ 0 & -3 & -6 \\ 0 & 0 & 0 \end{bmatrix} \quad \text{By applying E-row operation } \mathbf{R}_3 \to \mathbf{R}_3 + 2\mathbf{R}_2.$$

$$= \mathbf{B}.$$

The matrix **B** in above example is much simpler to handle than matrix **A**. Matrix **B** is in triangular form and contains a row of all zeroes. We have, thus, reduced matrix **A** to a triangular matrix **B** by elementary row operations. Such reductions are very helpful in finding the rank of a matrix and solving system of linear equations. The method used in reducing **A** to triangular matrix **B** in the above example can be described more generally by making sure that the first element in first column of matrix **A** is non-zero. If it is zero, we interchange rows to get a non-zero element as the leading element in the first column. All elements below this leading element of first column are made zero by adding the corresponding row to some multiple of first row. We now come to the second column and make all elements below the principal diagonal element of second column zero by E-row operations. And so, column by column, we make all elements below the principal diagonal zero. In this manner we obtain a triangular matrix. This method will always result in a triangular matrix.

THEOREM

Every matrix can be reduced to a triangular matrix by elementary row operations.

EXAMPLE 1.13 Reduce the matrix

$$\mathbf{A} = \begin{bmatrix} 1 & -2 & 3 & 9 \\ 2 & -3 & 1 & 8 \\ 3 & 0 & -1 & 3 \end{bmatrix}$$

to triangular form by elementary row operations.

SOLUTION

$$\mathbf{A} = \begin{bmatrix} 1 & -2 & 3 & 9 \\ 2 & -3 & 1 & 8 \\ 3 & 0 & -1 & 3 \end{bmatrix}$$

$$\sim \begin{bmatrix} 1 & -2 & 3 & 9 \\ 0 & 1 & -5 & -10 \\ 0 & 6 & -10 & -24 \end{bmatrix} \quad \text{By } \mathbf{R}_2 \to \mathbf{R}_2 - 2\mathbf{R}_1, \mathbf{R}_3 \to \mathbf{R}_3 - 3\mathbf{R}_1$$

$$\sim \begin{bmatrix} 1 & -2 & 3 & 9 \\ 0 & 1 & -5 & -10 \\ 0 & 0 & 20 & 36 \end{bmatrix} \quad \text{By } \mathbf{R}_3 \to \mathbf{R}_3 - 6\mathbf{R}_2$$

which is a triangular matrix.

EXAMPLE 1.14 Reduce the matrix $\mathbf{A} = \begin{bmatrix} 0 & 1 & 7 \\ 2 & 3 & 4 \\ -1 & 2 & 8 \end{bmatrix}$ to triangular form by E-row operations.

SOLUTION

$$\mathbf{A} = \begin{bmatrix} 0 & 1 & 7 \\ 2 & 3 & 4 \\ -1 & 2 & 8 \end{bmatrix}$$

$$\sim \begin{bmatrix} -1 & 2 & 8 \\ 2 & 3 & 4 \\ 0 & 1 & 7 \end{bmatrix} \quad \text{By } \mathbf{R}_1 \leftrightarrow \mathbf{R}_3$$

$$\sim \begin{bmatrix} -1 & 2 & 8 \\ 0 & 7 & 20 \\ 0 & 1 & 7 \end{bmatrix} \quad \text{By } \mathbf{R}_2 \to \mathbf{R}_2 + 2\mathbf{R}_1$$

$$\sim \begin{bmatrix} -1 & 2 & 8 \\ 0 & 7 & 20 \\ 0 & 0 & \frac{29}{7} \end{bmatrix} \quad \text{By } \mathbf{R}_3 \to \mathbf{R}_3 - \frac{1}{7}\mathbf{R}_2$$

which is a triangular matrix.

If **A** is a square matrix such that $|\mathbf{A}| \neq 0$, then A can be reduced to a unit matrix by elementary row operations. The method is simple: Make the first element in the first column 1 and all other elements in first column 0 by E-row operations. Now, make the second element of second column 1 and all other elements in second column 0 by E-row operations. Continue the process for all columns. We shall obtain the identity matrix of same order as **A.** The result can be stated as

THEOREM

Every non-singular matrix can be reduced to a unit matrix by elementary row operations.

The following example illustrates the method.

EXAMPLE 1.15 Reduce the matrix $\mathbf{A} = \begin{bmatrix} 2 & 4 & 6 \\ 3 & 1 & 2 \\ 0 & 1 & -1 \end{bmatrix}$ to a unit matrix of order 3.

SOLUTION

$$\mathbf{A} = \begin{bmatrix} 2 & 4 & 6 \\ 3 & 1 & 2 \\ 0 & 1 & -1 \end{bmatrix}$$

$$\sim \begin{bmatrix} 1 & 2 & 3 \\ 3 & 1 & 2 \\ 0 & 1 & -1 \end{bmatrix} \quad \text{By applying } \mathbf{R}_1 \to \frac{1}{2}\mathbf{R}_1$$

$$\sim \begin{bmatrix} 1 & 2 & 3 \\ 0 & -5 & -7 \\ 0 & 1 & -1 \end{bmatrix} \quad \text{By applying } \mathbf{R}_2 \to \mathbf{R}_2 - 3\mathbf{R}_1$$

$$\sim \begin{bmatrix} 1 & 2 & 3 \\ 0 & 1 & -1 \\ 0 & -5 & -7 \end{bmatrix} \quad \text{By applying } \mathbf{R}_2 \leftrightarrow \mathbf{R}_3$$

$$\sim \begin{bmatrix} 1 & 2 & 3 \\ 0 & 1 & -1 \\ 0 & 0 & -12 \end{bmatrix} \quad \text{By applying } \mathbf{R}_3 \to \mathbf{R}_3 + 5\mathbf{R}_2$$

$$\sim \begin{bmatrix} 1 & 0 & 5 \\ 0 & 1 & -1 \\ 0 & 0 & -12 \end{bmatrix} \quad \text{By applying } \mathbf{R}_1 \to \mathbf{R}_1 - 2\mathbf{R}_2$$

$$\sim \begin{bmatrix} 1 & 0 & 5 \\ 0 & 1 & -1 \\ 0 & 0 & 1 \end{bmatrix} \quad \text{By applying } \mathbf{R}_3 \to -\frac{1}{12}\mathbf{R}_3$$

$$\sim \begin{bmatrix} 1 & 0 & 0 \\ 0 & 1 & 0 \\ 0 & 0 & 1 \end{bmatrix} \quad \text{By applying } \mathbf{R}_2 \to \mathbf{R}_2 + \mathbf{R}_3, \mathbf{R}_1 \to \mathbf{R}_1 - 5\mathbf{R}_3,$$

We thus obtain a unit matrix from the matrix **A**.

We now state an important result which says that if we perform some E-row operation on product **AB**, then it is equivalent to performing the same operation on **A**. We first explain it through an example.

EXAMPLE 1.16 Let $\mathbf{A} = \begin{bmatrix} 2 & 1 \\ 0 & 3 \end{bmatrix}$ and $\mathbf{B} = \begin{bmatrix} 1 \\ 2 \end{bmatrix}$. We use the following notation:

If * denotes an elementary row operation, then by **A*** we denote the matrix obtained by performing the elementary row operation * on **A**.

First suppose that * denotes $\mathbf{R}_1 \leftrightarrow \mathbf{R}_2$, that is, interchanging rows $\mathbf{R}_1$ and $\mathbf{R}_2$.

Then $\quad \mathbf{A}^* = \begin{bmatrix} 0 & 3 \\ 2 & 1 \end{bmatrix}$

$$\mathbf{AB} = \begin{bmatrix} 2 & 1 \\ 0 & 3 \end{bmatrix}\begin{bmatrix} 1 \\ 2 \end{bmatrix} = \begin{bmatrix} 4 \\ 6 \end{bmatrix}$$

$$\therefore \quad (\mathbf{AB})^* = \begin{bmatrix} 6 \\ 4 \end{bmatrix}$$

$$\text{Also,} \quad \mathbf{A}^*\mathbf{B} = \begin{bmatrix} 0 & 3 \\ 2 & 1 \end{bmatrix}\begin{bmatrix} 1 \\ 2 \end{bmatrix} = \begin{bmatrix} 6 \\ 4 \end{bmatrix}$$

Thus, $(\mathbf{AB})^* = \mathbf{A}^*\mathbf{B}$. This means that the matrix obtained after interchanging the rows in **AB** is equivalent to the product obtained by post-multiplying **B** to the matrix obtained after interchanging rows of **A**.

Now, suppose that * denotes the E-row operation $\mathbf{R}_1 \to 2\mathbf{R}_1$, that is, multiplying $\mathbf{R}_1$ by 2.

$$\mathbf{A}^* = \begin{bmatrix} 4 & 2 \\ 0 & 3 \end{bmatrix}, \quad (\mathbf{AB})^* = \begin{bmatrix} 8 \\ 6 \end{bmatrix} \text{ and}$$

$$\mathbf{A}^*\mathbf{B} = \begin{bmatrix} 4 & 2 \\ 0 & 3 \end{bmatrix}\begin{bmatrix} 1 \\ 2 \end{bmatrix} = \begin{bmatrix} 8 \\ 6 \end{bmatrix} = (\mathbf{AB})^*$$

Next, suppose that * denotes the E-row operation

$$\mathbf{R}_1 \to \mathbf{R}_1 + 2\mathbf{R}_2$$

$$\text{Then} \quad \mathbf{A}^* = \begin{bmatrix} 2 & 7 \\ 0 & 3 \end{bmatrix}, \quad (\mathbf{AB})^* = \begin{bmatrix} 16 \\ 6 \end{bmatrix}$$

and $\mathbf{A}^*\mathbf{B} = \begin{bmatrix} 2 & 7 \\ 0 & 3 \end{bmatrix}\begin{bmatrix} 1 \\ 2 \end{bmatrix} = \begin{bmatrix} 16 \\ 6 \end{bmatrix}$. Thus, in this case also we have $(\mathbf{AB})^* = \mathbf{A}^*\mathbf{B}$.

From this example, we can conclude that if * denotes an elementary row operation, then $(\mathbf{AB})^* = \mathbf{A}^*\mathbf{B}$ where **A** and **B** are two matrices. This result can be stated as

THEOREM

An elementary row operation on the product of two matrices is equivalent to the same elementary row operation on the pre-factor of the product.

We can similarly discuss elementary column operations. By an elementary operation we mean either an elementary row operation or an elementary column operation. We end this section by giving a small definition.

DEFINITION: A matrix obtained from a unit matrix by a single elementary operation is called an 'elementary matrix'.

For example, $\begin{bmatrix} 0 & 1 \\ 1 & 0 \end{bmatrix}$ is an elementary matrix because it can be obtained from $\mathbf{I}_2$ by interchanging $\mathbf{R}_1$ and $\mathbf{R}_2$. The elementary

NOTE: Elementary matrices are always non-singular. Can you explain why?

matrix $\begin{bmatrix} 2 & 0 \\ 0 & 1 \end{bmatrix}$ is obtained from $\mathbf{I}_2$ by multiplying $\mathbf{R}_1$ by 2.

$\begin{bmatrix} 0 & 0 & 1 \\ 0 & 1 & 0 \\ 1 & 0 & 0 \end{bmatrix}$ is an elementary matrix because it is obtained from $\mathbf{I}_3$ by $\mathbf{R}_1 \leftrightarrow \mathbf{R}_3$.

EXERCISE 1.2

1. Reduce the following matrices to triangular matrices by elementary row operations.

(a) $\begin{bmatrix} 3 & 1 & 4 \\ 1 & 2 & -5 \\ 0 & 1 & 2 \end{bmatrix}$ **(b)** $\begin{bmatrix} 0 & 1 & 3 \\ 2 & 0 & -1 \\ 5 & 1 & 0 \end{bmatrix}$ **(c)** $\begin{bmatrix} -1 & 2 & 1 & 8 \\ 3 & 1 & -1 & 0 \\ 4 & 2 & 1 & 7 \end{bmatrix}$

(d) $\begin{bmatrix} 1 & 4 & 2 & -8 \\ 2 & 9 & 6 & 5 \\ -3 & -15 & 8 & 13 \end{bmatrix}$ **(e)** $\begin{bmatrix} 2 & 1 & 4 \\ -1 & 3 & -2 \\ 4 & 1 & 3 \end{bmatrix}$ **(f)** $\begin{bmatrix} 1 & 1 & 1 & 1 \\ 1 & 3 & -2 & 1 \\ 2 & 0 & -3 & 2 \\ 3 & 3 & 0 & 3 \end{bmatrix}$

2. Reduce the following non-singular matrices to identity matrices by elementary row operations.

(a) $\begin{bmatrix} 2 & 0 & -1 \\ 5 & 1 & 0 \\ 0 & 1 & 3 \end{bmatrix}$ **(b)** $\begin{bmatrix} 0 & 0 & 2 \\ 2 & 2 & 6 \\ 3 & 6 & 9 \end{bmatrix}$ **(c)** $\begin{bmatrix} 1 & 1 & 2 \\ 3 & 1 & 1 \\ 2 & 3 & 1 \end{bmatrix}$ **(d)** $\begin{bmatrix} 1 & 2 & 1 \\ 0 & 1 & -1 \\ 3 & -1 & 1 \end{bmatrix}$

3. Write down the elementary matrices obtained by applying each of the following operation to $\mathbf{I}_3$.

(a) $\mathbf{R}_2 \leftrightarrow \mathbf{R}_3$ **(b)** $\mathbf{R}_1 \rightarrow 3\mathbf{R}_1$ **(c)** $\mathbf{R}_1 \rightarrow \mathbf{R}_1 - 4\mathbf{R}_3$ **(d)** $\mathbf{R}_3 \rightarrow \mathbf{R}_3 - 2\mathbf{R}_2$

1.3 INVERSE OF A MATRIX

? DEFINITION: A non-zero square matrix **A** of order *n* is said to be 'invertible' if there exists a square matrix **B** of the same order such that

$$\mathbf{AB} = \mathbf{BA} = \mathbf{I}_n$$

where $\mathbf{I}_n$ is the identity matrix of order *n*.
The matrix **B** is called an 'inverse' of **A** (and **A** is called an inverse of **B**).

EXAMPLE 1.17 Let $\mathbf{A} = \begin{bmatrix} 1 & -1 \\ 1 & 1 \end{bmatrix}$ and $\mathbf{B} = \begin{bmatrix} \frac{1}{2} & \frac{1}{2} \\ -\frac{1}{2} & \frac{1}{2} \end{bmatrix}$.

Then $\mathbf{AB} = \begin{bmatrix} 1 & 0 \\ 0 & 1 \end{bmatrix} = \mathbf{I}_2$

and $\mathbf{BA} = \begin{bmatrix} 1 & 0 \\ 0 & 1 \end{bmatrix} = \mathbf{I}_2$

that is, $\mathbf{AB} = \mathbf{BA} = \mathbf{I}_2$

Thus, **B** is an inverse of **A**.

An invertible matrix has a unique inverse. It follows from the following.

THEOREM

If **A**, **B**, **C** are square matrices of order n, such that $\mathbf{AB} = \mathbf{BA} = \mathbf{I}_n$ and $\mathbf{AC} = \mathbf{CA} = \mathbf{I}_n$ then $\mathbf{B} = \mathbf{C}$.

Proof $\mathbf{B} = \mathbf{BI}_n = \mathbf{B(AC)} = \mathbf{(BA)C} = \mathbf{I}_n\mathbf{C} = \mathbf{C}$

Hence if A is invertible, it has a unique inverse, which we denote by $\mathbf{A}^{-1}$. Thus, we have

$$\mathbf{AA}^{-1} = \mathbf{A}^{-1}\mathbf{A} = \mathbf{I}$$

Every square matrix need not have an inverse. For example, the matrix $\begin{bmatrix} 2 & 0 \\ 1 & 0 \end{bmatrix}$ does not have an inverse. Because if $\mathbf{B} = \begin{bmatrix} a & b \\ c & d \end{bmatrix}$ is an inverse of **A**, then we must have $\mathbf{AB} = \mathbf{I}_2$.

That is, $\begin{bmatrix} 2 & 0 \\ 1 & 0 \end{bmatrix}\begin{bmatrix} a & b \\ c & d \end{bmatrix} = \begin{bmatrix} 1 & 0 \\ 0 & 1 \end{bmatrix}$

or $$\begin{bmatrix} 2a & 2b \\ a & b \end{bmatrix} = \begin{bmatrix} 1 & 0 \\ 0 & 1 \end{bmatrix} \tag{1}$$

which gives $2a = 1$, $a = 0$, $2b = 0$ and $b = 1$. However, if $2a = 1$, then $a \neq 0$. Hence, values of a, b, c and d that satisfy (1) do not exist, and the matrix $\begin{bmatrix} 2 & 0 \\ 1 & 0 \end{bmatrix}$ has no inverse.

EXAMPLE 1.18 Find the inverse of the matrix $\begin{bmatrix} 1 & 0 \\ 0 & 2 \end{bmatrix}$.

SOLUTION

Let $\mathbf{A} = \begin{bmatrix} 1 & 0 \\ 0 & 2 \end{bmatrix}$. Let $\mathbf{B} = \begin{bmatrix} a & b \\ c & d \end{bmatrix}$ be the inverse of **A**.

Then $\mathbf{AB} = \mathbf{I}_2$.

That is, $\begin{bmatrix} 1 & 0 \\ 0 & 2 \end{bmatrix}\begin{bmatrix} a & b \\ c & d \end{bmatrix} = \begin{bmatrix} 1 & 0 \\ 0 & 1 \end{bmatrix}$

or $\begin{bmatrix} a & b \\ 2c & 2d \end{bmatrix} = \begin{bmatrix} 1 & 0 \\ 0 & 1 \end{bmatrix}$

$\Rightarrow$ $a = 1, b = 0, c = 0, d = \frac{1}{2}$

So, $\mathbf{B} = \begin{bmatrix} 1 & 0 \\ 0 & \frac{1}{2} \end{bmatrix}$.

Also, $\mathbf{BA} = \begin{bmatrix} 1 & 0 \\ 0 & \frac{1}{2} \end{bmatrix}\begin{bmatrix} 1 & 0 \\ 0 & 2 \end{bmatrix} = \begin{bmatrix} 1 & 0 \\ 0 & 1 \end{bmatrix}$

Hence, $\mathbf{B} = \begin{bmatrix} 1 & 0 \\ 0 & \frac{1}{2} \end{bmatrix}$ is the inverse of **A**.

EXAMPLE 1.19 Find whether the matrix $\mathbf{A} = \begin{bmatrix} 1 & 2 \\ 2 & 4 \end{bmatrix}$ is invertible.

SOLUTION

Let $\mathbf{B} = \begin{bmatrix} a & b \\ c & d \end{bmatrix}$ be the inverse of **A**. Then $\mathbf{AB} = \mathbf{I}_2$

That is, $\begin{bmatrix} 1 & 2 \\ 2 & 4 \end{bmatrix}\begin{bmatrix} a & b \\ c & d \end{bmatrix} = \begin{bmatrix} 1 & 0 \\ 0 & 1 \end{bmatrix}$

which gives two pairs of system of linear equations.

$$\begin{aligned} a + 2c &= 1 \\ 2a + 4c &= 0 \end{aligned} \quad \text{and} \quad \begin{aligned} b + 2d &= 0 \\ 2b + 4d &= 1 \end{aligned}$$

These linear systems have no solutions, so no such matrix **B** can exist. Hence, **A** is not invertible.

From above examples, it is clear that a matrix may or may not have an inverse. Now the question arises: Under what conditions can we say that a given matrix is invertible? Before we answer this question, we make a small remark.

> **REMARK:** If **A** is invertible, and **B** is the inverse of **A**, then both **A** and **B** are square matrices of same order. This can be seen as follows:
> Suppose **A** is of order $p \times q$ and **B** is of order $r \times s$. Let $\mathbf{AB} = \mathbf{BA} = \mathbf{I}_n$ where $\mathbf{I}_n$ is the identity matrix of order n. For both **AB** and **BA** to be defined it is necessary that $q = r$ and $s = p$. Then **AB** is a $p \times s$ matrix and **BA** is $r \times q$ matrix. Since $\mathbf{AB} = \mathbf{BA} = \mathbf{I}_n$,
> $\therefore \quad p = q = r = s = n.$
> Thus, if **A** is not a square matrix, then there is no question of it being invertible.

Now, we state a theorem, without proof, which will enable us to find whether a given square matrix is invertible or not.

THEOREM

A square matrix **A** is invertible if and only if **A** is non-singular i.e., $|\mathbf{A}| \neq \mathbf{0}$.

Once we know that a matrix is invertible, we can find its inverse also. The method given in Example 1.18 is obviously not very practical for matrices of order greater than 2. We now discuss a method of obtaining inverse of a given square matrix using elementary row operations. Let **A** be the square matrix of order n whose inverse is to be found. Consider the identity $\mathbf{A} = \mathbf{IA}$ where **I** is the identity matrix. Reduce the matrix **A** on the L.H.S. to the identity matrix **I** by applying elementary row operations only. Apply all these operations (in the same order) to the pre-factor I on the R.H.S. of the identity. In this manner, the matrix **I** reduces to some matrix **B** such that $\mathbf{BA} = \mathbf{I}$. The matrix **B** so obtained is the inverse of **A**.

The method is illustrated in the following examples.

EXAMPLE 1.20 Find the inverse of the matrix $\begin{bmatrix} 1 & 2 \\ 3 & 4 \end{bmatrix}$.

SOLUTION

Let $\mathbf{A} = \begin{bmatrix} 1 & 2 \\ 3 & 4 \end{bmatrix}$. Consider the identity

$$\mathbf{A} = \mathbf{IA}$$

$$\begin{bmatrix} 1 & 2 \\ 3 & 4 \end{bmatrix} = \begin{bmatrix} 1 & 0 \\ 0 & 1 \end{bmatrix}\mathbf{A}$$

$$\Rightarrow \begin{bmatrix} 1 & 2 \\ 0 & -2 \end{bmatrix} = \begin{bmatrix} 1 & 0 \\ -3 & 1 \end{bmatrix}\mathbf{A} \qquad \text{By applying } \mathbf{R}_2 \to \mathbf{R}_2 - 3\mathbf{R}_1$$

$$\Rightarrow \begin{bmatrix} 1 & 2 \\ 0 & 1 \end{bmatrix} = \begin{bmatrix} 1 & 0 \\ \frac{3}{2} & -\frac{1}{2} \end{bmatrix}\mathbf{A} \qquad \text{By applying } \mathbf{R}_2 \to -\frac{1}{2}\mathbf{R}_2$$

$$\Rightarrow \begin{bmatrix} 1 & 0 \\ 0 & 1 \end{bmatrix} = \begin{bmatrix} -2 & 1 \\ \frac{3}{2} & -\frac{1}{2} \end{bmatrix}\mathbf{A} \qquad \text{By applying } \mathbf{R}_1 \to \mathbf{R}_1 - 2\mathbf{R}_2$$

that is, $\mathbf{I} = \mathbf{BA}$

where $\mathbf{B} = \begin{bmatrix} -2 & 1 \\ \frac{3}{2} & -\frac{1}{2} \end{bmatrix}$.

Thus, $\mathbf{A}^{-1} = \begin{bmatrix} -2 & 1 \\ \frac{3}{2} & -\frac{1}{2} \end{bmatrix}$.

EXAMPLE 1.21 Find the inverse of the matrix $\mathbf{A} = \begin{bmatrix} 1 & 1 & 1 \\ 0 & 2 & 3 \\ 5 & 5 & 1 \end{bmatrix}$ using elementary row operations.

[DU, B.Sc. (Ph. Sc.), 2006]

SOLUTION

Consider the identity $\mathbf{A} = \mathbf{IA}$.

$$\begin{bmatrix} 1 & 1 & 1 \\ 0 & 2 & 3 \\ 5 & 5 & 1 \end{bmatrix} = \begin{bmatrix} 1 & 0 & 0 \\ 0 & 1 & 0 \\ 0 & 0 & 1 \end{bmatrix}\mathbf{A}$$

$$\Rightarrow \begin{bmatrix} 1 & 1 & 1 \\ 0 & 2 & 3 \\ 0 & 0 & -4 \end{bmatrix} = \begin{bmatrix} 1 & 0 & 0 \\ 0 & 1 & 0 \\ -5 & 0 & 1 \end{bmatrix}\mathbf{A} \qquad \text{By applying } \mathbf{R}_3 \to \mathbf{R}_3 - 5\mathbf{R}_1$$

$$\Rightarrow \begin{bmatrix} 1 & 1 & 1 \\ 0 & 1 & 3/2 \\ 0 & 0 & -4 \end{bmatrix} = \begin{bmatrix} 1 & 0 & 0 \\ 0 & 1/2 & 0 \\ -5 & 0 & 1 \end{bmatrix}\mathbf{A} \qquad \text{By applying } \mathbf{R}_2 \to \frac{1}{2}\mathbf{R}_2$$

$$\Rightarrow \begin{bmatrix} 1 & 0 & -1/2 \\ 0 & 1 & 3/2 \\ 0 & 0 & -4 \end{bmatrix} = \begin{bmatrix} 1 & -1/2 & 0 \\ 0 & 1/2 & 0 \\ -5 & 0 & 1 \end{bmatrix}\mathbf{A} \qquad \text{By applying } \mathbf{R}_1 \to \mathbf{R}_1 - \mathbf{R}_2$$

$$\Rightarrow \begin{bmatrix} 1 & 0 & -1/2 \\ 0 & 1 & 3/2 \\ 0 & 0 & 1 \end{bmatrix} = \begin{bmatrix} 1 & -1/2 & 0 \\ 0 & 1/2 & 0 \\ 5/4 & 0 & -1/4 \end{bmatrix} \mathbf{A} \qquad \text{By applying } \mathbf{R}_3 \to -\frac{1}{4}\mathbf{R}_3$$

$$\Rightarrow \begin{bmatrix} 1 & 0 & 0 \\ 0 & 1 & 0 \\ 0 & 0 & 1 \end{bmatrix} = \begin{bmatrix} 13/8 & -1/2 & -1/8 \\ -15/8 & 1/2 & 3/8 \\ 5/4 & 0 & -1/4 \end{bmatrix} \mathbf{A} \qquad \text{By applying } \mathbf{R}_1 \to \mathbf{R}_1 + \frac{1}{2}\mathbf{R}_3,$$

$$\mathbf{R}_2 \to \mathbf{R}_2 - \frac{3}{2}\mathbf{R}_3$$

Thus, $$\mathbf{A}^{-1} = \begin{bmatrix} 13/8 & -1/2 & -1/8 \\ -15/8 & 1/2 & 3/8 \\ 5/4 & 0 & -1/4 \end{bmatrix}.$$

REMARK: Since the post-factor **A** on the R.H.S. of the identity **A** = **IA** remains unchanged, instead of considering the whole identity **A** = **IA**, we can just consider the $n \times 2n$ matrix [**A** : **I**] where **I** is the unit matrix of order n. By reducing **A** to unit matrix I by elementary row operations the matrix [**A** : **I**] reduces to the form [**I** : $\mathbf{A}^{-1}$] where A^{-1} gives the inverse of **A**. The following example illustrates this method.

EXAMPLE 1.22 Let $\mathbf{A} = \begin{bmatrix} 2 & 4 & 6 \\ 3 & 1 & 2 \\ 0 & 1 & -1 \end{bmatrix}$. Find $\mathbf{A}^{-1}$.

SOLUTION

Consider the 3×6 matrix $[\mathbf{A} : \mathbf{I}_3]$

That is, $$[\mathbf{A} : \mathbf{I}_3] = \begin{bmatrix} 2 & 4 & 6 & : & 1 & 0 & 0 \\ 3 & 1 & 2 & : & 0 & 1 & 0 \\ 0 & 1 & -1 & : & 0 & 0 & 1 \end{bmatrix}$$

$$\sim \begin{bmatrix} 1 & 2 & 3 & : & \frac{1}{2} & 0 & 0 \\ 3 & 1 & 2 & : & 0 & 1 & 0 \\ 0 & 1 & -1 & : & 0 & 0 & 1 \end{bmatrix} \qquad \left[\mathbf{R}_1 \to \frac{1}{2}\mathbf{R}_1\right]$$

$$\sim \begin{bmatrix} 1 & 2 & 3 & : & \frac{1}{2} & 0 & 0 \\ 0 & -5 & -7 & : & -\frac{3}{2} & 1 & 0 \\ 0 & 1 & -1 & : & 0 & 0 & 1 \end{bmatrix} \qquad [\mathbf{R}_2 \to \mathbf{R}_2 - 3\mathbf{R}_1]$$

$$\sim \begin{bmatrix} 1 & 2 & 3 & : & \frac{1}{2} & 0 & 0 \\ 0 & 1 & -1 & : & 0 & 0 & 1 \\ 0 & -5 & -7 & : & -\frac{3}{2} & 1 & 0 \end{bmatrix} \qquad [\mathbf{R}_2 \leftrightarrow \mathbf{R}_3]$$

$$\sim \begin{bmatrix} 1 & 2 & 3 & : & \frac{1}{2} & 0 & 0 \\ 0 & 1 & -1 & : & 0 & 0 & 1 \\ 0 & 0 & -12 & : & -\frac{3}{2} & 1 & 5 \end{bmatrix} \qquad [\mathbf{R}_3 \to \mathbf{R}_3 + 5\mathbf{R}_2]$$

$$\sim \begin{bmatrix} 1 & 0 & 5 & : & \frac{1}{2} & 0 & -2 \\ 0 & 1 & -1 & : & 0 & 0 & 1 \\ 0 & 0 & -12 & : & -\frac{3}{2} & 1 & 5 \end{bmatrix} \qquad [\mathbf{R}_1 \to \mathbf{R}_1 - 2\mathbf{R}_2]$$

$$\sim \begin{bmatrix} 1 & 0 & 5 & : & \frac{1}{2} & 0 & -2 \\ 0 & 1 & -1 & : & 0 & 0 & 1 \\ 0 & 0 & 1 & : & \frac{3}{24} & -\frac{1}{12} & -\frac{5}{12} \end{bmatrix} \qquad \left[\mathbf{R}_3 \to -\frac{1}{12}\mathbf{R}_3\right]$$

$$\sim \begin{bmatrix} 1 & 0 & 5 & : & \frac{1}{2} & 0 & -2 \\ 0 & 1 & 0 & : & \frac{3}{24} & -\frac{1}{12} & \frac{7}{12} \\ 0 & 0 & 1 & : & \frac{3}{24} & -\frac{1}{12} & -\frac{5}{12} \end{bmatrix} \qquad [\mathbf{R}_2 \to \mathbf{R}_2 + \mathbf{R}_3]$$

$$\sim \begin{bmatrix} 1 & 0 & 0 & : & -\frac{3}{24} & \frac{5}{12} & \frac{1}{12} \\ 0 & 1 & 0 & : & \frac{3}{24} & -\frac{1}{12} & \frac{7}{12} \\ 0 & 0 & 1 & : & \frac{3}{24} & -\frac{1}{12} & -\frac{5}{12} \end{bmatrix} \qquad [\mathbf{R}_1 \to \mathbf{R}_1 - 5\mathbf{R}_3]$$

$$= [\mathbf{I}_3 : \mathbf{A}^{-1}]$$

Hence,
$$\mathbf{A}^{-1} = \begin{bmatrix} -\frac{3}{24} & \frac{5}{12} & \frac{1}{12} \\ \frac{3}{24} & -\frac{1}{12} & \frac{7}{12} \\ \frac{3}{24} & -\frac{1}{12} & -\frac{5}{12} \end{bmatrix}.$$

EXAMPLE 1.23 Find $\mathbf{A}^{-1}$, if it exists, where $\mathbf{A} = \begin{bmatrix} 1 & 2 & -3 \\ 1 & -2 & 1 \\ 5 & -2 & -3 \end{bmatrix}$.

SOLUTION

We consider the 3×6 partitioned matrix $[\mathbf{A} : \mathbf{I}_3]$.

$$[\mathbf{A} : \mathbf{I}_3] = \begin{bmatrix} 1 & 2 & -3 & : & 1 & 0 & 0 \\ 1 & -2 & 1 & : & 0 & 1 & 0 \\ 5 & -2 & -3 & : & 0 & 0 & 1 \end{bmatrix}$$

$$\sim \begin{bmatrix} 1 & 2 & -3 & : & 1 & 0 & 0 \\ 0 & -4 & 4 & : & -1 & 1 & 0 \\ 5 & -2 & -3 & : & 0 & 0 & 1 \end{bmatrix} \quad [\mathbf{R}_2 \rightarrow \mathbf{R}_2 - \mathbf{R}_1]$$

$$\sim \begin{bmatrix} 1 & 2 & -3 & : & 1 & 0 & 0 \\ 0 & -4 & 4 & : & -1 & 1 & 0 \\ 0 & -12 & 12 & : & -5 & 0 & 1 \end{bmatrix} \quad [\mathbf{R}_3 \rightarrow \mathbf{R}_3 - 5\mathbf{R}_1]$$

$$\sim \begin{bmatrix} 1 & 2 & -3 & : & 1 & 0 & 0 \\ 0 & -4 & 4 & : & -1 & 1 & 0 \\ 0 & 0 & 0 & : & -2 & -3 & 1 \end{bmatrix}$$

At this point, we note that **A** is equivalent to the matrix $\begin{bmatrix} 1 & 2 & -3 \\ 0 & -4 & 4 \\ 0 & 0 & 0 \end{bmatrix}$ which has a row of zeros. This matrix cannot be reduced to an identity matrix. In fact, it is a singular matrix as its determinant is zero. Thus, **A** is not invertible.

EXERCISE 1.3

1. Find the inverse of the following matrices, if it exists (Use the method of Example 1.18 and Example 1.19).

   (a) $\begin{bmatrix} 1 & 2 \\ 0 & 0 \end{bmatrix}$ **(b)** $\begin{bmatrix} 3 & 1 \\ 4 & 2 \end{bmatrix}$

2. Find the inverse of the following 2×2 matrices using elementary row operations.

   (a) $\begin{bmatrix} 2 & 1 \\ 5 & 3 \end{bmatrix}$ **(b)** $\begin{bmatrix} 4 & 5 \\ 3 & 4 \end{bmatrix}$

3. Find the inverse, if it exists, of the following 3×3 matrices using elementary row operations.

   (a) $\begin{bmatrix} 1 & 2 & 1 \\ 3 & 2 & 3 \\ 1 & 1 & 2 \end{bmatrix}$ **(b)** $\begin{bmatrix} 2 & 0 & -1 \\ 5 & 1 & 0 \\ 0 & 1 & 3 \end{bmatrix}$ **(c)** $\begin{bmatrix} 0 & 0 & 2 \\ 1 & 2 & 6 \\ 3 & 6 & 9 \end{bmatrix}$ **(d)** $\begin{bmatrix} 1 & 2 & 3 \\ 1 & 1 & 2 \\ 0 & 1 & 2 \end{bmatrix}$

   (e) $\begin{bmatrix} 0 & 1 & 2 \\ 1 & 2 & 3 \\ 3 & 1 & 1 \end{bmatrix}$ **(f)** $\begin{bmatrix} 1 & -1 & 1 \\ 4 & 1 & 0 \\ 8 & 1 & 1 \end{bmatrix}$ **(g)** $\begin{bmatrix} 1 & 2 & 3 \\ 1 & 3 & 5 \\ 1 & 5 & 12 \end{bmatrix}$ **(h)** $\begin{bmatrix} 4 & -6 & 1 \\ -1 & -1 & 1 \\ -4 & 11 & -1 \end{bmatrix}$

4. Let $\mathbf{A} = \begin{bmatrix} 1 & 1 \\ 0 & 1 \end{bmatrix}$ and $\mathbf{B} = \begin{bmatrix} 1 & 0 \\ 2 & 1 \end{bmatrix}$. Find $\mathbf{A}^{-1}, \mathbf{B}^{-1}$ and $(\mathbf{AB})^{-1}$. Show that $(\mathbf{AB})^{-1} \neq \mathbf{A}^{-1}\mathbf{B}^{-1}$.

5. Let **A** and **B** be two invertible matrices of the same order. Prove that $(\mathbf{AB})^{-1} = \mathbf{B}^{-1}\mathbf{A}^{-1}$. Verify this result for the matrices of question 4.

6. If $\mathbf{A} = \begin{bmatrix} 2 & 2 & 1 \\ -2 & 1 & 2 \\ 1 & -2 & 2 \end{bmatrix}$ and $\mathbf{B} = \begin{bmatrix} 1 & 3 & 2 \\ 1 & 1 & 1 \\ 2 & -3 & -1 \end{bmatrix}$ verify that $(\mathbf{AB})^{-1} = \mathbf{B}^{-1}\mathbf{A}^{-1}$.

7. For the matrix $\mathbf{A} = \begin{bmatrix} 1 & 2 \\ 3 & 4 \end{bmatrix}$, verify that $(\mathbf{A}^{T})^{-1} = (\mathbf{A}^{-1})^{T}$.

8. Give a criterion for deciding whether the 2×2 matrix $\begin{bmatrix} a & b \\ c & d \end{bmatrix}$ has an inverse.

9. Give a formula for $\begin{bmatrix} a & b \\ c & d \end{bmatrix}^{-1}$, when it exists.

10. A non-singular matrix $\mathbf{A}$ has the property $\mathbf{AA}^{\mathrm{T}} = \mathbf{A}^{\mathrm{T}}\mathbf{A}$. Prove that $\mathbf{A}^{\mathrm{T}}\mathbf{A}^{-1} = \mathbf{A}^{-1}\mathbf{A}^{\mathrm{T}}$. Prove also that if $\mathbf{B} = \mathbf{A}^{-1}\mathbf{A}^{\mathrm{T}}$, then $\mathbf{BB}^{\mathrm{T}}$ is the identity matrix.

1.4 RANK OF A MATRIX

Given any matrix $\mathbf{A}$ of order $m \times n$, we can obtain square sub-matrices of order r ($0 < r \leq$ least of m and n) from $\mathbf{A}$ by selecting the elements in any r rows and r columns of the matrix. The determinants of these square sub-matrices of order r are called r-rowed minors of the matrix $\mathbf{A}$. For example, consider the matrix

$$\mathbf{A} = \begin{bmatrix} 1 & 2 & 4 & 1 \\ 2 & 1 & 3 & 0 \\ 3 & 0 & 2 & 1 \end{bmatrix}$$

This is a 3×4 matrix. Then deleting any one column gives a sub-matrix of order 3.

$\begin{bmatrix} 1 & 2 & 4 \\ 2 & 1 & 3 \\ 3 & 0 & 2 \end{bmatrix}$ is one such sub-matrix.

The determinant of this sub-matrix is a 3-rowed minor.

That is, $\begin{vmatrix} 1 & 2 & 4 \\ 2 & 1 & 3 \\ 3 & 0 & 2 \end{vmatrix}$ is a 3-rowed minor. Similarly,

$\begin{vmatrix} 1 & 2 \\ 2 & 1 \end{vmatrix}, \begin{vmatrix} 1 & 3 \\ 0 & 2 \end{vmatrix}$ and others are 2-rowed minors.

? DEFINITION: Let $\mathbf{A}$ be a matrix of order $m \times n$. The order of the largest square sub-matrix of $\mathbf{A}$ whose determinant has a non-zero value is called the 'rank' of the matrix $\mathbf{A}$.

The rank of the zero matrix is defined to be zero.

It is clear from the above definition that if r is the rank of a matrix $\mathbf{A}$ then

(i) $\mathbf{A}$ possesses at least one r-rowed non-zero minor.

(ii) $\mathbf{A}$ does not possess any non-zero $(r+1)$-rowed minor, that is, determinant of every square sub-matrix of order greater than r is zero.

EXAMPLE 1.24 Find the rank of the matrix $\begin{bmatrix} 4 & 2 & 3 \\ 8 & 5 & 2 \\ 12 & -4 & 5 \end{bmatrix}$.

SOLUTION

$$\text{Let } \mathbf{A} = \begin{bmatrix} 4 & 2 & 3 \\ 8 & 5 & 2 \\ 12 & -4 & 5 \end{bmatrix}$$

$$|\mathbf{A}| = \begin{vmatrix} 4 & 2 & 3 \\ 8 & 5 & 2 \\ 12 & -4 & 5 \end{vmatrix} = -176 \neq 0$$

Hence, rank of **A** is 3.

EXAMPLE 1.25 Find the rank of the matrix $\begin{bmatrix} 1 & 2 & 3 \\ 2 & 1 & 0 \\ 0 & 1 & 2 \end{bmatrix}$.

SOLUTION

$$\text{Let } \mathbf{A} = \begin{bmatrix} 1 & 2 & 3 \\ 2 & 1 & 0 \\ 0 & 1 & 2 \end{bmatrix}$$

$$|\mathbf{A}| = 1(2-0) - 2(4-0) + 3(2-0) = 0$$

So rank of **A** cannot be 3.

Now $\begin{bmatrix} 1 & 2 \\ 2 & 1 \end{bmatrix}$ is a square sub-matrix of **A** such that $\begin{vmatrix} 1 & 2 \\ 2 & 1 \end{vmatrix} = -3 \neq 0$.

$\therefore$ **A** has a 2-rowed non-zero minor. Hence, rank of **A** = 2.

EXAMPLE 1.26 Find the rank of the matrix $\mathbf{A} = \begin{bmatrix} 1 & -3 & 0 & 7 \\ 9 & 1 & 2 & 0 \end{bmatrix}$.

SOLUTION

The largest square sub-matrix of **A** is of order 2. So rank is less than or equal to 2. Since $\begin{vmatrix} 1 & -3 \\ 9 & 1 \end{vmatrix} = 28 \neq 0$, so rank of **A** = 2.

REMARK: We usually denote rank of a matrix **A** by $\rho(\mathbf{A})$. It is clear from above examples that

(1) If **A** is a non-singular matrix of order n, then $\rho(\mathbf{A}) = n$.
(2) If **A** is a singular matrix of order n, then $\rho(\mathbf{A}) < n$.
(3) If **A** is a matrix of order $m \times n$, then $\rho(\mathbf{A}) \leq$ least of m and n.
(4) If **A** is a non-zero matrix, then $\rho(\mathbf{A}) \neq 0$.

EXAMPLE 1.27 Find the rank of the matrix $\mathbf{A} = \begin{bmatrix} 4 & 2 & -1 & 3 \\ 0 & 5 & -1 & 2 \\ 12 & -4 & -1 & 5 \end{bmatrix}$.

SOLUTION

Since matrix **A** is of order 3×4, the rank of **A** cannot exceed 3.

It can be seen, however, that each of the 3-rowed minors of **A** has value zero, that is,

$$\begin{vmatrix} 4 & 2 & -1 \\ 0 & 5 & -1 \\ 12 & -4 & -1 \end{vmatrix} = 0 \qquad \begin{vmatrix} 4 & 2 & 3 \\ 0 & 5 & 2 \\ 12 & -4 & 5 \end{vmatrix} = 0$$

$$\begin{vmatrix} 4 & -1 & 3 \\ 0 & -1 & 2 \\ 12 & -1 & 5 \end{vmatrix} = 0 \quad \text{and} \quad \begin{vmatrix} 2 & -1 & 3 \\ 5 & -1 & 2 \\ -4 & -1 & 5 \end{vmatrix} = 0$$

Therefore, the rank of **A** cannot be three.

Since $\begin{vmatrix} 4 & 2 \\ 0 & 5 \end{vmatrix} = 20 \neq 0$, rank of $\mathbf{A} = 2$.

It is clear from the above example, that sometimes it is difficult to find the rank of a matrix of high order as we have to determine several minors of high order. The difficulty in determining the rank of a matrix of high order may be reduced by considering elementary row operations. We must note the following.

THEOREM

The rank of a matrix remains unchanged under elementary row operations.

This is because the order of the largest non-singular square sub-matrix of the matrix is not affected by the elementary row operations. In fact, under type I elementary row operation, only the signs of the values of some r-rowed minors are changed; under type II elementary row operation the values of some r-rowed minors are multiplied by a non-zero scalar; and under type III elementary row operation the value of every r-rowed minor is unchanged.

The above theorem provides an important tool for determining the rank of a matrix. By applying elementary row operations on a matrix **A**, we can obtain a matrix **B** whose rank can be easily found. This is illustrated in following examples.

EXAMPLE 1.28 Reduce the matrix

$$\mathbf{A} = \begin{bmatrix} 5 & 3 & 14 & 4 \\ 0 & 1 & 2 & 1 \\ 1 & -1 & 2 & 0 \end{bmatrix}$$

to triangular form by elementary row operations and hence determine its rank.

[DU, B.A.(P), 1990]

SOLUTION

$$\mathbf{A} = \begin{bmatrix} 5 & 3 & 14 & 4 \\ 0 & 1 & 2 & 1 \\ 1 & -1 & 2 & 0 \end{bmatrix}$$

$$\sim \begin{bmatrix} 1 & -1 & 2 & 0 \\ 0 & 1 & 2 & 1 \\ 5 & 3 & 14 & 4 \end{bmatrix} \quad \text{by } \mathbf{R}_1 \leftrightarrow \mathbf{R}_3$$

$$\sim \begin{bmatrix} 1 & -1 & 2 & 0 \\ 0 & 1 & 2 & 1 \\ 0 & 8 & 4 & 4 \end{bmatrix} \quad \text{by } \mathbf{R}_3 \to \mathbf{R}_3 - 5\mathbf{R}_1$$

$$\sim \begin{bmatrix} 1 & -1 & 2 & 0 \\ 0 & 1 & 2 & 1 \\ 0 & 0 & -12 & 4 \end{bmatrix} \quad \text{by } \mathbf{R}_3 \to \mathbf{R}_3 - 8\mathbf{R}_2$$

We have, thus, reduced **A** to triangular form by elementary row operations. The reduced matrix has a non-zero 3-rowed minor

$$\begin{vmatrix} 1 & -1 & 2 \\ 0 & 1 & 2 \\ 0 & 0 & -12 \end{vmatrix} = -12.$$

So rank of reduced matrix is 3. Hence, rank of **A** = 3.

EXAMPLE 1.29 Determine the rank of matrix **A** where

$$\mathbf{A} = \begin{bmatrix} 0 & 2 & 4 & 6 \\ 3 & -1 & 4 & -2 \\ 6 & -1 & 10 & -1 \end{bmatrix}$$

SOLUTION

$$\mathbf{A} = \begin{bmatrix} 0 & 2 & 4 & 6 \\ 3 & -1 & 4 & -2 \\ 6 & -1 & 10 & -1 \end{bmatrix}$$

$$\sim \begin{bmatrix} 3 & -1 & 4 & -2 \\ 0 & 2 & 4 & 6 \\ 6 & -1 & 10 & -1 \end{bmatrix} \quad \text{by } \mathbf{R}_1 \leftrightarrow \mathbf{R}_2$$

$$\sim \begin{bmatrix} 3 & -1 & 4 & -2 \\ 0 & 2 & 4 & 6 \\ 0 & 1 & 2 & 3 \end{bmatrix} \quad \text{by } \mathbf{R}_3 \to \mathbf{R}_3 - 2\mathbf{R}_1$$

$$\sim \begin{bmatrix} 3 & -1 & 4 & -2 \\ 0 & 2 & 4 & 6 \\ 0 & 0 & 0 & 0 \end{bmatrix} \quad \text{by } \mathbf{R}_3 \to \mathbf{R}_3 - \frac{1}{2}\mathbf{R}_2.$$

So the matrix **A** is transformed to the triangular matrix $\begin{bmatrix} 3 & -1 & 4 & -2 \\ 0 & 2 & 4 & 6 \\ 0 & 0 & 0 & 0 \end{bmatrix}$.

It is clear that every 3-rowed minor of this matrix has value zero.

Also since $\begin{vmatrix} 3 & -1 \\ 0 & 2 \end{vmatrix} = 6 \neq 0$, so rank of this matrix is 2.

Hence, rank **A** = 2.

EXAMPLE 1.30 Obtain the rank of the matrix:

$$\mathbf{A} = \begin{bmatrix} 1 & 1 & 1 & 1 \\ 1 & 3 & -2 & 1 \\ 2 & 0 & -3 & 2 \\ 3 & 3 & -3 & 3 \end{bmatrix}.$$

SOLUTION

We reduce matrix to triangular form

$$\mathbf{A} = \begin{bmatrix} 1 & 1 & 1 & 1 \\ 1 & 3 & -2 & 1 \\ 2 & 0 & -3 & 2 \\ 3 & 3 & -3 & 3 \end{bmatrix}$$

$$\sim \begin{bmatrix} 1 & 1 & 1 & 1 \\ 0 & 2 & -3 & 0 \\ 0 & -2 & -5 & 0 \\ 0 & 0 & -6 & 0 \end{bmatrix} \quad \text{By } \mathbf{R}_2 \to \mathbf{R}_2 - \mathbf{R}_1,\ \mathbf{R}_3 \to \mathbf{R}_3 - 2\mathbf{R}_1,\ \mathbf{R}_4 \to \mathbf{R}_4 - 3\mathbf{R}_1.$$

$$\sim \begin{bmatrix} 1 & 1 & 1 & 1 \\ 0 & 2 & -3 & 0 \\ 0 & 0 & -8 & 0 \\ 0 & 0 & 0 & 0 \end{bmatrix} \quad \text{By } \mathbf{R}_4 \to \mathbf{R}_4 - \frac{6}{8}\mathbf{R}_3$$

$= \mathbf{B}.$

Since A is 4×4 matrix, maximum possible rank of $\mathbf{A} = 4$. But last row of **B** consist of zeroes only.

$\therefore$ rank of **B** cannot be 4. Hence, rank of **A** also cannot be 4.

Also $\begin{vmatrix} 1 & 1 & 1 \\ 0 & 2 & -3 \\ 0 & 0 & -8 \end{vmatrix} = 1 \times 2 \times (-8) = -16 \neq 0.$

$\therefore$ rank of $\mathbf{B} = 3$ and, hence, rank of $\mathbf{A} = 3$.

EXERCISE 1.4

1. By finding a non-zero minor of largest order determine the rank of the following matrices.

(a) $\begin{bmatrix} 2 & 3 & 3 \\ 3 & 6 & 12 \\ 2 & 4 & 8 \end{bmatrix}$ **(b)** $\begin{bmatrix} 1 & 2 & 3 \\ 3 & 1 & 2 \\ 2 & 3 & 1 \end{bmatrix}$ **(c)** $\begin{bmatrix} 1 & 1 & -1 \\ 0 & 2 & 0 \\ 0 & 0 & 3 \end{bmatrix}$ **(d)** $\begin{bmatrix} 1 & 2 & 5 & 2 & 3 \\ 2 & 4 & 10 & 3 & 6 \end{bmatrix}$

2. Find the rank of the following matrices by reducing them to triangular form using elementary row operations.

(a) $\begin{bmatrix} 4 & 2 & 3 \\ 8 & 5 & 2 \\ 12 & -4 & 5 \end{bmatrix}$ **(b)** $\begin{bmatrix} 1 & 1 & 1 & 1 \\ 3 & 4 & 5 & 2 \\ 2 & 3 & 4 & 0 \end{bmatrix}$ *[DU, B.Sc. (G), 1992]*

(c) $\begin{bmatrix} 0 & 1 & 2 & 1 \\ 1 & 2 & 3 & 4 \\ 3 & 1 & 1 & 3 \end{bmatrix}$ **(d)** $\begin{bmatrix} 1 & 2 & 1 & 0 \\ -2 & 4 & 3 & 0 \\ 1 & 0 & 2 & -8 \end{bmatrix}$ *[DU, B.Sc. (G), 1995]*

(e) $\begin{bmatrix} 1 & 1 & -1 & 1 \\ 1 & -1 & 2 & -1 \\ 3 & 3 & 0 & 1 \end{bmatrix}$ **(f)** $\begin{bmatrix} 1 & 3 & 6 & -1 \\ 1 & 4 & 5 & 1 \\ 1 & 5 & 4 & 3 \end{bmatrix}$

(g) $\begin{bmatrix} 1 & 1 & 2 & 3 \\ 1 & 3 & 0 & 3 \\ 1 & -2 & -3 & 0 \\ 1 & 1 & 2 & 3 \end{bmatrix}$ *[DU, B.Sc. (G), 1997]*

(h) $\begin{bmatrix} -2 & -1 & -3 & -1 \\ 1 & 2 & 3 & -1 \\ 1 & 0 & 1 & 1 \\ 0 & 1 & 1 & -1 \end{bmatrix}$ *[DU, B.Sc. (G), 1996]*

(i) $\begin{bmatrix} 1 & -1 & 2 & -3 \\ 4 & 1 & 0 & 2 \\ 0 & 3 & 1 & 4 \\ 0 & 1 & 0 & 2 \end{bmatrix}$ *[DU, B.Sc. (G), 1994]*

(j) $\begin{bmatrix} 0 & 1 & -3 & -1 \\ 1 & 0 & 1 & 1 \\ 3 & 1 & 0 & 2 \\ 1 & 1 & -2 & 0 \end{bmatrix}$ *[DU, B.A. (P), 1994]*

1.5 SYSTEMS OF LINEAR EQUATIONS

Consider a system of m linear equations in n variables $x_1, x_2, \ldots, x_n$:

$$\begin{aligned} a_{11}x_1 + a_{12}x_2 + \cdots + a_{1n}x_n &= b_1 \\ a_{21}x_1 + a_{22}x_2 + \cdots + a_{2n}x_n &= b_2 \\ \cdots \quad \cdots \quad \cdots \quad &\cdots \\ a_{m1}x_1 + a_{m2}x_2 + \cdots + a_{mn}x_n &= b_m \end{aligned} \qquad (1)$$

This system may be written in matrix form as

$$\begin{bmatrix} a_{11} & a_{12} & \ldots & a_{1n} \\ a_{21} & a_{22} & \ldots & a_{2n} \\ \ldots & \ldots & \ldots & \ldots \\ a_{m1} & a_{m2} & \ldots & a_{mn} \end{bmatrix} \begin{bmatrix} x_1 \\ x_2 \\ \ldots \\ x_n \end{bmatrix} = \begin{bmatrix} b_1 \\ b_2 \\ \ldots \\ b_m \end{bmatrix}$$

The matrix $[a_{ij}]$ of order $m \times n$ is called the coefficient matrix of the given system of equations. We denote this matrix by A. If we denote matrices $\begin{bmatrix} x_1 \\ x_2 \\ \vdots \\ x_n \end{bmatrix}$ and $\begin{bmatrix} b_1 \\ b_2 \\ \vdots \\ b_m \end{bmatrix}$ by **X** and **B** respectively, the system (1) may be written as

$$\mathbf{AX} = \mathbf{B} \qquad (2)$$

If **B** is a zero-matrix, that is, $b_1 = b_2 = \cdots = b_m = 0$, then the system (2) $\big($as well as system (1)$\big)$ is called a 'homogeneous system' of linear equations. If $\mathbf{B} \neq \mathbf{O}$, that is, if the numbers $b_1, b_2, \ldots, b_m$ are not all zero, then the system is called 'linear non-homogeneous system.'

The matrix obtained by adding to the coefficient matrix **A** an additional column **B** is called the 'augmented matrix,' that is, the matrix

$$\begin{bmatrix} a_{11} & a_{12} & \dots & a_{1n} & b_1 \\ a_{21} & a_{22} & \dots & a_{2n} & b_2 \\ \dots & \dots & \dots & \dots & \dots \\ a_{m1} & a_{m2} & \dots & a_{mn} & b_m \end{bmatrix}$$

is the augmented matrix of the given system of equations. The augmented matrix of the system $\mathbf{AX} = \mathbf{B}$ is usually denoted by $[\mathbf{A} : \mathbf{B}]$.

If the given system of linear equations is such that the equations are all satisfied simultaneously by at least one set of values of the variables, then it is said to be 'consistent'. Such a set of values of the variables is called a 'solution'. The system is said to be 'inconsistent' if the equations are not satisfied simultaneously by any set of values of the variables, that is, if the system has no solution.

NOTE: A homogeneous system is always consistent. This is because a homogeneous system has at least one solution, namely, the trivial solution or zero solution given by $\mathbf{X} = \mathbf{O}$, that is, $x_1 = x_2 = \dots = x_n = 0$.

A linear non-homogeneous system, on the other hand, may or may not be consistent. For example, the system $x + y = 2$ and $x + y = 1$ has no solution because no numbers x and y can exist such that their sum is both 2 and 1. The system $x + y = 2$, $x - y = 0$ is consistent and has exactly one solution, $x = 1$ and $y = 1$. The system $x + y = 2$ is also consistent, but it has more than one solution. In fact, any two numbers whose sum is 2 is a solution of this system and we know that infinitely many pairs of such numbers exist.

The three systems of equations just discussed indicate that for a linear non-homogeneous system of equations one and only one of the following is true:

1. The system has a unique solution.
2. The system has infinitely many solutions.
3. The system has no solution.

Now the question is how to find whether a given system of equations is consistent or not, and if consistent, whether the solution is unique or not.

The answer to the former is given in the following theorem that we state without proof.

THEOREM 1

A necessary and sufficient condition for the system $\mathbf{AX} = \mathbf{B}$ of linear equations to be consistent is that the coefficient matrix **A** and the augmented matrix $[\mathbf{A} : \mathbf{B}]$ of the system are of the same rank.

Now, we come to the next question: If the system is consistent, how to determine whether the system has a unique solution or infinitely many solutions?

To answer this, we consider the system $\mathbf{AX} = \mathbf{B}$ having m equations and n unknowns.

(I) First, suppose that the system is consistent with $m = n$, that is, the number of equations is the same as the number of variables. Then the system of equation $\mathbf{AX} = \mathbf{B}$ has a unique solution if A is non-singular, and the unique solution is given by $\mathbf{X} = \mathbf{A}^{-1}\mathbf{B}$.

(II) Suppose the system is consistent with $m > n$ (that is, more equations than variables).

Case (i) If the rank of augmented matrix $[\mathbf{A} : \mathbf{B}]$ = the rank of coefficient matrix $\mathbf{A} = n$, then the solution is unique.

Case (ii) If the rank of augmented matrix $[\mathbf{A} : \mathbf{B}]$ = the rank of coefficient matrix $\mathbf{A} = r < n$, then the system has infinitely many solutions. In this case, the $(n - r)$ variables are assigned arbitrary values and the values of the remaining r variables are found in terms of the values of $n - r$ variables.

(III) Lastly, suppose that the system is consistent with $m < n$ (that is, more variables than equations).

Case (i) Let the rank of the augmented matrix $[\mathbf{A} : \mathbf{B}]$ = rank of the coefficient matrix $\mathbf{A} = m$. Then the system has infinite solutions. Here, $n - m$ variables are assigned arbitrary values and the remaining m variables are found in terms of these $n - m$ variables.

Case (ii) If the rank of augmented matrix $[\mathbf{A} : \mathbf{B}]$ = rank of the coefficient matrix $\mathbf{A} = r < m$, then the system has infinitely many solutions. In this case, $n - r$ variables are assigned arbitrary values and the values of the remaining r variables are found in terms of those assigned values.

Using (I), (II), and (III) we can clearly describe the nature of the solutions of a given system of linear non-homogeneous equations. For linear homogeneous system $\mathbf{AX} = \mathbf{O}$, we have the following result.

THEOREM 2

A necessary and sufficient condition that the system $\mathbf{AX} = \mathbf{O}$ of m homogeneous equations in n variables have non-trivial solutions is that the rank of its coefficient matrix is less than the number of variables, or, $\rho(\mathbf{A}) < n$.

The following two corollaries of this theorem are of special significance.

Cor. 1 If the homogeneous system $\mathbf{AX} = \mathbf{O}$ has more unknowns than equations or $n > m$, then the system has infinitely many solutions.

Cor. 2 A homogeneous system $\mathbf{AX} = \mathbf{O}$ of n linear equations in n unknowns has a non-trivial solution if and only if $\mathbf{A}$ is singular or $|\mathbf{A}| = 0$.

All the statements made above for determining the nature of solutions of a linear non-homogeneous and homogeneous system of equations may be proved, but we shall assume them, without proof, throughout this book. The following examples illustrate the above results.

EXAMPLE 1.31 Show that the following system of equations is not consistent.

$$x + y - z = 2$$

$$2x - y + z = 1$$

$$3x - y + z = 0$$

SOLUTION

The augmented matrix of the given system is

$$[\mathbf{A}:\mathbf{B}] = \begin{bmatrix} 1 & 1 & -1 & : & 2 \\ 2 & -1 & 1 & : & 1 \\ 3 & -1 & 1 & : & 0 \end{bmatrix}$$

By applying elementary row operations, we can show that the matrix $[\mathbf{A}:\mathbf{B}]$ is equivalent to the matrix $\begin{bmatrix} 1 & 1 & -1 & : & 2 \\ 0 & 1 & -1 & : & 1 \\ 0 & 0 & 0 & : & 1 \end{bmatrix}$ (verify)

It is clear that the rank of the transformed matrix is 3. So $\rho([\mathbf{A}:\mathbf{B}]) = 3$.

Also the matrix **A** is reduced to $\begin{bmatrix} 1 & 1 & 1 \\ 0 & 1 & -1 \\ 0 & 0 & 0 \end{bmatrix}$ by elementary row operations. So rank of $\mathbf{A} = 2$.

$\therefore \quad \rho([\mathbf{A}:\mathbf{B}]) \neq \rho(\mathbf{A})$.

Hence, the given system of equations is inconsistent.

EXAMPLE 1.32 Determine the nature of solution of the following system of equations:

$$x + 3y + 2z = 0$$
$$2x - y + 3z = 0$$
$$3x - 5y + 4z = 0$$
$$x + 17y + 4z = 0$$

SOLUTION

This is a homogeneous equation with more equations than variables. Here $m = 4$, $n = 3$.

$$\mathbf{A} = \begin{bmatrix} 1 & 3 & 2 \\ 2 & -1 & 3 \\ 3 & -5 & 4 \\ 1 & 17 & 4 \end{bmatrix}$$

$$\sim \begin{bmatrix} 1 & 3 & 2 \\ 0 & -7 & -1 \\ 0 & -14 & -2 \\ 0 & 14 & 2 \end{bmatrix} \quad \text{By } \mathbf{R}_2 \to \mathbf{R}_2 - 2\mathbf{R}_1,\ \mathbf{R}_3 \to \mathbf{R}_3 - 3\mathbf{R}_1,\ \mathbf{R}_4 \to \mathbf{R}_4 - \mathbf{R}_1.$$

$$\sim \begin{bmatrix} 1 & 3 & 2 \\ 0 & -7 & -1 \\ 0 & 0 & 0 \\ 0 & 0 & 0 \end{bmatrix} \quad \text{By } \mathbf{R}_3 \to \mathbf{R}_3 - 2\mathbf{R}_2,\ \mathbf{R}_4 \to \mathbf{R}_4 + 2\mathbf{R}_2.$$

So, rank $\mathbf{A} = 2 < 3$.

Hence, by Theorem 2, the system has infinitely many solutions.

EXAMPLE 1.33 Show that the only real value of λ for which the following equations have non-zero solution is 6,

$$x + 2y + 3z = \lambda x$$
$$3x + y + 2z = \lambda y$$
$$2x + 3y + z = \lambda z$$

SOLUTION

The given system can be written in the form

$$\begin{aligned}(1-\lambda)x + 2y + 3z &= 0\\ 3x + (1-\lambda)y + 2z &= 0\\ 2x + 3y + (1-\lambda)z &= 0\end{aligned} \qquad (1)$$

Hence, coefficient matrix $\mathbf{A} = \begin{bmatrix} 1-\lambda & 2 & 3 \\ 3 & 1-\lambda & 2 \\ 2 & 3 & 1-\lambda \end{bmatrix}$.

In order to have a non-zero solution of the given system of equations (1), we must have rank of **A** less than 3. Now the rank of **A** will be less than 3 if $|\mathbf{A}| = 0$.

$$\text{Now} \quad |\mathbf{A}| = \begin{vmatrix} 1-\lambda & 2 & 3 \\ 3 & 1-\lambda & 2 \\ 2 & 3 & 1-\lambda \end{vmatrix}$$

$$= \begin{vmatrix} 6-\lambda & 6-\lambda & 6-\lambda \\ 3 & 1-\lambda & 2 \\ 2 & 3 & 1-\lambda \end{vmatrix} \qquad \text{By } \mathbf{R}_1 \to \mathbf{R}_1 + (\mathbf{R}_2 + \mathbf{R}_3)$$

$$= (6-\lambda)\begin{vmatrix} 1 & 1 & 1 \\ 3 & 1-\lambda & 2 \\ 2 & 3 & 1-\lambda \end{vmatrix}$$

$$= (6-\lambda)\begin{vmatrix} 1 & 0 & 0 \\ 3 & -\lambda-2 & -1 \\ 2 & 1 & -\lambda-1 \end{vmatrix} \qquad \begin{matrix}\text{By } \mathbf{C}_2 \to \mathbf{C}_2 - \mathbf{C}_1 \\ \mathbf{C}_3 \to \mathbf{C}_3 - \mathbf{C}_1\end{matrix}$$

$$= (6-\lambda)\{(\lambda+2)(\lambda+1) + 1\}$$

$$= (6-\lambda)(\lambda^2 + 3\lambda + 3)$$

So, $|\mathbf{A}| = 0 \Rightarrow (6-\lambda) = 0$ or $\lambda^2 + 3\lambda + 3 = 0$

That is, $\lambda = 6$ or $\lambda = \dfrac{-3 \pm \sqrt{-3}}{2}$.

Hence, the only real value of λ for which the given system of equations has a non-zero solution is 6.

Our next aim is to find a method for solving a given system of linear equations. The method consists of transforming the given system into another system whose solution can be easily found and the solution of transformed system is same as that of the given system. This transformation is obtained by using elementary row operations. By applying elementary row operations to the coefficient matrix **A** of the system **AX** = **B**, and also applying the same elementary operations on **B**, we get another system **CX** = **D** which is equivalent to the given system **AX** = **B**. Note that the two systems of linear equations are called equivalent if they have exactly the same solutions. The method of solving a system of linear equations is given in the following steps.

Step 1. Put the given system of linear equations in the form **AX** = **B**.

Step 2. Transform the matrix **A** to triangular form by applying elementary row operations. Apply the same operations (in the same order) to matrix **B**, to get the transformed system **CX** = **D**.

Step 3. Express the system **CX** = **D** as linear equations. The resulting equivalent system of equations is simple to solve.

The method is explained through following examples.

EXAMPLE 1.34 Solve the system of equations

$$x + y + z = 6$$
$$2x + 3y + 4z = 20$$
$$x + y = z$$

SOLUTION

The given system of equations is

$$\begin{aligned} x + y + z &= 6 \\ 2x + 3y + 4z &= 20 \\ x + y - z &= 0 \end{aligned} \qquad (1)$$

The system of equations can be written in the form **AX** = **B**, where

$$\mathbf{A} = \begin{bmatrix} 1 & 1 & 1 \\ 2 & 3 & 4 \\ 1 & 1 & -1 \end{bmatrix}, \mathbf{X} = \begin{bmatrix} x \\ y \\ z \end{bmatrix} \text{ and } \mathbf{B} = \begin{bmatrix} 6 \\ 20 \\ 0 \end{bmatrix}$$

So (1) becomes

$$\begin{bmatrix} 1 & 1 & 1 \\ 2 & 3 & 4 \\ 1 & 1 & -1 \end{bmatrix} \begin{bmatrix} x \\ y \\ z \end{bmatrix} = \begin{bmatrix} 6 \\ 20 \\ 0 \end{bmatrix}$$

$$\Rightarrow \quad \begin{bmatrix} 1 & 1 & 1 \\ 0 & 1 & 2 \\ 1 & 1 & -1 \end{bmatrix} \begin{bmatrix} x \\ y \\ z \end{bmatrix} = \begin{bmatrix} 6 \\ 8 \\ 0 \end{bmatrix} \quad \text{By applying } \mathbf{R}_2 \to \mathbf{R}_2 - 2\mathbf{R}_1$$

$$\Rightarrow \quad \begin{bmatrix} 1 & 1 & 1 \\ 0 & 1 & 2 \\ 0 & 0 & -2 \end{bmatrix} \begin{bmatrix} x \\ y \\ z \end{bmatrix} = \begin{bmatrix} 6 \\ 8 \\ -6 \end{bmatrix} \quad \text{By } \mathbf{R}_3 \to \mathbf{R}_3 - \mathbf{R}_1 \qquad (2)$$

So, by reducing the matrix **A** to triangular form, we obtain system (2) which is equivalent to system (1). Now system (2) can be written as

$$x + y + z = 6$$
$$y + 2z = 8$$
$$-2z = -6$$
$$\Rightarrow \quad z = 3, y = 2, x = 1.$$

Hence, the given system of linear equations has a unique solution given by $x = 1$, $y = 2$ and $z = 3$.

EXAMPLE 1.35 Solve the system of equations:

$$x - 4y + 7z = 8$$
$$3x + 8y - 2z = 6$$
$$7x - 8y + 26z = 31$$

SOLUTION

The given system of equations can be written in matrix form as

$$\begin{bmatrix} 1 & -4 & 7 \\ 3 & 8 & -2 \\ 7 & -8 & 26 \end{bmatrix} \begin{bmatrix} x \\ y \\ z \end{bmatrix} = \begin{bmatrix} 8 \\ 6 \\ 31 \end{bmatrix}$$

By applying elementary row operations $\mathbf{R}_2 \rightarrow \mathbf{R}_2 - 3\mathbf{R}_1$, and $\mathbf{R}_3 \rightarrow \mathbf{R}_3 - 7\mathbf{R}_1$, we get

$$\begin{bmatrix} 1 & -4 & 7 \\ 0 & 20 & -23 \\ 0 & 20 & -23 \end{bmatrix} \begin{bmatrix} x \\ y \\ z \end{bmatrix} = \begin{bmatrix} 8 \\ -18 \\ -25 \end{bmatrix}$$

Again, applying elementary row operation $\mathbf{R}_3 \rightarrow \mathbf{R}_3 - \mathbf{R}_2$, we get

$$\begin{bmatrix} 1 & -4 & 7 \\ 0 & 20 & -23 \\ 0 & 0 & 0 \end{bmatrix} \begin{bmatrix} x \\ y \\ z \end{bmatrix} = \begin{bmatrix} 8 \\ -18 \\ -7 \end{bmatrix}$$

$\therefore$ the given system is equivalent to

$$x - 4y + 7z = 8$$
$$20y - 23z = -18$$
and
$$0 = -7$$

Since the conclusion $0 = -7$ is absurd, so the given system of equations is inconsistent.

REMARK: Note that in the above example, rank of coefficient matrix = 2, while rank of augmented matrix = 3. Since the two are not equal, the system is inconsistent.

EXAMPLE 1.36 Solve, if consistent, the system of equations:

$$x + y + 3z = 1$$
$$2x + 3y - z = 3$$
$$5x + 7y + z = 7$$

[DU, B.Sc.(Ph. Sc.), 2006]

SOLUTION

The given system of linear equations can be written in matrix form as

$$\begin{bmatrix} 1 & 1 & 3 \\ 2 & 3 & -1 \\ 5 & 7 & 1 \end{bmatrix} \begin{bmatrix} x \\ y \\ z \end{bmatrix} = \begin{bmatrix} 1 \\ 3 \\ 7 \end{bmatrix}$$

$$\Rightarrow \begin{bmatrix} 1 & 1 & 3 \\ 0 & 1 & -7 \\ 5 & 7 & 1 \end{bmatrix} \begin{bmatrix} x \\ y \\ z \end{bmatrix} = \begin{bmatrix} 1 \\ 1 \\ 7 \end{bmatrix}$$ By applying E-row operation $\mathbf{R}_2 \rightarrow \mathbf{R}_2 - 2\mathbf{R}_1$

$$\Rightarrow \quad \begin{bmatrix} 1 & 1 & 3 \\ 0 & 1 & -7 \\ 0 & 2 & -14 \end{bmatrix} \begin{bmatrix} x \\ y \\ z \end{bmatrix} = \begin{bmatrix} 1 \\ 1 \\ 2 \end{bmatrix}$$ By applying E-row operation $\mathbf{R}_3 \to \mathbf{R}_3 - 5\mathbf{R}_1$

$$\Rightarrow \quad \begin{bmatrix} 1 & 1 & 3 \\ 0 & 1 & -7 \\ 0 & 0 & 0 \end{bmatrix} \begin{bmatrix} x \\ y \\ z \end{bmatrix} = \begin{bmatrix} 1 \\ 1 \\ 0 \end{bmatrix}$$ By applying E-row operation $\mathbf{R}_3 \to \mathbf{R}_3 - 2\mathbf{R}_1$

$$\Rightarrow \quad \begin{aligned} x + y + 3z &= 1 \\ y - 7z &= 1 \\ 0z &= 0 \end{aligned}$$

$$\Rightarrow \quad \begin{aligned} y &= 1 + 7z \\ x &= 1 - y - 3z \\ &= 1 - (1 + 7z) - 3z \\ &= -10z \end{aligned}$$

and $\quad z = z.$

If we put $z = k$, we get $x = -10k$, $y = 1 + 7k$ where k is arbitrary. We can verify that these values of x, y and z satisfy the given system of equations. Thus, the given system of equations is consistent and the complete solution is given by

$$\left.\begin{aligned} x &= -10k \\ y &= 1 + 7k \\ z &= k \end{aligned}\right\} \quad \text{for all } k.$$

The system has infinitely many solutions.

> **NOTE:** In the above example, rank of the coefficient matrix = rank of the augmented matrix = 2, so the system is consistent. Moreover, 2 < 3 = number of variables, so the system has infinitely many solutions and the solution can be obtained by assigning 3 − 2 = 1 variable arbitrary values. In the above solution we have assigned an arbitrary value to *z*.

EXAMPLE 1.37 Show that the only real value of λ for which the following system of equations has a non-zero solution is 6 and then solve the equation.

$$\begin{aligned} x + 2y + 3z &= \lambda x \\ 3x + y + 2z &= \lambda y \\ 2x + 3y + z &= \lambda z \end{aligned}$$

[DU B.Sc.(G), 1998, 1995]

SOLUTION

We have already solved the first part of the question in Example 1.33. For $\lambda = 6$, the homogeneous system of linear equations becomes

$$\begin{aligned} -5x + 2y + 3z &= 0 \\ 3x - 5y + 2z &= 0 \\ 2x + 3y - 5z &= 0 \end{aligned}$$

Writing in matrix form, we get

$$\begin{bmatrix} -5 & 2 & 3 \\ 3 & -5 & 2 \\ 2 & 3 & -5 \end{bmatrix}\begin{bmatrix} x \\ y \\ z \end{bmatrix} = \begin{bmatrix} 0 \\ 0 \\ 0 \end{bmatrix}$$

$$\Rightarrow \quad \begin{bmatrix} -5 & 2 & 3 \\ 3 & -5 & 2 \\ 0 & 0 & 0 \end{bmatrix}\begin{bmatrix} x \\ y \\ z \end{bmatrix} = \begin{bmatrix} 0 \\ 0 \\ 0 \end{bmatrix} \qquad \text{By } \mathbf{R}_3 \to \mathbf{R}_3 + (\mathbf{R}_1 + \mathbf{R}_2)$$

$$\Rightarrow \quad \begin{bmatrix} -5 & 2 & 3 \\ 0 & -\frac{19}{5} & \frac{19}{5} \\ 0 & 0 & 0 \end{bmatrix}\begin{bmatrix} x \\ y \\ z \end{bmatrix} = \begin{bmatrix} 0 \\ 0 \\ 0 \end{bmatrix} \qquad \text{By } \mathbf{R}_2 \to \mathbf{R}_2 + \frac{3}{5}\mathbf{R}_1$$

$$-5x + 2y + 3z = 0 \tag{i}$$

$$\Rightarrow \quad -\frac{19}{5}y + \frac{19}{5}z = 0 \tag{ii}$$

$$0 = 0 \tag{iii}$$

From (ii), we have $y = z$. Substituting $y = z$ in (i) we get $x = z$.
So, we get $x = y = z = k$ (say)
Thus, the complete solution is

$$x = y = z = k, \text{ for all } k.$$

EXERCISE 1.5

Solve, if consistent, the following system of equations

1. $x + y + z = 7$
   $x + 2y + 3z = 16$
   $x + 3y + 4z = 22$

2. $2x + 3y + z = 9$
   $x + 2y + 3z = 6$
   $3x + y + 2z = 8$ *[DU, B.A. (P), 1997]*

3. $x + 2y - z = 3$
   $3x - y + 3z = 1$
   $2x - 2y + 3z = 2$
   $x - y + z = -1$ *[DU, B.Sc. (G), 1994]*

4. $x - 2y + 3z = 6$
   $3x + y - 4z = -7$
   $5x - 3y + 2z = 5$

5. $x - y + 2z = 4$
   $3x + y + 4z = 6$
   $x + y + z = 1$ *[DU, B.A. (P), 1995]*

6. $x - 3y + z = -1$
   $2x + y - 4z = -1$
   $6x - 7y + 8z = 7$ *[DU, B.A. (P), 1992]*

7. $x + 2y - z = 3$
   $3x - y + 2z = 1$
   $2x - 2y + 3z = 2$

8. $x + 3y + z = 6$
   $3x - 2y - 8z = 7$
   $4x + 5y - 3z = 17$

9. $2x + 6y + 11 = 0$
   $6x + 20y - 6z = 3$
   $6y - 18z + 1 = 0$

10. $x + 2y + 3z = 14$
    $3x + y + 2z = 11$
    $2x + 3y + z = 11$

11. $2x - 3y + 4z = 3$
    $x - 3z = -2$
    $x + y + z = 2$

12. $x + 2y + 3z = 5$
    $x + 3y + 6z = 11$
    $x + 4y + 10z = 21$

Solve the following system of homogeneous linear equations.

13. $x + 3y + 2z = 0$
$2x - y + 3z = 0$
$3x - 5y + 4z = 0$
$x + 17y + 4z = 0$ *[DU, B.Sc. (G), 1993]*

14. $x - y + z = 0$
$-3x + y - 4z = 0$
$7x - 3y - 9z = 0$
$4x - 2y + 5z = 0$

15. $3x - 4y + 5z = 0$
$x + y - 2z = 0$
$2x + 3y + z = 0$

16. $x + 3y - 2z = 0$
$2x - y + 4z = 0$
$x - 11y + 14z = 0$

17. For what values of λ does the following system of equations have a solution?

$$x + y + z = 1$$
$$x + 2y + 4z = \lambda$$
$$x + 4y + 10z = \lambda^2$$

Also find the solution in each case. *[DU, B.Sc. (G) 1997]*

18. For what value of λ, does the system of equations

$$\lambda x + 2y - 2z = 1$$
$$4x + 2\lambda y - z = 2$$
$$6x + 6y + \lambda z = 3$$

have a unique solution? Find the solution. *[DU, B.Sc. (G) 1992]*

19. For what values of λ and μ do the following system of equations

$$x + y + z = 6$$
$$x + 2y + 3z = 10$$
$$x + 2y + \lambda z = \mu$$

have (i) a unique solution, (ii) no solution, (iii) an infinite number of solutions ?

20. Discuss for all values of k, the nature of solutions of the following system of linear equations.

$$2x + 3ky + (3k + 4)z = 0$$
$$x + (k + 4)y + (4k + 2)z = 0$$
$$x + 2(k + 1)y + (3k + 4)z = 0$$

CHAPTER TWO

Vectors in $\mathbf{R}^2$ and $\mathbf{R}^3$

■ 2.1 INTRODUCTION

Many physical quantities, such as length, area, and mass are completely described by a single real number which indicates their magnitude. On the other hand, some quantities, such as force and velocity, require magnitude and direction to describe them. Such physical quantities are called vectors in mechanics.

A vector in a plane may be described using a directed line segment or an arrow. The tail of the arrow is called the *initial point* and the tip, or the head of the arrow, is called the *terminal point*.

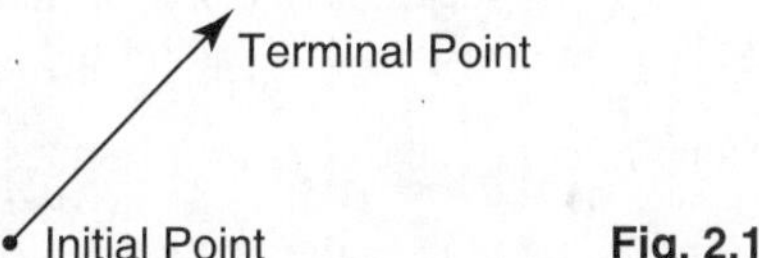

Fig. 2.1

Two directed line segments having the same direction and the same length represent *equivalent vectors*. Since we want a vector to be determined solely by its length and direction, equivalent vectors are regarded as equal even though they may be located in different positions.

To describe these geometric vectors algebraically, we consider a *rectangular coordinate system* in the plane. By this system we associate each point in a plane with a pair of numbers (x, y). For example, point A is point $(2, 3)$ in Fig. 2.2.

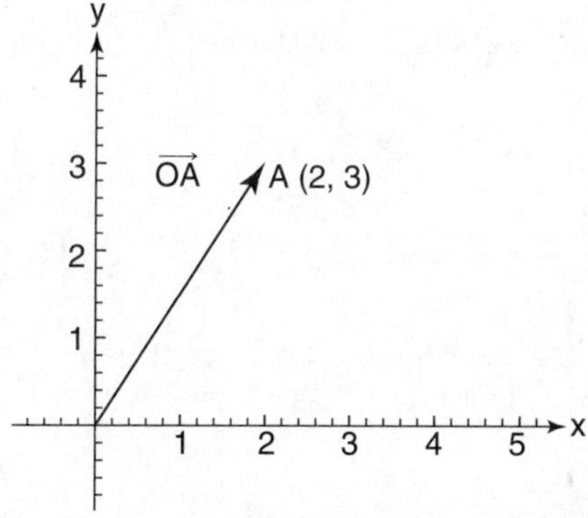

Fig. 2.2

Geometric vectors may be placed as directed line segments in the coordinate system. Since equivalent vectors are regarded as equal, we shall place the initial point of each vector at the *origin* of the coordinate system for the plane. If $P(x, y)$ is the terminal point of a vector, then this vector is completely described by the ordered pair (x, y). Thus, corresponding to every directed line segment $\overrightarrow{OP}$, there is a point in the plane. Conversely, given any point P on the plane, we can associate a directed line segment $\overrightarrow{OP}$ with its tail at origin and head at P. For example, vector $\overrightarrow{OA}$ in Fig. 2.2 is completely described by the point $A(2, 3)$ and conversely.

Thus, every vector in a plane may be regarded as a directed line segment or as an ordered pair. The first approach is geometric and the second, algebraic. We shall use this second approach for the study of vectors. We may write the coordinates of any point $P(x, y)$ in a column rather than in a row, that is, as a matrix $\begin{bmatrix} x \\ y \end{bmatrix}$. We formally define our vector in a plane in this manner.

> **? DEFINITION:** A 'vector in a plane' is a 2×1 matrix $\mathbf{X} = \begin{bmatrix} x \\ y \end{bmatrix}$, where x and y are real numbers, called the components of **X**.

Let **R** denote a set of real numbers. It can be visualized as a straight line taken in the horizontal position. By $\mathbf{R}^2$ we denote the set of all ordered pairs of real numbers. As discussed above, $\mathbf{R}^2$ may be regarded as a plane since we can associate a point in the plane with each ordered pair. $\mathbf{R}^2$ may also be taken as the set of all 2×1 matrices. Thus, elements of $\mathbf{R}^2$ may just be called vectors and $\mathbf{R}^2$ may be defined as the set of all the vectors in the plane. It must be obvious by now that by a 'vector in the plane' we may mean a point P in the plane, a directed line segment $\overrightarrow{OP}$, an ordered pair (x, y) or a 2×1 matrix $\begin{bmatrix} x \\ y \end{bmatrix}$. Thus, $\mathbf{X} = \begin{bmatrix} x \\ y \end{bmatrix}$ and $\mathbf{X} = (x, y)$ represent the same vector **X**, and we shall use the two representations of **X** interchangeably.

We shall denote vectors by capital letters, such as **X**, **Y** and so on, as is usually done to denote matrices. In many books small bold face letters **u**, **v** or $\vec{u}$, $\vec{v}$ or $\bar{u}$, $\bar{v}$ are also used to denote vectors. The point $\begin{bmatrix} 0 \\ 0 \end{bmatrix}$ in $\mathbf{R}^2$ with both coordinates zero is called a zero vector and is denoted by **O**.

Two vectors are equal if their components are equal. Thus, the vectors $\mathbf{X} = \begin{bmatrix} x_1 \\ y_1 \end{bmatrix}$ and $\mathbf{Y} = \begin{bmatrix} x_2 \\ y_2 \end{bmatrix}$ are equal if $x_1 = x_2$ and $y_1 = y_2$.

The length or magnitude of the vector $\mathbf{X} = \begin{bmatrix} x \\ y \end{bmatrix}$ is defined as

$$\|\mathbf{X}\| = \sqrt{x^2 + y^2}$$

This can be justified by Pythagoras Theorem (See Fig. 2.3).

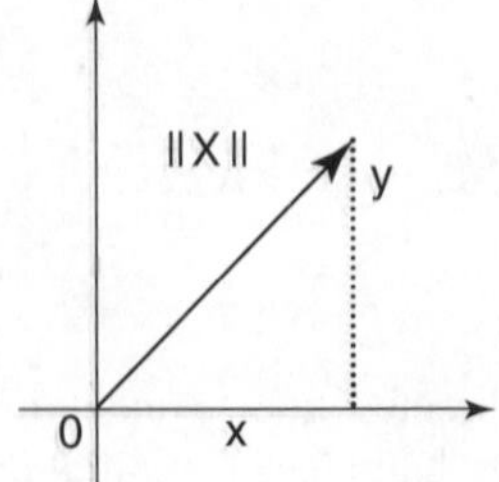

Fig. 2.3

The dot product of two vectors $\mathbf{X} = (x_1, y_1)$ and $\mathbf{Y} = (x_2, y_2)$ in $\mathbf{R}^2$ is defined to be

$$\mathbf{X} \cdot \mathbf{Y} = x_1 x_2 + y_1 y_2 \tag{1}$$

It is clear from this equation that

$$\|\mathbf{X}\| = \sqrt{\mathbf{X} \cdot \mathbf{X}}$$

for any vector $\mathbf{X}$ in $\mathbf{R}^2$.

The angle θ between two vectors $\mathbf{X}$ and $\mathbf{Y}$ is given by the relation

$$\cos\theta = \frac{\mathbf{X} \cdot \mathbf{Y}}{\|\mathbf{X}\|\|\mathbf{Y}\|} \tag{2}$$

From Equation 2, we may easily deduce that two non-zero vectors $\mathbf{X}$ and $\mathbf{Y}$ are perpendicular or orthogonal if and only if $\mathbf{X} \cdot \mathbf{Y} = 0$.

EXAMPLE 2.1 Let $\mathbf{X} = \begin{bmatrix} 2 \\ 4 \end{bmatrix}$ and $\mathbf{Y} = \begin{bmatrix} -1 \\ 2 \end{bmatrix}$. Then

$$\mathbf{X} \cdot \mathbf{Y} = 2 \cdot (-1) + 4 \cdot 2 = 6$$
$$\|\mathbf{X}\| = \sqrt{2^2 + 4^2} = \sqrt{20}$$
$$\|\mathbf{Y}\| = \sqrt{(-1)^2 + 2^2} = \sqrt{5}$$

Hence, $\cos\theta = \dfrac{6}{\sqrt{20}\sqrt{5}} = 0.6$.

We can obtain the approximate angle by using a table of cosines; we find that $\theta = 53^\circ 8'$.

EXAMPLE 2.2 Find the cosine of the angle between the vectors.

(a) $\mathbf{X} = (1, 2)$, $\quad \mathbf{Y} = (2, -3)$ $\qquad$ **(b)** $\mathrm{X} = (2, 1)$, $\quad \mathbf{Y} = (-2, -1)$

SOLUTION

(a) If θ is the angle between $\mathbf{X}$ and $\mathbf{Y}$ then

$$\cos\theta = \frac{\mathbf{X} \cdot \mathbf{Y}}{\|\mathbf{X}\|\|\mathbf{Y}\|} = \frac{1 \cdot 2 + 2 \cdot (-3)}{\sqrt{1^2 + 2^2}\sqrt{2^2 + (-3)^2}} = \frac{-4}{\sqrt{5}\sqrt{13}}$$

(b) It can easily be seen that

$$\cos\theta = \frac{2(-2) + 1 \cdot (-1)}{5} = \frac{-5}{5} = -1$$

whence $\theta = \pi$.

EXAMPLE 2.3 Find whether the following pair of vectors is orthogonal or not.

(a) $\mathbf{X} = (2, -4)$ and $\mathbf{Y} = (4, 2)$ $\qquad$ **(b)** $\mathbf{X} = (0, 2)$, and $\mathbf{Y} = (-3, 3)$

SOLUTION

(a) Here

$$\mathbf{X} \cdot \mathbf{Y} = 2 \cdot 4 + (-4) \cdot 2 = 0$$

So, $\mathbf{X}$ and $\mathbf{Y}$ are orthogonal.

(b) Here

$$\mathbf{X} \cdot \mathbf{Y} = 0(-3) + 2 \cdot 3 = 6 \neq 0$$

So, $\mathbf{X}$ and $\mathbf{Y}$ are not orthogonal.

Just as vectors in a plane can be described by pairs of real numbers, vectors in space can be described by triple of real numbers. By $\mathbf{R}^3$ we denote the set of all ordered triple (x, y, z) of real numbers. $\mathbf{R}^3$ may also be taken as the set of all 3×1 matrices $\begin{bmatrix} x \\ y \\ z \end{bmatrix}$ where x, y and z are real numbers. A vector in space or a 3-vector is an element of $\mathbf{R}^3$. More formally, we may give the following definition.

? DEFINITION: A 'vector in space' (or a '3-vector') is a 3×1 matrix $\mathbf{X} = \begin{bmatrix} x \\ y \\ z \end{bmatrix}$ where x, y, z are real numbers called components of $\mathbf{X}$.

Geometrically, any vector $\mathbf{X} = \begin{bmatrix} x \\ y \\ z \end{bmatrix}$ in $\mathbf{R}^3$ may be described using a 3-dimensional Cartesian coordinate system. See Fig. 2.4.

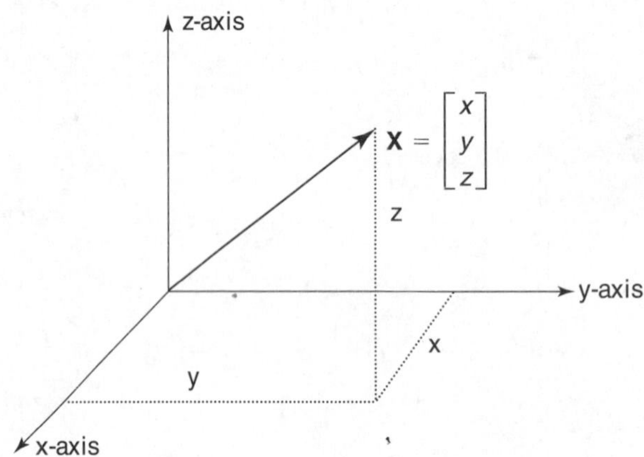

Fig. 2.4

If $\mathbf{X} = (x_1, y_1, z_1)$ and $\mathbf{Y} = (x_2, y_2, z_2)$ are two vectors in $\mathbf{R}^3$, their dot product in $\mathbf{R}^3$ is defined by

$$\mathbf{X} \cdot \mathbf{Y} = x_1x_2 + y_1y_2 + z_1z_2$$

Also, length of the vector $\mathbf{X}$ is defined as

$$\|\mathbf{X}\| = \sqrt{x_1^2 + y_1^2 + z_1^2}$$

As in the case of 2-vectors, the angle between the 3-vectors $\mathbf{X}$ and $\mathbf{Y}$ is given by

$$\cos\theta = \frac{\mathbf{X} \cdot \mathbf{Y}}{\|\mathbf{X}\|\|\mathbf{Y}\|}.$$

EXERCISE 2.1

1. Sketch a directed line segment in $\mathbf{R}^2$ representing each of the vectors

   (a) $\mathbf{X} = \begin{bmatrix} 2 \\ -3 \end{bmatrix}$ **(b)** $\mathbf{Y} = \begin{bmatrix} 3 \\ 2 \end{bmatrix}$

2. Find the length $\|\mathbf{X}\|$ of vector $\mathbf{X}$ where

   (a) $\mathbf{X} = \begin{bmatrix} 2 \\ -1 \end{bmatrix}$ **(b)** $\mathbf{X} = \begin{bmatrix} 0 \\ 0 \end{bmatrix}$ **(c)** $\mathbf{X} = \begin{bmatrix} 1 \\ 0 \\ -1 \end{bmatrix}$

3. Find the dot product of the following pair of vectors.

   (a) $(1, 0)$ and $(0, 1)$ **(b)** $\begin{bmatrix} 2 \\ 3 \end{bmatrix}$ and $\begin{bmatrix} 1 \\ -2 \end{bmatrix}$ **(c)** $\begin{bmatrix} 1 \\ -1 \\ 1 \end{bmatrix}$ and $\begin{bmatrix} 1 \\ -2 \\ 1 \end{bmatrix}$

4. Find the cosine of the angle between the vectors.

   (a) $\mathbf{X} = \begin{bmatrix} 1 \\ -1 \end{bmatrix}, \mathbf{Y} = \begin{bmatrix} -2 \\ 1 \end{bmatrix}$ **(b)** $\mathbf{X} = \begin{bmatrix} 1 \\ 0 \\ 0 \end{bmatrix}, \mathbf{Y} = \begin{bmatrix} 1 \\ 0 \\ 1 \end{bmatrix}$

5. Show that the following pair of vectors are orthogonal.

   (a) $(1, 0)$ and $(0, 1)$ **(b)** $(2, -3, 1)$ and $(1, 2, 4)$

6. Let (a, b) be a vector in $\mathbf{R}^2$. Prove that the vector $(-b, a)$ is orthogonal to (a, b).
7. Determine a vector in $\mathbf{R}^2$ that is orthogonal to $(3, -1)$. Show that there are many such vectors and that they all lie on a line.
8. Find all values of c such that $\|c\,(3, 0, 4)\| = 15$.

2.2 VECTOR OPERATIONS

Let $\mathbf{X} = (x_1, y_1)$ and $\mathbf{Y} = (x_2, y_2)$ be two vectors in a plane. The sum of the vectors $\mathbf{X}$ and $\mathbf{Y}$ is the vector $(x_1 + x_2, y_1 + y_2)$ and is denoted by $\mathbf{X} + \mathbf{Y}$. Thus, vectors are added by adding their components.

If $\mathbf{X} = (x, y)$ and c is a scalar (that is, a real number), then the scalar multiple c$\mathbf{X}$ of $\mathbf{X}$ by c is the vector (cx, cy). Thus, the scalar multiple c$\mathbf{X}$ of $\mathbf{X}$ by c is obtained by multiplying each component of $\mathbf{X}$ by c.

It must be noted here that if we write the vectors $\mathbf{X}$ and $\mathbf{Y}$ above as 2×1 matrices, then the operations of addition and scalar multiplication defined on $\mathbf{R}^2$ above are exactly the same as addition and scalar multiplication for matrices.

That is, if $\mathbf{X} = \begin{bmatrix} x_1 \\ y_1 \end{bmatrix}, \quad \mathbf{Y} = \begin{bmatrix} x_2 \\ y_2 \end{bmatrix}$

then
$$\mathbf{X} + \mathbf{Y} = \begin{bmatrix} x_1 + x_2 \\ y_1 + y_2 \end{bmatrix} \quad \text{and} \quad c\mathbf{X} = \begin{bmatrix} cx_1 \\ cy_1 \end{bmatrix}$$

We can interpret vector addition and scalar multiplication geometrically as follows:

Let $\mathbf{X} = \begin{bmatrix} x_1 \\ y_1 \end{bmatrix}$ and $\mathbf{Y} = \begin{bmatrix} x_2 \\ y_2 \end{bmatrix}$ be two vectors represented by arrows (directed line segments) from the origin $(0, 0)$ to the points (x_1, y_1) and (x_2, y_2) respectively. We represent the sum of $\mathbf{X}$ and $\mathbf{Y}$ by the arrow from $(0, 0)$ to the point $(x_1 + x_2, y_1 + y_2)$ as illustrated in Fig. 2.5.

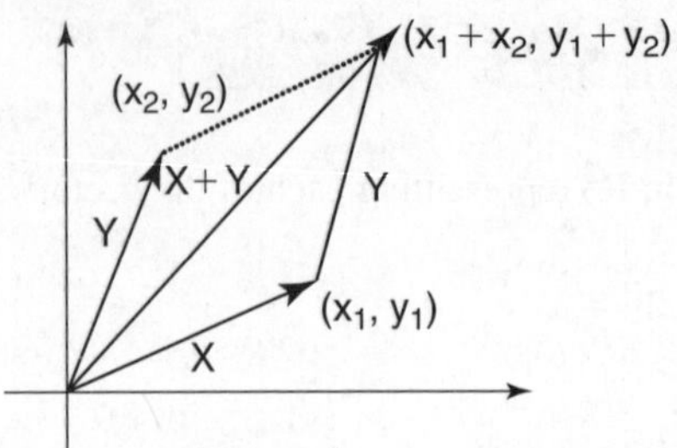

Fig. 2.5

In fact $\mathbf{X} + \mathbf{Y}$ is an arrow along the diagonal of the parallelogram with sides $\mathbf{X}$ and $\mathbf{Y}$.

Scalar multiplication can also be interpreted geometrically. Let c be a real number and let the vector $\mathbf{X} = \begin{bmatrix} x \\ y \end{bmatrix}$ be represented by the arrow from the origin $(0, 0)$ to the point (x, y). We represent the vector $c\mathbf{X} = \begin{bmatrix} cx \\ cy \end{bmatrix}$ by the arrow from $(0, 0)$ to (cx, cy). Note that if $c > 0$, $c\mathbf{X}$ points in the same direction as $\mathbf{X}$ but has length c times that of $\mathbf{X}$ and if $c < 0$, $c\mathbf{X}$ points in the direction opposite to $\mathbf{X}$ and has length $|c|$ times that of $\mathbf{X}$. See Fig. 2.6.

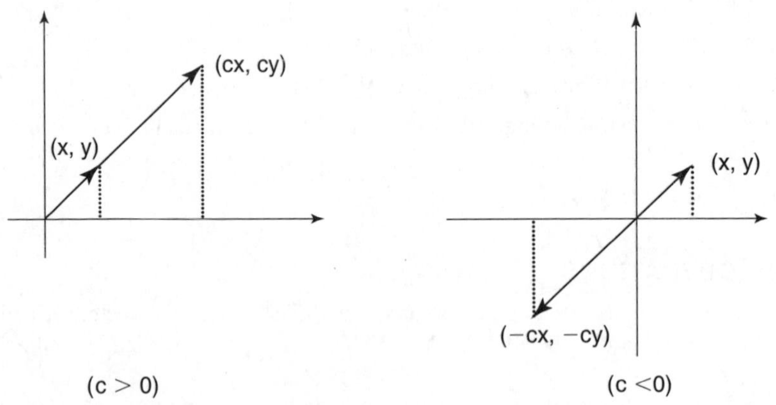

Fig. 2.6

We shall write $(-1)\mathbf{X} = -\mathbf{X}$ and $\mathbf{X} + (-1)\mathbf{Y} = \mathbf{X} - \mathbf{Y}$.

EXAMPLE 2.4 Let $\mathbf{X} = \begin{bmatrix} 1 \\ 2 \end{bmatrix}$ and $\mathbf{Y} = \begin{bmatrix} 2 \\ -3 \end{bmatrix}$. Then $\mathbf{X} + \mathbf{Y} = \begin{bmatrix} 1 + 2 \\ 2 - 3 \end{bmatrix} = \begin{bmatrix} 3 \\ -1 \end{bmatrix}$

If $c_1 = 2$, $c_1\mathbf{X} = 2\begin{bmatrix} 1 \\ 2 \end{bmatrix} = \begin{bmatrix} 2 \\ 4 \end{bmatrix}$

If $c_2 = -3$, $c_2\mathbf{X} = -3\begin{bmatrix} 1 \\ 2 \end{bmatrix} = \begin{bmatrix} -3 \\ -6 \end{bmatrix}$

The vectors $\mathbf{X} + \mathbf{Y}$, $c_1\mathbf{X}$ and $c_2\mathbf{X}$ are represented in Fig. 2.7 (a), (b) and (c) respectively.

EXAMPLE 2.5 Two or more forces acting simultaneously on a body are equivalent to a single force called the resultant force. If the forces lie in a plane, they can be represented by vectors in $\mathbf{R}^2$. The resultant force is represented by the sum of these vectors. This law is called the parallelogram of forces.

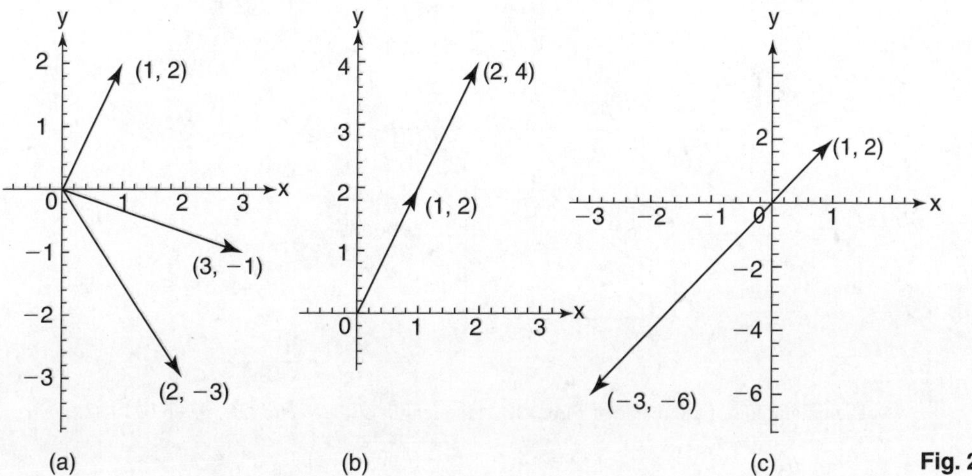

Fig. 2.7

Let a body be located at origin **O** of a rectangular coordinate system. Two forces $(1, 3)$ and $(2, 1)$ act on the body. In Fig. 2.8 we have represented the vectors by $\overrightarrow{OA}$ and $\overrightarrow{OB}$ respectively. The resultant force is the vector $\overrightarrow{OC} = \overrightarrow{OA} + \overrightarrow{OB} = (1, 3) + (2, 1) = (3, 4)$.

The magnitude of the resultant force is the length of the vector $(3, 4)$, given by

$$\|\overrightarrow{OC}\| = \sqrt{3^2 + 4^2} = \sqrt{25} = 5$$

So the magnitude of the resultant force is 5 and the direction is as indicated in Fig. 2.8.

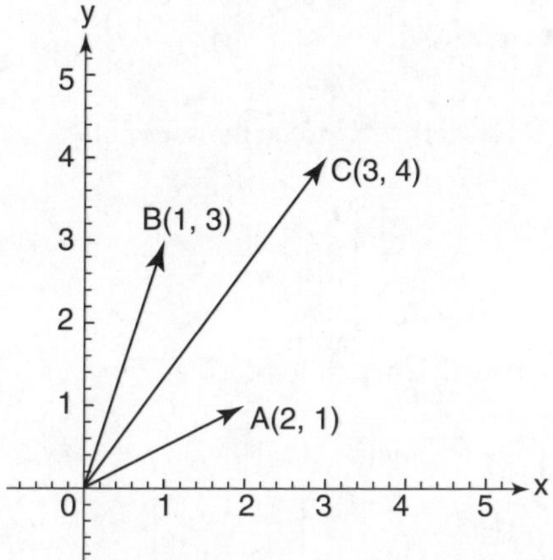

Fig. 2.8

EXAMPLE 2.6 A ship is being pushed by a tugboat with a force of 300 pounds along the negative y-axis while another tugboat is pushing the negative x-axis with a force of 400 pounds. Find the magnitude and sketch the direction of the resultant force.

[DU, B.Sc.(Phy. Sc.), 2006].

SOLUTION

Let us represent the force along the negative y-axis by $\overrightarrow{OA}$ and the force along the negative x-axis by $\overrightarrow{OB}$. Thus $\overrightarrow{OA} = (0, -300)$ and $\overrightarrow{OB} = (-400, 0)$. The resultant force is the vector $\overrightarrow{OC} = \overrightarrow{OA} + \overrightarrow{OB} = (-400, -300)$ which acts along the diagonal of the parallelogram formed by the vector $\overrightarrow{OA}$ and $\overrightarrow{OB}$. See Fig. 2.9.

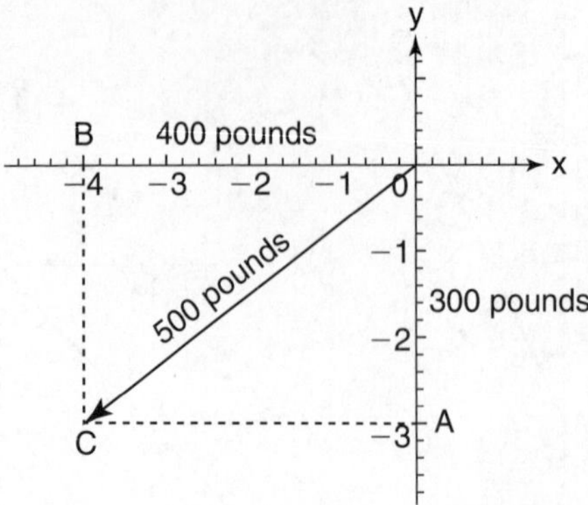

Fig. 2.9

The magnitude of the resultant force $\overrightarrow{OC}$ is the length of the vector $\overrightarrow{OC}$, given by

$$\|\overrightarrow{OC}\| = \sqrt{(400)^2 + (300)^2} = 500$$

Thus, the magnitude of the resultant force is 500 pounds and the direction is as indicated in Fig. 2.9.

Similarly we may define addition and scalar multiplication in $\mathbf{R}^3$ as follows:

If $\mathbf{X} = \begin{bmatrix} x_1 \\ y_1 \\ z_1 \end{bmatrix}$ and $\mathbf{Y} = \begin{bmatrix} x_2 \\ y_2 \\ z_2 \end{bmatrix}$, then

$$\mathbf{X} + \mathbf{Y} = \begin{bmatrix} x_1 + x_2 \\ y_1 + y_2 \\ z_1 + z_2 \end{bmatrix} \quad \text{and} \quad c\mathbf{X} = \begin{bmatrix} cx_1 \\ cy_1 \\ cz_1 \end{bmatrix} \text{ for any scalar c.}$$

We can now generalize $\mathbf{R}^2$ and $\mathbf{R}^3$ to define $\mathbf{R}^n$ to be the set of all $n \times 1$ matrices

$$\mathbf{X} = \begin{bmatrix} x_1 \\ x_2 \\ \vdots \\ x_n \end{bmatrix}$$

where $x_1, x_2, \ldots, x_n$ are real numbers, which are called the components of **X,** and **X** is called an n-vector.

We can define addition in $\mathbf{R}^n$ by the equation

$$\begin{bmatrix} x_1 \\ x_2 \\ \vdots \\ x_n \end{bmatrix} + \begin{bmatrix} y_1 \\ y_2 \\ \vdots \\ y_n \end{bmatrix} = \begin{bmatrix} x_1 + y_1 \\ x_2 + y_2 \\ \vdots \\ x_n + y_n \end{bmatrix}.$$

Similarly, scalar multiplication can be defined by the equation

$$c\begin{bmatrix} x_1 \\ x_2 \\ \vdots \\ x_n \end{bmatrix} = \begin{bmatrix} cx_1 \\ cx_2 \\ \vdots \\ cx_n \end{bmatrix}.$$

Since, an n-vector is an $n \times 1$ matrix and operations of addition and scalar multiplication in $\mathbf{R}^n$ are merely addition and multiplication of a matrix by a real number, it can be easily shown that the operations of vector addition and scalar multiplication satisfy the following properties:

THEOREM 1

Let **X**, **Y**, **Z** be vectors in $\mathbf{R}^n$; c and d be any scalars (real numbers). Then

(a) $\mathbf{X} + \mathbf{Y} = \mathbf{Y} + \mathbf{X}$

(b) $\mathbf{X} + (\mathbf{Y} + \mathbf{Z}) = (\mathbf{X} + \mathbf{Y}) + \mathbf{Z}$

(c) $\mathbf{X} + \mathbf{O} = \mathbf{O} + \mathbf{X} = \mathbf{X}$

(d) $\mathbf{X} + (-\mathbf{X}) = \mathbf{O}$, that is, $\mathbf{X} - \mathbf{X} = \mathbf{O}$

(e) $c(d\mathbf{X}) = (cd)\mathbf{X}$

(f) $c(\mathbf{X} + \mathbf{Y}) = c\mathbf{X} + c\mathbf{Y}$

(g) $(c + d)\mathbf{X} = c\mathbf{X} + d\mathbf{X}$

(h) $1\mathbf{X} = \mathbf{X}$

NOTE: As in the case of $\mathbf{R}^2$ and $\mathbf{R}^3$, we shall identify the vector $\begin{bmatrix} x_1 \\ x_2 \\ \vdots \\ x_n \end{bmatrix}$ with the point $(x_1, x_2, \ldots, x_n)$ so that points and vectors can be used interchangeably. Thus, we may view $\mathbf{R}^n$ as consisting of vectors or of points, and we write $(x_1, x_2, \ldots, x_n)$.The important point here is that no matter how we view $\mathbf{R}^n$—as n vectors, points, or $n \times 1$ matrices—its algebraic behaviour is always the same.

DEFINITION: A 'real vector space' is a set **V** of elements called vectors together with two operations

(i) *addition*: to each pair **X**, **Y** of vectors, there is a vector **X** + **Y** in **V**.

(ii) *scalar multiplication*: to each pair c and **X**, where c is a real number and **X** is a vector in V, there is a vector c**X** in **V** such that for any **X**, **Y**, **Z** $\in$ **V** and c, d $\in$ **R** the properties (a)–(h) of Theorem 1 are satisfied.

NOTE: A real vector space is also called a vector space over **R**. By a vector space we shall always mean a real vector space.

EXAMPLE 2.7 From Theorem 1, it is clear that $\mathbf{R}^n$ is a vector space over **R**. Let us verify in detail that $\mathbf{R}^2$ is a vector space over **R**, that is, we prove Theorem 1 for the case $n = 2$.

Let $\mathbf{X} = (x_1, x_2)$, $\mathbf{Y} = (y_1, y_2)$ and $\mathbf{Z} = (z_1, z_2)$ be any vectors in $\mathbf{R}^2$ and c and d be any real numbers.

(a)
$$\begin{aligned} \mathbf{X} + \mathbf{Y} = (x_1, x_2) + (y_1, y_2) &= (x_1 + y_1, x_2 + y_2) \\ &= (y_1 + x_1, y_2 + x_2) \\ &= (y_1, y_2) + (x_1, x_2) \\ &= \mathbf{Y} + \mathbf{X} \end{aligned}$$

(b) $\mathbf{X} + (\mathbf{Y} + \mathbf{Z}) = (x_1, x_2) + [(y_1, y_2) + (z_1, z_2)]$
$$= (x_1, x_2) + (y_1 + z_1, y_2 + z_2)$$
$$= (x_1 + (y_1 + z_1), x_2 + (y_2 + z_2))$$
$$= ((x_1 + y_1) + z_1, (x_2 + y_2) + z_2)$$
$$= (x_1 + y_1,\ x_2 + y_2) + (z_1, z_2)$$
$$= (\mathbf{X} + \mathbf{Y}) + \mathbf{Z}$$

(c) $\mathbf{X} + \mathbf{O} = (x_1, x_2) + (0, 0) = (x_1 + 0,\ x_2 + 0) = (x_1, x_2) = \mathbf{X}$

Similarly $\mathbf{O} + \mathbf{X} = \mathbf{X}$

(d) $\mathbf{X} + (-\mathbf{X}) = (x_1, x_2) + (-x_1, -x_2) = (x_1 - x_1,\ x_2 - x_2) = (0, 0) = \mathbf{O}$

(e) $\mathrm{c}(\mathrm{d}\mathbf{X}) = \mathrm{c}(\mathrm{d}(x_1, x_2)) = \mathrm{c}(\mathrm{d}x_1, \mathrm{d}x_2) = (\mathrm{c}(\mathrm{d}x_1), \mathrm{c}(\mathrm{d}x_2))$
$$= ((\mathrm{cd})x_1, (\mathrm{cd})x_2) = (\mathrm{cd})(x_1, x_2) = (\mathrm{cd})\mathbf{X}$$

(f) $\mathrm{c}(\mathbf{X} + \mathbf{Y}) = \mathrm{c}((x_1, x_2) + (y_1, y_2)) = \mathrm{c}(x_1 + y_1,\ x_2 + y_2)$
$$= (\mathrm{c}(x_1 + y_1), \mathrm{c}(x_2 + y_2))$$
$$= (\mathrm{c}x_1 + \mathrm{c}y_1,\ \mathrm{c}x_2 + \mathrm{c}y_2)$$
$$= (\mathrm{c}x_1, \mathrm{c}x_2) + (\mathrm{c}y_1, \mathrm{c}y_2)$$
$$= \mathrm{c}(x_1, x_2) + \mathrm{c}(y_1, y_2)$$
$$= \mathrm{c}\mathbf{X} + \mathrm{c}\mathbf{Y}$$

(g) $(\mathrm{c} + \mathrm{d})\mathbf{X} = (\mathrm{c} + \mathrm{d})(x_1, x_2) = ((\mathrm{c} + \mathrm{d})x_1, (\mathrm{c} + \mathrm{d})x_2)$
$$= (\mathrm{c}x_1 + \mathrm{d}x_1,\ \mathrm{c}x_2 + \mathrm{d}x_2)$$
$$= (\mathrm{c}x_1,\ \mathrm{c}x_2) + (\mathrm{d}x_1,\ \mathrm{d}x_2)$$
$$= \mathrm{c}(x_1, x_2) + \mathrm{d}(x_1, x_2) = \mathrm{c}\mathbf{X} + \mathrm{d}\mathbf{X}$$

(h) $1\mathbf{X} = 1(x_1, x_2) = (1x_1,\ 1x_2) = (x_1, x_2) = \mathbf{X}$

Thus, $\mathbf{R}^2$ is a real vector space.

> **NOTE:** In the following sections, we shall denote a vector space by **V**. Though the definitions and results are true for a general vector space, we shall assume that **V** is $\mathbf{R}^n$ for some value of *n*. In particular, we may take *n* to be 2 or 3, that is, **V** to be $\mathbf{R}^2$ or $\mathbf{R}^3$ as is done in examples and most of the exercises.

2.2.1 Standard Basis for $\mathbf{R}^2$ and $\mathbf{R}^3$

We have already discussed in previous section that for a vector $\mathbf{X} = \begin{bmatrix} x_1 \\ x_2 \end{bmatrix}$ in $\mathbf{R}^2$, $\|\mathbf{X}\| = \sqrt{x_1^2 + x_2^2}$ gives the length of vector **X**.

A unit vector is a vector whose length is 1.
If **X** is any non-zero vector then the vector

$$\mathrm{U} = \frac{1}{\|\mathbf{X}\|}\mathbf{X} \text{ is a unit vector in the direction of } \mathbf{X}.$$

For example, if $\mathbf{X} = \begin{bmatrix} -3 \\ 4 \end{bmatrix}$, then

$$U = \frac{1}{\|\mathbf{X}\|}\mathbf{X} = \frac{1}{\sqrt{(-3)^2 + 4^2}}\begin{bmatrix} -3 \\ 4 \end{bmatrix} = \frac{1}{5}\begin{bmatrix} -3 \\ 4 \end{bmatrix} = \begin{bmatrix} -\frac{3}{5} \\ \frac{4}{5} \end{bmatrix}$$

is a unit vector, since $\|U\| = 1$.

There are two vectors in $\mathbf{R}^2$ that are of special importance. They are $\hat{i} = (1, 0)$ and $\hat{j} = (0, 1)$, the unit vectors along the positive x- and y-axis respectively.

If $\mathbf{X} = (x_1, y_1)$ is any vector in $\mathbf{R}^2$, then we can write $\mathbf{X}$ in terms of $\hat{i}$ and $\hat{j}$ as

$$\mathbf{X} = x_1\hat{i} + y_1\hat{j}.$$

For example, if $\mathbf{X} = (2, 3)$ then $\mathbf{X} = 2\hat{i} + 3\hat{j}$.
We may note that $\hat{i} \cdot \hat{j} = 0$. Therefore $\hat{i}$ and $\hat{j}$ are orthogonal vectors.

We also denote $\hat{i}$ and $\hat{j}$ by e_1 and e_2 respectively.

In fact, in matrix form, $e_1 = \begin{bmatrix} 1 \\ 0 \end{bmatrix}$ and $e_2 = \begin{bmatrix} 0 \\ 1 \end{bmatrix}$ and for any $\mathbf{X} = \begin{bmatrix} x \\ y \end{bmatrix}$ we have

$$\begin{bmatrix} x \\ y \end{bmatrix} = x\begin{bmatrix} 1 \\ 0 \end{bmatrix} + y\begin{bmatrix} 0 \\ 1 \end{bmatrix}$$

or

$$\mathbf{X} = xe_1 + ye_2$$

Thus, every vector in $\mathbf{R}^2$ can be expressed in terms of e_1 and e_2. The set $\{e_1, e_2\}$ of orthogonal vectors e_1 and e_2 is called standard basis of $\mathbf{R}^2$.

Similarly, if $e_1 = \begin{bmatrix} 1 \\ 0 \\ 0 \end{bmatrix}$, $e_2 = \begin{bmatrix} 0 \\ 1 \\ 0 \end{bmatrix}$ and $e_3 = \begin{bmatrix} 0 \\ 0 \\ 1 \end{bmatrix}$, then for any vector $\mathbf{X} = \begin{bmatrix} x \\ y \\ z \end{bmatrix}$ in $\mathbf{R}^3$, we have

$$\begin{bmatrix} x \\ y \\ z \end{bmatrix} = x\begin{bmatrix} 1 \\ 0 \\ 0 \end{bmatrix} + y\begin{bmatrix} 0 \\ 1 \\ 0 \end{bmatrix} + z\begin{bmatrix} 0 \\ 0 \\ 1 \end{bmatrix}$$

or

$$\mathbf{X} = xe_1 + ye_2 + ze_3$$

The set $\{e_1, e_2, e_3\}$ is called standard basis of $\mathbf{R}^3$. As in case of $\mathbf{R}^2$, we may also denote e_1, e_2, e_3 by $\hat{i}, \hat{j}, \hat{k}$ respectively.

EXERCISE 2.2

1. Let $\mathbf{X} = \begin{bmatrix} 4 \\ 2 \end{bmatrix}$ and $\mathbf{Y} = \begin{bmatrix} -1 \\ 3 \end{bmatrix}$. Find $\mathbf{X} + \mathbf{Y}$, $3\mathbf{X}$, $2\mathbf{X} + 4\mathbf{Y}$, $3\mathbf{X} - \mathbf{Y}$, $\mathbf{Y} + (-\mathbf{Y})$. Show each combination graphically.

2. Find $\mathbf{X} + \mathbf{Y}$, $\mathbf{X} - \mathbf{Y}$, $2\mathbf{X}$ and $3\mathbf{X} - 2\mathbf{Y}$ if

(a) $\mathbf{X} = (2, 3)$, $\mathbf{Y} = (-2, 5)$ **(b)** $\mathbf{X} = (-1, 3)$, $\mathbf{Y} = (2, 4)$

(c) $\mathbf{X} = (1, 2, -3)$, $\mathbf{Y} = (0, 1, -2)$ **(d)** $\mathbf{X} = (2, 0, -4)$, $\mathbf{Y} = (3, 2, 1)$

3. Let $\mathbf{X} = (-1, 3)$ and $\mathbf{Y} = (4, 9)$. Find
(a) $\|\mathbf{X}\| + \|\mathbf{Y}\|$ **(b)** $\|-2\mathbf{X}\| + 2\|\mathbf{X}\|$ **(c)** $\|\mathbf{X} + \mathbf{Y}\|$ **(d)** $\|\mathbf{X} - \mathbf{Y}\|$

4. Verify (a)–(h) of Theorem 1 for the vectors $\mathbf{X} = \begin{bmatrix} 1 \\ 2 \end{bmatrix}$, $\mathbf{Y} = \begin{bmatrix} 3 \\ -1 \end{bmatrix}$, $\mathbf{Z} = \begin{bmatrix} -1 \\ 3 \end{bmatrix}$ and scalars $c = 2$, $d = 3$.

5. As done for $\mathbf{R}^2$, verify that $\mathbf{R}^3$ is a vector space over $\mathbf{R}$.

6. Show that the set consisting of the single vector $\mathbf{O}$ is a vector space.

7. Let $\mathbf{X}$, $\mathbf{Y}$ and $\mathbf{Z}$ be three vectors in $\mathbf{R}^2$. Show graphically that $(\mathbf{X}+\mathbf{Y})+\mathbf{Z} = \mathbf{X}+(\mathbf{Y}+\mathbf{Z})$.

8. Show that the following sets together with the given operations is a vector space.

(a) The set of all pairs of real numbers of the form $\begin{bmatrix} x \\ 0 \end{bmatrix}$ with the operations $\begin{bmatrix} x_1 \\ 0 \end{bmatrix} + \begin{bmatrix} x_2 \\ 0 \end{bmatrix} = \begin{bmatrix} x_1 + x_2 \\ 0 \end{bmatrix}$ and $c\begin{bmatrix} x_1 \\ 0 \end{bmatrix} = \begin{bmatrix} cx_1 \\ 0 \end{bmatrix}$.

(b) The set of all pairs of real numbers $\begin{bmatrix} x \\ y \end{bmatrix}$ with the operations $\begin{bmatrix} x_1 \\ y_1 \end{bmatrix} + \begin{bmatrix} x_2 \\ y_2 \end{bmatrix} = \begin{bmatrix} x_1 + x_2 \\ y_1 + y_2 \end{bmatrix}$ and $c\begin{bmatrix} x \\ y \end{bmatrix} = \begin{bmatrix} cx \\ y \end{bmatrix}$.

(c) The set of all pairs of real numbers of the form $\begin{bmatrix} x \\ 2x \end{bmatrix}$ with the standard operations defined on $\mathbf{R}^2$.

(d) The set of all ordered triples of real numbers of the form $(x, y, 0)$ with the operations

$$(x_1, y_1, 0) + (x_2, y_2, 0) = (x_1 + x_2, y_1 + y_2, 0)$$
$$c(x, y, 0) = (cx, cy, 0).$$

9. A boat is travelling east across a river at the rate of 4 miles per hour while the river's current is flowing south at a rate of 3 miles per hour. Find the resultant velocity of the boat.

10. A force of 12 pounds is applied to an object along the negative x-axis and a force of 5 pounds is applied to the object along the positive y-axis. Find the magnitude and direction of the resultant force.

11. Find the resultant of the force $(0, 3)$ and $(3, 1)$. How can the resultant be doubled in magnitude by changing $(0, 3)$ but keeping $(3, 1)$ unchanged ?

12. Show that $\hat{i} \cdot \hat{i} = \hat{j} \cdot \hat{j} = 1$ but $\hat{i} \cdot \hat{j} = 0$ where $\hat{i}$ and $\hat{j}$ are unit vectors of $\mathbf{R}^3$.

13. If $\mathbf{X}$, $\mathbf{Y}$ and $\mathbf{Z}$ are vectors in $\mathbf{R}^2$ and c is a scalar, then prove that

(a) $\mathbf{X} \cdot \mathbf{Y} = \mathbf{Y} \cdot \mathbf{X}$

(b) $(\mathbf{X}+\mathbf{Y}) \cdot \mathbf{Z} = \mathbf{X} \cdot \mathbf{Z} + \mathbf{Y} \cdot \mathbf{Z}$

(c) $(c\mathbf{X}) \cdot \mathbf{Y} = c(\mathbf{X} \cdot \mathbf{Y}) = \mathbf{X} \cdot (c\mathbf{Y})$

(d) $\mathbf{X} \cdot \mathbf{X} \geq 0$. Further $\mathbf{X} \cdot \mathbf{X} = 0$ if and only if $\mathbf{X} = \mathbf{O}$.

14. Repeat question 13, by taking vectors $\mathbf{X}$, $\mathbf{Y}$ and $\mathbf{Z}$ in $\mathbf{R}^3$.

2.3 LINEAR COMBINATION OF VECTORS

? DEFINITION: Let **V** be any vector space. Let $\mathbf{X}_1, \mathbf{X}_2, \ldots, \mathbf{X}_n$ be vectors from **V** and let $c_1, c_2, \ldots, c_n$ be a system of scalars (real numbers). Then the vector

$$\mathbf{X} = c_1\mathbf{X}_1 + c_2\mathbf{X}_2 + \cdots + c_n\mathbf{X}_n$$

is called a 'linear combination' of the vectors $\mathbf{X}_1, \mathbf{X}_2, \ldots, \mathbf{X}_n$. We also say that vector X is generated by the vectors $\mathbf{X}_1, \mathbf{X}_2, \ldots, \mathbf{X}_n$

EXAMPLE 2.8 The vector $2\begin{bmatrix}1\\3\end{bmatrix}$ is a linear combination of the vector $\begin{bmatrix}1\\3\end{bmatrix}$ of $\mathbf{R}^2$ with $c_1 = 2$. Here $n = 1$ since only one vector is involved.

EXAMPLE 2.9 The vector $2\begin{bmatrix}1\\3\end{bmatrix} + 3\begin{bmatrix}2\\1\end{bmatrix}$ is a linear combination of the vectors $\begin{bmatrix}1\\3\end{bmatrix}$ and $\begin{bmatrix}2\\1\end{bmatrix}$ of $\mathbf{R}^2$ with $c_1 = 2$, $c_2 = 3$, $\mathbf{X}_1 = \begin{bmatrix}1\\3\end{bmatrix}$ and $\mathbf{X}_2 = \begin{bmatrix}2\\1\end{bmatrix}$. Note that by applying addition and scalar multiplication defined on $\mathbf{R}^2$ we get

$$2\begin{bmatrix}1\\3\end{bmatrix} + 3\begin{bmatrix}2\\1\end{bmatrix} = \begin{bmatrix}8\\9\end{bmatrix}$$

Thus, we may also say that the vector $\begin{bmatrix}8\\9\end{bmatrix}$ is a linear combination of $\begin{bmatrix}1\\3\end{bmatrix}$ and $\begin{bmatrix}2\\1\end{bmatrix}$ because there exist $c_1 = 2$ and $c_2 = 3$ such that

$$2\begin{bmatrix}1\\3\end{bmatrix} + 3\begin{bmatrix}2\\1\end{bmatrix} = \begin{bmatrix}8\\9\end{bmatrix}.$$

EXAMPLE 2.10 Show that the vector $\begin{bmatrix}8\\7\end{bmatrix}$ is a linear combination of the vectors $\begin{bmatrix}1\\2\end{bmatrix}$ and $\begin{bmatrix}2\\1\end{bmatrix}$ of $\mathbf{R}^2$.

SOLUTION

We must show that there exist scalars c_1, c_2 such that

$$c_1\begin{bmatrix}1\\2\end{bmatrix} + c_2\begin{bmatrix}2\\1\end{bmatrix} = \begin{bmatrix}8\\7\end{bmatrix}$$

Since

$$c_1\begin{bmatrix}1\\2\end{bmatrix} + c_2\begin{bmatrix}2\\1\end{bmatrix} = \begin{bmatrix}c_1\\2c_1\end{bmatrix} + \begin{bmatrix}2c_2\\c_2\end{bmatrix} = \begin{bmatrix}c_1 + 2c_2\\2c_1 + c_2\end{bmatrix}$$

we get the vector equation

$$\begin{bmatrix}c_1 + 2c_2\\2c_1 + c_2\end{bmatrix} = \begin{bmatrix}8\\7\end{bmatrix}$$

Equating corresponding components , we get the system of equations

$$c_1 + 2c_2 = 8$$
$$2c_1 + c_2 = 7$$

the solution of which is given by $c_1 = 2$ and $c_2 = 3$. Thus, we may write

$$\begin{bmatrix}8\\7\end{bmatrix} = 2\begin{bmatrix}1\\2\end{bmatrix} + 3\begin{bmatrix}2\\1\end{bmatrix}$$

EXAMPLE 2.11 From the Sec. 2.2.1, we note that every vector $\mathbf{X} = \begin{bmatrix}x\\y\end{bmatrix}$ of $\mathbf{R}^2$ can be expressed as $\mathbf{X} = xe_1 + ye_2$ where $e_1 = \begin{bmatrix}1\\0\end{bmatrix}$ and $e_2 = \begin{bmatrix}0\\1\end{bmatrix}$.

Thus, we can say that every vector $\mathbf{X}$ of $\mathbf{R}^2$ can be expressed as a linear combination of vectors e_1 and e_2 of $\mathbf{R}^2$. In other words, every vector in $\mathbf{R}^2$ can be generated by vectors e_1 and e_2 of $\mathbf{R}^2$.

Similarly every vector $\mathbf{X} = \begin{bmatrix} x \\ y \\ z \end{bmatrix}$ of $\mathbf{R}^3$ can be expressed as a linear combination of vectors

$e_1 = \begin{bmatrix} 1 \\ 0 \\ 0 \end{bmatrix}, e_2 = \begin{bmatrix} 0 \\ 1 \\ 0 \end{bmatrix}$ and $e_3 = \begin{bmatrix} 0 \\ 0 \\ 1 \end{bmatrix}$ of $\mathbf{R}^3$.

> **? DEFINITION:** The vectors $\mathbf{X}_1, \mathbf{X}_2, \ldots, \mathbf{X}_n$ in a vector space **V** are said to '*span*' **V** if every vector in **V** is a linear combinaton of $\mathbf{X}_1, \mathbf{X}_2, \ldots, \mathbf{X}_n$. If $\mathbf{S} = \{\mathbf{X}_1, \mathbf{X}_2, \ldots, \mathbf{X}_n\}$, then we say that **S** spans **V**, or that **V** is spanned by **S**, if vectors $\mathbf{X}_1, \mathbf{X}_2, \ldots, \mathbf{X}_n$ span **V**. We write it as span **S** = **V**.

EXAMPLE 2.12 Let **V** be the vector space $\mathbf{R}^3$ and let $\mathbf{S} = \{\mathbf{X}_1, \mathbf{X}_2, \mathbf{X}_3\}$ where $\mathbf{X}_1 = (1, 2, 1)$, $\mathbf{X}_2 = (1, 0, 2)$ and $\mathbf{X}_3 = (1, 1, 0)$. To find out whether **S** spans $\mathbf{R}^3$, we pick any vector $\mathbf{X} = (x, y, z)$ in $\mathbf{R}^3$ (x, y and z are arbitrary real numbers) and must find out whether there are constants c_1, c_2 and c_3 such that

$$c_1\mathbf{X}_1 + c_2\mathbf{X}_2 + c_3\mathbf{X}_3 = \mathbf{X}$$

That is,

$$c_1(1, 2, 1) + c_2(1, 0, 2) + c_3(1, 1, 0) = (x, y, z)$$

This leads to the linear system,

$$\begin{aligned} c_1 + c_2 + c_3 &= x \\ 2c_1 \qquad + c_3 &= y \\ c_1 + 2c_2 \qquad &= z \end{aligned}$$

Solving them we get a solution as

$$c_1 = \frac{-2x + 2y + z}{3}, c_2 = \frac{x - y + z}{3}, c_3 = \frac{4x - y - 2z}{3}$$

Thus, we have obtained a solution for every choice of x, y and z.

For example, if $\mathbf{X} = (1, 2, 3)$ then $c_1 = \frac{5}{3}, c_2 = \frac{2}{3}$ and $c_3 = -\frac{4}{3}$.

So,
$$\mathbf{X} = \frac{5}{3}\mathbf{X}_1 + \frac{2}{3}\mathbf{X}_2 - \frac{4}{3}\mathbf{X}_3.$$

Hence, $\mathbf{X}_1, \mathbf{X}_2, \mathbf{X}_3$ span $\mathbf{R}^3$.

EXAMPLE 2.13 From Example 1.11, it is clear that the vectors $e_1 = \hat{i} = (1, 0)$ and $e_2 = \hat{j} = (0, 1)$ span $\mathbf{R}^2$ and the vector $e_1 = \hat{i} = (1, 0, 0), e_2 = \hat{j} = (0, 1, 0)$ and $e_3 = \hat{k} = (0, 0, 1)$ span $\mathbf{R}^3$. Similarly, the vectors $e_1 = (1, 0, 0, \ldots, 0)$, $e_2 = (0, 1, 0, \ldots, 0), \ldots, e_n = (0, 0, \ldots, 1)$ span $\mathbf{R}^n$, since any vector $\mathbf{X} = (x_1, x_2, \ldots, x_n)$ in $\mathbf{R}^n$ can be written as

$$\mathbf{X} = x_1e_1 + x_2e_2 + \cdots + x_ne_n.$$

EXAMPLE 2.14 Does the set $\left\{\begin{bmatrix}1\\0\\0\end{bmatrix}, \begin{bmatrix}0\\1\\0\end{bmatrix}\right\}$ span $\mathbf{R}^3$?

SOLUTION

In order to answer this question, we take an arbitrary vector $\begin{bmatrix}x\\y\\z\end{bmatrix}$ in $\mathbf{R}^3$ and try to determine if some linear combination of $\begin{bmatrix}1\\0\\0\end{bmatrix}$ and $\begin{bmatrix}0\\1\\0\end{bmatrix}$ will generate the vector $\begin{bmatrix}x\\y\\z\end{bmatrix}$. Thus, we write

$$c_1\begin{bmatrix}1\\0\\0\end{bmatrix} + c_2\begin{bmatrix}0\\1\\0\end{bmatrix} = \begin{bmatrix}x\\y\\z\end{bmatrix}$$

and try to determine c_1 and c_2 (in terms of x, y, z) such that the equation above is satisfied. Performing the operations on the left-hand side of the equation yields

$$\begin{bmatrix}c_1\\c_2\\0\end{bmatrix} = \begin{bmatrix}x\\y\\z\end{bmatrix}$$

The conclusion is, therefore, that if $z \neq 0$ then the equation $c_1\begin{bmatrix}1\\0\\0\end{bmatrix} + c_2\begin{bmatrix}0\\1\\0\end{bmatrix} = \begin{bmatrix}x\\y\\z\end{bmatrix}$ cannot be satisfied by any scalars c_1 and c_2.

Thus, none of the vectors $\begin{bmatrix}x\\y\\z\end{bmatrix}$ in $\mathbf{R}^3$ with $z \neq 0$ can be generated by linear combination of $\begin{bmatrix}1\\0\\0\end{bmatrix}$ and $\begin{bmatrix}0\\1\\0\end{bmatrix}$. We conclude that $\mathbf{R}^3$ is not spanned by the vectors $\begin{bmatrix}1\\0\\0\end{bmatrix}$ and $\begin{bmatrix}0\\1\\0\end{bmatrix}$.

EXERCISE 2.3

1. Show that the vector $\begin{bmatrix}3\\7\end{bmatrix}$ is a linear combination of the vectors $\begin{bmatrix}1\\3\end{bmatrix}$ and $\begin{bmatrix}0\\1\end{bmatrix}$ of $\mathbf{R}^2$.
2. Show that the vector $(5, 4, 2)$ is a linear combination of vectors $(1, 2, 0)$, $(3, 1, 4)$ and $(1, 0, 3)$.
3. Show that the vector $(3, -4, -6)$ cannot be expressed as a linear combination of the vectors $(1, 2, 3)$, $(-1, -1, -2)$ and $(1, 4, 5)$.
4. In the following sets of vectors, determine whether the first vector is a linear combination of the other vectors.
   (a) $(7, 5); (1, -1), (5, 1)$ **(b)** $(-1, 15); (-1, 4), (2, -8)$
   (c) $(4, 3, 8); (-1, 0, 1), (2, 1, 3), (0, 1, 5)$ **(d)** $(0, 10, 8); (-1, 2, 3), (1, 3, 1), (1, 8, 5)$
5. Which of the following sets of vectors span $\mathbf{R}^2$?
   (a) $(1, 2), (-1, 1)$ **(b)** $(1, 1), (1, -1), (1, 3)$
   (c) $(-3, 1), (6, -2)$ **(d)** $(0, 0), (1, 1), (-3, -3)$

6. Which of the following sets of vectors span $\mathbf{R}^3$?

(a) $(1, -1, 2), (0, 2, 2)$

(b) $(1, -1, 1), (1, 1, -1), (1, 1, 1)$

(c) $(1, 2, 1), (-1, 3, 0), (0, 5, 1)$

(d) $(1, -1, 1), (1, 2, -1), (2, 1, 0)$

7. Show that the following sets of vectors span $\mathbf{R}^3$. Express the vector $\begin{bmatrix} 1 \\ 3 \\ -2 \end{bmatrix}$ in terms of each spanning set.

(a) $\begin{bmatrix} 1 \\ 2 \\ 3 \end{bmatrix}, \begin{bmatrix} -1 \\ -1 \\ 0 \end{bmatrix}, \begin{bmatrix} 2 \\ 5 \\ 4 \end{bmatrix}$ **(b)** $\begin{bmatrix} 5 \\ 1 \\ 3 \end{bmatrix}, \begin{bmatrix} 2 \\ 0 \\ 1 \end{bmatrix}, \begin{bmatrix} -2 \\ -3 \\ -1 \end{bmatrix}$

2.4 LINEAR INDEPENDENCE

? DEFINITION: Let $\mathbf{S} = \{\mathbf{X}_1, \mathbf{X}_2, \ldots, \mathbf{X}_k\}$ be a set of vectors in a vector space $\mathbf{V}$. We say that the vectors $\mathbf{X}_1, \mathbf{X}_2 \ldots, \mathbf{X}_k$ are '*linearly dependent*,' or that the set of vectors S is a *linearly dependent set*, if there exist scalars $c_1, c_2, \ldots, c_k$, not all zero, such that

$$c_1\mathbf{X}_1 + c_2\mathbf{X}_2 + \cdots + c_k\mathbf{X}_k = \mathbf{O} \tag{1}$$

Otherwise, the vectors $\mathbf{X}_1, \mathbf{X}_2, \ldots, \mathbf{X}_k$ (or the set $\mathbf{S}$) are called 'linearly independent'.

In other words, the vectors $\mathbf{X}_1, \mathbf{X}_2, \ldots, \mathbf{X}_k$ are linearly independent if whenever

$$c_1\mathbf{X}_1 + c_2\mathbf{X}_2 + \cdots + c_k\mathbf{X}_k = \mathbf{O}$$

then

$$c_1 = c_2 = \cdots = c_k = 0$$

It must be noted that equation (1) holds if we choose all the scalars $c_1, c_2, \ldots, c_k$ equal to zero. In fact, the above definitions imply the following:

The set $\mathbf{S} = \{\mathbf{X}_1, \mathbf{X}_2, \ldots, \mathbf{X}_k\}$ is linearly independent if and only if equation (1) has the unique solution $c_1 = c_2 = \cdots = c_k = 0$; and $\mathbf{S}$ is linearly dependent if equation (1) has at least two different solutions for $c_1, c_2, \ldots, c_k$ (Obviously, one solution is always $c_1 = c_2 = \cdots = c_k = 0$).

EXAMPLE 2.15 Prove that the vectors $\begin{bmatrix} 1 \\ 2 \end{bmatrix}$ and $\begin{bmatrix} 2 \\ 1 \end{bmatrix}$ of $\mathbf{R}^2$ are linearly independent.

SOLUTION

We examine the equation

$$c_1\begin{bmatrix} 1 \\ 2 \end{bmatrix} + c_2\begin{bmatrix} 2 \\ 1 \end{bmatrix} = \begin{bmatrix} 0 \\ 0 \end{bmatrix}$$

for all possible solutions c_1 and c_2.

The above equation reduces to the system of equations

$$c_1 + 2c_2 = 0$$
$$2c_1 + c_2 = 0$$

Using matrix notation, we may write the system as

$$\begin{bmatrix} 1 & 2 \\ 2 & 1 \end{bmatrix}\begin{bmatrix} c_1 \\ c_2 \end{bmatrix} = \begin{bmatrix} 0 \\ 0 \end{bmatrix}$$

The matrix $\begin{bmatrix} 1 & 2 \\ 2 & 1 \end{bmatrix}$ is invertible and

$$\begin{bmatrix} 1 & 2 \\ 2 & 1 \end{bmatrix}^{-1} = \begin{bmatrix} -\frac{1}{3} & \frac{2}{3} \\ \frac{2}{3} & -\frac{1}{3} \end{bmatrix}$$

$$\therefore \quad \begin{bmatrix} c_1 \\ c_2 \end{bmatrix} = \begin{bmatrix} -\frac{1}{3} & \frac{2}{3} \\ \frac{2}{3} & -\frac{1}{3} \end{bmatrix}\begin{bmatrix} 0 \\ 0 \end{bmatrix}$$

or

$$\begin{bmatrix} c_1 \\ c_2 \end{bmatrix} = \begin{bmatrix} 0 \\ 0 \end{bmatrix}.$$

Thus, the equation $c_1\begin{bmatrix} 1 \\ 2 \end{bmatrix} + c_2\begin{bmatrix} 2 \\ 1 \end{bmatrix} = \begin{bmatrix} 0 \\ 0 \end{bmatrix}$ has the unique solution $c_1 = 0$ and $c_2 = 0$.

Hence, $\left\{\begin{bmatrix} 1 \\ 2 \end{bmatrix}, \begin{bmatrix} 2 \\ 1 \end{bmatrix}\right\}$ is linearly independent.

EXAMPLE 2.16 Consider the vectors $\begin{bmatrix} 1 \\ 2 \end{bmatrix}, \begin{bmatrix} 2 \\ 1 \end{bmatrix}$ and $\begin{bmatrix} 8 \\ 7 \end{bmatrix}$ in $\mathbf{R}^2$. Show that the set $\left\{\begin{bmatrix} 1 \\ 2 \end{bmatrix}, \begin{bmatrix} 2 \\ 1 \end{bmatrix}, \begin{bmatrix} 8 \\ 7 \end{bmatrix}\right\}$ is not linearly independent.

SOLUTION

Consider the equation

$$c_1\begin{bmatrix} 1 \\ 2 \end{bmatrix} + c_2\begin{bmatrix} 2 \\ 1 \end{bmatrix} + c_3\begin{bmatrix} 8 \\ 7 \end{bmatrix} = \begin{bmatrix} 0 \\ 0 \end{bmatrix}$$

Performing the operations of addition and scalar multiplication indicated on the left-hand side of the equation, we get

$$\begin{bmatrix} c_1 + 2c_2 + 8c_3 \\ 2c_1 + c_2 + 7c_3 \end{bmatrix} = \begin{bmatrix} 0 \\ 0 \end{bmatrix}$$

This is equivalent to the system

$$\left.\begin{aligned} c_1 + 2c_2 + 8c_3 &= 0 \\ 2c_1 + c_2 + 7c_3 &= 0 \end{aligned}\right\} \tag{1}$$

This is a homogeneous system of linear equations with more unknowns than equations. As we know that a homogeneous system $\mathbf{AX} = \mathbf{O}$ with more unknowns than the equations has infinitely many solutions, there exists a non-trivial solution for system (1).

In fact, $c_1 = 2, c_2 = 3$ and $c_3 = -1$ is one such solution (see Example 2.10).

Thus, the vectors $\begin{bmatrix} 1 \\ 2 \end{bmatrix}, \begin{bmatrix} 2 \\ 1 \end{bmatrix}$ and $\begin{bmatrix} 8 \\ 7 \end{bmatrix}$ are not linearly independent.

EXAMPLE 2.17 The set $\{e_1, e_2\}$ where $e_1 = \begin{bmatrix} 1 \\ 0 \end{bmatrix}$ and $e_2 = \begin{bmatrix} 0 \\ 1 \end{bmatrix}$ is a linearly independent set of vectors of $\mathbf{R}^2$ because the equation $c_1\begin{bmatrix} 1 \\ 0 \end{bmatrix} + c_2\begin{bmatrix} 0 \\ 1 \end{bmatrix} = \begin{bmatrix} 0 \\ 0 \end{bmatrix}$ implies $c_1 = 0$ and $c_2 = 0$.

Similarly, the equation $c_1\begin{bmatrix} 1 \\ 0 \\ 0 \end{bmatrix} + c_2\begin{bmatrix} 0 \\ 1 \\ 0 \end{bmatrix} + c_3\begin{bmatrix} 0 \\ 0 \\ 1 \end{bmatrix} = \begin{bmatrix} 0 \\ 0 \\ 0 \end{bmatrix}$ implies that $c_1 = c_2 = c_3 = 0$.

Therefore, the vectors $e_1 = \begin{bmatrix} 1 \\ 0 \\ 0 \end{bmatrix}$, $e_2 = \begin{bmatrix} 0 \\ 1 \\ 0 \end{bmatrix}$ and $e_3 = \begin{bmatrix} 0 \\ 0 \\ 1 \end{bmatrix}$ of $\mathbf{R}^3$ are linearly independent.

EXAMPLE 2.18 The vectors (1, 2, 3), (0, 1, 2) and (0, 0, 1) in $\mathbf{R}^3$ are linearly independent.

For, $\quad c_1(1, 2, 3) + c_2(0, 1, 2) + c_3(0, 0, 1) = (0, 0, 0)$

$\Leftrightarrow \quad (c_1, 2c_1 + c_2, 3c_1 + 2c_2 + c_3) = (0, 0, 0)$

$\Leftrightarrow \quad c_1 = 0, 2c_1 + c_2 = 0, 3c_1 + 2c_2 + c_3 = 0$

$\Leftrightarrow \quad c_1 = c_2 = c_3 = 0$

EXAMPLE 2.19 Are the vectors $\mathbf{X}_1 = (1, 0, 1, 2)$, $\mathbf{X}_2 = (0, 1, 1, 2)$ and $\mathbf{X}_3 = (1, 1, 1, 3)$ in $\mathbf{R}^4$ linearly dependent or linearly independent?

SOLUTION

We solve equation

$$c_1\mathbf{X}_1 + c_2\mathbf{X}_2 + c_3\mathbf{X}_3 = \mathbf{O}$$

for c_1, c_2 and c_3.

The resulting linear system is

$$\begin{aligned} c_1 \quad\quad\;\; + c_3 &= 0 \\ c_2 + c_3 &= 0 \\ c_1 + c_2 + c_3 &= 0 \\ 2c_1 + 2c_2 + 3c_3 &= 0 \end{aligned}$$

which has as its only solution $c_1 = c_2 = c_3 = 0$ (verify) showing that the given vectors are linearly independent.

EXAMPLE 2.20 Consider the vectors $\mathbf{X}_1 = (1, 2, -1)$, $\mathbf{X}_2 = (1, -2, 1)$, $\mathbf{X}_3 = (-3, 2, -1)$ and $\mathbf{X}_4 = (2, 0, 0)$ in $\mathbf{R}^3$. Is $\mathbf{S} = \{\mathbf{X}_1, \mathbf{X}_2, \mathbf{X}_3, \mathbf{X}_4\}$ linearly dependent or linearly independent?

SOLUTION

Consider the equation

$$c_1\mathbf{X}_1 + c_2\mathbf{X}_2 + c_3\mathbf{X}_3 + c_4\mathbf{X}_4 = \mathbf{O}$$

This leads to the system

$$\begin{aligned} c_1 + c_2 - 3c_3 + 2c_4 &= 0 \\ 2c_1 - 2c_2 + 2c_3 \quad\quad &= 0 \\ -c_1 + c_2 - c_3 \quad\quad &= 0 \end{aligned}$$

which is a homogeneous system of three equations in four unknowns. Thus, we are sure of existence of a non-trivial solution. In fact, two of the infinitely many solutions are

$$c_1 = 1, \qquad c_2 = 2, \qquad c_3 = 1, \qquad c_4 = 0$$

and

$$c_1 = 1, \qquad c_2 = 1, \qquad c_3 = 0, \qquad c_4 = -1$$

Hence **S** is linearly dependent.

EXAMPLE 2.21 Prove that the set $\{\mathbf{X}\}$ consisting of a non-zero vector **X** of $\mathbf{R}^n$ is a linearly independent set.

SOLUTION

For all possible solutions for c.

$$c\mathbf{X} = \mathbf{O}$$

If $c \neq 0$, then multiplying both sides of the equation by scalar $\frac{1}{c}$, we get

$$\frac{1}{c}(c\mathbf{X}) = \frac{1}{c}\mathbf{O}$$

which leads to $\mathbf{X} = \mathbf{O}$, a contradiction. Hence, we must have $c = 0$. Thus, the only solution for $c\mathbf{X} = \mathbf{O}$, when $\mathbf{X} \neq \mathbf{O}$ is $c = 0$. Therefore, every non-zero vector constitutes a linearly independent set.

It is also obvious from this argument that the equation $c\mathbf{X} = \mathbf{O}$ has infinitely many solutions when $\mathbf{X} = \mathbf{O}$ or $c\mathbf{O} = \mathbf{O}$ is satisfied for non-zero values of c. Thus, the set consisting of the zero vector of any vector space is not linearly independent.

We can go one step further with zero vector in following example.

EXAMPLE 2.22 Prove that any set of vectors containing the zero vector is linearly dependent.

SOLUTION

Consider the set $\{\mathbf{O}, \mathbf{X}_2, \ldots, \mathbf{X}_n\}$ which contains the zero vector. Let us examine the equation

$$c_1\mathbf{O} + c_2\mathbf{X}_2 + \cdots + c_n\mathbf{X}_n = \mathbf{O}$$

If we choose $c_2 = c_3 = \cdots = c_n = 0$ and $c_1 = 1$ (or any non-zero number), then the equation is satisfied. Thus, we get scalars $c_1, c_2, \ldots, c_n$ not all zero such that $c_1\mathbf{O} + c_2\mathbf{X}_2 + \cdots + c_n\mathbf{X}_n = \mathbf{O}$. Thus $\{\mathbf{O}, \mathbf{X}_2, \ldots, \mathbf{X}_n\}$ is linearly dependent.

From Example 2.21, it is clear that the set $\{\mathbf{X}\}$ is linearly dependent if and only if $\mathbf{X} = \mathbf{O}$. In following examples we discuss the linear dependence of set consisting of two and three vectors.

EXAMPLE 2.23 Prove that the set $\{\mathbf{X}_1, \mathbf{X}_2\}$ of vectors in $\mathbf{R}^n$ is linearly dependent if $\mathbf{X}_1$ and $\mathbf{X}_2$ are collinear, that is, one of them is a scalar multiple of the other.

SOLUTION

Suppose that $\mathbf{X}_1$ and $\mathbf{X}_2$ are linearly dependent. Then there exist scalars c_1 and c_2, not both zero, such that

$$c_1\mathbf{X}_1 + c_2\mathbf{X}_2 = \mathbf{O}$$

If $c_1 \neq 0$, then $\mathbf{X}_1 = \left(-\frac{c_2}{c_1}\right)\mathbf{X}_2$

If $c_2 \neq 0$, then $\mathbf{X}_2 = \left(-\frac{c_1}{c_2}\right)\mathbf{X}_1$

Thus one of the vectors is a scalar multiple of the other.
Conversely, suppose that $\mathbf{X}_1 = c\mathbf{X}_2$. Then

$$1\mathbf{X}_1 - c\mathbf{X}_2 = \mathbf{O}$$

and since for both $\mathbf{X}_1$ and $\mathbf{X}_2$ the coefficients are not zero, it follows that $\mathbf{X}_1$ and $\mathbf{X}_2$ are linearly dependent. Fig. 2.10 (a) and (b) clarify this when $\mathbf{X}_1$ and $\mathbf{X}_2$ are vectors in $\mathbf{R}^2$.

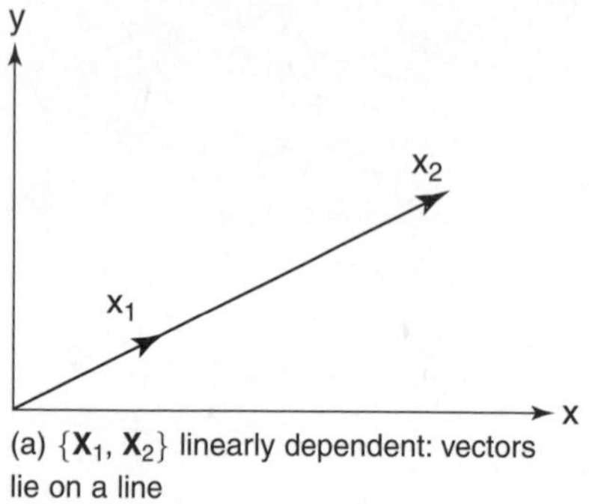

(a) $\{\mathbf{X}_1, \mathbf{X}_2\}$ linearly dependent: vectors lie on a line

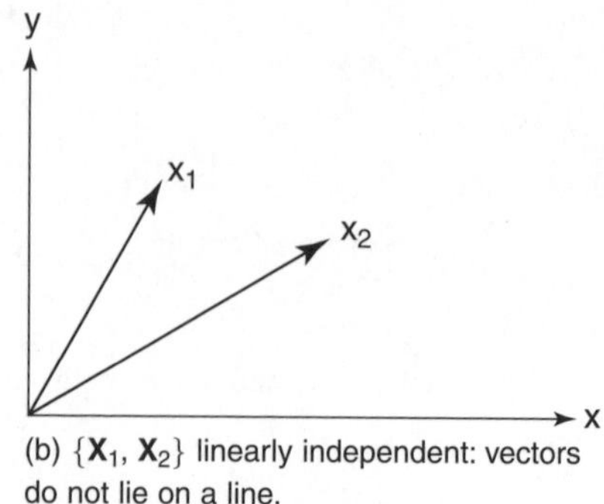

(b) $\{\mathbf{X}_1, \mathbf{X}_2\}$ linearly independent: vectors do not lie on a line.

Fig. 2.10

EXAMPLE 2.24 As in Example 2.23, it can similarly be proved that the set $\{\mathbf{X}_1, \mathbf{X}_2, \mathbf{X}_3\}$ of three vectors is linearly dependent if $\mathbf{X}_1$, $\mathbf{X}_2$ and $\mathbf{X}_3$ are coplanar, that is, one of them is a linear combination of the other two.

More generally, we have the following result for a set of more than three vectors.

The set $\mathbf{S} = \{\mathbf{X}_1, \mathbf{X}_2, \ldots, \mathbf{X}_n\}$ of non-zero vectors is linearly dependent if and only if **S** contains at least one vector which is a linear combination of the remaining vectors of **S**.

We state one more result to determine linear dependence in $\mathbf{R}^n$.

THEOREM 1

Let $\mathbf{S} = \{x_1, x_2, \ldots, x_r\}$ be a set of vectors in $\mathbf{R}^n$. If $r > n$, then **S** is linearly dependent.

In particular, this theorem tells us that a set in $\mathbf{R}^2$ with more than two vectors is linearly dependent, and a set in $\mathbf{R}^3$ with more than three vectors is linearly dependent.

EXERCISE 2.4

1. By inspection only, explain why the following are linearly dependent set of vectors.

(a) $\mathbf{X} = (1, 3)$ and $\mathbf{Y} = (-2, -6)$ in $\mathbf{R}^2$

(b) $\mathbf{X} = (1, 2)$, $\mathbf{Y} = (4, 5)$, $\mathbf{Z} = (2, 3)$ in $\mathbf{R}^2$

(c) $\mathbf{X} = (-1, 2, 5)$ and $\mathbf{Y} = (-3, 6, 15)$ in $\mathbf{R}^3$

2. Decide whether each of the following sets of vectors is linearly dependent or linearly independent.

(a) $\left\{\begin{bmatrix}1\\1\end{bmatrix}, \begin{bmatrix}1\\-1\end{bmatrix}\right\}$ in $\mathbf{R}^2$ **(b)** $\left\{\begin{bmatrix}-2\\3\end{bmatrix}, \begin{bmatrix}6\\-9\end{bmatrix}\right\}$ in $\mathbf{R}^2$

(c) $\left\{\begin{bmatrix}1\\2\end{bmatrix},\begin{bmatrix}-1\\4\end{bmatrix},\begin{bmatrix}-1\\16\end{bmatrix}\right\}$ in $\mathbf{R}^2$

(d) $\left\{\begin{bmatrix}1\\2\\3\end{bmatrix},\begin{bmatrix}3\\-1\\0\end{bmatrix}\right\}$ in $\mathbf{R}^3$

(e) $\left\{\begin{bmatrix}1\\1\\1\end{bmatrix},\begin{bmatrix}1\\-1\\1\end{bmatrix},\begin{bmatrix}1\\1\\-1\end{bmatrix}\right\}$ in $\mathbf{R}^3$

(f) $\left\{\begin{bmatrix}1\\1\\0\\0\end{bmatrix},\begin{bmatrix}0\\1\\-1\\0\end{bmatrix}\right\}$ in $\mathbf{R}^4$

3. Express vector $\mathbf{X} = (3, 1, -4)$ as a linear combination of the vectors $\mathbf{X}_1 = (1, 1, 1)$, $\mathbf{X}_2 = (0, 1, 1)$, $\mathbf{X}_3 = (0, 0, 1)$. Is the set $\mathbf{S} = \{\mathbf{X}, \mathbf{X}_1, \mathbf{X}_2, \mathbf{X}_3\}$ linearly dependent or linearly independent?

4. Express $\begin{bmatrix}1\\2\\3\end{bmatrix}$ as a linear combination of vectors of the set $\mathbf{S} = \left\{\begin{bmatrix}1\\0\\0\end{bmatrix},\begin{bmatrix}0\\1\\0\end{bmatrix},\begin{bmatrix}1\\0\\1\end{bmatrix},\begin{bmatrix}0\\0\\1\end{bmatrix}\right\} \subset \mathbf{R}^3$ in two ways. Show that **S** is linearly dependent set.

5. Find the values of c for which the following sets of vectors in $\mathbf{R}^2$ are linearly dependent.

(a) $\{(1, -2), (c, -4)\}$ **(b)** $\{(2, -c), (2c + 6, 4c)\}$

6. For what real values of λ do the following sets of vectors form a linearly dependent set in $\mathbf{R}^3$?

(a) $\mathbf{X}_1 = (-1, 0, -1)$, $\mathbf{X}_2 = (2, 1, 2)$, $\mathbf{X}_3 = (1, 1, \lambda)$

(b) $\mathbf{X}_1 = \left(\lambda, -\frac{1}{2}, -\frac{1}{2}\right)$, $\mathbf{X}_2 = \left(-\frac{1}{2}, \lambda, -\frac{1}{2}\right)$, $\mathbf{X}_3 = \left(-\frac{1}{2}, -\frac{1}{2}, \lambda\right)$

7. (a) Let the set $\{\mathbf{X}, \mathbf{Y}\}$ be linearly independent in a vector space **V**. Prove that $\{\mathbf{X} + \mathbf{Y}, \mathbf{X} - \mathbf{Y}\}$ is also linearly independent.

(b) Let the set $\{\mathbf{X}, \mathbf{Y}\}$ be linearly dependent in a vector space **V**. Prove that $\{\mathbf{X} + \mathbf{Y}, \mathbf{X} - \mathbf{Y}\}$ is also linearly dependent.

8. Prove that for any vectors **X**, **Y** and **Z**, the vectors $\mathbf{X} - \mathbf{Y}$, $\mathbf{Y} - \mathbf{Z}$ and $\mathbf{Z} - \mathbf{X}$ form a linearly dependent set.

9. Prove that any subset of a linearly independent set of vectors is linearly independent.

10. (a) If $\{\mathbf{X}_1, \mathbf{X}_2, \mathbf{X}_3\}$ is a linearly dependent set of vectors in $\mathbf{R}^n$, show that $\{\mathbf{X}_1, \mathbf{X}_2, \mathbf{X}_3, \mathbf{X}_4\}$ is also linearly dependent, where $\mathbf{X}_4$ is any other vector in $\mathbf{R}^n$.

(b) Prove that any set which contains a linearly dependent set is linearly dependent.

11. (a) Show that any set of three vectors from $\mathbf{R}^2$ is linearly dependent.

(b) Let $\mathbf{S} = \{\mathbf{X}_1, \mathbf{X}_2, \ldots, \mathbf{X}_m\}$ be a set of vectors in $\mathbf{R}^n$. Show that if $m > n$ then **S** is linearly dependent.

2.5 BASIS AND DIMENSION

? DEFINITION: Let **V** be a vector space. Let $\mathbf{S} = \{\mathbf{X}_1, \mathbf{X}_2, \ldots, \mathbf{X}_n\}$ be a finite subset of **V**. Then **S** is called a 'basis' of **V** if

(i) **S** spans **V**

(ii) **S** is linearly independent.

We have already discussed standard basis for $\mathbf{R}^2$ and $\mathbf{R}^3$. The following example clarifies why we call them 'basis'.

EXAMPLE 2.25 Let $e_1 = \begin{bmatrix} 1 \\ 0 \end{bmatrix}$ and $e_2 = \begin{bmatrix} 0 \\ 1 \end{bmatrix}$. Then

(i) $\{e_1, e_2\}$ spans $\mathbf{R}^2$. (proved in Example 2.13)

(ii) $\{e_1, e_2\}$ is linearly independent. (proved in Example 2.17)

$\therefore$ The set $\{e_1, e_2\}$ is a basis for $\mathbf{R}^2$.

Similarly, $\{e_1, e_2, e_3\}$ is a basis for $\mathbf{R}^3$ and in general $\{e_1, e_2, \ldots, e_n\}$ is a basis for $\mathbf{R}^n$. As already noted, each of these is called the standard basis or the natural basis for $\mathbf{R}^2$, $\mathbf{R}^3$ and $\mathbf{R}^n$ respectively.

A vector space may have more than one basis. Following are examples of distinct bases for $\mathbf{R}^2$ and $\mathbf{R}^3$.

EXAMPLE 2.26 The set $\left\{\begin{bmatrix} 1 \\ 1 \end{bmatrix}, \begin{bmatrix} 1 \\ 2 \end{bmatrix}\right\}$ is a basis for $\mathbf{R}^2$.

Consider the equation

$$c_1\begin{bmatrix} 1 \\ 1 \end{bmatrix} + c_2\begin{bmatrix} 1 \\ 2 \end{bmatrix} = \begin{bmatrix} 0 \\ 0 \end{bmatrix}$$

This reduces to the system

$$\begin{aligned} c_1 + c_2 &= 0 \\ c_1 + 2c_2 &= 0 \end{aligned}$$

which has the unique solution $c_1 = 0$ and $c_2 = 0$.

$\therefore \left\{\begin{bmatrix} 1 \\ 1 \end{bmatrix}, \begin{bmatrix} 1 \\ 2 \end{bmatrix}\right\}$ is linearly independent.

To prove that $\left\{\begin{bmatrix} 1 \\ 1 \end{bmatrix}, \begin{bmatrix} 1 \\ 2 \end{bmatrix}\right\}$ spans $\mathbf{R}^2$, we let $\begin{bmatrix} x \\ y \end{bmatrix}$ to be an arbitrary vector in $\mathbf{R}^2$ and write

$$a_1\begin{bmatrix} 1 \\ 1 \end{bmatrix} + a_2\begin{bmatrix} 1 \\ 2 \end{bmatrix} = \begin{bmatrix} x \\ y \end{bmatrix}.$$

This leads to the system of equations

$$\begin{aligned} a_1 + a_2 &= x \\ a_1 + 2a_2 &= y \end{aligned}$$

Solving for a_1 and a_2, we get

$$a_1 = 2x - y \quad \text{and} \quad a_2 = y - x$$

which shows that $\left\{\begin{bmatrix} 1 \\ 1 \end{bmatrix}, \begin{bmatrix} 1 \\ 2 \end{bmatrix}\right\}$ spans $\mathbf{R}^2$. Hence, it is a basis for $\mathbf{R}^2$.

EXAMPLE 2.27 Verify that the set $\left\{\begin{bmatrix} 1 \\ 1 \end{bmatrix}, \begin{bmatrix} 1 \\ -1 \end{bmatrix}\right\}$ is linearly independent in $\mathbf{R}^2$ and spans $\mathbf{R}^2$. Thus it is a basis of $\mathbf{R}^2$.

SOLUTION

In fact, it can be shown that any two vectors $\begin{bmatrix} a \\ b \end{bmatrix}$ and $\begin{bmatrix} c \\ d \end{bmatrix}$ in $\mathbf{R}^2$ will form a basis for $\mathbf{R}^2$ provided that neither of the two is a scalar multiple of the other vector.

EXAMPLE 2.28 The set of vectors

$$\mathbf{S} = \left\{ \begin{bmatrix} 0 \\ 1 \\ 0 \end{bmatrix}, \begin{bmatrix} 1 \\ 0 \\ 1 \end{bmatrix}, \begin{bmatrix} 1 \\ 1 \\ 0 \end{bmatrix} \right\}$$

constitutes a basis for $\mathbf{R}^3$.

(i) Prove that **S** is linearly independent.

SOLUTION

If $c_1, c_2, c_3 \in \mathbf{R}$ and

$$c_1 \begin{bmatrix} 0 \\ 1 \\ 0 \end{bmatrix} + c_2 \begin{bmatrix} 1 \\ 0 \\ 1 \end{bmatrix} + c_3 \begin{bmatrix} 1 \\ 1 \\ 0 \end{bmatrix} = \begin{bmatrix} 0 \\ 0 \\ 0 \end{bmatrix}$$

then

$$\begin{aligned} c_2 + c_3 &= 0 \\ c_1 \quad + c_3 &= 0 \\ c_2 \quad &= 0 \end{aligned}$$

or

$$c_1 = c_2 = c_3 = 0$$

so that the set $\left\{ \begin{bmatrix} 0 \\ 1 \\ 0 \end{bmatrix}, \begin{bmatrix} 1 \\ 0 \\ 1 \end{bmatrix}, \begin{bmatrix} 1 \\ 1 \\ 0 \end{bmatrix} \right\}$ is linearly independent.

(ii) Prove that **S** spans $\mathbf{R}^3$.

SOLUTION

Let $\begin{bmatrix} x \\ y \\ z \end{bmatrix}$ be an arbitrary vector of $\mathbf{R}^3$. We have to find real numbers a_1, a_2, a_3 such that

$$a_1 \begin{bmatrix} 0 \\ 1 \\ 0 \end{bmatrix} + a_2 \begin{bmatrix} 1 \\ 0 \\ 1 \end{bmatrix} + a_3 \begin{bmatrix} 1 \\ 1 \\ 0 \end{bmatrix} = \begin{bmatrix} x \\ y \\ z \end{bmatrix}$$

or

$$\begin{aligned} a_2 + a_3 &= x \\ a_1 \quad + a_3 &= y \\ a_2 \quad &= z \end{aligned}$$

or

$$a_1 = -x + y + z, a_2 = z, a_3 = x - z$$

$\therefore$ **S** spans $\mathbf{R}^3$.

Hence, **S** is a basis of $\mathbf{R}^3$.

EXAMPLE 2.29 Show that the set $\{(1, 0, -1), (1, 1, 1), (1, 2, 4)\}$ is a basis for $\mathbf{R}^3$.

SOLUTION

We first show that the set spans $\mathbf{R}^3$. Let (x, y, z) be an arbitrary element of $\mathbf{R}^3$. We try to find scalars a_1, a_2, a_3 such that

$$(x, y, z) = a_1(1, 0, -1) + a_2(1, 1, 1) + a_3(1, 2, 4).$$

This identity leads to the system of equations

$$\left. \begin{aligned} a_1 + a_2 + \ a_3 &= x \\ a_2 + 2a_3 &= y \\ -a_1 + a_2 + 4a_3 &= z \end{aligned} \right\} \qquad (1)$$

That is,
$$\begin{bmatrix} 1 & 1 & 1 \\ 0 & 1 & 2 \\ -1 & 1 & 4 \end{bmatrix}\begin{bmatrix} a_1 \\ a_2 \\ a_3 \end{bmatrix} = \begin{bmatrix} x \\ y \\ z \end{bmatrix}$$

Since $\begin{vmatrix} 1 & 1 & 1 \\ 0 & 1 & 2 \\ -1 & 1 & 4 \end{vmatrix} = 1 \neq 0$, therefore for all values of x, y, z, the system of equations (1) has a unique solution.

In fact, the solution is given by

$$a_1 = 2x - 3y + z, \quad a_2 = -2x + 5y - 2z, \quad a_3 = x - 2y + z$$

Thus, the set spans $\mathbf{R}^3$.

We now show the set is linearly independent. Consider the equation

$$b_1(1, 0, -1) + b_2(1, 1, 1) + b_3(1, 2, 4) = (0, 0, 0)$$

This equation leads to the system of equations

$$\begin{aligned} b_1 + b_2 + b_3 &= 0 \\ b_2 + 2b_3 &= 0 \\ -b_1 + b_2 + 4b_3 &= 0 \end{aligned}$$

This system has the unique solution $b_1 = 0, b_2 = 0$ and $b_3 = 0$. Thus, the set is linearly independent. We have shown that the set $\{(1, 0, -1), (1, 1, 1), (1, 2, 4)\}$ spans $\mathbf{R}^3$ and is linearly independent. It, thus, forms a basis of $\mathbf{R}^3$.

EXAMPLE 2.30 Show that the set $\mathbf{S} = \{\mathbf{X}_1, \mathbf{X}_2, \mathbf{X}_3, \mathbf{X}_4\}$ where $\mathbf{X}_1 = (1, 0, 1, 0)$, $\mathbf{X}_2 = (0, 1, -1, 2)$, $\mathbf{X}_3 = (0, 2, 2, 1)$ and $\mathbf{X}_4 = (1, 0, 0, 1)$ is a basis for $\mathbf{R}^4$.

SOLUTION

To show that **S** is linearly independent, we consider the equation

$$c_1\mathbf{X}_1 + c_2\mathbf{X}_2 + c_3\mathbf{X}_3 + c_4\mathbf{X}_4 = \mathbf{O}$$

or
$$c_1(1, 0, 1, 0) + c_2(0, 1, -1, 2) + c_3(0, 2, 2, 1) + c_4(1, 0, 0, 1) = (0, 0, 0, 0)$$

or
$$\begin{aligned} c_1 + c_4 &= 0 \\ c_2 + 2c_3 &= 0 \\ c_1 - c_2 + 2c_3 &= 0 \\ 2c_2 + c_3 + c_4 &= 0 \end{aligned}$$

This system has the unique solution

$$c_1 = c_2 = c_3 = c_4 = 0 \qquad \text{(verify)}$$

To show that **S** spans $\mathbf{R}^4$, we let $\mathbf{X} = (a, b, \mathrm{c}, \mathrm{d})$ be any vector in $\mathbf{R}^4$. We now seek constants k_1, k_2, k_3 and k_4 such that

$$k_1\mathbf{X}_1 + k_2\mathbf{X}_2 + k_3\mathbf{X}_3 + k_4\,\mathbf{X}_4 = \mathbf{X}$$

Substituting for $\mathbf{X}_1$, $\mathbf{X}_2$, $\mathbf{X}_3$, $\mathbf{X}_4$ and $\mathbf{X}$, we find a solution (verify) for k_1, k_2, k_3 and k_4 to the resulting linear system for any a, b, c and d.

Hence, **S** spans $\mathbf{R}^4$ and is a basis for $\mathbf{R}^4$.

The above examples show that a vector space may not have a unique basis. We also note that the bases $\left\{\begin{bmatrix}1\\0\end{bmatrix},\begin{bmatrix}0\\1\end{bmatrix}\right\}$, $\left\{\begin{bmatrix}1\\1\end{bmatrix},\begin{bmatrix}1\\2\end{bmatrix}\right\}$ and $\left\{\begin{bmatrix}1\\1\end{bmatrix},\begin{bmatrix}1\\-1\end{bmatrix}\right\}$ for $\mathbf{R}^2$ in Examples 2.25, 2.26 and 2.27 respectively have two vectors each. Again the bases for $\mathbf{R}^3$ in Examples 2.25, 2.28 and 2.29 have three vectors each. In Example 2.30, the basis of $\mathbf{R}^4$ contains four vectors. Can we thus conclude: A basis of $\mathbf{R}^n$ must contain n vectors? In fact, this is true. The following is a more general result:

THEOREM 1

Although a vector space may have many bases, all bases have the same number of vectors.

In other words, if $\mathbf{S}_1 = \{\mathbf{X}_1, \mathbf{X}_2, \ldots, \mathbf{X}_n\}$ and $\mathbf{S}_2 = \{\mathbf{Y}_1, \mathbf{Y}_2, \ldots, \mathbf{Y}_m\}$ are two bases for a vector space, then $n = m$.

Thus, the number of vectors in a basis is uniquely determined and this number is given a special name.

? DEFINITION: The 'dimension' of a non-zero vector space **V** is the number of vectors in a basis for **V**. We often write dim **V** for the dimension of **V**. If dim $\mathbf{V} = n$, we say that **V** is an n-dimensional vector space.

We define dimension of the zero vector space to be 0. It is in agreement with the fact that zero vector space has no basis as the set $\{\mathbf{O}\}$ is linearly dependent.

In order to determine the dimension of a non-zero vector space **V**, we need to find only one basis and then count the number of vectors in it.

EXAMPLE 2.31 The dimension of $\mathbf{R}^2$ is 2 since $\left\{\begin{bmatrix}1\\0\end{bmatrix},\begin{bmatrix}0\\1\end{bmatrix}\right\}$ is a basis for $\mathbf{R}^2$.

The set $\left\{\begin{bmatrix}1\\0\\0\end{bmatrix},\begin{bmatrix}0\\1\\0\end{bmatrix},\begin{bmatrix}0\\0\\1\end{bmatrix}\right\}$ is a basis for $\mathbf{R}^3$, therefore $\dim\mathbf{R}^3 = 3$. In general, $\dim\mathbf{R}^n = n$.

EXAMPLE 2.32 Let $\mathbf{R}^1$ be the vector space of all real numbers under standard addition and scalar multiplication. In this case, the numbers play the dual roles of vectors and scalars. It can be shown that the set $\{b\}$ where b is a non-zero real number is a basis for $\mathbf{R}^1$. In particular, $\{1\}$ is standard basis for $\mathbf{R}^1$. Therefore,

$$\dim\mathbf{R}^1 = 1.$$

It may be noted here that the definition of dimension is compatible with our intuitive ideas when we talk of line as a one-dimensional object and plane as a two-dimensional object and so on.

From the definition of a basis, a set of vectors in a vector space **V** is a basis for **V** if it spans **V** and is linearly independent. However, if we are given the additional information that the dimension of **V** is n, we need only verify one of the two conditions. We state this result in the following theorem called the basis theorem.

BASIS THEOREM

Let **V** be a vector space with dim **V** = n, and let **S** = $\{X_1, X_2, \ldots, X_n\}$ be a set of n vectors in **V**.

(a) If **S** is linearly independent, then it is a basis for **V**.

(b) If **S** spans **V**, then it is a basis for **V**.

EXERCISE 2.5

1. Explain, by inspection only, why the following sets of vectors are not bases for the indicated vector spaces

(a) $\mathbf{X}_1 = (1, 2), \mathbf{X}_2 = (0, 3), \mathbf{X}_3 = (2, 7)$ for $\mathbf{R}^2$

(b) $\mathbf{X}_1 = (1, 1), \mathbf{X}_2 = (-2, -2)$ for $\mathbf{R}^2$

(c) $\mathbf{X}_1 = (-1, 3, 2), \mathbf{X}_2 = (6, 1, 1)$ for $\mathbf{R}^3$

2. Which of the following sets of vectors are bases for $\mathbf{R}^2$?

(a) $\begin{bmatrix}1\\2\end{bmatrix}, \begin{bmatrix}2\\1\end{bmatrix}$ **(b)** $\begin{bmatrix}0\\0\end{bmatrix}, \begin{bmatrix}1\\2\end{bmatrix}$ **(c)** $\begin{bmatrix}3\\4\end{bmatrix}, \begin{bmatrix}-4\\-12\end{bmatrix}$

(d) $\begin{bmatrix}\pi\\0\end{bmatrix}, \begin{bmatrix}1\\e\end{bmatrix}$ **(e)** $\begin{bmatrix}1\\3\end{bmatrix}, \begin{bmatrix}1\\-1\end{bmatrix}$ **(f)** $\begin{bmatrix}1\\2\end{bmatrix}, \begin{bmatrix}1\\-3\end{bmatrix}, \begin{bmatrix}3\\2\end{bmatrix}$

3. Which of the following sets of vectors are bases for $\mathbf{R}^3$?

(a) $\begin{bmatrix}1\\2\\3\end{bmatrix}, \begin{bmatrix}0\\1\\1\end{bmatrix}, \begin{bmatrix}1\\0\\0\end{bmatrix}$ **(b)** $\begin{bmatrix}2\\1\\0\end{bmatrix}, \begin{bmatrix}-1\\1\\1\end{bmatrix}, \begin{bmatrix}3\\3\\1\end{bmatrix}$ **(c)** $(1, 2, 1), (2, 1, 0), (1, -1, 2)$

(d) $(1, 2, 1), (1, 0, -1), (0, -3, 2)$

4. Let $\begin{bmatrix}a\\b\end{bmatrix}$ and $\begin{bmatrix}c\\d\end{bmatrix}$ be two vectors in $\mathbf{R}^2$. Show that if neither of the two is a scalar multiple of the other, then the set $\left\{\begin{bmatrix}a\\b\end{bmatrix}, \begin{bmatrix}c\\d\end{bmatrix}\right\}$ is a basis for $\mathbf{R}^2$.

5. Prove that any non-zero real number is a basis for **R**.

6. Let $\{\mathbf{X}_1, \mathbf{X}_2, \ldots, \mathbf{X}_n\}$ be a basis for a vector space **V**. Let c be a non-zero scalar. Show that the set $\{c\mathbf{X}_1, c\mathbf{X}_2, \ldots, c\mathbf{X}_n\}$ is also a basis for **V**.

7. Let $\{\mathbf{X}_1, \mathbf{X}_2, \mathbf{X}_3\}$ be a basis for a vector space **V**. Show that $\{\mathbf{Y}_1, \mathbf{Y}_2, \mathbf{Y}_3\}$ is also a basis, where $\mathbf{Y}_1 = \mathbf{X}_1$, $\mathbf{Y}_2 = \mathbf{X}_1 + \mathbf{X}_2$, $\mathbf{Y}_3 = \mathbf{X}_1 + \mathbf{X}_2 + \mathbf{X}_3$.

8. Examine whether there exists a real number λ for which the set $\{(1+\lambda, 1, 1), (2+\lambda, 2+\lambda, 2), (3+\lambda, 3+\lambda, 3+\lambda)\}$ forms a basis for $\mathbf{R}^3$.

9. State (with a brief explanation) whether the following statements are true or false.

(i) The set $\{(-1, 2), (3, -6)\}$ is a basis for $\mathbf{R}^2$.

(ii) Since dim $\mathbf{R}^2 = 2$, therefore $\{(1, 2), (-1, 3), (5, 2)\}$ are linearly dependent in $\mathbf{R}^2$.

(iii) The maximum number of linearly dependent vectors in $\mathbf{R}^2$ is two.

(iv) A single vector can be added to any two vectors in $\mathbf{R}^3$ to get a basis for $\mathbf{R}^3$.

(v) A basis for a vector space **V** can include the zero vector.

(vi) If there are three linearly independent vectors in a vector space, the dimension must be greater than or equal to three.

(vii) There exists a set that spans **V** but that is not linearly independent.

(viii) There exists a set that is linearly independent but does not span **V**.

(ix) If dim $\mathbf{V} > 1$, there is more than one basis for **V**.

(x) If dim $\mathbf{V} = n$ and if **S** spans **V**, then **S** is a basis for **V**.

10. Let $\{\mathbf{X}_1, \mathbf{X}_2, \ldots, \mathbf{X}_n\}$ be a basis for a vector space **V**. Then each vector in **V** can be expressed uniquely as a linear combination of these vectors.

11. Prove that any two basis of a finitely generated vector space **V** contain the same number of vectors.

12. Prove the Basis Theorem.

2.6 SUBSPACES

? DEFINITION: Let **V** be a vector space and **W** a non-empty subset of **V**. If **W** is a vector space with respect to the operations in **V**, then **W** is called a '*subspace*' of **V**.

EXAMPLE 2.33 If **V** is a vector space, then **V** is a subspace of itself. Also, it can easily be seen that the set $\{\mathbf{O}\}$ consisting of zero vector of **V** is a subspace of **V**. Thus, every vector space **V** has at least two subspaces: **V** itself and the subspace consisting of the zero vector of **V**.

Before giving some more examples, let us note the following simple observation.

THEOREM 1

If **W** is a subspace of a vector space **V**, then **W** contains the zero vector of **V**.

PROOF

Let **X** be an arbitrary vector in **W** and **O** be zero vector of **V**. Let 0 be the zero scalar. Since **W** is a subspace of **V**, and **W** is itself a vector space and thus **W** is closed under scalar multiplication.

$\therefore \quad 0\mathbf{X} \in \mathbf{W}$

But $\quad 0\mathbf{X} = \mathbf{O} \quad \therefore \quad \mathbf{O} \in \mathbf{W}$.

Thus this theorem tells us that all subspaces of a vector space contain the zero vector.

EXAMPLE 2.34 Let $\mathbf{W} = \left\{\begin{bmatrix} x \\ 0 \end{bmatrix} : x \text{ is a real number}\right\}$. Then **W** is a subset of $\mathbf{R}^2$. We show that **W** is a subspace of $\mathbf{R}^2$.

It is clear that **W** contains $\begin{bmatrix} 0 \\ 0 \end{bmatrix}$ which serves as the zero vector. Also **W** is closed under addition and scalar multiplication. To see this let $\mathbf{X} = \begin{bmatrix} x \\ 0 \end{bmatrix}$ and $\mathbf{Y} = \begin{bmatrix} y \\ 0 \end{bmatrix}$ be vectors in **W**.

Then $\mathbf{X} + \mathbf{Y} = \begin{bmatrix} x \\ 0 \end{bmatrix} + \begin{bmatrix} y \\ 0 \end{bmatrix} = \begin{bmatrix} x + y \\ 0 \end{bmatrix}$ is in **W**.

And for any real number c,

$$c\mathbf{X} = c\begin{bmatrix} x \\ 0 \end{bmatrix} = \begin{bmatrix} cx \\ 0 \end{bmatrix} \text{ is also in } \mathbf{W}.$$

For every vector $\begin{bmatrix} x \\ 0 \end{bmatrix} \in \mathbf{W}$, we also have $\begin{bmatrix} -x \\ 0 \end{bmatrix} \in \mathbf{W}$. All other properties of being a vector space are satisfied automatically for **W** because they hold for $\mathbf{R}^2$ (which includes **W**) [In fact, we can easily verify them also]. Hence, **W** is a subspace of $\mathbf{R}^2$.

Geometrically, **W** is the set of vectors that lie on the *x*-axis. Note that the sum of two such vectors lies on the *x*-axis and so does the scalar multiple of any such vector. Thus, *x*-axis is a subspace of plane ($\mathbf{R}^2$). It is a one-dimensional subspace of $\mathbf{R}^2$.

We can similarly show that the subset $U = \left\{\begin{bmatrix} 0 \\ y \end{bmatrix} : y \in \mathbf{R}\right\}$ is also a subspace of $\mathbf{R}^2$. Clearly U represents *y*-axis on the plane. In general, it can be shown that any line passing through origin is a subspace of $\mathbf{R}^2$.

All possible subspaces of $\mathbf{R}^2$ are listed below:

(1) Origin is a subspace of $\mathbf{R}^2$. The dimension of this subspace is zero.

(2) The one-dimensional subspaces of $\mathbf{R}^2$ are lines through origin.

(3) The two-dimensional subspace of $\mathbf{R}^2$ is the whole space $\mathbf{R}^2$ itself.

Statements (1) and (3) are obvious while a proof of (2) will be given later.

At this point, we wish to answer the following question: Given a non-empty subset **W** of a vector space **V**, how can one determine in the most efficient way whether or not **W** is a subspace of **V**?

The answer is that it is enough merely to check that **W** is closed under addition and scalar multiplication. Thus, we have following:

THEOREM 2

Let **W** be a subset of $\mathbf{R}^n$. Then **W** is a subspace of $\mathbf{R}^n$ if and only if the following conditions hold:

(a) If **X** and **Y** are any vectors in **W**, then **X** + **Y** is in **W**.

(b) If c is a real number and **X** is a vector in **W** then c**X** is in **W**.

EXAMPLE 2.35 Let $\mathbf{W} = \left\{\begin{bmatrix} x \\ 0 \\ 0 \end{bmatrix} : x \in \mathbf{R}\right\}$ be a subset of $\mathbf{R}^3$. Then **W** is closed under addition and scalar multiplication as can be seen from the equations

$$\begin{bmatrix} x_1 \\ 0 \\ 0 \end{bmatrix} + \begin{bmatrix} x_2 \\ 0 \\ 0 \end{bmatrix} = \begin{bmatrix} x_1 + x_2 \\ 0 \\ 0 \end{bmatrix} \text{ and } c\begin{bmatrix} x \\ 0 \\ 0 \end{bmatrix} = \begin{bmatrix} cx \\ 0 \\ 0 \end{bmatrix}.$$

Thus, **W** is a subspace of $\mathbf{R}^3$.

Similarly, the subsets $\left\{\begin{bmatrix} 0 \\ y \\ 0 \end{bmatrix} : y \in \mathbf{R}\right\}$ and $\left\{\begin{bmatrix} 0 \\ 0 \\ z \end{bmatrix} : z \in \mathbf{R}\right\}$ are subspaces of $\mathbf{R}^3$.

EXAMPLE 2.36 Let $\mathbf{W} = \left\{ \begin{bmatrix} x \\ y \\ 0 \end{bmatrix} : x, y \in \mathbf{R} \right\}$. Show that **W** is a subspace of $\mathbf{R}^3$.

[D.U. B.Sc. (Ph.Sc) 2008]

SOLUTION

Let $\mathbf{X} = \begin{bmatrix} x_1 \\ y_1 \\ 0 \end{bmatrix}$ and $\mathbf{Y} = \begin{bmatrix} x_2 \\ y_2 \\ 0 \end{bmatrix}$ be two vectors in **W**. Then

$$\mathbf{X} + \mathbf{Y} = \begin{bmatrix} x_1 \\ y_1 \\ 0 \end{bmatrix} + \begin{bmatrix} x_2 \\ y_2 \\ 0 \end{bmatrix} = \begin{bmatrix} x_1 + x_2 \\ y_1 + y_2 \\ 0 \end{bmatrix} \text{ is in } \mathbf{W}.$$

$\therefore$ **W** is closed under addition.
Let c be any real number. Then

$$c\mathbf{X} = c\begin{bmatrix} x_1 \\ y_1 \\ 0 \end{bmatrix} = \begin{bmatrix} cx_1 \\ cy_1 \\ 0 \end{bmatrix} \text{ is also in } \mathbf{W}.$$

So **W** is closed under scalar multiplication. Thus, **W** is a subspace of $\mathbf{R}^3$.

Note that **W** of Example 2.35 represents x-axis geometrically and the subspaces $\left\{ \begin{bmatrix} 0 \\ y \\ 0 \end{bmatrix} : y \in \mathbf{R} \right\}$ and $\left\{ \begin{bmatrix} 0 \\ 0 \\ z \end{bmatrix} : z \in \mathbf{R} \right\}$ represent y- and z-axes respectively. These are one dimensional subspaces of $\mathbf{R}^3$.

The subspace $\mathbf{W} = \left\{ \begin{bmatrix} x \\ y \\ 0 \end{bmatrix} : x, y \in \mathbf{R} \right\}$ of Example 2.36 represents xy-plane geometrically. This is a 2-dimensional subspace of $\mathbf{R}^3$.

Similarly the yz-plane $\left\{ \begin{bmatrix} 0 \\ y \\ z \end{bmatrix} : y, z \in \mathbf{R} \right\}$ and xz-plane $\left\{ \begin{bmatrix} x \\ 0 \\ z \end{bmatrix} : x, z \in \mathbf{R} \right\}$ are also 2-dimenasional subspaces of $\mathbf{R}^3$ (verify).

All subspaces of $\mathbf{R}^3$ are listed below:

(1) The origin is subspace of $\mathbf{R}^3$. The dimension of this subspace is defined to be zero.

(2) The one-dimensional subspaces of $\mathbf{R}^3$ are lines through the origin.

(3) The two-dimensional subspaces of $\mathbf{R}^3$ are planes through the origin.

(4) The whole space $\mathbf{R}^3$ is three-dimensional subspace of itself.

Statements (1) and (4) are obvious. We come to statement (2). We first prove using Theorem 2 that any line passing through origin is a subspace of $\mathbf{R}^3$ (and also of $\mathbf{R}^2$ as we have noted already). Let l be a line passing through origin in $\mathbf{R}^3$. Then, the sum of any two vectors on this line also lies on the line l and that the scalar multiple of any vector on line l lies on l. Therefore, l is closed with respect to addition and scalar multiplication and so it is a subspace of $\mathbf{R}^3$.

Moreover, the lines through origin are the only one-dimensional subspaces of $\mathbf{R}^3$. The proof is simple:

Let $\mathbf{S}$ be a one-dimensional subspace of $\mathbf{R}^3$. Since dim $\mathbf{S} = 1$, any basis of $\mathbf{S}$ must contain only one linearly independent vector. Let $\{\mathbf{X}\}$ where $\mathbf{X} \neq \mathbf{O}$ be a basis of $\mathbf{S}$. Then, any vector in $\mathbf{S}$ must be of the form $c\mathbf{X}$ for some scalar c. Clearly, these vectors form a line through the origin.

We can now similarly prove statement (3), that is, the two-dimensional subspaces of $\mathbf{R}^3$ are the planes through origin.

EXAMPLE 2.37 Let $\mathbf{V}$ be a subset of $\mathbf{R}^3$ consisting of vectors of the form (a, a, b). $\mathbf{V}$ consist of all the elements of $\mathbf{R}^3$ that have the first two components same. Show that $\mathbf{V}$ is a subspace of $\mathbf{R}^3$.

SOLUTION

Let (a, a, b) and (c, c, d) be two elements of $\mathbf{V}$ and let k be a scalar (real number).

Then
$$(a, a, b) + (\mathrm{c}, \mathrm{c}, \mathrm{d}) = (a + \mathrm{c}, a + \mathrm{c}, b + \mathrm{d}) \in \mathbf{V}$$
and
$$k(a, a, b) = (ka, ka, b) \in \mathbf{V}$$

So $\mathbf{V}$ is closed under operations of addition and scalar multiplication.

Thus, $\mathbf{V}$ is a subspace of $\mathbf{R}^3$.

Note that geometrically $\mathbf{V}$ is a plane perpendicular to the xy-plane, through the line $y = x$, $z = 0$.

EXAMPLE 2.38 Let $\mathbf{S} = \left\{ \begin{bmatrix} x \\ y \\ z \end{bmatrix} : x + y = z, \text{ and } x, y, z \in \mathbf{R} \right\}$. Show that $\mathbf{S}$ is a subspace of $\mathbf{R}^3$.

SOLUTION

We must show that $\mathbf{S}$ is closed under addition and scalar multiplication defined on $\mathbf{R}^3$.

Let $\begin{bmatrix} x_1 \\ y_1 \\ z_1 \end{bmatrix}$ and $\begin{bmatrix} x_2 \\ y_2 \\ z_2 \end{bmatrix}$ be elements of $\mathbf{S}$.

Then,
$$\begin{bmatrix} x_1 \\ y_1 \\ z_1 \end{bmatrix} + \begin{bmatrix} x_2 \\ y_2 \\ z_2 \end{bmatrix} = \begin{bmatrix} x_1 + x_2 \\ y_1 + y_2 \\ z_1 + z_2 \end{bmatrix}$$

which will belong to $\mathbf{S}$ if

$$(x_1 + x_2) + (y_1 + y_2) = z_1 + z_2$$

Since $x_1 + y_1 = z_1$ and $x_2 + y_2 = z_2$,

$\therefore$
$$(x_1 + x_2) + (y_1 + y_2) = (x_1 + y_1) + (x_2 + y_2) = z_1 + z_2$$

Thus, $\mathbf{S}$ is closed under addition.

Let c be a scalar (real number) and $\begin{bmatrix} x \\ y \\ z \end{bmatrix}$ be any vector in $\mathbf{S}$.

Then
$$\mathrm{c}\begin{bmatrix} x \\ y \\ z \end{bmatrix} = \begin{bmatrix} \mathrm{c}x \\ \mathrm{c}y \\ \mathrm{c}z \end{bmatrix}$$

Now $$cx + cy = c(x + y) = cz$$

$\therefore \quad c\begin{bmatrix} x \\ y \\ z \end{bmatrix}$ belongs to **S**.

Thus **S** is closed under scalar multiplication. We conclude that **S** is a subspace of $\mathbf{R}^3$.

EXAMPLE 2.39 Determine which of the following subsets of $\mathbf{R}^2$ are subspaces.

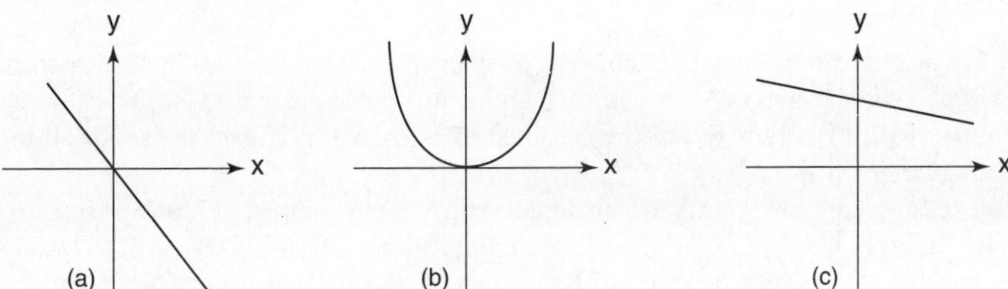

Fig. 2.11

SOLUTION

Clearly only (a) is a subspace of $\mathbf{R}^2$ being a line through origin. (b) cannot be a subspace as it is not a line. In fact, if we take any two points on this parabola, their sum may not lie on the parabola. (c) is not a line through origin and therefore does not contain the zero vector and, hence, by Theorem 1, it cannot be a subspace.

EXAMPLE 2.40 Show that $\mathbf{W} = \left\{ \begin{bmatrix} 1 \\ y \end{bmatrix} : y \text{ is a real number} \right\}$ is not a subspace of $\mathbf{R}^2$.

SOLUTION

Let $\begin{bmatrix} 1 \\ y_1 \end{bmatrix}$ and $\begin{bmatrix} 1 \\ y_2 \end{bmatrix}$ be two elements of **W**. Then

$$\begin{bmatrix} 1 \\ y_1 \end{bmatrix} + \begin{bmatrix} 1 \\ y_2 \end{bmatrix} = \begin{bmatrix} 2 \\ y_1 + y_2 \end{bmatrix}$$

But $\begin{bmatrix} 2 \\ y_1 + y_2 \end{bmatrix} \notin \mathbf{W}$ from defintion of **W**.

Thus, **W** is not a subspace of $\mathbf{R}^2$.

EXAMPLE 2.41 Let **W** be the subset of $\mathbf{R}^3$ consisting of all vectors of the form $(a, b, 1)$, where a and b are real numbers. We show that **W** is not a subspace of $\mathbf{R}^3$.

Let $(a_1, b_1, 1)$ and $(a_2, b_2, 1)$ be two elements of $\mathbf{R}^3$. Then
$(a_1, b_1, 1) + (a_2, b_2, 1) = (a_1 + a_2, b_1 + b_2, 2)$ is not an element of **W**.

Hence, **W** is not a subspace of $\mathbf{R}^3$.

In the above two examples, we could also have used Theorem 1. If a given subset does not contain the zero vector, it cannot be a subspace. $\mathbf{W} = \left\{ \begin{bmatrix} 1 \\ y \end{bmatrix} : y \in \mathbf{R} \right\}$ cannot contain the zero vector as the first component is always 1 and similarly **W** of Example 2.41 cannot contain zero vector. Therefore, they are not subspaces of $\mathbf{R}^2$ and $\mathbf{R}^3$ respectively. The following example also illustrates this approach.

EXAMPLE 2.42 Let **W** be the set of vectors of the form $(a, a, a + 2)$. Show that **W** is not a subspace of $\mathbf{R}^3$.

SOLUTION

We check to see if $(0, 0, 0)$ is in **W**. Suppose there is some a for which

$$(a, a, a + 2) = (0, 0, 0)$$

Equating corresponding components, we get

$$a = 0 \text{ and } a + 2 = 0$$

But this system of equation has no solution. Thus, no such a exists. So $(0, 0, 0)$ is not an element of W. Hence, **W** is not a subspace of $\mathbf{R}^3$.

In Section 2.4, we have introduced the concept of span. We have defined that a set **S** spans a vector space **V** if every vector in **V** is a linear combination of elements in **S**. For example, $\{(1, 0, 0), (0, 1, 0), (0, 0, 1)\}$ spans $\mathbf{R}^3$. If **V** is spanned by **S**, then **V** is in fact the set of all linear combinations of elements of **S**.

We have also seen certain subsets of **V** which do not span **V**. For example, $\{(1, 0, 0), (0, 1, 0)\}$ does not span $\mathbf{R}^3$. In other words, the set of all linear combinations of $\{(1, 0, 0), (0, 1, 0)\}$ is not equal to $\mathbf{R}^3$. But such a set clearly is a subset of $\mathbf{R}^3$. In fact, it is also a subspace of $\mathbf{R}^3$. Such subspaces which are generated by a subset **S** are called (linear) span of **S**.

? **DEFINITION:** Let **S** be a non-empty subset of a vector space **V**. Then the set of all linear combinations of elements of **S** is called the *'span'* of **S** denoted by *span* **S**. If $\mathbf{S} = \phi$, we define the span **S** to be $\{\mathbf{O}\}$.

THEOREM

Let **S** be a non-empty finite subset of a vector space **V**. Then span **S** is a subspace of **V**.

PROOF

Let $\mathbf{S} = \{\mathbf{X}_1, \mathbf{X}_2, \ldots, \mathbf{X}_n\}$. Let **X** and **Y** be linear combination of vectors in S. Suppose $\mathbf{X} = a_1\mathbf{X}_1 + a_2\mathbf{X}_2 + \cdots + a_n\mathbf{X}_n$ and $\mathbf{Y} = b_1\mathbf{X}_1 + b_2\mathbf{X}_2 + \cdots + b_n\mathbf{X}_n$ for some scalars $a_1, a_2, \ldots, a_n, b_1, b_2, \ldots, b_n$.

Then
$$\begin{aligned}\mathbf{X} + \mathbf{Y} &= a_1\mathbf{X}_1 + a_2\mathbf{X}_2 + \cdots + a_n\mathbf{X}_n + b_1\mathbf{X}_1 + b_2\mathbf{X}_2 + \cdots + b_n\mathbf{X}_n \\ &= (a_1 + b_1)\mathbf{X}_1 + (a_2 + b_2)\mathbf{X}_2 + \cdots + (a_n + b_n)\mathbf{X}_n\end{aligned}$$

which is a linear combination of vectors $\mathbf{X}_1, \mathbf{X}_2, \ldots, \mathbf{X}_n$.

$\therefore \mathbf{X} + \mathbf{Y} \in$ span **S**.

Let c be an arbitrary real number. Then

$$\begin{aligned}c\mathbf{X} &= c(a_1\mathbf{X}_1 + a_2\mathbf{X}_2 + \cdots + a_n\mathbf{X}_n) \\ &= ca_1\mathbf{X}_1 + ca_2\mathbf{X}_2 + \cdots + ca_n\mathbf{X}_n\end{aligned}$$

So c**X** is a linear combination of $\mathbf{X}_1, \mathbf{X}_2, \ldots, \mathbf{X}_n$.

$\therefore c\mathbf{X} \in$ span **S**.

Thus, span **S** is closed under addition and scalar multiplication.

Hence, span **S** is a subspace of **V**.

It must be noted that span **S** contains the set **S**. Moreover, it is the smallest subspace of **V** containing **S**.

EXAMPLE 2.43 Let **X** be a non-zero vector in $\mathbf{R}^2$. The set $\{\mathbf{X}\}$ is not a subspace of $\mathbf{R}^2$. Span of $\{\mathbf{X}\}$ is the set of all scalar multiples of **X**, i.e., $\{c\mathbf{X} : c \in \mathbf{R}\}$. It contains X and is a subspace of $\mathbf{R}^2$. In fact, the set is the smallest subspace of $\mathbf{R}^2$ which contains **X**. Geometrically, the set is a line through the origin **O** in the direction of the vector **X**. It is also clear that $\{\mathbf{X}\}$ acts as a basis for the subspace and that its dimension is 1.

EXAMPLE 2.44 If $\mathbf{S} = \left\{\begin{bmatrix} 1 \\ 0 \end{bmatrix}\right\}$, then span $\mathbf{S} = \left\{\begin{bmatrix} x \\ 0 \end{bmatrix} : x \in \mathbf{R}\right\}$. Clearly it is the one dimensional subspace x-axis of $\mathbf{R}^2$.

EXAMPLE 2.45 If **X** and **Y** are two non collinear vectors in $\mathbf{R}^2$, then span $\{\mathbf{X}, \mathbf{Y}\} = \mathbf{R}^2$. Thus, the smallest subspace of $\mathbf{R}^2$ containing both **X** and **Y** is $\mathbf{R}^2$ itself.

EXERCISE 2.6

1. Prove that the subset **W** of all vectors in $\mathbf{R}^2$ of the form $c\begin{bmatrix} 1 \\ 2 \end{bmatrix}$ for $c \in \mathbf{R}$ is a subspace of $\mathbf{R}^2$.
[Geometrically it is a line through the origin and the point $(1, 2)$]

2. Examine which of the following is a subspace of $\mathbf{R}^2$. If it is a subsapce, give its geometrical interpretations.

(a) all vectors of the form $\begin{bmatrix} x \\ x \end{bmatrix}$ where $x \in \mathbf{R}$.

(b) all vectors of the form $\begin{bmatrix} 2x \\ 3x \end{bmatrix}$ where $x \in \mathbf{R}$

(c) all vectors of the form $\begin{bmatrix} x \\ y \end{bmatrix}$ where $x + y = 0, x, y \in \mathbf{R}$

(d) all vectors of the form $\begin{bmatrix} x \\ 2 \end{bmatrix}$ where $x \in \mathbf{R}$

3. Examine whether **V** is a subspace of $\mathbf{R}^2$.

(a) $\mathbf{V} = \{(a, 2a) : a \in \mathbf{R}\}$
(b) $\mathbf{V} = \{(0, k) : k \in \mathbf{R}\}$
(c) $\mathbf{V} = \{(2k + 1, 3k + 1) : k \in \mathbf{R}\}$
(d) $\mathbf{V} = \{(a, b^2) : a, b \in \mathbf{R}\}$
(e) $\mathbf{V} = \{(a, b) : a > 0,\ a, b \in \mathbf{R}\}$
(f) $\mathbf{V} = \{(a, b^3) : a, b \in \mathbf{R}\}$
(g) $\mathbf{V} = \{(a, b) : a + b = 1, a, b \in \mathbf{R}\}$

Give the geometrical interpretation of each subspace **V**.

4. Let **V** be the first quadrant in the xy-plane, that is, let $\mathbf{V} = \left\{\begin{bmatrix} x \\ y \end{bmatrix} : x \geq 0, y \geq 0\right\}$

(a) If **X** and **Y** are in **V**, is $\mathbf{X} + \mathbf{Y}$ in **V**? Why?
(b) Find a specific vector **X** in **V** and a specific scalar c such that c**X** is not in **V**.

Is (b) enough to conclude that **V** is not a subspace of $\mathbf{R}^2$?

5. Let **W** be the set of all points inside and on the unit circle in the xy-plane, that is, let $\mathbf{W} = \left\{ \begin{bmatrix} x \\ y \end{bmatrix} : x^2 + y^2 \leq 1 \right\}$. Is **W** a subsapce of xy-plane (that is, $\mathbf{R}^2$)?

6. **(a)** Let $\mathbf{W}_1$ be the subset of $\mathbf{R}^3$ consisting of all vectors of the form $(a, 0, 0)$. Show that $\mathbf{W}_1$ is a subspace of $\mathbf{R}^3$. Describe it geometrically.

(b) Let $\mathbf{W}_2$ be the set of vectors of the form (a, a^2, b). Show that $\mathbf{W}_2$ is not a subspace of $\mathbf{R}^3$.

7. Determine which of the following are subspaces of $\mathbf{R}^3$. If it is a subspace, describe it geometrically.

(a) all vectors $\begin{bmatrix} x \\ 2x \\ 3x \end{bmatrix}$ where $x \in \mathbf{R}$

(b) all vectors $\begin{bmatrix} x \\ 1 \\ 0 \end{bmatrix}$ where $x \in \mathbf{R}$

(c) all vectors $\begin{bmatrix} x \\ y \\ z \end{bmatrix}$ where $x + y = 0;\ x, y, z \in \mathbf{R}$

(d) all vectors $\begin{bmatrix} x \\ y \\ z \end{bmatrix}$ where $x + y = 1;\ x, y, z \in \mathbf{R}$

(e) all vectors $\begin{bmatrix} x \\ x \\ x \end{bmatrix}$ where $x \in \mathbf{R}$.

8. Examine which of the following subsets of $\mathbf{R}^3$ are subspaces of $\mathbf{R}^3$. If it is a subspace, give its geometrical interpretation.

[DU, B.Sc. (Ph. Sc.), 2006]

(a) $\mathbf{W}_1 = \{(a, b, 2) : a, b \in \mathbf{R}\}$
(b) $\mathbf{W}_2 = \{(a, b, \text{c}) : \text{c} = a + b;\ a, b, \text{c} \in \mathbf{R}\}$
(c) $\mathbf{W}_3 = \{(a, 2a, b) : a, b \in \mathbf{R}\}$
(d) $\mathbf{W}_4 = \{(a, b, \text{c}) : b = 2a + 1;\ a, b, \text{c} \in \mathbf{R}\}$
(e) $\mathbf{W}_5 = \{(a, b, \text{c}) : \text{c} > 0, \text{c} \in \mathbf{R}\}$
(f) $\mathbf{W}_6 = \{(a, b, \text{c}) : a = -\text{c};\ a, b, \text{c} \in \mathbf{R}\}$
(g) $\mathbf{W}_7 = \{(a, a + b, 3a) : a, b \in \mathbf{R}\}$
(h) $\mathbf{W}_8 = \{(a, a^2, 5a) : a \in \mathbf{R}\}$

9. Show that the set $\mathbf{W} = \{(a, b, \text{c}) : a + b + \text{c} = 0;\ a, b, \text{c} \in \mathbf{R}\}$ is a subspace of $\mathbf{R}^3$. Describe it geometrically. Also give dimension of **W**.

10. Draw the graphs of following equations. Hence, find whether they are subspaces of $\mathbf{R}^2$ or not.

(a) $x + y = 1$
(b) $x^2 + y^2 = 1$
(c) $x - 2y = 0$
(d) $y^2 = 4x$.

11. Let **H** and **K** be subspaces of a vector space **V**. The *intersection* of **H** and **K**, written as $\mathbf{H} \cap \mathbf{K}$, is the set of all vectors **X** in **V** that belong to both **H** and **K**. Show that $\mathbf{H} \cap \mathbf{K}$ is a subspace of **V**. Give an example in $\mathbf{R}^2$ to show that the union of two subspaces is not, in general, a subspace.

12. If $\mathbf{W}_1$ is the line $y = x$ and $\mathbf{W}_2$ is the line $y = -x$, then $\mathbf{W}_1$ and $\mathbf{W}_2$ are subspaces of $\mathbf{R}^2$. Is the union $\mathbf{W}_1 \cup \mathbf{W}_2$ a subspace of $\mathbf{R}^2$?

13. Let **W** be the set whose only element is the vector $\begin{bmatrix} 1 \\ 2 \end{bmatrix}$. Define addition $\oplus$ and scalar multiplication $\odot$ by $\begin{bmatrix} 1 \\ 2 \end{bmatrix} \oplus \begin{bmatrix} 1 \\ 2 \end{bmatrix} = \begin{bmatrix} 1 \\ 2 \end{bmatrix}$ and $c \odot \begin{bmatrix} 1 \\ 2 \end{bmatrix} = \begin{bmatrix} 1 \\ 2 \end{bmatrix}$ for all $c \in \mathbf{R}$. Prove that **W** is a vector space and that **W** is not a subspace of $\mathbf{R}^2$.

14. Find the span of the vectors $\begin{bmatrix} 1 \\ 0 \\ 0 \end{bmatrix}$ and $\begin{bmatrix} 0 \\ 1 \\ 1 \end{bmatrix}$.

15. Let **S** be the subset of $\mathbf{R}^2$ consisting of all unit vectors : $\mathbf{S} = \{\mathbf{X} \in \mathbf{R}^2 : \|\mathbf{X}\| = 1\}$. Find the span of **S**.

16. Let **S** be a subset of a vector space **V**. Prove that span **S** is the smallest subspace of **V** containing **S**.

17. We know that a line is one-dimensional and a plane is two-dimensional. Find the dimension of subspaces of $\mathbf{R}^2$ in problems 2 and 3, and the subspaces of $\mathbf{R}^3$ in problems 7 and 8.

CHAPTER THREE

Linear Transformations

3.1 INTRODUCTION

The reader must be familiar with the concept of a function.

A function f from a set U (called domain) into a set V (called co-domain) is a rule which associates, to each member u of U, a unique member of V, denoted by $f(u)$.

In this chapter, we shall study vector-valued functions, whose domain (and co-domain) are vector spaces. Thus, if we write $\mathbf{Y} = \mathrm{T}(\mathbf{X})$, we mean that T is a function which assigns, to each vector $\mathbf{X}$ in a vector space V, a unique vector $\mathbf{Y}$ in a vector space W. $\mathrm{T}(\mathbf{X})$ is called image of $\mathbf{X}$ under T.

We shall take V and W to be the vector spaces $\mathbf{R}^n$ and $\mathbf{R}^m$ for some values of n and m. It is customary to refer to the function T as a transformation. We now define a special class of transformations called linear transformations.

Linear transformations have many applications in physics, engineering, social sciences, economics and various branches of mathematics. Linear transformations play a fundamental role in computer graphics, image processing and manipulation, harvesting of animal population, genetics and other areas.

? DEFINITION: A function $\mathrm{T}:\mathbf{R}^n \to \mathbf{R}^m$ is called a 'linear transformation' from $\mathbf{R}^n$ into $\mathbf{R}^m$ if the following conditions hold:

(a) $\mathrm{T}(\mathrm{X} + \mathrm{Y}) = \mathrm{T}(\mathrm{X}) + \mathrm{T}(\mathrm{Y})$ for all vectors X and Y in $\mathbf{R}^n$.

(b) $\mathrm{T}(c\mathrm{X}) = c\mathrm{T}(\mathrm{X})$ for all vectors X in $\mathbf{R}^n$ and all real numbers c.

If T satisfies the above two properties, we say that T preserves addition and multiplication by scalars. The two properties can be combined together to get one formula, $\mathrm{T}(c\mathrm{X} + d\mathrm{Y}) = c\mathrm{T}(\mathrm{X}) + d\mathrm{T}(\mathrm{Y})$.

EXAMPLE 3.1 Let $\mathrm{T}: \mathbf{R}^2 \to \mathbf{R}^2$ be defined by

$$\mathrm{T}\left(\begin{bmatrix} x_1 \\ x_2 \end{bmatrix}\right) = \begin{bmatrix} x_1 \\ 0 \end{bmatrix}$$

Then T is a linear transformation. To verify this, let

$$\mathbf{X} = \begin{bmatrix} x_1 \\ x_2 \end{bmatrix} \quad \text{and} \quad \mathbf{Y} = \begin{bmatrix} y_1 \\ y_2 \end{bmatrix}.$$

Then
$$\begin{aligned}
\mathrm{T}(\mathbf{X}+\mathbf{Y}) &= \mathrm{T}\left(\begin{bmatrix} x_1 \\ x_2 \end{bmatrix} + \begin{bmatrix} y_1 \\ y_2 \end{bmatrix}\right) \\
&= \mathrm{T}\left(\begin{bmatrix} x_1 + y_1 \\ x_2 + y_2 \end{bmatrix}\right) \\
&= \begin{bmatrix} x_1 + y_1 \\ 0 \end{bmatrix} \\
&= \begin{bmatrix} x_1 \\ 0 \end{bmatrix} + \begin{bmatrix} y_1 \\ 0 \end{bmatrix} \\
&= \mathrm{T}\left(\begin{bmatrix} x_1 \\ x_2 \end{bmatrix}\right) + \mathrm{T}\left(\begin{bmatrix} y_1 \\ y_2 \end{bmatrix}\right)
\end{aligned}$$

Also, if c is a real number, then
$$\begin{aligned}
\mathrm{T}(\mathrm{c}\mathbf{X}) &= \mathrm{T}\left(\mathrm{c}\begin{bmatrix} x_1 \\ x_2 \end{bmatrix}\right) \\
&= \mathrm{T}\left(\begin{bmatrix} \mathrm{c}x_1 \\ \mathrm{c}x_2 \end{bmatrix}\right) \\
&= \begin{bmatrix} \mathrm{c}x_1 \\ 0 \end{bmatrix} \\
&= \mathrm{c}\begin{bmatrix} x_1 \\ 0 \end{bmatrix} \\
&= \mathrm{cT}\left(\begin{bmatrix} x_1 \\ x_2 \end{bmatrix}\right)
\end{aligned}$$

Hence, T is a linear transformation.

EXAMPLE 3.2 Let $\mathrm{T}: \mathbf{R}^3 \rightarrow \mathbf{R}^2$ be defined by
$$\mathrm{T}\left(\begin{bmatrix} x \\ y \\ z \end{bmatrix}\right) = \begin{bmatrix} x \\ y \end{bmatrix}.$$

To verify that T is a linear transformation, we let
$$\mathbf{X} = \begin{bmatrix} x_1 \\ y_1 \\ z_1 \end{bmatrix} \quad \text{and} \quad \mathbf{Y} = \begin{bmatrix} x_2 \\ y_2 \\ z_2 \end{bmatrix}.$$

Then
$$\begin{aligned}
\mathrm{T}(\mathbf{X}+\mathbf{Y}) &= \mathrm{T}\left(\begin{bmatrix} x_1 \\ y_1 \\ z_1 \end{bmatrix} + \begin{bmatrix} x_2 \\ y_2 \\ z_2 \end{bmatrix}\right) = \mathrm{T}\left(\begin{bmatrix} x_1 + x_2 \\ y_1 + y_2 \\ z_1 + z_2 \end{bmatrix}\right) \\
&= \begin{bmatrix} x_1 + x_2 \\ y_1 + y_2 \end{bmatrix} = \begin{bmatrix} x_1 \\ y_1 \end{bmatrix} + \begin{bmatrix} x_2 \\ y_2 \end{bmatrix} \\
&= \mathrm{T}\left(\begin{bmatrix} x_1 \\ y_1 \\ z_1 \end{bmatrix}\right) + \mathrm{T}\left(\begin{bmatrix} x_2 \\ y_2 \\ z_2 \end{bmatrix}\right) \\
&= \mathrm{T}(\mathbf{X}) + \mathrm{T}(\mathbf{Y})
\end{aligned}$$

Also, if c is a real number

$$T(c\mathbf{X}) = T\left(\begin{bmatrix} cx_1 \\ cy_1 \\ cz_1 \end{bmatrix}\right) = \begin{bmatrix} cx_1 \\ cy_1 \end{bmatrix} = c\begin{bmatrix} x_1 \\ y_1 \end{bmatrix} = cT(\mathbf{X})$$

Hence, T is a linear transformation.
T in the above two examples is called a projection.

Geometrically, projections are very important. We shall study their geometrical significance in the next section.

EXAMPLE 3.3 Let $T: \mathbf{R}^3 \to \mathbf{R}^2$ be defined by

$$T\left(\begin{bmatrix} x \\ y \\ z \end{bmatrix}\right) = \begin{bmatrix} x + y \\ x + z \end{bmatrix}$$

Let us now show that T is a linear transformation.

Let $\mathbf{X} = \begin{bmatrix} x_1 \\ y_1 \\ z_1 \end{bmatrix}$ and $\mathbf{Y} = \begin{bmatrix} x_2 \\ y_2 \\ z_2 \end{bmatrix}$. Then

$$\begin{aligned} T(\mathbf{X} + \mathbf{Y}) &= T\left(\begin{bmatrix} x_1 + x_2 \\ y_1 + y_2 \\ z_1 + z_2 \end{bmatrix}\right) \\ &= \begin{bmatrix} (x_1 + x_2) + (y_1 + y_2) \\ (x_1 + x_2) + (z_1 + z_2) \end{bmatrix} \\ &= \begin{bmatrix} (x_1 + y_1) + (x_2 + y_2) \\ (x_1 + z_1) + (x_2 + z_2) \end{bmatrix} \\ &= \begin{bmatrix} x_1 + y_1 \\ x_1 + z_1 \end{bmatrix} + \begin{bmatrix} x_2 + y_2 \\ x_2 + z_2 \end{bmatrix} \\ &= T\left(\begin{bmatrix} x_1 \\ y_1 \\ z_1 \end{bmatrix}\right) + T\left(\begin{bmatrix} x_2 \\ y_2 \\ z_2 \end{bmatrix}\right) \\ &= T(\mathbf{X}) + T(\mathbf{Y}) \end{aligned}$$

Also, let c be a real number. Then

$$\begin{aligned} T(c\mathbf{X}) &= T\left(\begin{bmatrix} cx_1 \\ cy_1 \\ cz_1 \end{bmatrix}\right) \\ &= \begin{bmatrix} cx_1 + cy_1 \\ cx_1 + cz_1 \end{bmatrix} \\ &= c\begin{bmatrix} x_1 + y_1 \\ x_1 + z_1 \end{bmatrix} \\ &= cT\left(\begin{bmatrix} x_1 \\ y_1 \\ z_1 \end{bmatrix}\right) = cT(\mathbf{X}) \end{aligned}$$

Thus, we conclude that T is a linear transformation.

EXAMPLE 3.4 Let $T: \mathbf{R}^n \to \mathbf{R}^n$ be defined by

$$T(\mathbf{X}) = \mathbf{X}, \text{ for every } \mathbf{X} \text{ in } \mathbf{R}^n.$$

Then,
$$T(\mathbf{X} + \mathbf{Y}) = \mathbf{X} + \mathbf{Y} = T(\mathbf{X}) + T(\mathbf{Y})$$
and
$$T(c\mathbf{X}) = c\mathbf{X} = cT(\mathbf{X})$$

Thus, T is a linear transformation and is called the identity transformation on $\mathbf{R}^n$.

Before discussing some properties of linear transformations, we give some examples of transformations which are not linear. Such transformations are called non-linear transformations.

EXAMPLE 3.5 Let a transformation T from $\mathbf{R}^2$ into $\mathbf{R}^3$ be defined by the formula

$$T(x, y) = (x, y, 1)$$

Show that T is not linear. *[D.U. B.Sc. (Ph. Sc.) 2008]*

SOLUTION

Let (x_1, y_1) and (x_2, y_2) be vectors in $\mathbf{R}^2$. Then, we have

$$T[(x_1, y_1) + (x_2, y_2)] = T(x_1 + x_2, y_1 + y_2) = (x_1 + x_2, y_1 + y_2, 1)$$

and
$$T(x_1, y_1) + T(x_2, y_2) = (x_1, y_1, 1) + (x_2, y_2, 1) = (x_1 + x_2, y_1 + y_2, 2)$$

Since
$$(x_1 + x_2, y_1 + y_2, 1) \neq (x_1 + x_2, y_1 + y_2, 2)$$

it follows that

$$T[(x_1, y_1) + (x_2, y_2)] \neq T(x_1, y_1) + T(x_2, y_2)$$

and T is not linear.

EXAMPLE 3.6 Let $T: \mathbf{R}^3 \to \mathbf{R}^2$ be defined by

$$T\left(\begin{bmatrix} x_1 \\ x_2 \\ x_3 \end{bmatrix}\right) = \begin{bmatrix} x_1 + 1 \\ x_2 - x_3 \end{bmatrix}$$

[D.U. B.Sc. (Ph. Sc.) 2008]

Determine whether T is a linear transformation, let

SOLUTION

Let
$$\mathbf{X} = \begin{bmatrix} x_1 \\ x_2 \\ x_3 \end{bmatrix} \text{ and } \mathbf{Y} = \begin{bmatrix} y_1 \\ y_2 \\ y_3 \end{bmatrix}$$

Then,

$$T(\mathbf{X} + \mathbf{Y}) = T\left(\begin{bmatrix} x_1 + y_1 \\ x_2 + y_2 \\ x_3 + y_3 \end{bmatrix}\right) = \begin{bmatrix} (x_1 + y_1) + 1 \\ (x_2 + y_2) - (x_3 + y_3) \end{bmatrix}$$

On the other hand,

$$\mathrm{T}(\mathbf{X}) + \mathrm{T}(\mathbf{Y}) = \begin{bmatrix} x_1 + 1 \\ x_2 - x_3 \end{bmatrix} + \begin{bmatrix} y_1 + 1 \\ y_2 - y_3 \end{bmatrix} = \begin{bmatrix} (x_1 + y_1) + 2 \\ (x_2 - x_3) + (y_2 - y_3) \end{bmatrix}$$

Clearly, $\mathrm{T}(\mathbf{X} + \mathbf{Y}) \neq \mathrm{T}(\mathbf{X}) + \mathrm{T}(\mathbf{Y})$.
We conclude that T is non-linear transformation.

We now state following two important properties of linear transformations. Under a linear transformation,

1. zero vector goes to zero vector
2. lines are mapped to lines

Let us discuss them one by one.

If $\mathrm{T}: \mathbf{R}^n \to \mathbf{R}^m$ is a linear transformation, then T takes the zero vector of $\mathbf{R}^n$ to zero vector of $\mathbf{R}^m$. The proof is simple.

If $\mathbf{O}_{\mathbf{R}^n}$ is the zero vector of $\mathbf{R}^n$ and $\mathbf{O}_{\mathbf{R}^m}$ denotes the zero vector of $\mathbf{R}^m$, then

$$\mathbf{O}_{\mathbf{R}^m} + \mathrm{T}(\mathbf{X}) = \mathrm{T}(\mathbf{X}) = \mathrm{T}(\mathbf{X} + \mathbf{O}_{\mathbf{R}^n}) = \mathrm{T}(\mathbf{X}) + \mathrm{T}(\mathbf{O}_{\mathbf{R}^n})$$

whence $\mathrm{T}(\mathbf{O}_{\mathbf{R}^n}) = \mathbf{O}_{\mathbf{R}^m}$.

It may well happen that $\mathrm{T}(\mathbf{X}) = \mathbf{O}$ even when $\mathbf{X} \neq \mathbf{O}$. In fact, $\mathrm{T}: \mathbf{R}^n \to \mathbf{R}^n$ defined by $\mathrm{T}(\mathbf{X}) = \mathbf{O}$ for every $\mathbf{X}$ in $\mathbf{R}^n$ is a linear transformation, called the *zero transformation*.

REMARK: It is also clear from the above property that if T does not map the zero vector to zero vector, then T cannot be linear.

EXAMPLE 3.7 The transformation $\mathrm{T}: \mathbf{R}^n \to \mathbf{R}^n$ defined by

$$\mathrm{T}(x, y) = (x + 1, y)$$

is not linear, because

$$\mathrm{T}(\mathrm{O}) = \mathrm{T}(0, 0) = (1, 0) \neq \mathrm{O}$$

This example clearly shows that the above property is sometimes very effective in proving a transformation to be non-linear.

The second property that lines are mapped to lines is even more important. We state this property more precisely in the following theorem.

THEOREM

If $\mathrm{T}: \mathbf{R}^2 \to \mathbf{R}^2$ is a linear transformation, then

(i) lines through origin are mapped to lines through origin.
(ii) lines are mapped to lines (or a single point).
(iii) line segment joining origin and a vector $\mathbf{X}$ is mapped to line segment joining origin and the vector $\mathrm{T}(\mathbf{X})$.
(iv) line segment joining two vectors $\mathbf{X}$ and $\mathbf{Y}$ is mapped to line segment joining $\mathrm{T}(\mathbf{X})$ and $\mathrm{T}(\mathbf{Y})$.

We prove part (i).

Let l be a line through origin. Let $\mathbf{X}$ be a non-zero vector on l. Then all other vectors on l are of the form $c\mathbf{X}$, that is, scalar multiple of $\mathbf{X}$.

Since T is linear $T(c\mathbf{X}) = cT(\mathbf{X})$.

So, the image of l under T is the set $\{cT(\mathbf{X}) : c \in \mathbf{R}\}$ which is the set of all scalar multiples of vector $T(\mathbf{X})$. Clearly, it is a line passing through origin and the vector $T(\mathbf{X})$. Note that if $T(\mathbf{X}) = \mathbf{O}$, the image set is trivially the single point origin, which we regard as a degenerate line through origin.

It is clear from above theorem that if the image of a line under a transformation T is neither a line nor a single point, it cannot be linear.

EXAMPLE 3.8 Let $T : \mathbf{R}^2 \to \mathbf{R}^2$ be defined by $T(x, y) = (x, y^2)$. Prove that T is not linear.

SOLUTION

Note that $T(\mathbf{O}) = \mathbf{O}$, so we cannot use the method of Example 3.7.
Consider the subset $U = \{(x, x) : x \in \mathbf{R}\}$ of $\mathbf{R}^2$.

Geometrically it is the line $y = x$ through origin. Also, $T(U) = \{(x, x^2) : x \in \mathbf{R}\}$ or, the image $T(U)$ of line U under T is the set of all points of the from (x, x^2). Clearly all these points lie on the parabola $y = x^2$. Thus, the image of line $y = x$ under T is the parabola $y = x^2$, which is neither a line nor a point, and hence T is not linear (See Fig. 3.1).

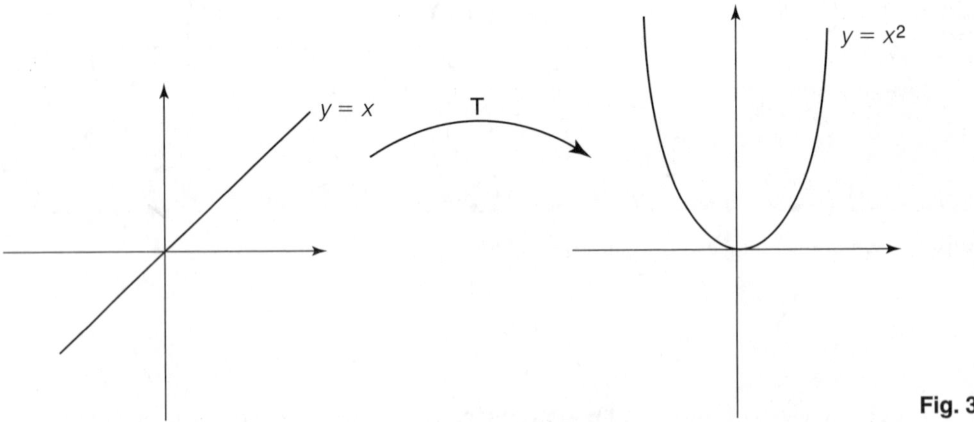

Fig. 3.1

We conclude this section by noting the following remarks.

REMARK 1: If $T : \mathbf{R}^3 \to \mathbf{R}^3$ is linear, then planes through origin are mapped to planes through origin.

REMARK 2: From above properties and the Remark 1, we can observe the following. If $T : \mathbf{R}^n \to \mathbf{R}^n$ is linear and U is a subspace of $\mathbf{R}^n$, then $T(U)$ is a subspace of $\mathbf{R}^n$.

EXERCISE 3.1

1. Prove that the transformation $T : \mathbf{R}^2 \rightarrow \mathbf{R}^2$ defined by $T\left(\begin{bmatrix} x \\ y \end{bmatrix}\right) = \begin{bmatrix} 2x \\ x - y \end{bmatrix}$ is linear. Find the images of the elements $\begin{bmatrix} 1 \\ 2 \end{bmatrix}$ and $\begin{bmatrix} -1 \\ 4 \end{bmatrix}$ under this transformation.

2. Which of the following functions $T : \mathbf{R}^2 \rightarrow \mathbf{R}^2$ are linear transformations?

(a) $T\left(\begin{bmatrix} x \\ y \end{bmatrix}\right) = \begin{bmatrix} -x \\ -y \end{bmatrix}$

(b) $T\left(\begin{bmatrix} x \\ y \end{bmatrix}\right) = \begin{bmatrix} x \\ 1 - y \end{bmatrix}$

(c) $T\left(\begin{bmatrix} x \\ y \end{bmatrix}\right) = \begin{bmatrix} x^2 \\ y \end{bmatrix}$

(d) $T\left(\begin{bmatrix} x \\ y \end{bmatrix}\right) = \begin{bmatrix} 2x \\ 3y \end{bmatrix}$

3. Which of the following transformations are linear?

(a) $T : \mathbf{R}^3 \rightarrow \mathbf{R}^3;\ T(x, y, z) = (x + y, x + z, x)$

(b) $T : \mathbf{R}^2 \rightarrow \mathbf{R}^3;\ T(x, y) = (0, x - y, x + y)$

(c) $T : \mathbf{R}^3 \rightarrow \mathbf{R}^2;\ T(x_1, x_2, x_3) = (x_1 - x_3, x_1 + x_2)$

(d) $T : \mathbf{R}^2 \rightarrow \mathbf{R};\ T(x, y) = x + y$

(e) $T : \mathbf{R} \rightarrow \mathbf{R}^2;\ T(x) = (x, x^2)$

(f) $T : \mathbf{R}^2 \rightarrow \mathbf{R}^2;\ T(x_1, x_2) = (x_1 + x_2 + 1, x_1 - x_2)$

(g) $T : \mathbf{R}^2 \rightarrow \mathbf{R}^3;\ T(x, y) = (x, y, xy)$

4. Which of the following transformations $T : \mathbf{R}^2 \rightarrow \mathbf{R}$ are linear transformations?

(a) $T(x, y) = x$

(b) $T(x, y) = 0$

(c) $T(x, y) = y^2$

(d) $T(x, y) = 2x - y$

5. Prove that the following are not linear transformations.

(a) $T : \mathbf{R}^2 \rightarrow \mathbf{R}^2;\ T(x, y) = (1, 2)$

(b) $T : \mathbf{R} \rightarrow \mathbf{R};\ T(x) = ax^2$

(c) $T : \mathbf{R} \rightarrow \mathbf{R};\ T(x) = x^3$

(d) $T : \mathbf{R}^3 \rightarrow \mathbf{R}^2;\ T(x, y, z) = (x + 2, 4y)$

6. Suppose that $T : \mathbf{R}^2 \rightarrow \mathbf{R}^2$ maps lines through the origin to lines through the origin. Must f be linear? Justify your answer.

7. Let $T : \mathbf{R}^3 \rightarrow \mathbf{R}^2$ be a linear transformation for which $T(1, 0, 0) = (2, -1)$, $T(0, 1, 0) = (3, 1)$ and $T(0, 0, 1) = (-1, 2)$. Find $T(-3, 4, 2)$.

8. Let $T : \mathbf{R}^3 \rightarrow \mathbf{R}^2$ be a linear transformation such that $T(1, 0, 1) = (2, -1)$, $T(0, 1, 1) = (1, 1)$ and $T(1, 1, 0) = (-1, 4)$. Find

(a) $T(2, -1, 1)$

(b) $T(1, 1, 1)$

(c) $T(1, 2, 3)$

9. Let $T: \mathbf{R}^n \rightarrow \mathbf{R}^m$ be a transformation. Prove that T is a linear transformation if and only if $T(c\mathbf{X} + d\mathbf{Y}) = cT(\mathbf{X}) + dT(\mathbf{Y})$ for all $\mathbf{X}, \mathbf{Y} \in \mathbf{R}^n$ and scalars c, d.

10. Prove that $T: \mathbf{R}^n \rightarrow \mathbf{R}^m$ defined by $T(\mathbf{X}) = \mathbf{O}$ is a linear transformation.

11. If $T: \mathbf{R}^n \rightarrow \mathbf{R}^m$ is such that $T(\mathbf{O}) \neq \mathbf{O}$, then T is not linear.

3.2 SOME SPECIAL TRANSFORMATIONS

In this section, we give a variety of examples of linear and non-linear transformations of geometric importance.

3.2.1 Projection

EXAMPLE 3.9 In Examples 3.1 and 3.2 we have considered two transformations:

(1) $T: \mathbf{R}^2 \rightarrow \mathbf{R}^2$ defined by $T(x, y) = (x, 0)$

(2) $T: \mathbf{R}^3 \rightarrow \mathbf{R}^2$ defined by $T(x, y, z) = (x, y)$

We have shown that these transformations are linear and are called projections.

It is simple and helpful to describe geometrically the effect of T in (1) and (2) above:

In (1), T transforms each vector (x, y) in the plane to the nearest point $(x, 0)$ on the horizontal axis. So the transformation projects points of plane on the x-axis (See Fig. 3.2(a)).

In (2), the image of a point (x, y, z) in $\mathbf{R}^3$ under T is the point (x, y) in $\mathbf{R}^2$ which is obtained by drawing perpendicular from (x, y, z) onto xy-plane (See Fig. 3.2(b)).

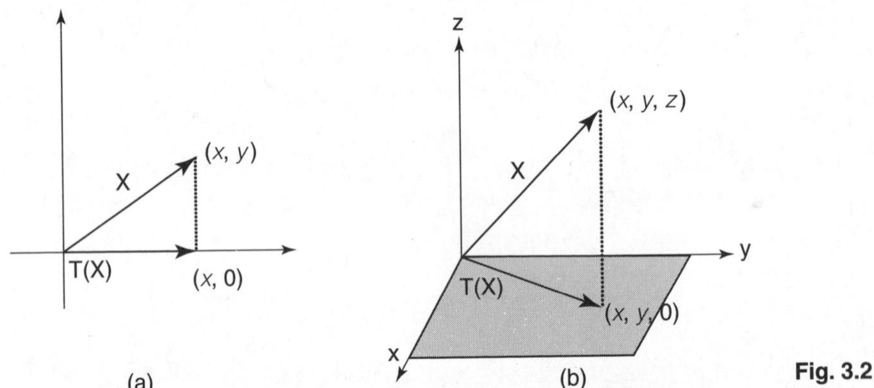

Fig. 3.2

Note that the transformation in (2) is equivalent to the transformation $T: \mathbf{R}^3 \rightarrow \mathbf{R}^3$ given by $T(x, y, z) = (x, y, 0)$. Both have same geometrical effect and are called projection on xy-plane.

3.2.2 Dilation and Contraction

EXAMPLE 3.10 Let $T_r: \mathbf{R}^3 \rightarrow \mathbf{R}^3$ be defined by $T_r(\mathbf{X}) = r\mathbf{X}$ where r is a real number. Then T_r is a linear transformation for every real number r.

In fact, $$T(\mathbf{X} + \mathbf{Y}) = r(\mathbf{X} + \mathbf{X}) = r\mathbf{X} + r\mathbf{Y} = T(\mathbf{X}) + T(\mathbf{Y})$$

and $$T(c\mathbf{X}) = r(c\mathbf{X}) = c(r\mathbf{X}) = cT(\mathbf{X}).$$

If $r > 1$, T_r is called dilation and if $0 < r < 1$, T_r is called a contraction. If $r = 1$, T_r is just an identity transformation. In Fig. 3.3(a) we show the vector $T_2(\mathbf{X}) = 2\mathbf{X}$ and in Fig. 3.3(b), the vector $T_{\frac{1}{2}}(\mathbf{X}) = \frac{1}{2}\mathbf{X}$.

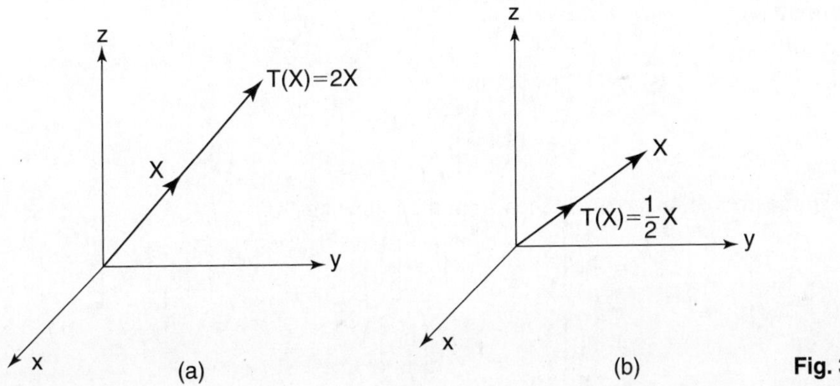

Fig. 3.3

The transformation T_2 stretches everything by a factor of 2. Note, that the square region ABCD in $\mathbf{R}^2$ is transformed into the square region T(A)T(B)T(C)T(D) by the transformation $T(\mathbf{X}) = 2\mathbf{X}$. (See Fig. 3.4).

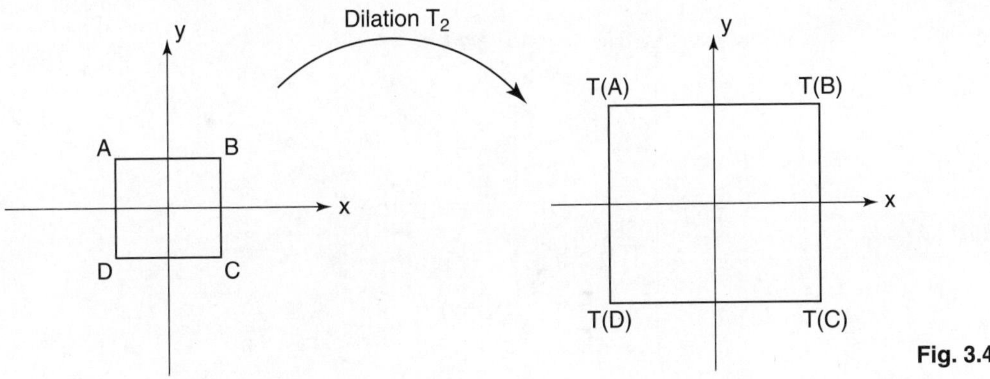

Fig. 3.4

With the transformation $T_{\frac{1}{2}}(\mathbf{X}) = \frac{1}{2}\mathbf{X}$, the circle of radius 2 will contract to a circle of radius 1. See Fig. 3.5 below.

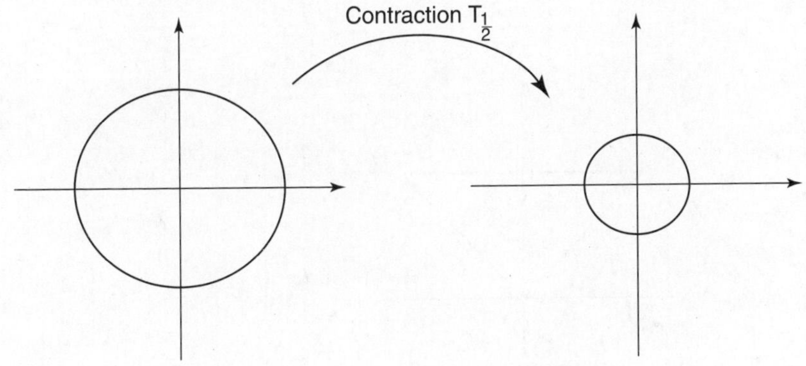

Fig. 3.5

3.2.3 Reflection

EXAMPLE 3.11 Let $T : \mathbf{R}^2 \to \mathbf{R}^2$ be defined by

$$T\left(\begin{bmatrix} x \\ y \end{bmatrix}\right) = \begin{bmatrix} x \\ -y \end{bmatrix}$$

T is a linear transformation as can be seen from the following:

$$T\left(\begin{bmatrix} x_1 \\ y_1 \end{bmatrix} + \begin{bmatrix} x_2 \\ y_2 \end{bmatrix}\right) = T\left(\begin{bmatrix} x_1 + x_2 \\ y_1 + y_2 \end{bmatrix}\right) = \begin{bmatrix} x_1 + x_2 \\ -(y_1 + y_2) \end{bmatrix} = \begin{bmatrix} x_1 \\ -y_1 \end{bmatrix} + \begin{bmatrix} x_2 \\ -y_2 \end{bmatrix}$$
$$= T\left(\begin{bmatrix} x_1 \\ y_1 \end{bmatrix}\right) + T\left(\begin{bmatrix} x_2 \\ y_2 \end{bmatrix}\right)$$

and

$$T\left(c\begin{bmatrix} x \\ y \end{bmatrix}\right) = T\left(\begin{bmatrix} cx_1 \\ cy_1 \end{bmatrix}\right) = \begin{bmatrix} cx \\ -cy \end{bmatrix} = c\begin{bmatrix} x \\ -y \end{bmatrix} = cT\left(\begin{bmatrix} x \\ y \end{bmatrix}\right)$$

This linear transformation is called a reflection with respect to the x-axis, and is shown in Fig. 3.6.

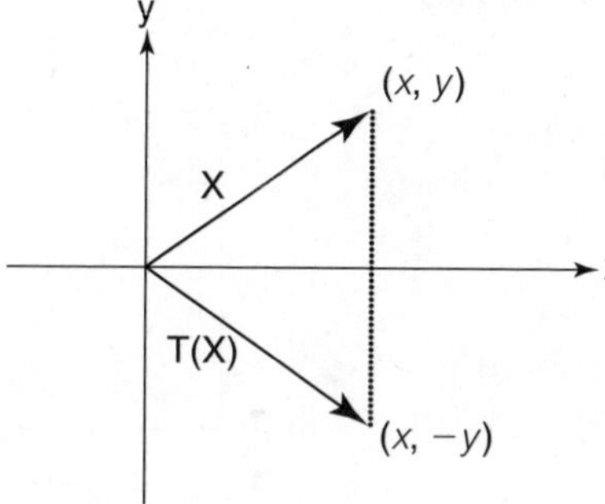

Fig. 3.6

Similarly, the transformation $T : \mathbf{R}^2 \to \mathbf{R}^2$ given by $T\left(\begin{bmatrix} x \\ y \end{bmatrix}\right) = \begin{bmatrix} -x \\ y \end{bmatrix}$ maps each point in the plane into its reflection about the y-axis (See Fig. 3.7).

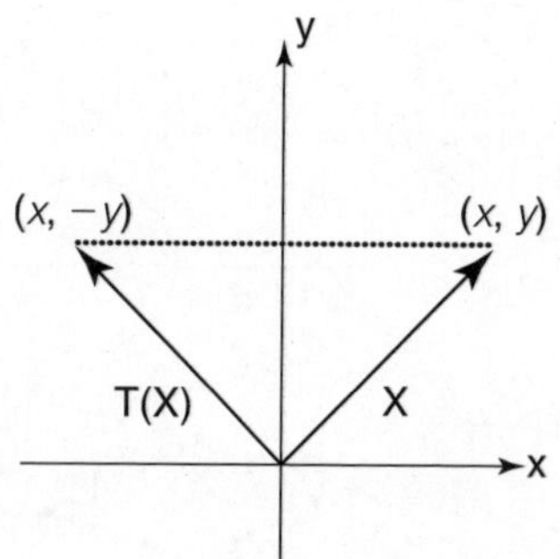

Fig. 3.7

EXAMPLE 3.12 Let T: $\mathbf{R}^3 \to \mathbf{R}^3$ be defined by $T(x, y, z) = (-x, y, z)$. Then T is a linear transformation. (Verify).

This transformation sends each point (x, y, z) to its mirror image $(-x, y, z)$ with respect to yz-plane. It is called reflection with respect to yz-plane.

Similarly, we can define reflections about xy-plane and xz-plane, and with respect to coordinate axes.

EXAMPLE 3.13 Consider the following transformation from $\mathbf{R}^3$ into $\mathbf{R}^3$:

$$T(x, y, z) = (-x, -y, -z)$$

This is a linear transformation (verify) known as *reflection in origin*.

The Fig. 3.8 shows reflection in origin in $\mathbf{R}^2$:

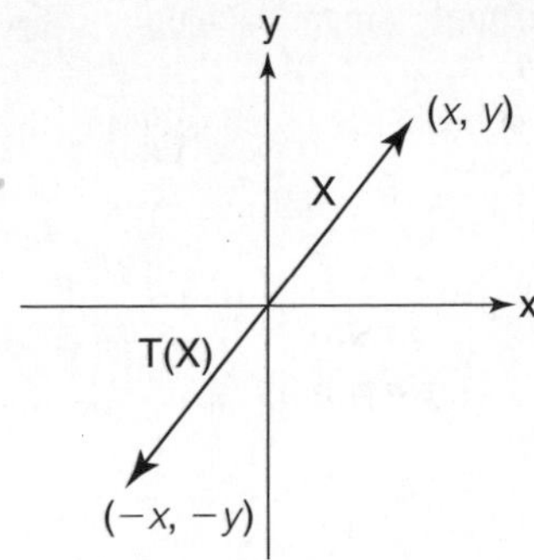

Fig. 3.8

3.2.4 Rotation

EXAMPLE 3.14 In this example we show that rotation of the plane about the origin is a linear transformation. Suppose that we rotate every vector in $\mathbf{R}^2$ counterclockwise through an angle θ about the origin of a Cartesian coordinate system. Let $P(x, y)$ be any point on the plane. After rotating, it is mapped to the point P′ with coordinates (x', y'). Let $\overrightarrow{OP}$ makes an angle ϕ with the x-axis. See Fig. 3.9.

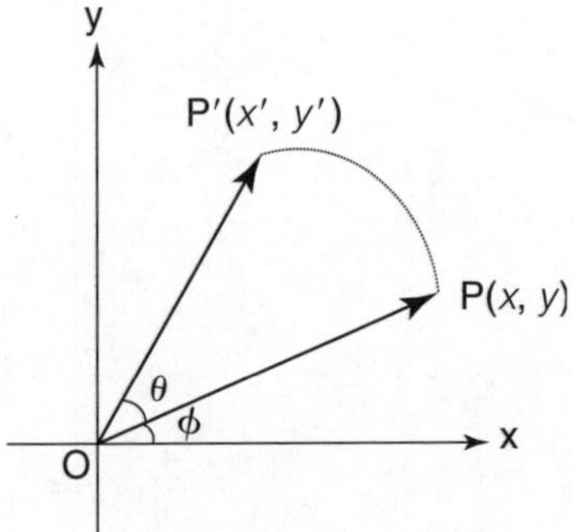

Fig. 3.9

If r denotes the length of vector $\overrightarrow{OP}$, we see from Fig. 3.9 that

$$x = r\cos\phi, \qquad y = r\sin\phi \tag{1}$$

and

$$x' = r\cos(\phi + \theta), \qquad y' = r\sin(\phi + \theta) \tag{2}$$

Then

$$\begin{aligned} x' &= r\cos(\phi + \theta) \\ &= r(\cos\phi\cos\theta - \sin\phi\sin\theta) \\ &= r\cos\phi\cos\theta - r\sin\phi\sin\theta \\ &= x\cos\theta - y\sin\theta \qquad \text{[using (1)]} \end{aligned}$$

and

$$\begin{aligned} y' &= r\sin(\phi + \theta) \\ &= r[\sin\phi\cos\theta + \cos\phi\sin\theta] \\ &= r\sin\phi\cos\theta + r\cos\phi\sin\theta \\ &= x\sin\theta + y\cos\theta \qquad \text{[using (1)]} \end{aligned}$$

Thus, the counterclockwise rotation by an angle θ maps each point (x, y) on the plane onto its image point $(x\cos\theta - y\sin\theta, x\sin\theta + y\cos\theta)$.

We may also perform this change of coordinates by considering the transformation

$$T_\theta : \mathbf{R}^2 \to \mathbf{R}^2 \quad \text{defined by}$$

$$T_\theta\left[\begin{pmatrix} x \\ y \end{pmatrix}\right] = \begin{bmatrix} x\cos\theta - y\sin\theta \\ x\sin\theta + y\cos\theta \end{bmatrix} \tag{3}$$

Then T_θ is a linear transformation (verify). The linear transformation T_θ is called a rotation through the angle θ.

NOTE: To see that T_θ is linear we may also note that rotation preserves the length of each vector and also the angle between any two vectors. See Fig 3.10(a) & 3.10(b).

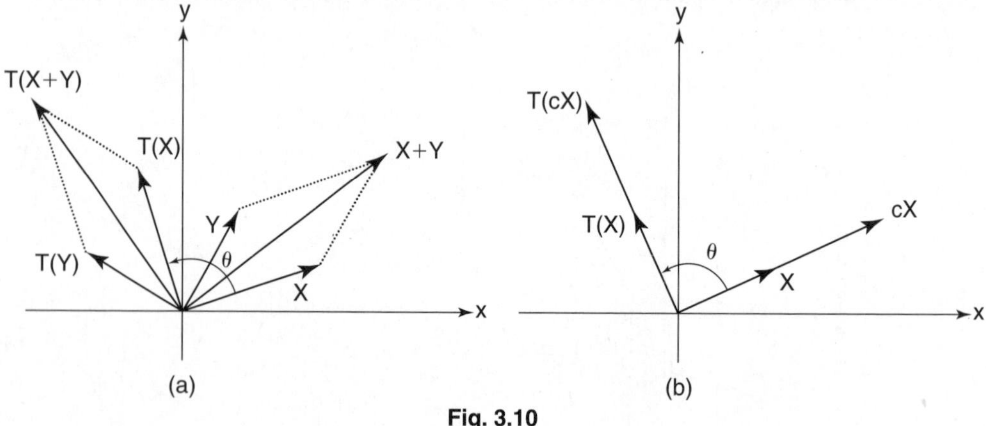

Fig. 3.10

The linear transformation T_θ in Equation (3) can also be written as

$$T\left(\begin{bmatrix} x \\ y \end{bmatrix}\right) = \begin{bmatrix} \cos\theta & -\sin\theta \\ \sin\theta & \cos\theta \end{bmatrix}\begin{bmatrix} x \\ y \end{bmatrix} \tag{4}$$

which suggests that rotation is a matrix multiplication. The matrix $\begin{bmatrix} \cos\theta & -\sin\theta \\ \sin\theta & \cos\theta \end{bmatrix}$ is called a rotation matrix.

The image of a point $\begin{bmatrix} 3 \\ 2 \end{bmatrix}$ under a rotation of 90° about the origin is given by

$$T\left(\begin{bmatrix} 3 \\ 2 \end{bmatrix}\right) = \begin{bmatrix} \cos 90^\circ & -\sin 90^\circ \\ \sin 90^\circ & \cos 90^\circ \end{bmatrix}\begin{bmatrix} 3 \\ 2 \end{bmatrix}$$

$$= \begin{bmatrix} 0 & -1 \\ 1 & 0 \end{bmatrix}\begin{bmatrix} 3 \\ 2 \end{bmatrix} = \begin{bmatrix} -2 \\ 3 \end{bmatrix}$$

Thus a vector $\begin{bmatrix} 3 \\ 2 \end{bmatrix}$ in $\mathbf{R}^2$ is transformed to the vector $\begin{bmatrix} -2 \\ 3 \end{bmatrix}$ under rotation of 90°. (See Fig. 3.11 (a))

The same vector $\begin{bmatrix} 3 \\ 2 \end{bmatrix}$ is transformed to the vector $\begin{bmatrix} \frac{3\sqrt{3}-2}{2} \\ \frac{3+2\sqrt{3}}{2} \end{bmatrix}$ (See Fig. 3.11 (b)) under a rotation of 30° as can be seen from the equation

$$T\left(\begin{bmatrix} 3 \\ 2 \end{bmatrix}\right) = \begin{bmatrix} \cos 30^\circ & -\sin 30^\circ \\ \sin 30^\circ & \cos 30^\circ \end{bmatrix}\begin{bmatrix} 3 \\ 2 \end{bmatrix}$$

$$= \begin{bmatrix} \frac{\sqrt{3}}{2} & -\frac{1}{2} \\ \frac{1}{2} & \frac{\sqrt{3}}{2} \end{bmatrix}\begin{bmatrix} 3 \\ 2 \end{bmatrix} = \begin{bmatrix} \frac{3\sqrt{3}-2}{2} \\ \frac{3+2\sqrt{3}}{2} \end{bmatrix}$$

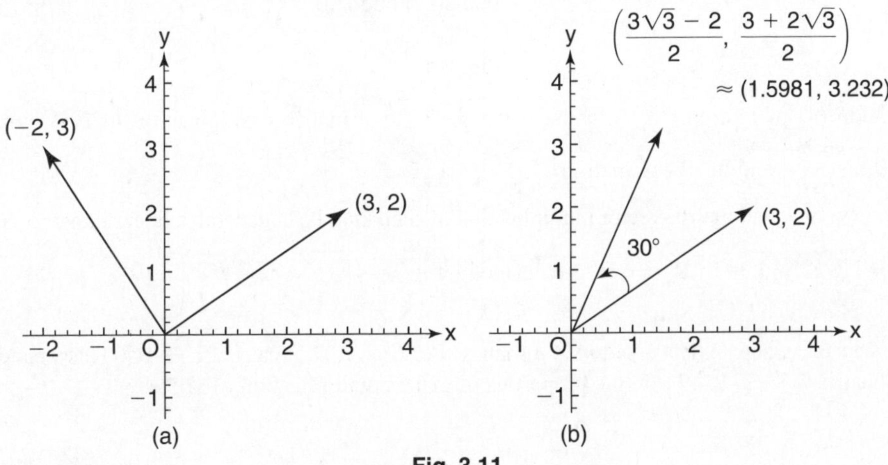

Fig. 3.11

EXAMPLE 3.15 Determine the equation satisfied by the set of image points of the locus of $x = y^2$ under a rotation of the plane about the origin through an angle of 90°.

SOLUTION

Since the angle of rotation $\theta = 90°$, $\cos\theta = 0$ and $\sin\theta = 1$. The image point (x', y') of each point (x, y) is given by the matrix equation

$$\begin{bmatrix} x' \\ y' \end{bmatrix} = \begin{bmatrix} x\cos\theta - y\sin\theta \\ x\sin\theta + y\cos\theta \end{bmatrix}$$

$$= \begin{bmatrix} x\cdot 0 - y\cdot 1 \\ x\cdot 1 + y\cdot 0 \end{bmatrix}$$

$$= \begin{bmatrix} -y \\ x \end{bmatrix}$$

or $x = y'$ and $y = -x'$.

Hence, replacing x by y' and y by $-x'$ in $x = y^2$, we find that the equation satisfied by the set of image points becomes

$$y' = (-x')^2$$

or

$$y' = x'^2$$

That is, $y = x^2$ since (x', y') are the coordinate of each image point with reference to xy-axes. $x = y^2$ is a parabola on x-axis and $y = x^2$ is a parabola on y-axis. See Fig. 3.12.

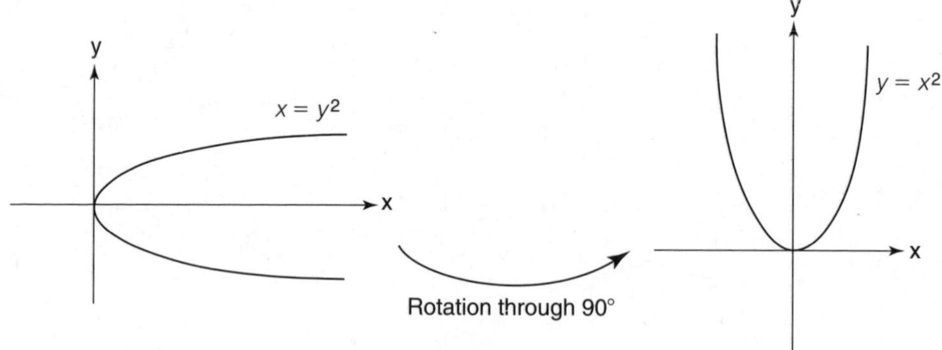

Fig. 3.12

In equation (4) of Example 3.14, we see that $T_\theta(\mathbf{X})$ is a matrix multiplication of 2×2 matrix $\begin{bmatrix} \cos\theta & -\sin\theta \\ \sin\theta & \cos\theta \end{bmatrix}$ and the 2×1 matrix $\begin{bmatrix} x \\ y \end{bmatrix}$.

We now show that matrix-vector multiplication of such kind is, in general, a linear transformation.

EXAMPLE 3.16 Let $T: \mathbf{R}^n \to \mathbf{R}^m$ be defined by

$$T(\mathbf{X}) = \mathbf{AX} \tag{1}$$

for all $\mathbf{X}$ in $\mathbf{R}^n$, where $\mathbf{A}$ is a fixed $m \times n$ matrix. For each $n \times 1$ matrix $\mathbf{X} \in \mathbf{R}^n$, T associates an $m \times 1$ matrix $\mathbf{AX} \in \mathbf{R}^m$. From the properties of matrix multiplication, we note that

$$\mathbf{A}(\mathbf{X} + \mathbf{Y}) = \mathbf{AX} + \mathbf{AY} \tag{2}$$

and

$$\mathbf{A}(c\mathbf{X}) = c(\mathbf{AX}) \tag{3}$$

From Equations (1), (2) and (3) it follows that

$$T(\mathbf{X} + \mathbf{Y}) = T(\mathbf{X}) + T(\mathbf{Y})$$

and

$$T(c\mathbf{X}) = cT(\mathbf{X})$$

Thus, T is a linear transformation. This transformation is sometimes referred to as a matrix transformation.

3.2.5 Shear

EXAMPLE 3.17 The transformation $T: \mathbf{R}^2 \to \mathbf{R}^2$ defined by $T\left(\begin{bmatrix} x \\ y \end{bmatrix}\right) = \begin{bmatrix} x + ky \\ y \end{bmatrix}$ is a linear transformation (verify) called shear in the x-direction with factor k. We discuss its geometrical effect using the unit square (See Fig. 3.13 (a)). The square OABC with vertices $\mathbf{O}(0, 0)$, $\mathbf{A}(1, 0)$, $\mathbf{B}(1, 1)$ and $\mathbf{C}(0, 1)$ is called the unit square. Under shear in the x-direction with factor $k = 2$, the square OABC will transform to the parallelogram $T(\mathbf{O})T(\mathbf{A})T(\mathbf{B})T(\mathbf{C})$ where

$$T(\mathbf{O}) = T\left(\begin{bmatrix} 0 \\ 0 \end{bmatrix}\right) = \begin{bmatrix} 0 \\ 0 \end{bmatrix}$$

$$T(\mathbf{A}) = T\left(\begin{bmatrix} 1 \\ 0 \end{bmatrix}\right) = \begin{bmatrix} 1 \\ 0 \end{bmatrix}$$

$$T(\mathbf{B}) = T\left(\begin{bmatrix} 1 \\ 1 \end{bmatrix}\right) = \begin{bmatrix} 1 + 2 \\ 1 \end{bmatrix} = \begin{bmatrix} 3 \\ 1 \end{bmatrix}$$

$$T(\mathbf{C}) = T\left(\begin{bmatrix} 0 \\ 1 \end{bmatrix}\right) = \begin{bmatrix} 0 + 2 \\ 1 \end{bmatrix} = \begin{bmatrix} 2 \\ 1 \end{bmatrix}$$

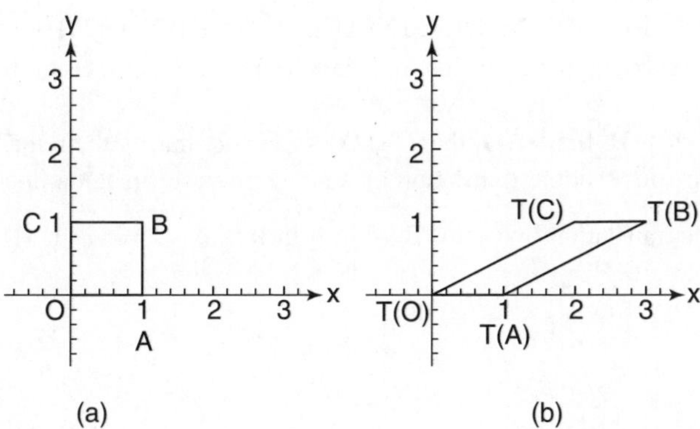

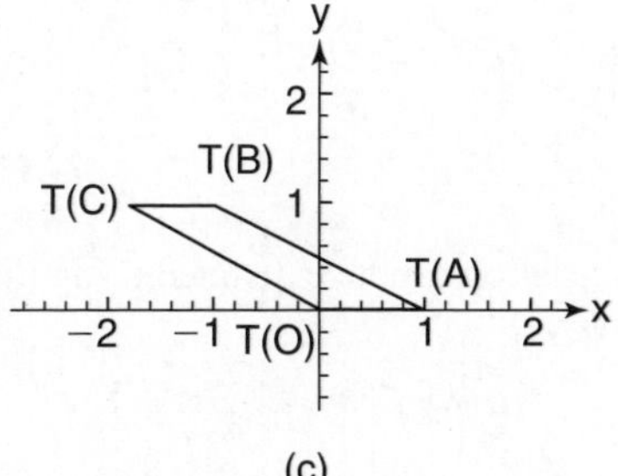

Fig. 3.13

Note that the points on x-axis remain fixed, and as we move away from the x-axis, the points further from the x-axis move a greater distance than those closer.

The effect of shear in the x-direction with factor $k = -2$ on the unit square is shown in Fig. 3.13 (c).

EXAMPLE 3.18 A shear in the y-direction with factor k is defined by the linear transformation $T\left(\begin{bmatrix} x \\ y \end{bmatrix}\right) = \begin{bmatrix} x \\ y + kx \end{bmatrix}$. When $k = 2$, the image of unit square under T is shown in Fig. 3.14.

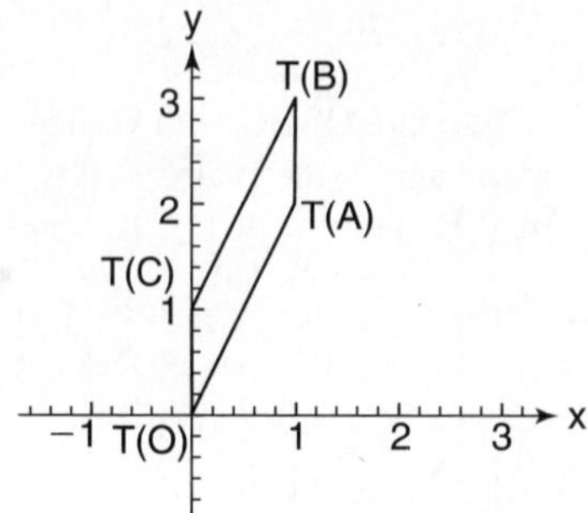

Fig. 3.14

Upto now we have discussed examples of linear transformations. Some more linear transformations will be discussed in the next section and the exercises that follow. We now give examples of two non-linear transformations of geometric importance.

3.2.6 Translation

EXAMPLE 3.19 Let $T : \mathbf{R}^2 \rightarrow \mathbf{R}^2$ be defined by $T(\mathbf{X}) = \mathbf{X} + \mathbf{B}$, where $\mathbf{B}$ is a fixed vector in $\mathbf{R}^2$. We call this a translation by vector $\mathbf{B}$ (or just translation) and use the notation tran_B or T_B, that is, $T_B(\mathbf{X}) = \mathbf{X} + \mathbf{B}$.

We note that $T_B(\mathbf{O}) = \mathbf{B}$. If $\mathbf{B} \neq \mathbf{O}$, then $T_B(\mathbf{O}) \neq \mathbf{O}$ and thus by the remark provided before Example 3.7 of the previous section, translation by a non-zero vector is not a linear transformation.

Fig. 3.15 shows the translation by vector $\mathbf{B} = \begin{bmatrix} a \\ b \end{bmatrix}$ in $\mathbf{R}^2$.

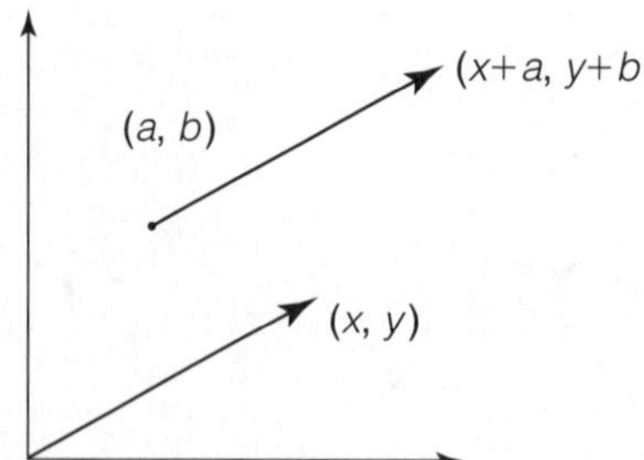

Fig. 3.15

Geometric importance of translation can be explained using the following particular case. Let $T : \mathbf{R}^2 \rightarrow \mathbf{R}^2$ be defined by

$$T\left(\begin{bmatrix} x \\ y \end{bmatrix}\right) = \begin{bmatrix} x + 1 \\ y + 1 \end{bmatrix}$$

Then T is essentially the translation by vector $\begin{bmatrix} 1 \\ 1 \end{bmatrix}$. This is the familiar translation of axes where the origin is moved to Q : (1, 1) and the point P : (−1, −1) in the domain becomes the origin O′ : (0, 0) in the range. See Fig. 3.16.

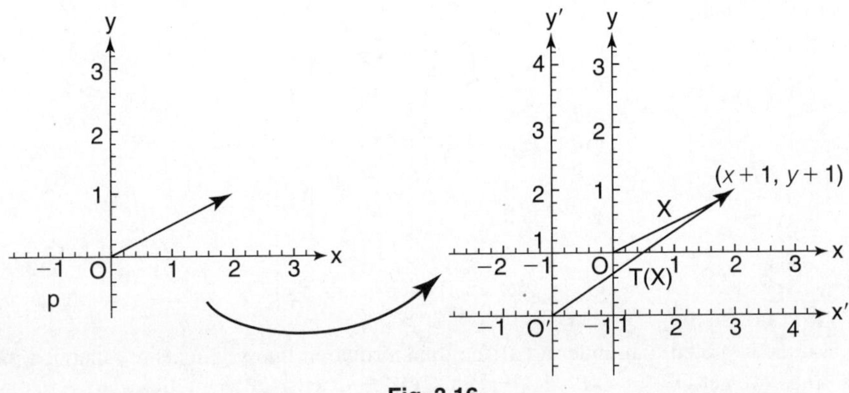

Fig. 3.16

It must be noted that under a translation, distance between two points remains the same, only the position of points is changed. Thus, line segments are mapped to line segments and a triangle remains the same triangle but shifts its position. For example the triangle with vertices A(0, 0), B(2, 0) and C(1, 2) under the above translation by vector (1, 1) is mapped to a triangle with vertices A′(1, 1), B′(3, 1) and C′(2, 3) (See Fig. 3.17).

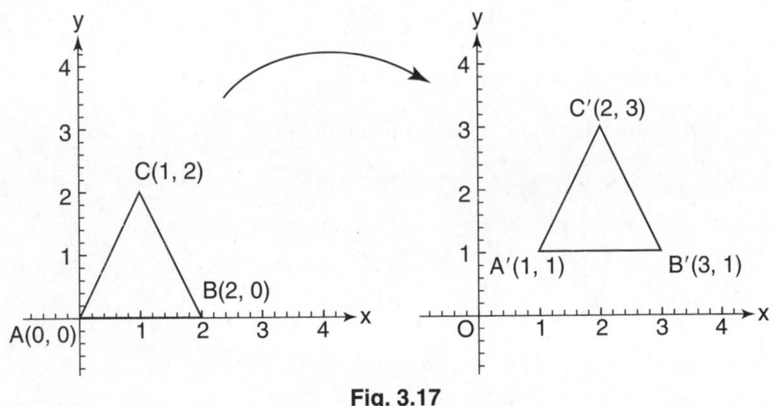

Fig. 3.17

EXAMPLE 3.20 An affine transformation is a transformation

$$T : \mathbf{R}^n \rightarrow \mathbf{R}^n, \text{ defined by}$$

$$T(\mathbf{X}) = \mathbf{AX} + \mathbf{B}$$

where $\mathbf{A}$ is an $n \times n$ matrix and $\mathbf{B}$ is a fixed vector in $\mathbf{R}^n$.

It is clear, that for $\mathbf{B} \neq \mathbf{O}$ affine transformation T is non-linear.

An affine transformation can be interpreted as a matrix transformation followed by a translation. Affine transformations are very important in computer graphics. Note that all linear transformations and translations are particular cases of affine transformations.

As an example, consider the following affine transformation on $\mathbf{R}^2$.

$$T\left(\begin{bmatrix} x \\ y \end{bmatrix}\right) = \begin{bmatrix} 2 & 1 \\ 1 & 1 \end{bmatrix}\begin{bmatrix} x \\ y \end{bmatrix} + \begin{bmatrix} 1 \\ 2 \end{bmatrix}$$

Let us find the image of the unit square OPQR, of Fig 3.18.

Here,
$$P' = T(P) = T\left(\begin{bmatrix}1\\0\end{bmatrix}\right) = \begin{bmatrix}3\\3\end{bmatrix}.$$
$$Q' = T(Q) = T\left(\begin{bmatrix}1\\1\end{bmatrix}\right) = \begin{bmatrix}4\\4\end{bmatrix}$$
$$R' = T(R) = T\left(\begin{bmatrix}0\\1\end{bmatrix}\right) = \begin{bmatrix}2\\3\end{bmatrix}$$
$$O' = T(O) = T\left(\begin{bmatrix}0\\0\end{bmatrix}\right) = \begin{bmatrix}1\\2\end{bmatrix}$$

Also, it must be noted that under an affine transformation line segments are mapped into line segments. Thus, we get OP $\rightarrow$ O′P′, PQ $\rightarrow$ P′Q′, QR $\rightarrow$ Q′R′, OR $\rightarrow$ O′R′

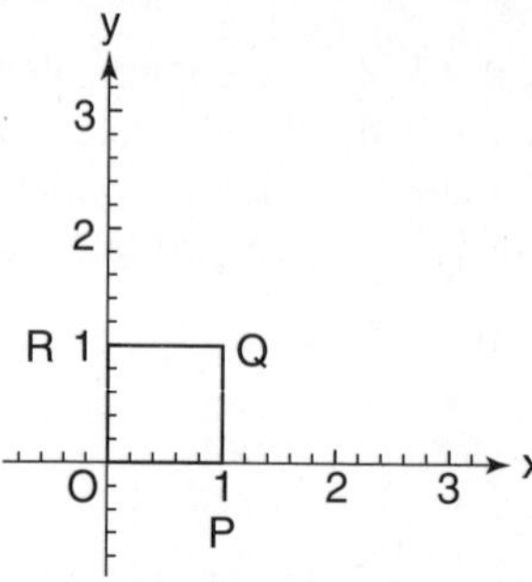

Fig. 3.18

The square OPQR is transformed into the parallelogram O′P′Q′R′ (See Fig. 3.19).

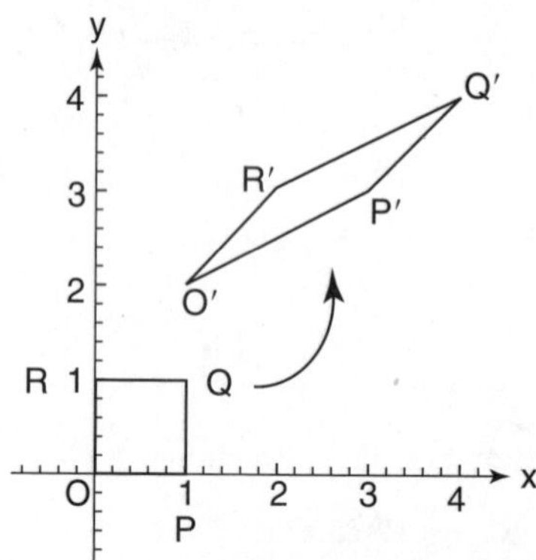

Fig. 3.19

EXERCISE 3.2

1. Construct a transformation that projects $\mathbf{R}^2$, that is, plane on *y*-axis. Show that it is linear.
2. Construct a transformation that projects $\mathbf{R}^3$ on **(a)** *z*-axis **(b)** *yz*-plane. In each case, show that the transformation is linear.
3. Prove that $T: \mathbf{R}^2 \rightarrow \mathbf{R}^2$ defined by $T\left(\begin{bmatrix}x\\y\end{bmatrix}\right) = \begin{bmatrix}2x\\2y\end{bmatrix}$ is a linear transformation. Describe the effect of this transformation on a triangle ABC with vertices A(0, 0), B(2, 0) and C(1, 2).

4. Show that the following transformations $T: \mathbf{R}^2 \to \mathbf{R}^2$ are linear. Describe them geometrically.

(a) $T(x, y) = (y, x)$ **(b)** $T(x, y) = (-y, -x)$

(c) $T(x, y) = (-y, x)$ **(d)** $T(x, y) = (-x, -y)$

5. Describe the geometrical effects of following transformations from $\mathbf{R}^3$ into $\mathbf{R}^3$.

(a) $T(x, y, z) = (x, -y, z)$ **(b)** $T(x, y, z) = (x, y, -z)$

(c) $T(x, y, z) = (x, -y, -z)$ **(d)** $T(x, y, z) = (-x, -y, z)$

6. Determine the image point of the point P under a rotation of the plane about the origin through the angle θ where

(a) $P: (5, \sqrt{3}); \theta = 30°$ **(b)** $P: (2, -3); \theta = 90°$

(c) $P: (1, 2); \theta = 45°$ **(d)** $P: (\sqrt{3}, 1); \theta = -60°$

7. Use Equation (4) of Example 3.14 to find the matrix of a counterclockwise rotation of $\mathbf{R}^2$ through the angle θ if θ is given by

(a) $\frac{\pi}{6}$ **(b)** $\frac{\pi}{4}$ **(c)** $\frac{\pi}{3}$ **(d)** $\frac{\pi}{2}$ **(e)** $\frac{2\pi}{3}$ **(f)** π

8. Let $T: \mathbf{R}^2 \to \mathbf{R}^2$ be the counterclockwise rotation through θ. Find θ if T is given by

(a) $T\left(\begin{bmatrix} x \\ y \end{bmatrix}\right) = \begin{bmatrix} -y \\ x \end{bmatrix}$ **(b)** $T\left(\begin{bmatrix} x \\ y \end{bmatrix}\right) = \begin{bmatrix} -x \\ -y \end{bmatrix}$

9. Let T be the rotation through 60°. If ABC is a triangle with vertices $A(1, 1)$, $B(-3, 3)$ and $C(2, -1)$, find and sketch the image of ABC under T.

10. Find and sketch the image of the unit square with vertices $(0, 0)$, $(0, 1)$, $(1, 1)$, $(1, 0)$ under the following transformations of the plane:

(a) a reflection about x-axis **(b)** a reflection about y-axis

(c) a dilation of factor 3 **(d)** a contraction of factor 1/2

(e) a counterclockwise rotation through $\pi/2$ **(f)** a counterclockwise rotation through $\pi/4$

(g) a translation by vector $(1, 1)$ **(h)** a reflection about origin

(i) a shear of factor 2 in x-direction **(j)** a shear of factor 0.5 in y-direction

11. Let $T: \mathbf{R}^2 \to \mathbf{R}^2$ be given by

$$T\left(\begin{bmatrix} x \\ y \end{bmatrix}\right) = \begin{bmatrix} ax + by \\ cx + dy \end{bmatrix}$$

where a, b, c and d are real numbers.

(a) Prove that T is a linear transformation.

(b) This transformation is of special importance for various values of a, b, c and d. For example, when $a = 1, b = 0, c = 0$ and $d = -1$, it represents reflection with respect to x-axis. Identify the transformation when

(i) $a = 1, b = c = d = 0$ **(ii)** $a = -1, b = c = 0, d = 1$

(iii) $a = b = c = 0, d = 1$ **(iv)** $a = 0, b = c = 1, d = 0$

(v) $a = -1, b = c = 0, d = -1$ **(vi)** $a = 0, b = -1, c = 1, d = 0$

12. Interpret geometrically the transformation of square ABCD, $A: (0, 0)$, $B: (2, 0)$, $C: (0, 2)$, $D: (2, 2)$ under $T(x, y) = (x + 1, y + 2)$.

13. Find the image of the triangle having vertices $(1, 2)$, $(3, 4)$ and $(4, 6)$ under the translation that takes the point $(1, 2)$ to $(2, -3)$.

14. Find the equation of the image of the ellipse $\frac{x^2}{4} + \frac{y^2}{9} = 1$ under a rotation through an angle of $\frac{\pi}{2}$.

15. Find the equation of the image of the unit circle $x^2 + y^2 = 1$ under a dilation of factor 3.

16. Find the equation of the image of the line $y = 2x$ under

(a) a reflection about the y-axis

(b) a reflection about $y = x$

(c) a rotation of 60° about the origin

(d) a contraction of factor 1/2

17. Find the image of the line $y = 3x + 1$ under the translation $T\left(\begin{bmatrix} x \\ y \end{bmatrix}\right) = \begin{bmatrix} x \\ y \end{bmatrix} + \begin{bmatrix} a \\ b \end{bmatrix}$ where

(a) $a = 2, b = 5$ **(b)** $a = -1, b = 1$.

18. Let R be the rectangle with vertices $(1, 1)$, $(1, 4)$, $(3, 1)$ and $(3, 4)$. Determine and sketch the image of R under

(a) shear in the x-direction with $k = 3$

(b) shear in the y-direction with $k = 2$

(c) dilation with $r = 3$

19. Let $T : \mathbf{R}^2 \to \mathbf{R}^2$ be a transformation. A point (x, y) is called a fixed point under T if it remains same under T, that is, $T(x, y) = (x, y)$. State which of the following statements are true.

(a) Every point on x-axis is a fixed point under reflection about x-axis.

(b) Every point on x-axis is a fixed point under reflection about y-axis.

(c) Origin is the only fixed point under rotation of $\pi/2$.

(d) No point can be a fixed point under translation by a non-zero vector.

(e) All points inside the unit square are fixed points under dilation of factor 2.

20. A geometric property that is unchanged under a given transformation is called an *invariant* under that transformation. Under which of the following transformations is the distance between two points on a plane invariant?

(a) reflection about x-axis

(b) rotation through an angle θ

(c) dilation of factor 2

(d) contraction of factor 1/3

(e) translation by a non-zero vector

(f) reflection about the line $x = y$

(g) projection on x-axis.

Under which of the above transformations is the area of a triangle invariant?

21. Let $T : \mathbf{R}^3 \to \mathbf{R}^3$ be defined by

$$T\left(\begin{bmatrix} x \\ y \end{bmatrix}\right) = \begin{bmatrix} 1 & 1 \\ 0 & 1 \\ 1 & -2 \end{bmatrix}\begin{bmatrix} x \\ y \end{bmatrix}$$

Prove that T is linear.

22. Let **A** be a 7×5 matrix. If $T : \mathbf{R}^m \to \mathbf{R}^n$ be such that $T(\mathbf{X}) = \mathbf{AX}$, what are the values of m and n?

23. Consider the linear transformation $\mathrm{T} : \mathbf{R}^2 \rightarrow \mathbf{R}^2$ defined by $\mathrm{T}(\mathbf{X}) = \mathbf{AX}$ where $\mathbf{A} = \begin{bmatrix} 1 & 2 \\ -1 & 3 \\ 1 & 2 \end{bmatrix}$.

Determine the images of the vectors $\mathbf{X} = \begin{bmatrix} -1 \\ 1 \end{bmatrix}$, $\mathbf{Y} = \begin{bmatrix} 2 \\ 3 \end{bmatrix}$, $\mathbf{Z} = \begin{bmatrix} 5 \\ 2 \end{bmatrix}$.

3.3 MATRIX REPRESENTATION OF A LINEAR TRANSFORMATION

In Example 3.16 we have shown that if **A** is an $m \times n$ matrix, then the transformation $\mathrm{T} : \mathbf{R}^n \rightarrow \mathbf{R}^m$ given by

$$\mathrm{T}(\mathbf{X}) = \mathbf{AX} \tag{1}$$

is a linear transformation. We now investigate whether the converse of the statement is true. Therefore, we raise the following question.

Question Suppose T is a linear transformation from $\mathbf{R}^n$ into $\mathbf{R}^m$. Can T be represented by an equation like Equation (1) with a properly chosen matrix **A**?

The answer to this question is in the affirmative and is stated in the following theorem.

THEOREM

Let $\mathrm{T}: \mathbf{R}^n \rightarrow \mathbf{R}^m$ be a linear transformation. Then there exists a unique $m \times n$ matrix A such that

$$\mathrm{T}(\mathbf{X}) = \mathbf{AX}$$

for all **X** in $\mathbf{R}^n$.

PROOF

It can be seen that if T is a linear transformation then

$$\mathrm{T}(c_1\mathbf{X}_1 + c_2\mathbf{X}_2 + \cdots + c_n\mathbf{X}_n) = c_1\mathrm{T}(\mathbf{X}_1) + c_2\mathrm{T}(\mathbf{X}_2) + \cdots + c_n\mathrm{T}(\mathbf{X}_n) \tag{1}$$

where $c_1, c_2, \ldots, c_n$ are real numbers and $\mathbf{X}_1, \mathbf{X}_2, \ldots, \mathbf{X}_n$ are vectors in $\mathbf{R}^n$.

Let $\mathbf{X} = \begin{bmatrix} x_1 \\ x_2 \\ \vdots \\ x_n \end{bmatrix}$ be a vector in $\mathbf{R}^n$ and $\{e_1, e_2, \ldots, e_n\}$ be the standard basis of $\mathbf{R}^n$.

Then $$\mathbf{X} = x_1e_1 + x_2e_2 + \cdots + x_ne_n$$

Since T is linear,

$$\therefore \quad \mathrm{T}(\mathbf{X}) = x_1\mathrm{T}(e_1) + x_2\mathrm{T}(e_2) + \cdots + x_n\mathrm{T}(e_n) \tag{2}$$

using Equation (1).

Here, $\mathrm{T}(e_j)$ is an $m \times 1$ matrix for each j. Therefore, if we denote

$$\mathrm{T}(e_j) = \begin{bmatrix} a_{1j} \\ a_{2j} \\ \vdots \\ a_{mj} \end{bmatrix},$$

then equation (2) becomes

$$\mathrm{T}(\mathbf{X}) = x_1\begin{bmatrix} a_{11} \\ a_{21} \\ \vdots \\ a_{m1} \end{bmatrix} + x_2\begin{bmatrix} a_{12} \\ a_{22} \\ \vdots \\ a_{m2} \end{bmatrix} + \cdots + x_n\begin{bmatrix} a_{1n} \\ a_{2n} \\ \vdots \\ a_{mn} \end{bmatrix}$$

$$= \begin{bmatrix} a_{11} & a_{12} & \cdots & a_{1n} \\ a_{21} & a_{22} & \cdots & a_{2n} \\ \vdots & \vdots & \vdots & \vdots \\ a_{m1} & a_{m2} & \cdots & a_{mn} \end{bmatrix}\begin{bmatrix} x_1 \\ x_2 \\ \vdots \\ x_n \end{bmatrix}$$

$$= \mathbf{AX}$$

where **A** is the matrix whose j^{th} column is $\mathrm{T}(e_j)$. Now, we show **A** is unique. Suppose that we also have $\mathrm{T}(\mathbf{X}) = \mathbf{BX}$ for all **X** in $\mathbf{R}^n$. Then

$$\mathrm{T}(e_j) = \mathrm{B}e_j = j^{\text{th}} \text{ column of } \mathbf{B}$$

and $$\mathrm{T}(e_j) = \mathrm{A}e_j = j^{\text{th}} \text{ column of } \mathbf{A}.$$

Thus, the columns of A and B agree, so $\mathbf{A} = \mathbf{B}$.

In the following examples we find matrix representation for some of the transformations discussed in the previous section. The above theorem says that the matrix representation of a given transformation $\mathrm{T}: \mathbf{R}^n \to \mathbf{R}^m$ is the matrix $\mathbf{A} = [\mathrm{T}(e_1) : \mathrm{T}(e_2) : \ldots : \mathrm{T}(e_n)]$ whose columns are the vectors $\mathrm{T}(e_j)$ where e_j is a standard basis vector. The matrix **A** is called standard matrix representing T.

EXAMPLE 3.21 Let T be the transformation given by

$$\mathrm{T}\left(\begin{bmatrix} x \\ y \\ z \end{bmatrix}\right) = \begin{bmatrix} x + y \\ x + z \end{bmatrix} \tag{1}$$

It was shown in Example 3.3 that T is a linear transformation from $\mathbf{R}^3$ to $\mathbf{R}^2$. Find a matrix representation for T.

SOLUTION

Let **A** be the matrix such that

$$\mathrm{T}\left(\begin{bmatrix} x \\ y \\ z \end{bmatrix}\right) = \mathbf{A}\begin{bmatrix} x \\ y \\ z \end{bmatrix} \quad \text{for all} \begin{bmatrix} x \\ y \\ z \end{bmatrix} \text{ in } \mathbf{R}^3.$$

The columns of **A** are given by the vectors

$$\mathrm{T}\left(\begin{bmatrix} 1 \\ 0 \\ 0 \end{bmatrix}\right), \mathrm{T}\left(\begin{bmatrix} 0 \\ 1 \\ 0 \end{bmatrix}\right) \text{ and } \mathrm{T}\left(\begin{bmatrix} 0 \\ 0 \\ 1 \end{bmatrix}\right).$$

Using (1) we get,

$$T\left(\begin{bmatrix}1\\0\\0\end{bmatrix}\right) = \begin{bmatrix}1\\1\end{bmatrix}$$

$$T\left(\begin{bmatrix}0\\1\\0\end{bmatrix}\right) = \begin{bmatrix}1\\0\end{bmatrix}$$

$$T\left(\begin{bmatrix}0\\0\\1\end{bmatrix}\right) = \begin{bmatrix}0\\1\end{bmatrix}$$

Thus the matrix **A** has the form

$$\mathbf{A} = \begin{bmatrix}1 & 1 & 0\\1 & 0 & 1\end{bmatrix}$$

It can easily be checked that

$$T\left(\begin{bmatrix}x\\y\\z\end{bmatrix}\right) = \begin{bmatrix}1 & 1 & 0\\1 & 0 & 1\end{bmatrix}\begin{bmatrix}x\\y\\z\end{bmatrix}$$

EXAMPLE 3.22 Obtain the matrix representation for the projection $T : \mathbf{R}^2 \to \mathbf{R}^2$ defined by

$$T\left(\begin{bmatrix}x\\y\end{bmatrix}\right) = \begin{bmatrix}x\\0\end{bmatrix}.$$

SOLUTION

The standard basis of $\mathbf{R}^2$ is $\{e_1, e_2\}$ where $e_1 = \begin{bmatrix}1\\0\end{bmatrix}$ and $e_2 = \begin{bmatrix}0\\1\end{bmatrix}$. If **A** is the matrix representation of T, then the columns of **A** are the vectors $T(e_1)$ and $T(e_2)$.

$$T(e_1) = T\left(\begin{bmatrix}1\\0\end{bmatrix}\right) = \begin{bmatrix}1\\0\end{bmatrix}$$

and $$T(e_2) = T\left(\begin{bmatrix}0\\1\end{bmatrix}\right) = \begin{bmatrix}0\\0\end{bmatrix}$$

$\therefore$ $$\mathbf{A} = \begin{bmatrix}1 & 0\\0 & 0\end{bmatrix}.$$

EXAMPLE 3.23 Find a matrix representation for the reflection about x-axis in $\mathbf{R}^2$.

SOLUTION

If $T : \mathbf{R}^2 \to \mathbf{R}^2$ is the reflection about x-axis then

$$T\left(\begin{bmatrix}x\\y\end{bmatrix}\right) = \begin{bmatrix}x\\-y\end{bmatrix}$$

Let **A** be the matrix representation of T. Then, the columns of **A** are given by

$$T\left(\begin{bmatrix}1\\0\end{bmatrix}\right) = \begin{bmatrix}1\\0\end{bmatrix}$$

and $$\mathrm{T}\left(\begin{bmatrix}0\\1\end{bmatrix}\right)=\begin{bmatrix}0\\-1\end{bmatrix}$$

$\therefore$ $$\mathbf{A}=\begin{bmatrix}1&0\\0&-1\end{bmatrix}$$

EXAMPLE 3.24 Let $\mathrm{T}:\mathbf{R}^2\to\mathbf{R}^2$ be defined by

$$\mathrm{T}\left(\begin{bmatrix}x\\y\end{bmatrix}\right)=\begin{bmatrix}-y\\-x\end{bmatrix}.$$

Show that T is a linear transformation known as reflection with respect to the line $y=-x$. Also find the standard matrix representing T.

SOLUTION

T is a linear transformation follows from the following equations

$$\mathrm{T}\left(\begin{bmatrix}x_1\\y_1\end{bmatrix}+\begin{bmatrix}x_2\\y_2\end{bmatrix}\right)=\mathrm{T}\left(\begin{bmatrix}x_1+x_2\\y_1+y_2\end{bmatrix}\right)=\begin{bmatrix}-y_1-y_2\\-x_1-x_2\end{bmatrix}$$
$$=\begin{bmatrix}-y_1\\-x_1\end{bmatrix}+\begin{bmatrix}-y_2\\-x_2\end{bmatrix}=\mathrm{T}\left(\begin{bmatrix}x_1\\y_1\end{bmatrix}\right)+\mathrm{T}\left(\begin{bmatrix}x_2\\y_2\end{bmatrix}\right)$$

and

$$\mathrm{T}\left(\mathrm{c}\begin{bmatrix}x\\y\end{bmatrix}\right)=\mathrm{T}\left(\begin{bmatrix}\mathrm{c}x\\\mathrm{c}y\end{bmatrix}\right)=\begin{bmatrix}-\mathrm{c}x\\-\mathrm{c}y\end{bmatrix}=\mathrm{c}\begin{bmatrix}-y\\-x\end{bmatrix}=\mathrm{cT}\begin{bmatrix}x\\y\end{bmatrix}$$

Let A be the standard matrix representing T. The standard basis of $\mathbf{R}^2$ is $\{e_1,e_2\}$ where $e_1=\begin{bmatrix}1\\0\end{bmatrix}$ and $e_2=\begin{bmatrix}0\\1\end{bmatrix}$.

Now $$\mathrm{T}(e_1)=\mathrm{T}\left(\begin{bmatrix}1\\0\end{bmatrix}\right)=\begin{bmatrix}0\\-1\end{bmatrix}$$

and $$\mathrm{T}(e_2)=\mathrm{T}\left(\begin{bmatrix}0\\1\end{bmatrix}\right)=\begin{bmatrix}-1\\0\end{bmatrix}$$

$\therefore$ $$\mathbf{A}=[\mathrm{T}(e_1)\quad\mathrm{T}(e_2)]$$
$$=\begin{bmatrix}0&-1\\-1&0\end{bmatrix}$$

EXAMPLE 3.25 Find a matrix representation for counterclockwise rotation of the plane about origin through 90°.

SOLUTION

Let $\mathrm{T}:\mathbf{R}^2\to\mathbf{R}^2$ be the linear transformation denoting rotation through 90°.

Let $\mathbf{A}=\begin{bmatrix}a&b\\c&d\end{bmatrix}$ be a 2×2 matrix such that $\mathrm{T}(\mathbf{X})=\mathbf{AX}$ (1)

for every vector $\mathbf{X}$ in $\mathbf{R}^2$.

It is clear that the rotation through 90° will transform the vector $\begin{bmatrix}1\\0\end{bmatrix}$ on the x-axis to $\begin{bmatrix}0\\1\end{bmatrix}$, and sends $\begin{bmatrix}0\\1\end{bmatrix}$ on the y-axis to $\begin{bmatrix}-1\\0\end{bmatrix}$. See Fig. 3.20.

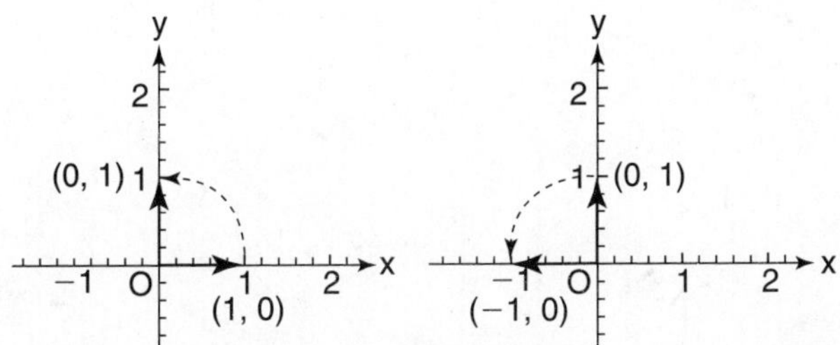

Rotation through 90°
Fig. 3.20

Thus $\quad T\left(\begin{bmatrix}1\\0\end{bmatrix}\right) = \begin{bmatrix}0\\1\end{bmatrix}$ and $T\left(\begin{bmatrix}0\\1\end{bmatrix}\right) = \begin{bmatrix}-1\\0\end{bmatrix}$

Using in Equation (1)

$$\begin{bmatrix}0\\1\end{bmatrix} = \begin{bmatrix}a & b\\c & d\end{bmatrix}\begin{bmatrix}1\\0\end{bmatrix}$$

and

$$\begin{bmatrix}-1\\0\end{bmatrix} = \begin{bmatrix}a & b\\c & d\end{bmatrix}\begin{bmatrix}0\\1\end{bmatrix}$$

That is, $\begin{bmatrix}0\\1\end{bmatrix} = \begin{bmatrix}a\\c\end{bmatrix}$ and $\begin{bmatrix}-1\\0\end{bmatrix} = \begin{bmatrix}b\\d\end{bmatrix}$

$\therefore \qquad a = 0, b = -1, c = 1, d = 0$

Hence, the matrix $\mathbf{A} = \begin{bmatrix}0 & -1\\1 & 0\end{bmatrix}$ represents rotation of the plane through 90°.

EXAMPLE 3.26 Let T be a counterclockwise rotation of vectors in $\mathbf{R}^2$ through the angle θ. Find a matrix representation for the rotation T.

SOLUTION

Let **A** be the 2×2 matrix such that

$$T\left(\begin{bmatrix}x\\y\end{bmatrix}\right) = \mathbf{A}\begin{bmatrix}x\\y\end{bmatrix} \tag{1}$$

for all $\mathbf{X} = \begin{bmatrix}x\\y\end{bmatrix}$ in $\mathbf{R}^2$.

In order to determine the columns of **A**, we must determine $T\left(\begin{bmatrix}1\\0\end{bmatrix}\right)$ and $T\left(\begin{bmatrix}0\\1\end{bmatrix}\right)$. Using elementary trigonometry (See Fig. 3.21), we obtain

$$T\left(\begin{bmatrix}1\\0\end{bmatrix}\right) = \begin{bmatrix}\cos\theta\\ \sin\theta\end{bmatrix}, \qquad T\left(\begin{bmatrix}0\\1\end{bmatrix}\right) = \begin{bmatrix}\cos\left(\theta + \dfrac{\pi}{2}\right)\\ \sin\left(\theta + \dfrac{\pi}{2}\right)\end{bmatrix}$$

or

$$T\left(\begin{bmatrix}1\\0\end{bmatrix}\right) = \begin{bmatrix}\cos\theta\\ \sin\theta\end{bmatrix} \text{ and } T\left(\begin{bmatrix}0\\1\end{bmatrix}\right) = \begin{bmatrix}-\sin\theta\\ \cos\theta\end{bmatrix}$$

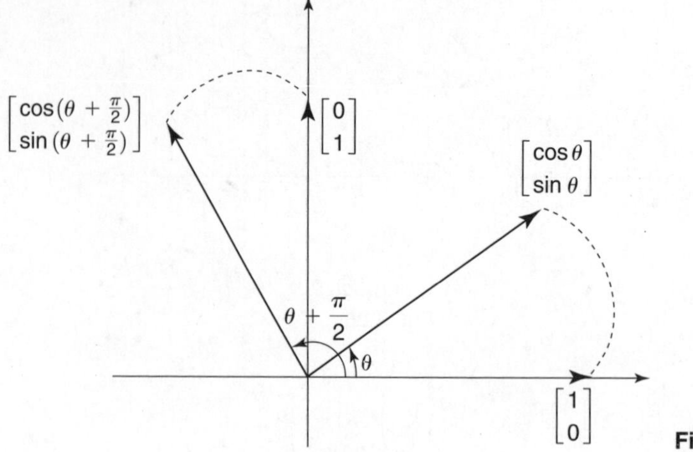

Fig. 3.21

Thus, matrix **A** of Equation (1) has the form

$$\mathbf{A} = \begin{bmatrix} \cos\theta & -\sin\theta \\ \sin\theta & \cos\theta \end{bmatrix}$$

Note that for this **A**, equation (1) becomes

$$\mathrm{T}\left(\begin{bmatrix} x \\ y \end{bmatrix}\right) = \begin{bmatrix} \cos\theta & -\sin\theta \\ \sin\theta & \cos\theta \end{bmatrix}\begin{bmatrix} x \\ y \end{bmatrix}$$

which is same as Equation (4) of Example 3.19.

EXAMPLE 3.27 Let l be the line in the xy plane that passes through the origin and makes an angle θ with the positive x-axis where $0 \le \theta \le \pi$. Let $\mathrm{T} : \mathbf{R}^2 \to \mathbf{R}^2$ be the linear transformation that reflects each vector about l. (See Fig. 3.22). Find the standard matrix of T.

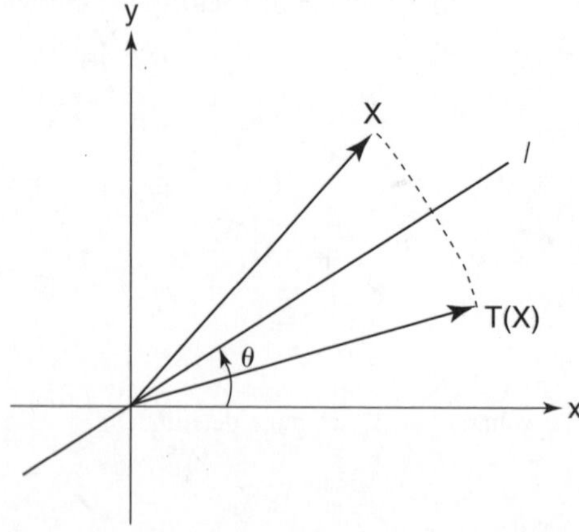

Fig. 3.22

SOLUTION

If **A** is the standard matrix of T, then $\mathbf{A} = [\mathrm{T}(e_1), \mathrm{T}(e_2)]$, where e_1, e_2 are standard basis vectors of T. It is clear from Fig. 3.23(a), that

$$\mathrm{T}(e_1) = \begin{bmatrix} ||\mathrm{T}(e_1)|| \cos 2\theta \\ ||\mathrm{T}(e_1)|| \sin 2\theta \end{bmatrix} = \begin{bmatrix} \cos 2\theta \\ \sin 2\theta \end{bmatrix}$$

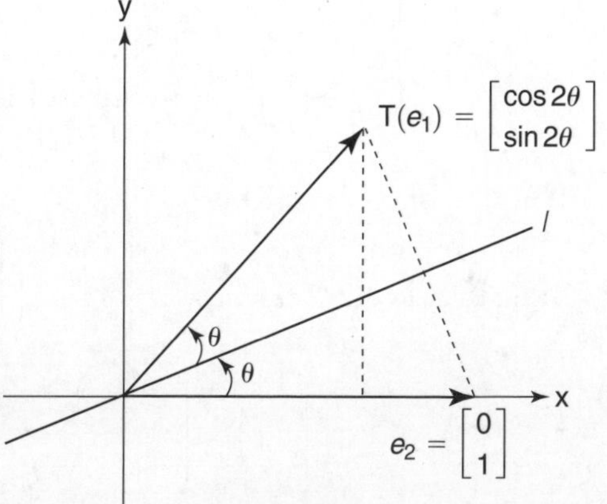

Fig. 3.23(a)

It can also be easily seen that $T(e_2)$ makes an angle $(2\theta - \frac{\pi}{2})$ with the x-axis.

$$\therefore \quad T(e_2) = \begin{bmatrix} \cos\left(2\theta - \dfrac{\pi}{2}\right) \\ \sin\left(2\theta - \dfrac{\pi}{2}\right) \end{bmatrix} = \begin{bmatrix} \sin 2\theta \\ -\cos 2\theta \end{bmatrix}$$

$$\therefore \quad \mathbf{A} = \begin{bmatrix} \cos 2\theta & \sin 2\theta \\ \sin 2\theta & -\cos 2\theta \end{bmatrix} \qquad (1)$$

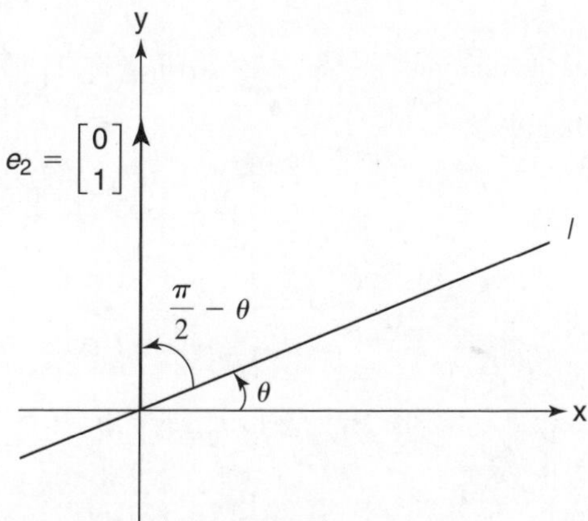

Fig. 3.23(b)

Thus, the reflection about any line l that passes through the origin and makes an angle θ with x-axis is given by $T(\mathbf{X}) = \mathbf{AX}$ where $\mathbf{A}$ is given by Equation (1).

EXAMPLE 3.28 Find a matrix representation for $T : \mathbf{R}^3 \rightarrow \mathbf{R}^3$ given by

$$T\left(\begin{bmatrix} x \\ y \\ z \end{bmatrix}\right) = \begin{bmatrix} -x \\ y \\ z \end{bmatrix}$$

Note that T is a linear transformation called reflection in yz-plane.

SOLUTION

Standard basis for $\mathbf{R}^3$ is $\left\{\begin{bmatrix}1\\0\\0\end{bmatrix}, \begin{bmatrix}0\\1\\0\end{bmatrix}, \begin{bmatrix}0\\0\\1\end{bmatrix}\right\}$. If **A** is a matrix such that

$$\mathbf{T}(\mathbf{X}) = \mathbf{AX}$$

for every **X** in $\mathbf{R}^3$, then the columns of **A** are given by $T\left(\begin{bmatrix}1\\0\\0\end{bmatrix}\right)$, $T\left(\begin{bmatrix}0\\1\\0\end{bmatrix}\right)$ and $T\left(\begin{bmatrix}0\\1\\0\end{bmatrix}\right)$.

Now $$T\left(\begin{bmatrix}1\\0\\0\end{bmatrix}\right) = \begin{bmatrix}-1\\0\\0\end{bmatrix}$$

$$T\left(\begin{bmatrix}0\\1\\0\end{bmatrix}\right) = \begin{bmatrix}0\\1\\0\end{bmatrix}$$

and $$T\left(\begin{bmatrix}0\\0\\1\end{bmatrix}\right) = \begin{bmatrix}0\\0\\1\end{bmatrix}$$

∴ $$\mathbf{A} = \begin{bmatrix}-1 & 0 & 0\\0 & 1 & 0\\0 & 0 & 1\end{bmatrix}.$$

EXAMPLE 3.29 Consider the dilation $T : \mathbf{R}^2 \to \mathbf{R}^2$ defined by $T\left(\begin{bmatrix}x\\y\end{bmatrix}\right) = \begin{bmatrix}2x\\2y\end{bmatrix}$. We find the standard matrix representation of T.

Note that $T\left(\begin{bmatrix}1\\0\end{bmatrix}\right) = \begin{bmatrix}2\\0\end{bmatrix}$ and $T\left(\begin{bmatrix}0\\1\end{bmatrix}\right) = \begin{bmatrix}0\\2\end{bmatrix}$ which give columns of **A**, the standrad matrix representation of T.

Thus, $$\mathbf{A} = \begin{bmatrix}2 & 0\\0 & 2\end{bmatrix}.$$

In general, it can be shown that for the transformation $T_r : \mathbf{R}^2 \to \mathbf{R}^2$ given by $T_r(X) = rX$, the matrix representation is given by

$$\mathbf{A} = \begin{bmatrix}r & 0\\0 & r\end{bmatrix}.$$

EXAMPLE 3.30 Consider the transformation $T : \mathbf{R}^2 \to \mathbf{R}^2$ defined by

$$T\left(\begin{bmatrix}x\\y\end{bmatrix}\right) = \begin{bmatrix}kx\\y\end{bmatrix}$$

where k is a positive constant. Since the x-coordinate of every vector is multiplied by a positive constant k, the effect of T is to expand or compress each plane figure in the x-direction. This transformation is therefore called dilation in the x-direction if $k > 1$ and called a contraction in the x-direction if $0 < k < 1$. We can easily verify that T is linear. Also since $T(e_1) = T\left(\begin{bmatrix}1\\0\end{bmatrix}\right) = \begin{bmatrix}k\\0\end{bmatrix}$ and $T(e_2) = T\left(\begin{bmatrix}0\\1\end{bmatrix}\right) = \begin{bmatrix}0\\1\end{bmatrix}$, the standard matrix of T is $\begin{bmatrix}k & 0\\0 & 1\end{bmatrix}$.

The Fig. 3.24 shows the effect of T on the unit square with factor $k = 2$ and $k = \frac{1}{2}$.

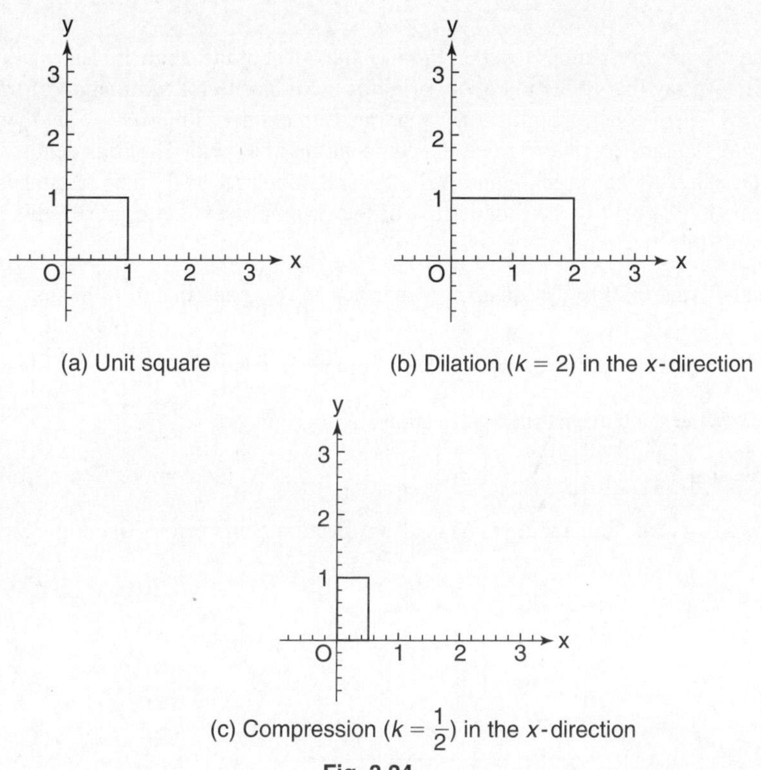

Fig. 3.24

We can similarly define dilation or compression in the y-direction (See *Example 3.29*)

In fact, $\begin{bmatrix}1 & 0\\0 & k\end{bmatrix}$ is the standard matrix for dilation or compression in the y-direction.

3.3.1. Composition of Two Transformations

Let $T_1 : \mathbf{R}^2 \to \mathbf{R}^2$ be the reflection with respect to the y-axis and $T_2 : \mathbf{R}^2 \to \mathbf{R}^2$ be the reflection with respect to the x-axis. Let $\mathbf{X} = (x, y)$ be a vector in $\mathbf{R}^2$. If we apply T_1 on $\mathbf{X}$ followed by T_2, we get the vector $(-x, -y)$. See Fig. 3.25 below.

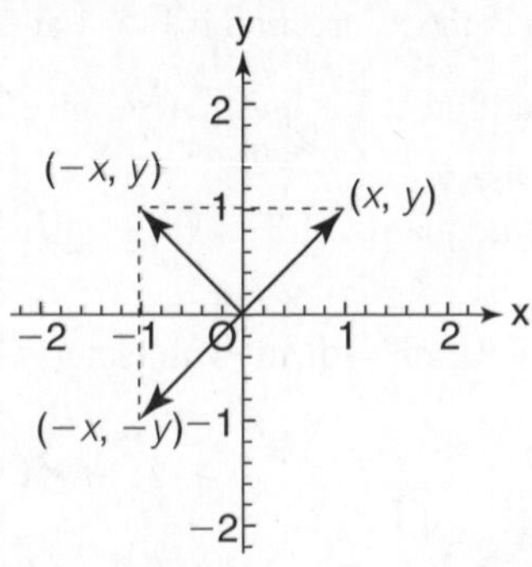

Fig. 3.25

Thus, the two transformations together produce the effect of the single transformation 'reflection about origin'. We say that reflection about origin is a composite transformation of transformation T_1 followed by T_2. In general, application of a transformation T_1 followed by another transformation T_2 produces a transformation called the composition of T_2 with T_1 and is denoted by $T_2 \text{ o } T_1$.

If **A** is the standard matrix of T_1 and **B** is the standard matrix of T_2, then the standard matrix of the transformation $T_2 \text{ o } T_1$ is the product **BA** of the two matrices. Let us verify this result for the above example of reflections.

The standard matrix of reflection about y-axis is $\begin{bmatrix} -1 & 0 \\ 0 & 1 \end{bmatrix}$ and that of reflection about x-axis is $\begin{bmatrix} 1 & 0 \\ 0 & -1 \end{bmatrix}$, and hence the standard matrix of their composition is $\begin{bmatrix} -1 & 0 \\ 0 & 1 \end{bmatrix}\begin{bmatrix} 1 & 0 \\ 0 & -1 \end{bmatrix} = \begin{bmatrix} -1 & 0 \\ 0 & -1 \end{bmatrix}$ which is clearly the standard matrix of reflection about origin.

Also note that $\begin{bmatrix} 1 & 0 \\ 0 & -1 \end{bmatrix}\begin{bmatrix} -1 & 0 \\ 0 & 1 \end{bmatrix} = \begin{bmatrix} -1 & 0 \\ 0 & -1 \end{bmatrix}$, that is, the composition of reflection about x-axis followed by reflection about y-axis is also reflection about origin. See Fig. 3.26.

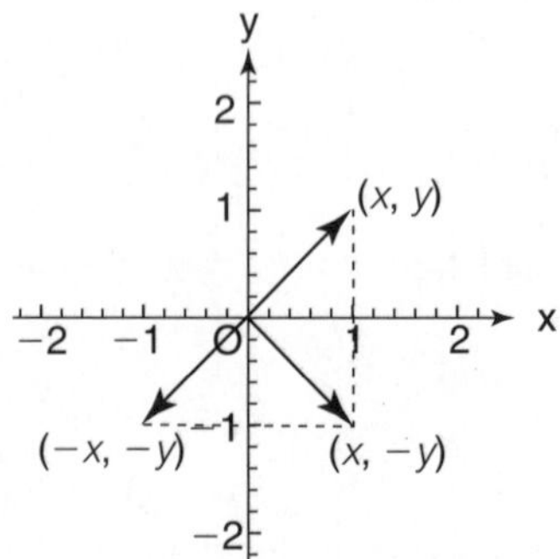

Fig. 3.26

In other words, $T_1 \text{ o } T_2 = T_2 \text{ o } T_1$.

But, in general, the composition of two transformations is not commutative because matrix multiplication is not commutative. Thus, the order in which the linear transformations are composed makes a difference.

EXAMPLE 3.31 Let $T_1 : \mathbf{R}^2 \to \mathbf{R}^2$ be reflection about the line $y = x$ and $T_2 : \mathbf{R}^2 \to \mathbf{R}^2$ be the projection on x-axis. Find the standard matrix of $T_2 \text{ o } T_1$. Is this matrix same as the standard matrix of $T_1 \text{ o } T_2$? Illustrate your result graphically.

SOLUTION

Standard matrix of $T_1 = \begin{bmatrix} 0 & 1 \\ 1 & 0 \end{bmatrix}$ and standard matrix of $T_2 = \begin{bmatrix} 1 & 0 \\ 0 & 0 \end{bmatrix}$.

$\therefore$ standard matrix of $T_2 \circ T_1 = \begin{bmatrix} 1 & 0 \\ 0 & 0 \end{bmatrix}\begin{bmatrix} 0 & 1 \\ 1 & 0 \end{bmatrix} = \begin{bmatrix} 0 & 1 \\ 0 & 0 \end{bmatrix}$

and standard matrix of $T_1 \circ T_2 = \begin{bmatrix} 0 & 1 \\ 1 & 0 \end{bmatrix}\begin{bmatrix} 1 & 0 \\ 0 & 0 \end{bmatrix} = \begin{bmatrix} 0 & 0 \\ 1 & 0 \end{bmatrix}$

Clearly, the two matrices are different and

hence $T_1 \circ T_2 \neq T_2 \circ T_1$.

This fact is also clear from Fig. 3.27 (a) and (b).

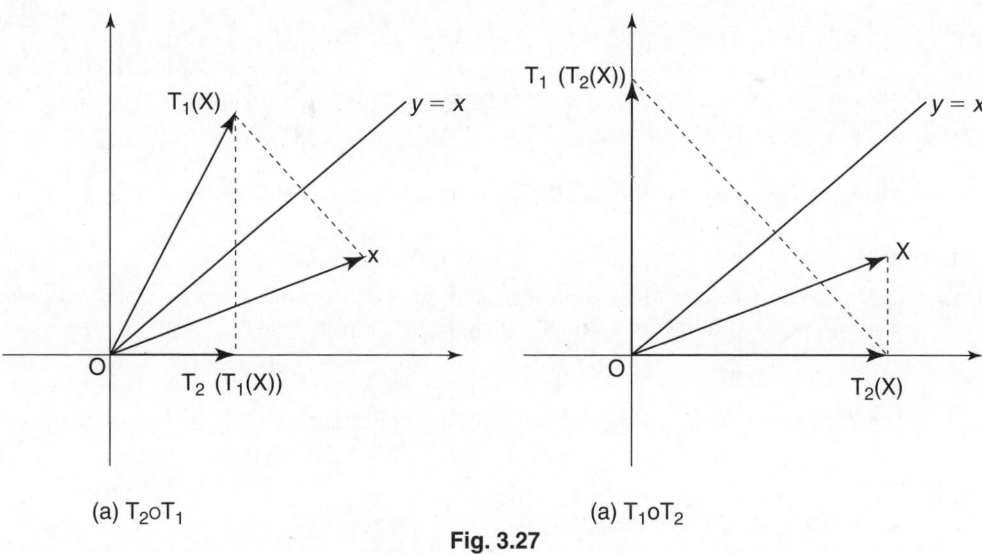

Fig. 3.27

3.3.2. Geometric Effect of Multiplication by a Matrix

We have shown that every linear transformation from $\mathbf{R}^n \to \mathbf{R}^m$ has a matrix representation and conversely, every matrix represents a linear transformation. In most cases, we can easily describe the linear transformations geometrically from their matrix representation. In the following examples, we discuss the geometric effect of multiplication by a matrix.

EXAMPLE 3.32 Consider the matrix $\begin{bmatrix} 1 & 0 \\ 0 & 0 \end{bmatrix}$. If we multiply this matrix with the vector $\begin{bmatrix} x \\ y \end{bmatrix}$, we get the vector $\begin{bmatrix} x \\ 0 \end{bmatrix}$. Thus, the matrix represents a transformation which maps the vector $\begin{bmatrix} x \\ y \end{bmatrix}$ onto the vector $\begin{bmatrix} x \\ 0 \end{bmatrix}$. Clearly, the matrix represents the projection on the x-axis.

EXAMPLE 3.33 Find the geometric description of the linear transformation with matrix

$$\begin{bmatrix} -1 & 0 & 0 \\ 0 & -1 & 0 \\ 0 & 0 & -1 \end{bmatrix}.$$

SOLUTION

If $T : \mathbf{R}^3 \to \mathbf{R}^3$ is the transformation whose matrix representation is $\begin{bmatrix} -1 & 0 & 0 \\ 0 & -1 & 0 \\ 0 & 0 & -1 \end{bmatrix}$, then

$$T\left(\begin{bmatrix} x \\ y \\ z \end{bmatrix}\right) = \begin{bmatrix} -1 & 0 & 0 \\ 0 & -1 & 0 \\ 0 & 0 & -1 \end{bmatrix}\begin{bmatrix} x \\ y \\ z \end{bmatrix} = \begin{bmatrix} -x \\ -y \\ -z \end{bmatrix}$$

Thus, under the transformation, each point (x, y, z) of $\mathbf{R}^3$ is mapped onto the point with coordinates $(-x, -y, -z)$. Therefore, the matrix represents reflection in origin.

EXAMPLE 3.34 Determine the effect of the transformation of the plane represented by $\begin{pmatrix} 1 & 0 \\ 1 & 0 \end{pmatrix}$.

SOLUTION

Under this transformation of the plane, each point (x, y) on the plane is mapped onto a point with coordinates (x, x). The matrix represents a projection of the plane onto the line $y = x$.

EXAMPLE 3.35 Describe the geometrical effect of multiplication by a 2×2 diagonal matrix.

SOLUTION

A 2×2 diagonal matrix is of the form

$$\mathbf{A} = \begin{bmatrix} k_1 & 0 \\ 0 & k_2 \end{bmatrix}$$

where k_1 and k_2 are constants.

If $T : \mathbf{R}^2 \to \mathbf{R}^2$ be the transformation such that

$$T(\mathbf{X}) = \mathbf{AX}$$

then
$$T\left(\begin{bmatrix} x \\ y \end{bmatrix}\right) = \begin{bmatrix} k_1 & 0 \\ 0 & k_2 \end{bmatrix}\begin{bmatrix} x \\ y \end{bmatrix} = \begin{bmatrix} k_1x \\ k_2y \end{bmatrix}$$

Thus, under this transformation x-coordinate is multiplied by k_1 and y-coordinate by k_2. If k_1 and k_2 are both positive, the transformation will expand or contract a vector in the x-direction by a factor k_1 and in the y-direction by a factor k_2. This fact is more clear if we consider composition of transformations. In fact, we can write

$$\begin{bmatrix} k_1 & 0 \\ 0 & k_2 \end{bmatrix} = \begin{bmatrix} 1 & 0 \\ 0 & k_2 \end{bmatrix}\begin{bmatrix} k_1 & 0 \\ 0 & 1 \end{bmatrix}$$

If $k_1 > 0$, $\begin{bmatrix} k_1 & 0 \\ 0 & 1 \end{bmatrix}$ represents dilation or contraction in the x-direction with factor k_1 (see Example 3.30). Similarly, $\begin{bmatrix} 1 & 0 \\ 0 & k_2 \end{bmatrix}$ is the dilation or contraction in the y-direction with factor k_2 if $k_2 > 0$. Thus the diagonal matrix $\begin{bmatrix} k_1 & 0 \\ 0 & k_2 \end{bmatrix}$ where k_1 and k_2 are positive constants represents the composition of a dilation or contraction in the x-direction by a factor k_1, followed by a dilation or contraction in the y-direction by a factor k_2.

It is clear that if $k_1 = k_2$, $\begin{bmatrix} k_1 & 0 \\ 0 & k_2 \end{bmatrix}$, will be a scalar matrix and represents simply a dilation or contraction of the plane. If $k_1 \neq k_2$, then the transformation represented by $\begin{bmatrix} k_1 & 0 \\ 0 & k_2 \end{bmatrix}$ is called a *magnification* or *scaling*. Since x and y do not change in the same manner, so a magnification usually distorts a plane figure. Note the effect of the magnification $\begin{bmatrix} 2 & 0 \\ 0 & 3 \end{bmatrix}$ on the unit square in Fig. 3.28.

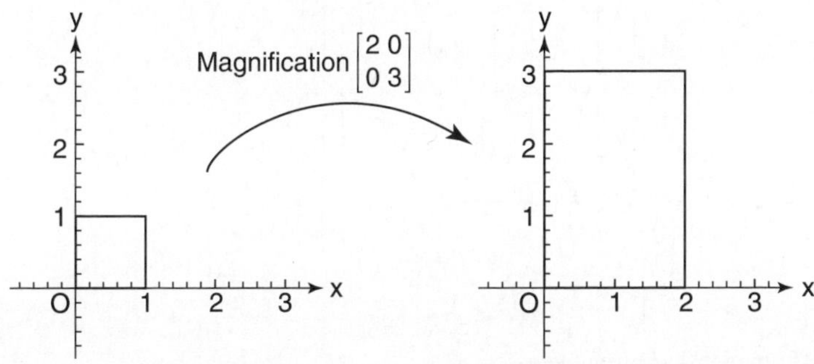

Fig. 3.28

REMARK: If $k_1 < 0$, that is, $k_1 = -k$ for some $k > 0$, then

$$\begin{bmatrix} k_1 & 0 \\ 0 & 1 \end{bmatrix} = \begin{bmatrix} -1 & 0 \\ 0 & 1 \end{bmatrix}\begin{bmatrix} k & 0 \\ 0 & 1 \end{bmatrix}$$

which represents the composition of a dilation or contraction in the x-direction with factor k, followed by a reflection about y-axis. Based on this, one can easily deduce the geometric effect of the diagonal matrix $\begin{bmatrix} k_1 & 0 \\ 0 & k_2 \end{bmatrix}$ if either k_1 or k_2 is negative.

Upto now, we have seen that the matrix representation of a linear transformation can be easily obtained if we know its action on the standard basis vectors. What if we know its action only on some other basis? The following example shows that we can still find the transformation.

EXAMPLE 3.36 Let $T : \mathbf{R}^2 \to \mathbf{R}^2$ be a linear transformation such that $T\left(\begin{bmatrix} 1 \\ 1 \end{bmatrix}\right) = \begin{bmatrix} 3 \\ 2 \end{bmatrix}$ and $T\left(\begin{bmatrix} 1 \\ 2 \end{bmatrix}\right) = \begin{bmatrix} 4 \\ 5 \end{bmatrix}$. Find a matrix representation for T.

SOLUTION

It can be easily verified that $\begin{bmatrix}1\\1\end{bmatrix}$ and $\begin{bmatrix}1\\2\end{bmatrix}$ form a basis of $\mathbf{R}^2$. Thus, we can write vectors $\begin{bmatrix}1\\0\end{bmatrix}$ and $\begin{bmatrix}0\\1\end{bmatrix}$ as linear combination of the vectors $\begin{bmatrix}1\\1\end{bmatrix}$ and $\begin{bmatrix}1\\2\end{bmatrix}$.

In fact,
$$\begin{bmatrix}1\\0\end{bmatrix} = 2\begin{bmatrix}1\\1\end{bmatrix} - \begin{bmatrix}1\\2\end{bmatrix}$$

and
$$\begin{bmatrix}0\\1\end{bmatrix} = -\begin{bmatrix}1\\1\end{bmatrix} + \begin{bmatrix}1\\2\end{bmatrix} \quad \text{(verify)}$$

Since T is linear,

$$\therefore \quad T\left(\begin{bmatrix}1\\0\end{bmatrix}\right) = 2T\left(\begin{bmatrix}1\\1\end{bmatrix}\right) - T\left(\begin{bmatrix}1\\2\end{bmatrix}\right)$$

and
$$T\left(\begin{bmatrix}0\\1\end{bmatrix}\right) = -T\left(\begin{bmatrix}1\\1\end{bmatrix}\right) + T\left(\begin{bmatrix}1\\2\end{bmatrix}\right)$$

Using
$$T\left(\begin{bmatrix}1\\1\end{bmatrix}\right) = \begin{bmatrix}3\\2\end{bmatrix} \quad \text{and} \quad T\left(\begin{bmatrix}1\\2\end{bmatrix}\right) = \begin{bmatrix}4\\5\end{bmatrix}$$

we get
$$T\left(\begin{bmatrix}1\\0\end{bmatrix}\right) = \begin{bmatrix}2\\-1\end{bmatrix} \quad \text{and} \quad T\left(\begin{bmatrix}0\\1\end{bmatrix}\right) = \begin{bmatrix}1\\3\end{bmatrix}$$

If **A** is the matrix representation of T, then

$$\mathbf{A} = \begin{bmatrix}2 & 1\\-1 & 3\end{bmatrix}$$

Then, the transformation T is given by

$$T\left(\begin{bmatrix}x\\y\end{bmatrix}\right) = \begin{bmatrix}2 & 1\\-1 & 3\end{bmatrix}\begin{bmatrix}x\\y\end{bmatrix}.$$

We conclude this chapter by observing the following remark based on the above example.

REMARK: If $T : \mathbf{R}^2 \rightarrow \mathbf{R}^2$ is linear, then T can be uniquely described by its action on any basis of $\mathbf{R}^2$.

EXERCISE 3.3

1. Find a matrix representing the transformation of the plane that maps the points $(1, 0)$ and $(0, 1)$ onto $(2, 1)$ and $(3, 2)$ respectively.
2. Find the standard matrix for each of the following linear transformations from $\mathbf{R}^2$ into $\mathbf{R}^2$

(a) $T\left(\begin{bmatrix}x\\y\end{bmatrix}\right) = \begin{bmatrix}x+y\\x-y\end{bmatrix}$ **(b)** $T\left(\begin{bmatrix}x\\y\end{bmatrix}\right) = \begin{bmatrix}x\\y\end{bmatrix}$

(c) $T\left(\begin{bmatrix} x \\ y \end{bmatrix}\right) = \begin{bmatrix} 2x + 3y \\ 3x - 2y \end{bmatrix}$ **(d)** $T\left(\begin{bmatrix} x \\ y \end{bmatrix}\right) = \begin{bmatrix} 0 \\ 0 \end{bmatrix}$

(e) $T\left(\begin{bmatrix} x \\ y \end{bmatrix}\right) = \begin{bmatrix} 2x \\ -y \end{bmatrix}$ **(f)** $T\left(\begin{bmatrix} x \\ y \end{bmatrix}\right) = \begin{bmatrix} x + 2y \\ y \end{bmatrix}$

3. Find the standard matrix for each of the following linear transformations of the plane:

(a) projection onto y-axis

(b) reflection with respect to y-axis

(c) reflection with respect to the line $y = x$

(d) rotation through the angle $\pi/3$

(e) rotation through the angle $\pi/4$

(f) dilation of factor 3

(g) contraction of factor 1/3

(h) reflection about origin

4. Let $T : \mathbf{R}^2 \to \mathbf{R}^2$ be a linear transformation. Find a matrix representation for T where T is

(a) a counterclockwise rotation through $\pi/6$

(b) a shear along the x-direction with $k = 2$

(c) a dilation in the x-direction with a factor 5

(d) a shear of factor 3 in the y-direction

5. Find the standard matrix for the linear transformation $T : \mathbf{R}^3 \to \mathbf{R}^3$ defined by

(a) $T\left(\begin{bmatrix} x \\ y \\ z \end{bmatrix}\right) = \begin{bmatrix} x \\ -y \\ 0 \end{bmatrix}$ **(b)** $T\left(\begin{bmatrix} x \\ y \\ z \end{bmatrix}\right) = \begin{bmatrix} 2x \\ 3y \\ 4z \end{bmatrix}$

(c) $T\left(\begin{bmatrix} x \\ y \\ z \end{bmatrix}\right) = \begin{bmatrix} x - y \\ y + z \\ z - x \end{bmatrix}$ **(d)** $T\left(\begin{bmatrix} x \\ y \\ z \end{bmatrix}\right) = \begin{bmatrix} x - y \\ x + z \\ y - z \end{bmatrix}$

(e) $T\left(\begin{bmatrix} x \\ y \\ z \end{bmatrix}\right) = \begin{bmatrix} y - x \\ 0 \\ z - x \end{bmatrix}$ **(f)** $T\left(\begin{bmatrix} x \\ y \\ z \end{bmatrix}\right) = \begin{bmatrix} y \\ z \\ x \end{bmatrix}$

6. Find the standard matrix representation for the following linear transformations on $\mathbf{R}^3$.

(a) a dilation of factor 4.

(b) a projection onto the xy-plane

(c) a projection onto the z-axis

(d) a reflection about the xy-plane

(e) a reflection about the xz-plane

(f) a reflection about the x-axis

(g) a reflection about the origin.

7. Find the matrix of each of the following linear transformations

(a) $T : \mathbf{R}^2 \to \mathbf{R}^3$, $T(x, y) = (x, x + y, y)$

(b) $T : \mathbf{R}^2 \to \mathbf{R}^3$, $T(x, y) = (x + y, y, x - y)$

(c) $T: \mathbf{R}^3 \to \mathbf{R}^2, \quad T(x, y, z) = (x + y, 2z - x)$
(d) $T: \mathbf{R}^3 \to \mathbf{R}^2, \quad T(x, y, z) = (x - y + z, x + 2y + z)$
(e) $T: \mathbf{R}^3 \to \mathbf{R}, \quad T(x, y, z) = y$
(f) $T: \mathbf{R} \to \mathbf{R}^3, \quad T(x) = (x, -x, 0)$
(g) $T: \mathbf{R}^2 \to \mathbf{R}^2, \quad T(x, y) = (ax + by, cx + dz)$

8. Determine the matrix of each of the following transformations from $\mathbf{R}^n$ into $\mathbf{R}^n$:
(a) the identity transformation
(b) the zero transformation

9. Let $T: \mathbf{R}^2 \to \mathbf{R}^2$ be defined by

$T\left(\begin{bmatrix} x \\ y \end{bmatrix}\right) = \begin{bmatrix} x \\ ky \end{bmatrix}$ where k is a positive constant.

T is called a dilation in the y-direction if $k > 1$, and called a contraction in the y-direction if $0 < k < 1$.
(a) Prove that T is a linear transformation.
(b) Determine the standard matrix representing T.
(c) Find and sketch the image of the unit square under T with $k = 2$ and $k = 1/2$.

10. Illustrate geometrically the effect of the linear transformation $\mathbf{A} = \begin{bmatrix} 3 & 1 \\ 2 & 5 \end{bmatrix}$ on the square whose vertices are (0, 0), (1, 0), (1, 1) and (0, 1).

11. Determine the geometric effect of multiplication by the matrix

(a) $\begin{bmatrix} 0 & 1 \\ 1 & 0 \end{bmatrix}$ **(b)** $\begin{bmatrix} 1 & 1 \\ 0 & 0 \end{bmatrix}$ **(c)** $\begin{bmatrix} 0 & 0 \\ 0 & 1 \end{bmatrix}$

(d) $\begin{bmatrix} 2 & 0 \\ 0 & 1 \end{bmatrix}$ **(e)** $\begin{bmatrix} -1 & 0 \\ 0 & 1 \end{bmatrix}$ **(f)** $\begin{bmatrix} 0.5 & 0 \\ 0 & 0.5 \end{bmatrix}$

12. Consider the transformation on $\mathbf{R}^2$ defined by each of the following matrices. Find the image of the unit square under each transformation.

(a) $\begin{bmatrix} 0 & -1 \\ 1 & 0 \end{bmatrix}$ **(b)** $\begin{bmatrix} 2 & 0 \\ 0 & 2 \end{bmatrix}$ **(c)** $\begin{bmatrix} 3 & 0 \\ 1 & 4 \end{bmatrix}$

(d) $\begin{bmatrix} -2 & -3 \\ 0 & 4 \end{bmatrix}$ **(e)** $\begin{bmatrix} 0 & -2 \\ 2 & 0 \end{bmatrix}$ **(f)** $\begin{bmatrix} 1 & 0 \\ 0 & 1 \end{bmatrix}$

13. Determine a rotation matrix that maps the points (3, 4) and (1, –2) onto the points (–4, 3) and (2, 1) respectively.

14. Determine the effect of the dilation of the plane represented by $\begin{pmatrix} 4 & 0 \\ 0 & 4 \end{pmatrix}$ upon the

(a) line $y = 3x + 2$
(b) triangle with vertices (0, 0), (1, 2) and (3, 1). Illustrate graphically.

15. Find the image of the triangle with vertices (5, 0), (0, 3) and (–2, 1) under the linear transformation $T: \mathbf{R}^2 \to \mathbf{R}^2$ defined by $T(\mathbf{X}) = \mathbf{AX}$, where $\mathbf{A} = \begin{bmatrix} -2 & 1 \\ 3 & 4 \end{bmatrix}$.

16. Determine the effect of the magnification of the plane represented by $\begin{pmatrix} 1 & 0 \\ 0 & 3 \end{pmatrix}$ upon the ellipse $\dfrac{x^2}{9} + \dfrac{y^2}{1} = 1$. Illustrate graphically.

17. Find a single matrix that defines each of the following transformations on $\mathbf{R}^2$.

(a) a counterclockwise rotation through $\pi/2$, followed by a dilation of factor 2.

(b) a dilation of factor 3, followed by a reflection in the x-axis.

(c) a rotation of $\pi/2$, followed by a reflection about the line $y = x$.

(d) a projection onto the x-axis, followed by a reflection about y-axis.

(e) a projection onto the y-axis, followed by a contraction with factor 1/2.

18. Construct single 2×2 matrices that define the following transformations on $\mathbf{R}^2$. Find the image of the point $\begin{bmatrix} 1 \\ 2 \end{bmatrix}$ under each transformation.

(a) a reflection about the line $y = x$, then a rotation through π, then a dilation of factor 2.

(b) a dilation of factor 3, then a shear of factor 2 in the x-direction.

(c) a contraction by a factor of 1/2 in the x-direction, then a dilation by a factor 5 in the y-direction.

(d) dilation by a factor of 5 in the y-direction, then a shear by a factor of 2 in the y-direction.

(e) reflection about the y-axis, the a dilation by a factor of 5 in the x-direction, and then a reflection about the line $y = x$.

19. Find the standard matrix for the stated composition of linear transformations on $\mathbf{R}^3$:

(a) a reflection about the yz-plane, followed by a projection on the xz-plane.

(b) a projection on the xy-plane, followed by a reflection about the yz-plane.

(c) a reflection about the xz-plane, followed by a contraction by a factor of 1/5.

(d) a reflection about the xy-plane, then a reflection about the xz-plane, and then a reflection about the yz-plane.

20. Determine whether $T_1 \circ T_2 = T_2 \circ T_1$.

(a) $T_1 : \mathbf{R}^2 \to \mathbf{R}^2$ is the projection on the x-axis and $T_2 : \mathbf{R}^2 \to \mathbf{R}^2$ is the projection on the y-axis.

(b) $T_1 : \mathbf{R}^2 \to \mathbf{R}^2$ is the rotation through an angle θ, and $T_2 : \mathbf{R}^2 \to \mathbf{R}^2$ is the projection on the x-axis.

(c) $T_1 : \mathbf{R}^2 \to \mathbf{R}^2$ is the dilation in the x-direction with factor 2 and $T_2 : \mathbf{R}^2 \to \mathbf{R}^2$ is the dilation in the y-direction with factor 3.

(d) $T_1 : \mathbf{R}^2 \to \mathbf{R}^2$ be reflection with respect to the x-axis and $T_2 : \mathbf{R}^2 \to \mathbf{R}^2$ be reflection with respect to the line $y = x$.

(e) $T_1 : \mathbf{R}^3 \to \mathbf{R}^3$ be dilation of factor k, and $T_2 : \mathbf{R}^3 \to \mathbf{R}^3$ be the reflection about origin.

21. Discuss the geometric effect of multiplication by the following matrices. Find the image of the vector $X = \begin{bmatrix} 1 \\ 2 \\ 3 \end{bmatrix}$ under the transformation using matrix multiplication. Check the geometric effect by plotting the vector X and its image.

(a) $\begin{bmatrix} 1 & 0 & 0 \\ 0 & 1 & 0 \\ 0 & 0 & 0 \end{bmatrix}$ **(b)** $\begin{bmatrix} 3 & 0 & 0 \\ 0 & 3 & 0 \\ 0 & 0 & 3 \end{bmatrix}$

(c) $\begin{bmatrix} 1 & 0 & 0 \\ 0 & -1 & 0 \\ 0 & 0 & 1 \end{bmatrix}$ **(d)** $\begin{bmatrix} -1 & 0 & 0 \\ 0 & -1 & 0 \\ 0 & 0 & 1 \end{bmatrix}$

22. Let $T : \mathbf{R}^2 \to \mathbf{R}^2$ be a linear transformation such that $T\left(\begin{bmatrix}1\\1\end{bmatrix}\right) = \begin{bmatrix}3\\2\end{bmatrix}$ and $T\left(\begin{bmatrix}1\\2\end{bmatrix}\right) = \begin{bmatrix}4\\5\end{bmatrix}$. Find a matrix representation for T.

23. Let $T : \mathbf{R}^2 \to \mathbf{R}^2$ be translation by vector $\begin{bmatrix}a\\b\end{bmatrix}$, that is, $T\left(\begin{bmatrix}x\\y\end{bmatrix}\right) = \begin{bmatrix}x\\y\end{bmatrix} + \begin{bmatrix}a\\b\end{bmatrix}$. If $T\left(\begin{bmatrix}x\\y\end{bmatrix}\right) = \begin{bmatrix}x'\\y'\end{bmatrix}$, then

$$\left.\begin{aligned} x' &= x + a \\ y' &= y + b \end{aligned}\right\} \tag{1}$$

Equation (1) can be written in matrix form as

$$\begin{pmatrix}x'\\y'\\1\end{pmatrix} = \begin{pmatrix}1 & 0 & a\\0 & 1 & b\\0 & 0 & 1\end{pmatrix}\begin{pmatrix}x\\y\\1\end{pmatrix}$$

The matrix $\begin{bmatrix}1 & 0 & a\\0 & 1 & b\\0 & 0 & 1\end{bmatrix}$ is called the translation matrix.

(a) Use the translation matrix to find the image of the point $(3, -4)$ under a translation of the plane that maps the origin onto $(5, 2)$.

(b) Find the image of $P(3, 2)$ under a translation that maps the origin onto $(4, -1)$.

CHAPTER FOUR

Eigenvalues and Eigenvectors

■ 4.1 EIGENVALUES AND EIGENVECTORS

Consider a linear transformation $T : \mathbf{R}^n \to \mathbf{R}^n$ that maps the vector space $\mathbf{R}^n$ onto itself. Such transformations are usually known as Linear Operators. The simplest such linear operators are the maps $T_\lambda : \mathbf{R}^n \to \mathbf{R}^n$ that multiply each vector $\mathbf{X} \in \mathbf{R}^n$ by some real number λ. If $\lambda > 1$, then T_λ is a dilation and if $0 < \lambda < 1$, T_λ is a contraction. If $\lambda < 0$, T_λ reverses the direction of X. Note that if $\lambda = 1$, T_λ is identity operator and if $\lambda = -1$, T_λ is reflection in origin. If $\lambda < 0$, T_λ can be viewed as a composition, $T_\lambda = T_{-1} \circ T_{|\lambda|}$ of a dilation or contraction of factor $|\lambda|$ followed by a reflection in origin. However, most linear operators are not so simple. But sometimes, it is possible to find certain vectors on which a given linear operator acts like T_λ for some λ.

In this section, we discuss the problem of finding scalars (real numbers) λ for which there exist non-zero vectors $\mathbf{X}$ such that

$$T(\mathbf{X}) = \lambda\mathbf{X} \tag{1}$$

for a given linear operator T. Such a real number λ is called an *eigenvalue* of T and the associated vector $\mathbf{X}$ satisfying Equation (1) is called an *eigenvector* of T.

It must be noted that $\mathbf{X} = \mathbf{O}$ is always a solution of (1) for any λ. But we do not refer to it as an eigenvector of T, since we are interested only in non-zero vectors satisfying Equation (1). We again insist that λ is a real number to exclude the possibility of complex eigenvalues.

It is important to realize that there exist linear operators having no eigenvectors at all. The following example provides such a linear operator.

EXAMPLE 4.1 Find a linear operator $T : \mathbf{R}^2 \to \mathbf{R}^2$ such that T has no eigenvectors.

SOLUTION

Let T be the counterclockwise rotation of vectors of $\mathbf{R}^2$ through 30°. Then, it is clear from the geometrical nature of T that the vector $T(\mathbf{X})$ cannot be a scalar multiple of $\mathbf{X}$ unless $\mathbf{X}$ is a zero vector of $\mathbf{R}^2$. Thus, T has no eigenvectors.

In the previous chapter we proved that every linear transformation can be represented by a matrix. Thus, if $T : \mathbf{R}^n \to \mathbf{R}^n$ is a linear operator we get $n \times n$ square matrix $\mathbf{A}$ such that $T(\mathbf{X}) = \mathbf{AX}$ for every $\mathbf{X} \in \mathbf{R}^n$. We can thus define eigenvalue and eigenvector for a general square matrix.

DEFINITION: Let A be an $n \times n$ matrix. A scalar (real number) λ is called an 'eigenvalue' of **A** if there exists a non-zero vector **X** in R^n such that

$$\mathbf{AX} = \lambda\mathbf{X}$$

The vector **X** is called an 'eigenvector' corresponding to λ.

Eigenvalues are also called characteristic values, proper values, or latent values, and eigenvectors are also called characteristic vectors, proper vectors, or latent vectors.

It must be noted here that the eigenvalues of a linear operator T are precisely the eigenvalues of its standard matrix and if **X** is an eigenvector of T corresponding to λ, then X is an eigienvector of **A** corresponding to λ, and vice-versa.

Let us look at the geometrical significance of an eigenvector that corresponds to a non-zero eigenvalue. The vector $\mathbf{AX} = \lambda\mathbf{X}$ is a vector parallel to the vector **X**. It is in the same or opposite direction as **X**, depending on the sign of λ. An eigenvector of **A** is thus a vector whose direction is unchanged or reversed when multiplied by **A**.

EXAMPLE 4.2 Let $\mathbf{A} = \begin{bmatrix} -1 & -2 \\ 0 & 3 \end{bmatrix}$ and $\mathbf{X}_1 = \begin{bmatrix} 1 \\ -2 \end{bmatrix}$. Then

$$\mathbf{AX}_1 = \begin{bmatrix} -1 & -2 \\ 0 & 3 \end{bmatrix}\begin{bmatrix} 1 \\ -2 \end{bmatrix} = \begin{bmatrix} 3 \\ -6 \end{bmatrix} = 3\begin{bmatrix} 1 \\ -2 \end{bmatrix} = 3\mathbf{X}_1.$$

Thus, $\lambda = 3$ is an eigenvalue of **A** and $\mathbf{X}_1 = \begin{bmatrix} 1 \\ -2 \end{bmatrix}$ is an eigenvector corresponding to $\lambda = 3$.

Let $\mathbf{X}_2 = \begin{bmatrix} 1 \\ 0 \end{bmatrix}$. Then

$$\mathbf{AX}_2 = \begin{bmatrix} -1 & -2 \\ 0 & 3 \end{bmatrix}\begin{bmatrix} 1 \\ 0 \end{bmatrix} = \begin{bmatrix} -1 \\ 0 \end{bmatrix} = -\begin{bmatrix} 1 \\ 0 \end{bmatrix} = -\mathbf{X}_2 = (-1)\mathbf{X}_2.$$

Thus, $\lambda = -1$ is an eigenvalue of **A** and $\mathbf{X}_2 = \begin{bmatrix} 1 \\ 0 \end{bmatrix}$ is the corresponding eigenvector. Fig. 4.1 shows that $\mathbf{X}_1$ and $\mathbf{AX}_1$ are parallel to each other and are in the same direction while $\mathbf{X}_2$ and $\mathbf{AX}_2$ are in the opposite direction.

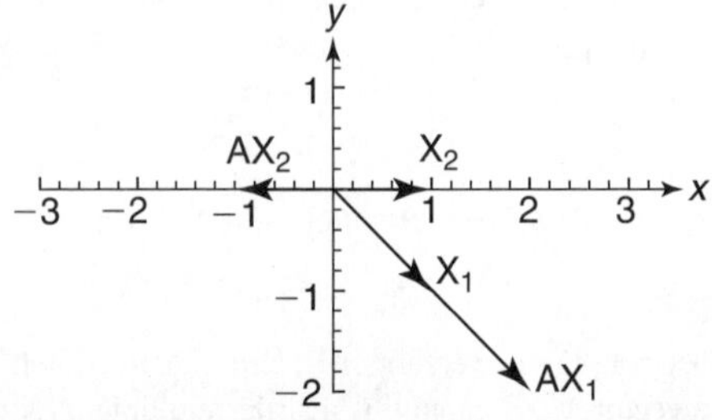

Fig. 4.1

REMARK: There is exactly one eigenvalue corresponding to a given eigenvector **X**. In fact, if we have $T(\mathbf{X}) = \lambda\mathbf{X}$ and $T(\mathbf{X}) = \mu\mathbf{X}$ for some $\mathbf{X} \neq \mathbf{O}$, then $\lambda\mathbf{X} = \mu\mathbf{X}$, so $\lambda = \mu$. Corresponding to a given eigenvalue, there can be many eigenvectors.

EXAMPLE 4.3 Let $\mathbf{A}$ be the identity matrix $\mathbf{I}_n$. Then

$$\mathbf{AX} = \mathbf{I}_n\mathbf{X} = 1\mathbf{X}, \text{ for every non-zero vector } \mathbf{X}.$$

Thus, the only eigenvalue is $\lambda = 1$ and every non-zero vector in $\mathbf{R}^n$ is an eigenvector of $\mathbf{A}$ associated with eigenvalue $\lambda = 1$.

In the following example we develop a general technique of finding eigenvalues and associated eigenvectors of any given $n \times n$ matrix.

EXAMPLE 4.4 Let $\mathbf{A} = \begin{bmatrix} 1 & 2 \\ 4 & 3 \end{bmatrix}$. Find the eigenvalues of $\mathbf{A}$ and their associated eigenvectors.

SOLUTION

We wish to find all real numbers λ and all non-zero vectors $\mathbf{X} = \begin{bmatrix} x_1 \\ x_2 \end{bmatrix}$ satisfying the equation

$$\mathbf{AX} = \lambda\mathbf{X}$$

or

$$\begin{bmatrix} 1 & 2 \\ 4 & 3 \end{bmatrix}\begin{bmatrix} x_1 \\ x_2 \end{bmatrix} = \lambda\begin{bmatrix} x_1 \\ x_2 \end{bmatrix} \tag{1}$$

Using $\lambda\begin{bmatrix} x_1 \\ x_2 \end{bmatrix} = \lambda\begin{bmatrix} 1 & 0 \\ 0 & 1 \end{bmatrix}\begin{bmatrix} x_1 \\ x_2 \end{bmatrix}$, we rewrite Equation (1) in the form

$$\begin{bmatrix} 1 & 2 \\ 4 & 3 \end{bmatrix}\begin{bmatrix} x_1 \\ x_2 \end{bmatrix} - \lambda\begin{bmatrix} 1 & 0 \\ 0 & 1 \end{bmatrix}\begin{bmatrix} x_1 \\ x_2 \end{bmatrix} = \begin{bmatrix} 0 \\ 0 \end{bmatrix}$$

Using the distribution property of matrix multiplication, we have

$$\left(\begin{bmatrix} 1 & 2 \\ 4 & 3 \end{bmatrix} - \lambda\begin{bmatrix} 1 & 0 \\ 0 & 1 \end{bmatrix}\right)\begin{bmatrix} x_1 \\ x_2 \end{bmatrix} = \begin{bmatrix} 0 \\ 0 \end{bmatrix} \tag{2}$$

Equation (2) represents a linear system with a matrix of coefficients given by

$$\mathbf{A} - \lambda\mathbf{I} = \begin{bmatrix} 1 & 2 \\ 4 & 3 \end{bmatrix} - \lambda\begin{bmatrix} 1 & 0 \\ 0 & 1 \end{bmatrix} = \begin{bmatrix} 1-\lambda & 2 \\ 4 & 3-\lambda \end{bmatrix}$$

Also, Equation (2) may be written as

$$(1-\lambda)x_1 + 2x_2 = 0$$
$$4x_1 + (3-\lambda)x_2 = 0$$

This is a homogeneous system and has a non-trivial solution if and only if the matrix of coefficients $\mathbf{A} - \lambda\mathbf{I}$ is singular.

That is,

$$|\mathbf{A} - \lambda\mathbf{I}| = 0$$

or

$$\begin{vmatrix} 1-\lambda & 2 \\ 4 & 3-\lambda \end{vmatrix} = 0$$

This means that

$$(3-\lambda)(1-\lambda) - 4 \cdot 2 = 0$$

or

$$\lambda^2 - 4\lambda - 5 = 0 \tag{3}$$

The solutions of Equation (3) are given by

$$\lambda_1 = -1 \text{ and } \lambda_2 = 5.$$

We now have two different eigenvalues of the matrix $\begin{bmatrix} 1 & 2 \\ 4 & 3 \end{bmatrix}$ given by $\lambda = -1$ and $\lambda = 5$. In order to determine the associated eigenvectors, we use Equation (1) first with $\lambda = -1$ and then with $\lambda = 5$.

For $\lambda = -1$, we obtain the system

$$\begin{bmatrix} 1 & 2 \\ 4 & 3 \end{bmatrix}\begin{bmatrix} x_1 \\ x_2 \end{bmatrix} = -\begin{bmatrix} x_1 \\ x_2 \end{bmatrix}$$

This gives

$$\begin{aligned} x_1 + 2x_2 &= -x_1 \\ 4x_1 + 3x_2 &= -x_2 \end{aligned}$$

or

$$\begin{aligned} 2x_1 + 2x_2 &= 0 \\ 4x_1 + 4x_2 &= 0 \end{aligned} \qquad (4)$$

A non-trivial solution of this system is $\begin{bmatrix} x_1 \\ x_2 \end{bmatrix} = \begin{bmatrix} 1 \\ -1 \end{bmatrix}$. Thus, the vector $\begin{bmatrix} 1 \\ -1 \end{bmatrix}$ is an eigenvector of matrix $\begin{bmatrix} 1 & 2 \\ 4 & 3 \end{bmatrix}$ corresponding to the eigenvalue $\lambda = -1$.

Substituting $\lambda = 5$ into Equation (1), we obtain

$$\begin{bmatrix} 1 & 2 \\ 4 & 3 \end{bmatrix}\begin{bmatrix} x_1 \\ x_2 \end{bmatrix} = 5\begin{bmatrix} x_1 \\ x_2 \end{bmatrix}$$

which gives

$$\begin{aligned} x_1 + 2x_2 &= 5x_1 \\ 4x_1 + 3x_2 &= 5x_2 \end{aligned}$$

or

$$\left.\begin{aligned} -4x_1 + 2x_2 &= 0 \\ 4x_1 - 2x_2 &= 0 \end{aligned}\right\} \qquad (5)$$

A non-zero solution of this last system is given by $\begin{bmatrix} x_1 \\ x_2 \end{bmatrix} = \begin{bmatrix} 1 \\ 2 \end{bmatrix}$. Thus, the vector $\begin{bmatrix} 1 \\ 2 \end{bmatrix}$ is an eigenvector of the matrix $\begin{bmatrix} 1 & 2 \\ 4 & 3 \end{bmatrix}$ corresponding to the eigenvalue $\lambda = 5$.

REMARK: It must be noted that the vector $\begin{bmatrix} x_1 \\ x_2 \end{bmatrix} = \begin{bmatrix} 1 \\ -1 \end{bmatrix}$ is not a unique solution of system (4). It can be easily seen that all the vectors having the form $k\begin{bmatrix} 1 \\ -1 \end{bmatrix}$ are also solutions of system (4). Thus all the vectors $k\begin{bmatrix} 1 \\ -1 \end{bmatrix}$, where $k \neq 0$, can serve as eigenvectors of the matrix $\begin{bmatrix} 1 & 2 \\ 4 & 3 \end{bmatrix}$ associated with the eigenvalue $\lambda = -1$.

A similar remark holds for the vector $\begin{bmatrix} 1 \\ 2 \end{bmatrix}$ in the example above.

Guided by the above example, we shall now derive a general procedure for finding the eigenvalues and their associated eigenvectors for $n \times n$ matrices.

Let $\mathbf{A}$ be an $n \times n$ matrix. A non zero vector $\mathbf{X}$ is an eigenvector of $\mathbf{A}$ corresponding to the eigenvalue λ if it satisfies the equation

$$\mathbf{AX} = \lambda\mathbf{X}$$

or

$$\mathbf{AX} - \lambda\mathbf{X} = 0 \tag{1}$$

We can write $\mathbf{X} = \mathbf{I}_n\mathbf{X}$ where $\mathbf{I}_n$ is the identity matrix of order n.

So Equation (1) gives us

$$(\mathbf{A} - \lambda\mathbf{I}_n)\mathbf{X} = 0$$

This matrix equation represents a system of homogeneous linear equations having matrix of coefficient $\mathbf{A} - \lambda\mathbf{I}_n$. This system has non-zero solution if and only if $\mathbf{A} - \lambda\mathbf{I}_n$ is singular, that is,

$$\det(\mathbf{A} - \lambda\mathbf{I}_n) = 0$$

or

$$|\mathbf{A} - \lambda\mathbf{I}_n| = \mathbf{0}$$

The expansion of this determinant results in a polynomial $f(\lambda)$ of degree n in λ. The zeros of the polynomial are the eigenvalues of the matrix $\mathbf{A}$. This leads to the following definition.

> **? DEFINITION:** Let $\mathbf{A}$ be an $n \times n$ matrix. The determinant $|\mathbf{A} - \lambda\mathbf{I}_n|$ is called the 'characteristic polynomial' of the matrix $\mathbf{A}$, and the equation $|\mathbf{A} - \lambda\mathbf{I}_n| = 0$ is called the 'characteristic equation' of the matrix $\mathbf{A}$.

The equation $f(\lambda) = |\mathbf{A} - \lambda\mathbf{I}_n| = 0$ is an nth degree equation and, thus, from Fundamental Theorem of algebra, it has n roots. The roots may be real or complex and also not all these roots may be different. It is clear that the eigenvalues of $\mathbf{A}$ are the real roots of characteristic equation $f(\lambda) = 0$. Therefore, the eigenvalues are sometimes also known as characteristic roots.

In the following, we shall write $\mathbf{I}$ in place of $\mathbf{I}_2$, $\mathbf{I}_3$, or $\mathbf{I}_n$ for our convenience.

EXAMPLE 4.5 Find the characteristic polynomial of a 2×2 matrix.

SOLUTION

Let $\mathbf{A} = \begin{bmatrix} a_{11} & a_{12} \\ a_{21} & a_{22} \end{bmatrix}$ be any 2×2 matrix.

$$f(\lambda) = |\mathbf{A} - \lambda\mathbf{I}|$$

Here

$$\mathbf{A} - \lambda\mathbf{I} = \begin{bmatrix} a_{11} & a_{12} \\ a_{21} & a_{22} \end{bmatrix} - \lambda\begin{bmatrix} 1 & 0 \\ 0 & 1 \end{bmatrix}$$

$$= \begin{bmatrix} a_{11} - \lambda & a_{12} \\ a_{21} & a_{22} - \lambda \end{bmatrix}$$

So

$$|\mathbf{A} - \lambda\mathbf{I}| = \begin{vmatrix} a_{11} - \lambda & a_{12} \\ a_{21} & a_{22} - \lambda \end{vmatrix}$$

Expanding the determinant, we get

$$f(\lambda) = \lambda^2 - (a_{11} + a_{22})\lambda + (a_{11}a_{22} - a_{12}a_{21}) \tag{1}$$

Note that the constant term in Equation (1) is exactly $|\mathbf{A}|$.

EXAMPLE 4.6 Find the characteristic equation of the matrix $\begin{bmatrix} 1 & 2 & 0 \\ 2 & 2 & 2 \\ 0 & 2 & 3 \end{bmatrix}$

SOLUTION

$$f(\lambda) = |\mathbf{A} - \lambda\mathbf{I}| = \begin{vmatrix} 1-\lambda & 2 & 0 \\ 2 & 2-\lambda & 2 \\ 0 & 2 & 3-\lambda \end{vmatrix}$$

$$= (1 - \lambda)(\lambda^2 - 5\lambda + 2) - 2(6 - 2\lambda)$$

$$= \lambda^3 - 6\lambda^2 + 3\lambda + 10$$

So, the characteristic equation is

$$f(\lambda) = 0$$

or

$$\lambda^3 - 6\lambda^2 + 3\lambda + 10 = 0$$

EXAMPLE 4.7 Find the characteristic roots of the matrix **A** where $\mathbf{A} = \begin{bmatrix} 1 & 2 & 3 \\ 0 & -4 & 2 \\ 0 & 0 & 7 \end{bmatrix}$

SOLUTION

$$|\mathbf{A} - \lambda\mathbf{I}| = \begin{vmatrix} 1-\lambda & 2 & 3 \\ 0 & -4-\lambda & 2 \\ 0 & 0 & 7-\lambda \end{vmatrix}$$

$$= (1 - \lambda)(-4 - \lambda)(7 - \lambda)$$

Characteristic equation is $|\mathbf{A} - \lambda\mathbf{I}| = 0$

$$(1 - \lambda)(-4 - \lambda)(7 - \lambda) = 0$$

which gives

$$\lambda = 1 \quad \text{or} \quad \lambda = -4 \quad \text{or} \quad \lambda = 7$$

Thus the characteristic roots of the matrix A are 1, –4 and 7.

The discussion before example 4.5 leads to the following general method of finding eigenvalus and their associated eigenvectors for a given $n \times n$ matrix.

Step 1. Construct the characteristic equation of **A**, namely, $|\mathbf{A} - \lambda\mathbf{I}_n| = 0$.

Step 2. Find all the roots of the equation $|\mathbf{A} - \lambda\mathbf{I}_n| = 0$. The roots obtained are the eigenvalues of the matrix **A**.

Step 3. For each eigenvalue λ found in Step 2, solve the system $\mathbf{AX} = \lambda\mathbf{X}$ to obtain the eigenvectors **X** corresponding to λ.

EXAMPLE 4.8 Let $\mathbf{A} = \begin{bmatrix} 1 & 2 & -1 \\ 1 & 0 & 1 \\ 4 & -4 & 5 \end{bmatrix}$. Find the eigenvalues and the corresponding eigenvectors of the matrix **A**.

SOLUTION

The characteristic equation of **A** is given by

$$|\mathbf{A} - \lambda \mathbf{I}| = 0$$

or
$$\begin{vmatrix} 1-\lambda & 2 & -1 \\ 1 & -\lambda & 1 \\ 4 & -4 & 5-\lambda \end{vmatrix} = 0$$

or
$$\lambda^3 - 6\lambda^2 + 11\lambda - 6 = 0$$

or
$$(\lambda - 1)(\lambda^2 - 5\lambda + 6) = 0$$

or
$$(\lambda - 1)(\lambda - 2)(\lambda - 3) = 0$$

$\therefore \quad \lambda = 1, 2, 3$

The eigenvalues of **A** are then

$$\lambda_1 = 1, \ \lambda_2 = 2, \ \lambda_3 = 3.$$

Eigenvectors corresponding to $\lambda_1 = 1$

To find the eigenvectors corresponding to $\lambda = 1$, we have to solve the system of linear equations

$$\mathbf{AX} = \lambda \mathbf{X} \text{ corresponding to } \lambda = 1.$$

$$(\mathbf{A} - \lambda \mathbf{I})\mathbf{X} = \begin{pmatrix} 1-1 & 2 & -1 \\ 1 & -1 & 1 \\ 4 & -4 & 5-1 \end{pmatrix} \begin{pmatrix} x \\ y \\ z \end{pmatrix} = \begin{pmatrix} 0 \\ 0 \\ 0 \end{pmatrix}$$

or
$$\begin{bmatrix} 0 & 2 & -1 \\ 1 & -1 & 1 \\ 4 & -4 & 4 \end{bmatrix} \begin{bmatrix} x \\ y \\ z \end{bmatrix} = \begin{bmatrix} 0 \\ 0 \\ 0 \end{bmatrix} \qquad (1)$$

Here
$$\begin{vmatrix} 0 & 2 & -1 \\ 1 & -1 & 1 \\ 4 & -4 & 4 \end{vmatrix} = 0$$

So the homogeneous system (1) has infinity many solutions. Taking $z = k$, matrix Equation (1) reduces to the system of equations.

$$2y - k = 0$$
$$x - y + k = 0$$

Solving them we get $x = -\dfrac{1}{2}k, \ y = \dfrac{1}{2}k$.

Thus, a solution of Equation (1) is $\begin{bmatrix} -\frac{1}{2}k \\ \frac{1}{2}k \\ k \end{bmatrix}$ for any real number k.

In particular, for $k = 2$,

$$\mathbf{X}_1 = \begin{bmatrix} -1 \\ 1 \\ 2 \end{bmatrix}$$

is an eigenvector of **A** associated with $\lambda_1 = 1$

Eigenvectors corresponding to $\lambda_2 = 2$

To find the eigenvectors associated with $\lambda_2 = 2$, we form the system

$$(\mathbf{A} - 2\mathbf{I})\mathbf{X} = 0$$

or
$$\begin{bmatrix} -1 & 2 & -1 \\ 1 & -2 & 1 \\ 4 & -4 & 3 \end{bmatrix} \begin{bmatrix} x \\ y \\ z \end{bmatrix} = \begin{bmatrix} 0 \\ 0 \\ 0 \end{bmatrix} \tag{2}$$

A solution is $\begin{bmatrix} -\frac{1}{2}k \\ \frac{1}{4}k \\ k \end{bmatrix}$ for any real number k. (verify)

Thus, for $k = 4$, $\mathbf{X}_2 = \begin{bmatrix} -2 \\ 1 \\ 4 \end{bmatrix}$ is an eigenvector of **A** associated with $\lambda_2 = 2$.

Eigenvectors corresponding to $\lambda_3 = 3$

We can similarly verify that a solution of the system $(\mathbf{A} - 3\mathbf{I})\mathbf{X} = 0$ is $\begin{bmatrix} -\frac{1}{4}k \\ \frac{1}{4}k \\ k \end{bmatrix}$ for any real number k.

Thus, for $k = 4$, $\mathbf{X}_3 = \begin{bmatrix} -1 \\ 1 \\ 4 \end{bmatrix}$ is an eigenvector of **A** associated with $\lambda_3 = 3$.

REMARK: In above example, any vector of the form $\begin{bmatrix} -\frac{1}{2}k \\ \frac{1}{2}k \\ k \end{bmatrix}$ or $k\begin{bmatrix} -1 \\ 1 \\ 2 \end{bmatrix}$ where $k \neq 0$, $k \in R$ is an eigenvector corresponding to $\lambda_1 = 1$. A similar remark can be made for $\lambda_2 = 2$ and $\lambda_3 = 3$.

For $k = 0$, the solution corresponds to the zero vector which we do not regard as an eigenvector. It must be noted, however, that the number zero can be an eigenvalue.

For example, consider the matrix $\mathbf{A} = \begin{bmatrix} 0 & 0 \\ 0 & 1 \end{bmatrix}$ and vector $\mathbf{X} = \begin{bmatrix} 1 \\ 0 \end{bmatrix}$.

Then, $\mathbf{AX} = \begin{bmatrix} 0 & 0 \\ 0 & 1 \end{bmatrix}\begin{bmatrix} 1 \\ 0 \end{bmatrix} = \begin{bmatrix} 0 \\ 0 \end{bmatrix} = 0\begin{bmatrix} 1 \\ 0 \end{bmatrix} = 0\mathbf{X}$.

Thus, $\lambda = 0$ is an eigenvalue of **A** and $\mathbf{X} = \begin{bmatrix} 1 \\ 0 \end{bmatrix}$ is an eigenvector of **A** associated with $\lambda = 0$. In fact, zero will be an eigenvalue of a matrix **A** if and only if the equation $\mathbf{AX} = 0\mathbf{X}$ has a nontrivial solution. But this equation is same as $\mathbf{AX} = 0$, which has a nontrivial solution if and only if **A** is not invertible, that is, it is singular. Thus, we observe the following.

'**0** is an eigenvalue of **A** if and only if **A** is singular'.

In the following examples, we find the eigenvalues and the corresponding eigenvectors for some of the transformations discussed earlier and try to interpret them geometrically.

EXAMPLE 4.9 Let $T : \mathbf{R}^2 \to \mathbf{R}^2$ be the projection onto the x-axis. Find the eigenvalues and the corresponding eigenvectors for T. Interpret them geometrically.

SOLUTION

It is clear that the eigenvalues of T are the roots of the characteristic equation of the standard matrix representation of T.

The projection T is given by $T\left(\begin{bmatrix} x \\ y \end{bmatrix}\right) = \begin{bmatrix} x \\ 0 \end{bmatrix}$ and its standard matrix is $\mathbf{A} = \begin{bmatrix} 1 & 0 \\ 0 & 0 \end{bmatrix}$.

The characteristic equation of **A** is given by

$$|\mathbf{A} - \lambda\mathbf{I}| = 0$$

or

$$\begin{vmatrix} 1-\lambda & 0 \\ 0 & -\lambda \end{vmatrix} = 0$$

or

$$\lambda(1-\lambda) = 0$$

$\Rightarrow$

$$\lambda = 0 \text{ or } \lambda = 1.$$

Thus the two eigenvectors of A are $\lambda_1 = 0$ and $\lambda_2 = 1$.

The eigenvectors corresponding to $\lambda_1 = 0$ are the solutions of the equation

$$(\mathbf{A} - 0\mathbf{I})\mathbf{X} = \mathbf{0}$$

That is,

$$\mathbf{AX} = \mathbf{0}$$

or

$$\begin{pmatrix} 1 & 0 \\ 0 & 0 \end{pmatrix}\begin{pmatrix} x \\ y \end{pmatrix} = \begin{pmatrix} 0 \\ 0 \end{pmatrix}$$

which gives

$$x = 0$$

$$0 \cdot y = 0$$

Solution of this system is $\begin{bmatrix} 0 \\ k \end{bmatrix}$, where k is a real number.

Thus, the eigenvectors corresponding to $\lambda_1 = 0$ are the vectors of the form $\begin{bmatrix} 0 \\ k \end{bmatrix}$ where $k \neq 0$.

It can similarly be seen that the eigenvectors corresponding to $\lambda_2 = 1$ are the vectors of the form $\begin{bmatrix} k \\ 0 \end{bmatrix}$, where k is a non-zero real number.

Geometrical interpretation

If **X** is a vector along x-axis, it is of the form $\begin{bmatrix} k \\ 0 \end{bmatrix}$ for some real number k. And

$$\mathrm{T}\left(\begin{bmatrix} k \\ 0 \end{bmatrix}\right) = \begin{bmatrix} k \\ 0 \end{bmatrix} = 1\begin{bmatrix} k \\ 0 \end{bmatrix}.$$

Geometrically, the projection of any vector along x-axis is the vector itself. Thus, 1 is an eigenvalue of T and the corresponding eigenvectors are the non-zero vectors along x-axis.

If **X** is a vector along y-axis, it is of the form $\begin{bmatrix} 0 \\ k \end{bmatrix}$, for some real number k. And

$$\mathrm{T}\left(\begin{bmatrix} 0 \\ k \end{bmatrix}\right) = \begin{bmatrix} 0 \\ 0 \end{bmatrix} = 0 \cdot \begin{bmatrix} 0 \\ k \end{bmatrix}.$$

In other words, T projects the vectors along y-axis onto the zero vector. It is clear that 0 is an eigenvalue and the non-zero vectors along y-axis are the corresponding eigenvectors.

If **X** is any other vector, that is, **X** is neither parallel to x-axis nor parallel to y-axis, then the projection $\mathrm{T}(\mathrm{X})$ of X on x-axis is not a scalar multiple of the given vector. Thus, such a vector's projection is neither zero nor parallel to x-axis. So it cannot be an eigenvector of T. See Fig. 4.2.

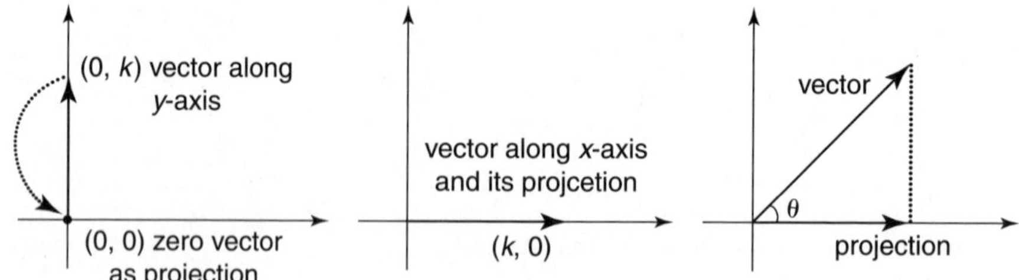

Fig. 4.2

EXAMPLE 4.10 Find the eigenvalues and eigenvectors of the reflection about x-axis in $\mathbf{R}^2$.

SOLUTION

We try to find the eigenvalues and eigenvectors directly from the geometry of the situation. If $\mathrm{T}: \mathbf{R}^2 \to \mathbf{R}^2$ is the reflection along x-axis, then it is given by

$$\mathrm{T}\left(\begin{bmatrix} x \\ y \end{bmatrix}\right) = \begin{bmatrix} x \\ -y \end{bmatrix}$$

Geometrically, we can see that the points on x-axis do not move under the action of T. In other words, the vectors along x-axis are reflected onto themselves. It is also clear from the equation $\mathrm{T}\left(\begin{bmatrix} x \\ 0 \end{bmatrix}\right) = \begin{bmatrix} x \\ 0 \end{bmatrix}$, as the vectors along x-axis are of the form $\begin{bmatrix} x \\ 0 \end{bmatrix}$. Thus, 1 is an eigenvalue of T and the non-zero vectors along x-axis are the corresponding eigenvectors.

If **X** is a vector along y-axis, under reflection about x-axis, its direction is reversed. See Fig. 4.3. This fact is also clear from the equation

$$\mathrm{T}(\mathbf{X}) = \mathrm{T}\left(\begin{bmatrix} 0 \\ y \end{bmatrix}\right) = \begin{bmatrix} 0 \\ -y \end{bmatrix} = -\begin{bmatrix} 0 \\ y \end{bmatrix} = -\mathbf{X}$$

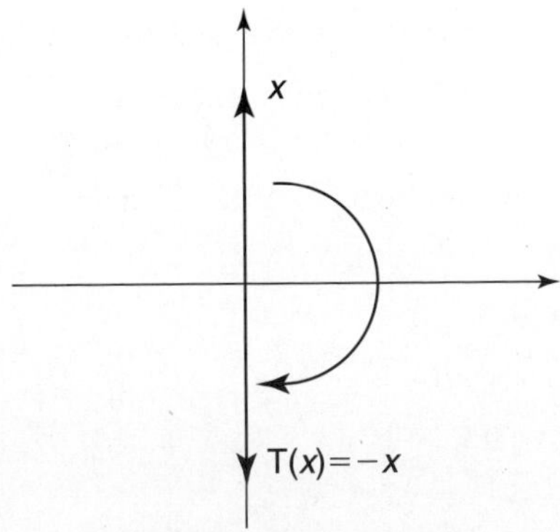

Fig. 4.3 Reflection about x-axis

as vector **X** along y-axis is of the form $\begin{bmatrix} 0 \\ y \end{bmatrix}$. Thus −1 is an eigenvector of T and the corresponding eigenvectos are the non-zero vectors along y-axis.

It can be verified that 1 and −1 are indeed the roots of the characteristic equation of the standard matrix of T.

EXAMPLE 4.11 Let $\mathrm{T} : \mathbf{R}^2 \rightarrow \mathbf{R}^2$ be the transformation given by $\mathrm{T}\left(\begin{bmatrix} x \\ y \end{bmatrix}\right) = \begin{bmatrix} y \\ x \end{bmatrix}$. Find the eigenvalues and eigenvectors of T without finding the standard matrix of T.

SOLUTION

It is clear that T is reflection through the line $y = x$. It is also clear that the points on this line do not move under the action of T. That is, if **X** is a non-zero vector along the line $y = x$, then $\mathrm{T}(\mathbf{X}) = \mathbf{X}$. Then **X** is an eigenvector and 1 is an eigenvalue. If **X** is a vector perpendicular to the line $y = x$, we have $\mathrm{T}(\mathbf{X}) = -\mathbf{X}$ and **X** is an eigenvector corresponding to the eigenvalue $\lambda = -1$. No other vector can be an eigenvector.

You can easily check these results using the standard matrix of T.

EXAMPLE 4.12 Find the eigenvalues and eigenvectors for the transformation $\mathrm{T} : \mathbf{R}^2 \rightarrow \mathbf{R}^2$ given by

$$\mathrm{T}\left(\begin{bmatrix} x \\ y \end{bmatrix}\right) = \begin{bmatrix} 3x \\ 3y \end{bmatrix}$$

SOLUTION

Note that T is dilation by a factor 3. It is clear that if **X** is any vector, then $\mathrm{T}(\mathbf{X}) = 3\mathbf{X}$.

Thus, every non-zero vector $\mathbf{R}^2$ is an eigenvector of T corresponding to the eigenvalue $\lambda = 3$. This fact is also clear if we consider the standard matrix **A** of T. The standard matrix A is given by $\begin{bmatrix} 3 & 0 \\ 0 & 3 \end{bmatrix}$.

The characteristic equation of **A** is

$$|\mathbf{A} - \lambda\mathbf{I}| = 0$$

or
$$\begin{vmatrix} 3-\lambda & 0 \\ 0 & 3-\lambda \end{vmatrix} = 0$$

or
$$(3-\lambda)^2 = 0$$

$\Rightarrow$
$$\lambda = 3$$

Thus 3 is an eigenvalue of A and since the equation $(\mathbf{A} - 3\mathbf{I})\mathbf{X} = \mathbf{O}$ is satisfied by every vector **X**, any non-zero vector of $\mathbf{R}^2$ is an eigenvector corresponding to $\lambda = 3$.

We have seen in Example 4.1 that rotation about 30° in $\mathbf{R}^2$ has no eigenvector. Following example considers the case for 90° rotation.

EXAMPLE 4.13 Let T be the counterclockwise rotation about 90° of $\mathbf{R}^2$. Find the eigenvalues and eigenvector of T. Give its geometrical interpretation.

SOLUTION

The standard matrix representation of rotation about 90° is given by

$$\mathbf{A} = \begin{bmatrix} 0 & -1 \\ 1 & 0 \end{bmatrix}$$

Characteristic equation of **A** is given by

$$|\mathbf{A} - \lambda \mathbf{I}| = 0$$

or

$$\begin{vmatrix} -\lambda & -1 \\ 1 & -\lambda \end{vmatrix} = 0$$

or

$$\lambda^2 + 1 = 0$$

This equation has no real roots.
Thus, **A** has no eigenvalues and hence no eigenvectors.

Geometrically, if **X** is a non-zero vector, $\mathrm{T}(\mathbf{X})$ is orthogonal to **X**. In other words, $\mathrm{T}(\mathbf{X})$ cannot be parallel to **X** if **X** is non-zero. Thus T possesses no eigenvectors.

The next example clarifies the situation for rotations in general.

EXAMPLE 4.14 Find the eigenvalues and eigenvectors for the linear transformation T_θ where T_θ is the counterclockwise rotation of plane about the origin through an angle θ.

SOLUTION

To find the eigenvalues and eigenvectors we must look for the non-zero vectors X and real number λ such that

$$\mathrm{T}_\theta(\mathbf{X}) = \lambda \mathbf{X}.$$

In other words, we must search for the non-zero vectors **X** such that **X** and $\mathrm{T}_\theta(\mathbf{X})$ are parallel. It is clear from the geometry of the situation that the vector **X** and the vector $\mathrm{T}_\theta(\mathbf{X})$ obtained after rotating **X** through an angle θ are parallel only when θ is an integral multiple of π. In fact, if θ is an odd multiple of π, then $\mathrm{T}_\theta(\mathbf{X}) = -\mathbf{X}$ and if θ is an even multiple of π, then $\mathrm{T}_\theta(\mathbf{X}) = \mathbf{X}$. Thus, T_θ has eigenvalues 1 or -1 according as θ is even or odd multiple of π. If θ is not an integral multiple of π, T_θ cannot have any eigenvalues and thus no eigenvectors. See Fig. 4.4.

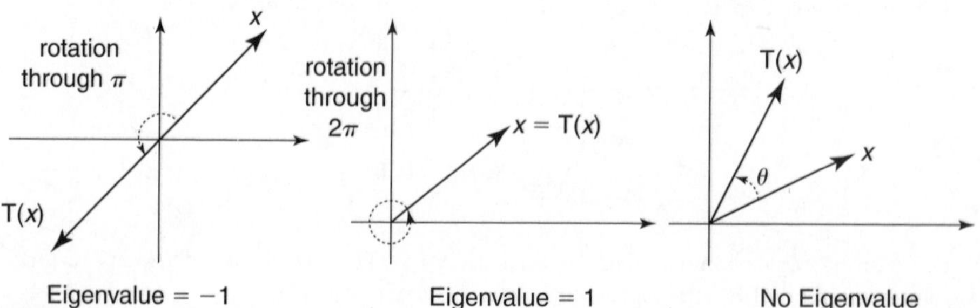

Fig. 4.4

Every non-zero vector is an eigenvector corresponding to the eigenvalues 1 or −1.

Let us now verify the above observations using the standard matrix A of rotation T_θ. We know that

$$\mathbf{A} = \begin{bmatrix} \cos\theta & -\sin\theta \\ \sin\theta & \cos\theta \end{bmatrix}$$

The characteristic equation of **A** is given by

$$|\mathbf{A} - \lambda \mathbf{I}| = 0$$

or
$$\begin{vmatrix} \cos\theta - \lambda & -\sin\theta \\ \sin\theta & \cos\theta - \lambda \end{vmatrix} = 0$$

or
$$(\cos\theta - \lambda)^2 + \sin^2\theta = 0$$

or
$$\lambda^2 - 2\cos\theta\lambda + 1 = 0 \quad (1)$$

Now the eigenvalues of **A** are the real roots of characteristic Equation (1) which has real roots if and only if

$$4\cos^2\theta - 4 \geq 0$$

or
$$\cos^2\theta - 1 \geq 0$$

or
$$-\sin^2\theta \geq 0$$

or
$$\sin^2\theta \leq 0$$

$\Rightarrow$
$$\sin^2\theta = 0 \qquad (\because \sin^2\theta \geq 0)$$

or
$$\sin\theta = 0$$

$\Rightarrow$
$$\theta = k\pi \qquad \text{where } k \text{ is an integer.}$$

Thus Equation (1) has real roots if and only if θ is an integral multiple of π and the real roots are given by $\lambda_1 = \lambda_2 = \cos\theta$ where $\theta = k\pi$, $k \in Z$.

Thus if θ is an odd multiple of π, $\lambda_1 = \lambda_2 = -1$ and if θ is an even multiple of π, $\lambda_1 = \lambda_2 = 1$. The corresponding eigenvectors are all non-zero vectors.

REMARK: Note that θ = 30° in Example 4.1 and θ = 90° in Example 4.13, both are not integral multiples of π and thus the linear transformations have no eigenvectors.

EXAMPLE 4.15 Find the eigenvalues and eigenvectors for reflection in the *xy*-plane in $\mathbf{R}^3$.

SOLUTION

It can easily be seen geometrically that every non-zero vector in the *xy*-plane is an eigenvector with eigenvalue 1. The remaining eigenvectors are the vectors of the form $\begin{bmatrix} 0 \\ 0 \\ z \end{bmatrix}$, $z \neq 0$, that is, non-zero vectors along *z*-axis. Each of them has eigenvalue −1.

EXERCISE 4.1

1. Show that the vectors $\begin{bmatrix} 1 \\ 2 \end{bmatrix}$ and $\begin{bmatrix} 1 \\ -1 \end{bmatrix}$ are eigenvectors for the matrix $\mathbf{A} = \begin{bmatrix} 3 & -1 \\ -2 & 2 \end{bmatrix}$. What are the eigenvalues associated with each of these vectors?
2. Find the characteristic equation, eigenvalues and corresponding eigenvectors for each of the following matrices:

   (a) $\begin{bmatrix} 3 & 4 \\ 2 & 1 \end{bmatrix}$ **(b)** $\begin{bmatrix} 2 & 0 \\ 0 & 0 \end{bmatrix}$ **(c)** $\begin{bmatrix} 3 & 1 \\ 2 & 2 \end{bmatrix}$ **(d)** $\begin{bmatrix} 1 & 1 \\ 1 & 1 \end{bmatrix}$

   (e) $\begin{bmatrix} 0 & 0 \\ 0 & 0 \end{bmatrix}$ **(f)** $\begin{bmatrix} 3 & 0 \\ 0 & 3 \end{bmatrix}$ **(g)** $\begin{bmatrix} 1 & -1 \\ -1 & 1 \end{bmatrix}$ **(h)** $\begin{bmatrix} 2 & \sqrt{2} \\ \sqrt{2} & 1 \end{bmatrix}$

3. Show that the matrix $\begin{bmatrix} 1 & 1 \\ -2 & -1 \end{bmatrix}$ has no eigenvalues.
4. Find the characteristic equation, eigenvalues and corresponding eigenvectors for each of the following matrices:

   (a) $\begin{bmatrix} 2 & 2 & 1 \\ 1 & 3 & 1 \\ 1 & 2 & 2 \end{bmatrix}$ **(b)** $\begin{bmatrix} 2 & 1 & 0 \\ 0 & 2 & 1 \\ 0 & 0 & 2 \end{bmatrix}$ *[DU, B.Sc (Ph.Sc), 2008]* **(c)** $\begin{bmatrix} -1 & 0 & 0 \\ -1 & 3 & 0 \\ 3 & 2 & -2 \end{bmatrix}$

   (d) $\begin{bmatrix} 8 & -6 & 2 \\ -6 & 7 & -4 \\ 2 & -4 & 3 \end{bmatrix}$ **(e)** $\begin{bmatrix} 2 & 2 & 3 \\ 1 & 2 & 1 \\ 2 & -2 & 1 \end{bmatrix}$ **(f)** $\begin{bmatrix} 3 & 0 & 0 \\ 0 & 4 & \sqrt{3} \\ 0 & \sqrt{3} & 6 \end{bmatrix}$

5. Find the eigenvalues and eigenvectors for the following transformations by first finding the characteristic equation from their standard matrix. Give a geometrical interpretation for the eigenvalues and eigenvectors. (Follow the type of arguments used in Example 4.9)
   (a) $T:\mathbf{R}^2 \to \mathbf{R}^2$ is a projection onto the y-axis.
   (b) $T:\mathbf{R}^2 \to \mathbf{R}^2$ is a reflection about the line $y = -x$.
   (c) $T:\mathbf{R}^2 \to \mathbf{R}^2$ is a rotation through $\pi/4$.
   (d) $T:\mathbf{R}^3 \to \mathbf{R}^3$ is a dilation of factor 3.
   (e) $T:\mathbf{R}^3 \to \mathbf{R}^3$ is a reflection about the origin.
6. Use the type of arguments used in Example 4.10, 4.11 and 4.12 to find the eigenvalues and corresponding eigenvectors of the following linear transformations T from the geometry of the situation only. Check your conclusions by calculating the eigenvalues and corresponding eigenvectors from the standard matrix of T
   (a) $T:\mathbf{R}^2 \to \mathbf{R}^2$ is the reflection about y-axis.
   (b) $T:\mathbf{R}^2 \to \mathbf{R}^2$ is a contraction by a factor of 1/2.
   (c) $T:\mathbf{R}^2 \to \mathbf{R}^2$ is a rotation through π.
   (d) $T:\mathbf{R}^3 \to \mathbf{R}^3$ is a projection onto the xy-plane.
   (e) $T:\mathbf{R}^3 \to \mathbf{R}^3$ is a reflection about yz-plane.
7. Show that if **A** is a diagonal matrix or a triangular matrix, then the eigenvalues of **A** are the elements on the main diagonal of **A**.
8. Show that the square matrices **A** and $\mathbf{A}^T$ have the same eigenvalues, However, the same is not true for the corresponding eigenvectors of **A** and $\mathbf{A}^T$. Give an example of a 2×2 matrix such that an eigenvector of **A** is not an eigenvector of $\mathbf{A}^T$.

9. **(a)** If λ is an eigenvalue of a matrix $\mathbf{A}$ corresponding to an eigenvector $\mathbf{X}$, then prove that λ^2 is an eigenvalue of $\mathbf{A}^2$ corresponding to the same eigenvector $\mathbf{X}$.

(b) If λ is an eigenvalue of $\mathbf{A}$, then prove that λ^m is an eigenvalue of $\mathbf{A}^m$ where m is a positive integer.

10. Prove that $\lambda = 0$ is an eigenvalue of $\mathbf{A}$ if and only if $\mathbf{A}$ is singular.

11. If the eigenvalues of a square matrix $\mathbf{A}$ are $k\lambda_1, k\lambda_2, \ldots, k\lambda_n$ then prove that the eigenvalues of $k\mathbf{A}$ are $k\lambda_1, k\lambda_2, \ldots, k\lambda_n$.

12. Let $\mathrm{T}: \mathbf{R}^2 \to \mathbf{R}^2$ be the rotation through an angle of $\pi/2$ radians. Although T has no eigenvectors, prove that every non-zero vector is an eigenvector of T^2. [Hint: We define T^2 as the composition T o T. Thus, if $\mathbf{A}$ is the standard matrix of T, then the standard matrix of T^2 is $\mathbf{A}^2$].

13. Show that it is impossible for any symmetric matrix of the form

$$\mathbf{A} = \begin{bmatrix} a & b \\ b & c \end{bmatrix} \qquad (b \neq 0)$$

to have two identical eigenvalues.

14. Find the condition under which the matrix

$$\mathbf{A} = \begin{bmatrix} a & b \\ c & d \end{bmatrix}$$

has two distinct (real) eigenvalues.

15. The Cayley-Hamilton Theorem states that "Every square matrix satisfies its characteristic equation". That is, if the characteristic equation of a matrix $\mathbf{A}$ is

$$\lambda^n + a_1\lambda^{n-1} + \cdots + a_n = 0 \quad \text{then} \quad \mathbf{A}^n + a_1\mathbf{A}^{n-1} + \cdots + a_n\mathrm{I} = \mathbf{O}$$

Verify the Cayley-Hamilton theorem for the following matrices:

(a) $\begin{bmatrix} 3 & 3 \\ 2 & 4 \end{bmatrix}$ **(b)** $\begin{bmatrix} 0 & 2 \\ -1 & 3 \end{bmatrix}$ **(c)** $\begin{bmatrix} 6 & -8 \\ 4 & -6 \end{bmatrix}$ **(d)** $\begin{bmatrix} 1 & 2 & 3 \\ 0 & 2 & 2 \\ 0 & 0 & -3 \end{bmatrix}$

4.2 EIGEN SPACE

Let $\mathrm{T}: \mathbf{R}^n \to \mathbf{R}^n$ be a linear transformation and λ be any real number. Let S_λ be a subset of $\mathbf{R}^n$ consisting of all vectors $\mathbf{X}$ such that

$$\mathrm{T}(\mathbf{X}) = \lambda\mathbf{X} \tag{1}$$

Then S_λ is a subspace of $\mathbf{R}^n$. In order to prove it, we just need to show that S_λ is closed with respect to vector addition and scalar multiplication. Let $\mathbf{X}, \mathbf{Y} \in \mathrm{S}_\lambda$. Then $\mathrm{T}(\mathbf{X}) = \lambda\mathbf{X}$ and $\mathrm{T}(\mathbf{Y}) = \lambda\mathbf{Y}$.

$$\mathrm{T}(\mathbf{X} + \mathbf{Y}) = \mathrm{T}(\mathbf{X}) + \mathrm{T}(\mathbf{Y}) = \lambda\mathbf{X} + \lambda\mathbf{Y} = \lambda(\mathbf{X} + \mathbf{Y})$$

Thus $\mathbf{X} + \mathbf{Y} \in \mathrm{S}_\lambda$.

If c is a real number, then

$$\mathrm{T}(\mathrm{c}\mathbf{X}) = \mathrm{c}\mathrm{T}(\mathbf{X}) = \mathrm{c}(\lambda\mathbf{X}) = \lambda(\mathrm{c}\mathbf{X}).$$

Thus $\mathrm{c}\mathbf{X} \in \mathrm{S}_\lambda$

Hence, S_λ is a subspace of $\mathbf{R}^n$.

Clearly Equation (1) is satisfied by the zero vector of $\mathbf{R}^n$. And most of the time $\mathrm{S}_\lambda = \{\mathbf{O}\}$. (Indeed, if λ is not an eigenvalue of T).

If λ is an eigenvalue, then Equation (1) is satisfied by non-zero vectors (eigenvectors) also.

In fact, the subspace S_λ consists of the zero vector and all the eigenvectors corresponding to λ. Thus, we have just proved that the set consisting of the eigenvectors of T with eigenvalue λ together with the zero vector forms a subspace of $\mathbf{R}^n$. This subspace is called *eigenspace* of T corresponding to λ.

REMARK: The set consisting of all eigenvectors corresponding to a given eigenvalue cannot form a subspace as every subspace must contain the zero vector. Thus it is necessary to include the zero vector.

Since a matrix represents a linear transformation, we define eigenspace for a matrix as follows:

? DEFINITION: Let **A** be an $n \times n$ matrix and λ an eigenvalue of **A**. The set of all eigenvectors corresponding to λ, together with the zero vector, is a subspace of $\mathbf{R}^n$. This subspace is called the 'eigenspace' of **A** corresponding to λ.

EXAMPLE 4.16 Find the eigenspaces of the matrix $\mathbf{A} = \begin{bmatrix} 3 & 0 \\ 0 & 2 \end{bmatrix}$

SOLUTION

The characteristic equation of matrix **A** is given by

$$\begin{vmatrix} 3 - \lambda & 0 \\ 0 & 2 - \lambda \end{vmatrix} = 0$$

which gives $\lambda = 2$ and $\lambda = 3$.

Thus, the eigenvalues of **A** are $\lambda_1 = 3$ and $\lambda_2 = 2$. The eigenvectors associated with $\lambda_1 = 3$ are given by the equation

$$\begin{pmatrix} 3 - 3 & 0 \\ 0 & 2 - 3 \end{pmatrix} \begin{pmatrix} x \\ y \end{pmatrix} = \begin{pmatrix} 0 \\ 0 \end{pmatrix}$$

A solution of this equation is $x = k$ and $y = 0$ where k is any real number.

Thus the eigenvectors of **A** corresponding to $\lambda_1 = 3$ are the vectors of the form $\begin{bmatrix} k \\ 0 \end{bmatrix}$, where k is a non-zero real number.

The eigenspace corresponding to $\lambda_1 = 3$ is the set of vectors of the form $\begin{bmatrix} k \\ 0 \end{bmatrix}$, where k is a real number. Notice that geometrically it represents x-axis . We can similarly show that the eigenvectors corresponding to $\lambda_2 = 2$ are the vectors of the form $\begin{bmatrix} 0 \\ k \end{bmatrix}$ where $k \neq 0$. Thus the eigenspace corresponding to $\lambda_2 = 2$ is the set of vectors of the form $\begin{bmatrix} 0 \\ k \end{bmatrix}$ where k is a real number. Again notice that any such vector lies along y-axis.

It must be noted that the vectors $\begin{bmatrix} k \\ 0 \end{bmatrix}$ under the transformation represented by the matrix $\mathbf{A} = \begin{bmatrix} 3 & 0 \\ 0 & 2 \end{bmatrix}$ is mapped to the vector $\begin{bmatrix} 3k \\ 0 \end{bmatrix}$. In fact,

$$\begin{bmatrix} 3 & 0 \\ 0 & 2 \end{bmatrix} \begin{bmatrix} k \\ 0 \end{bmatrix} = 3 \begin{bmatrix} k \\ 0 \end{bmatrix}$$

The transformed vector $\begin{bmatrix} 3k \\ 0 \end{bmatrix}$ is also a vector lying along x-axis. Thus the vectors along the x-axis (that is, the vectors of the eigenspace corresponding to $\lambda_1 = 3$) are mapped onto the x-axis itself. Similarly the vectors of the subspace y-axis (that is, the eigenspace corresponding to $\lambda_2 = 2$) are mapped onto the y-axis itself. Such subspaces which, under the given transformation, are mapped onto themselves are called invariant subsapces. We have thus shown that the eigenspaces of matrix A are invariant subspaces. This is an important property of eigenspaces which we shall discuss in Example 4.17 below.

> ? **DEFINITION:** A subspace U of $\mathbf{R}^n$ is called 'invariant' under a linear transformation T if T maps each element of U onto an element of U.
> In other words, U is invariant under T if $\mathbf{X} \in U \Rightarrow T(\mathbf{X}) \in U$, that is, $T(U) \subseteq U$.

EXAMPLE 4.17 Let $T : \mathbf{R}^n \rightarrow \mathbf{R}^n$ be a linear transformation and λ be an eigenvalue of T. Then the eigenspace corresponding to λ is an invariant subspace of $\mathbf{R}^n$.

PROOF

Let $\mathbf{X}$ be a vector in the eigenspace corresponding to λ. Then $T(\mathbf{X}) = \lambda\mathbf{X}$. If $\mathbf{X}$ is a zero vector, then $T(\mathbf{X}) = \mathbf{O}$ and hence belongs to the eigenspace. If $\mathbf{X} \neq \mathbf{O}$, $\mathbf{X}$ is an eigenvector of T. Then $\mathbf{Y} = \lambda\mathbf{X}$ is also an eigenvector of T because

$$T(\mathbf{Y}) = T(\lambda\mathbf{X}) = \lambda T(\mathbf{X}) = \lambda(\lambda\mathbf{X}) = \lambda\mathbf{Y}.$$

Thus, $\lambda\mathbf{X} = T(\mathbf{X})$ belongs to the eigenspace. Thus, eigenspace corresponding to λ is an invariant subsapce of $\mathbf{R}^n$.

Alternately we can also use the following argument:
Since eigenspace is a subspace of $\mathbf{R}^n$, it is closed with respect to scalar multiplication and thus if $\mathbf{X}$ is an element of eigenspace, $\lambda\mathbf{X}$ is also an element of eigenspace.

The matrix $\mathbf{A} = \begin{bmatrix} 3 & 0 \\ 0 & 2 \end{bmatrix}$ in Example 4.16 is a magnification of the plane and under this transformation x-axis and y-axis are mapped onto themselves respectively. They are thus the invariant eigenspaces of $\mathbf{R}^2$. See Fig. 4.5 below.

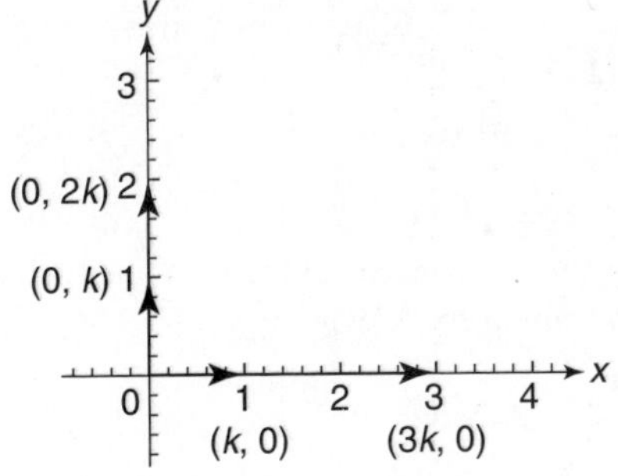

Fig. 4.5

EXAMPLE 4.18 Determine the eigenspaces of the matrix

$$\mathbf{A} = \begin{bmatrix} -4 & -6 \\ 3 & 5 \end{bmatrix}$$

SOLUTION

The characteristic equation of **A** is given by

$$\lambda^2 - \lambda - 2 = 0$$

and the eigenvalues of **A** are $\lambda = 2$ or -1 (verify).

The corresponding eigenvectors are the non-zero solutions of the equation $(\mathbf{A} - \lambda\mathbf{I})\mathbf{X} = \mathbf{O}$ for these values of λ. For $\lambda = 2$, we get

$$\begin{bmatrix} -6 & -6 \\ 3 & 3 \end{bmatrix} \begin{bmatrix} x \\ y \end{bmatrix} = \begin{bmatrix} 0 \\ 0 \end{bmatrix}$$

which gives $x = -y$.

The solution of this system of equations are $x = -r$, $y = r$ where r is a scalar. Thus the eigenvectors of **A** corresponding to $\lambda = 2$ are non-zero vectors of the form

$$r\begin{bmatrix} -1 \\ 1 \end{bmatrix}$$

The set of these eigenvectors together with the zero vector is the eigenspace corresponding to $\lambda = 2$. It is a one-dimensional subspace of $\mathbf{R}^2$ with basis $\left\{\begin{bmatrix} -1 \\ 1 \end{bmatrix}\right\}$. Geometrically, it is a line through $(-1, 1)$ and the origin. We can similarly show that the eigenspace corresponding to $\lambda = -1$ is the set of all multiples of $\begin{bmatrix} -2 \\ 1 \end{bmatrix}$ which is a line through $(-2, 1)$ and the origin. See Fig. 4.6.

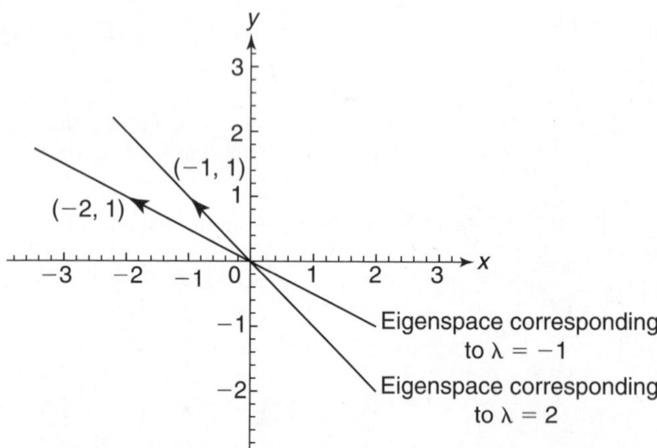

Fig. 4.6

EXAMPLE 4.19 Determine the invariant subspaces of the xy-plane (that is, $\mathbf{R}^2$) under a reflection with respect to x-axis represented by the matrix **A**, where

$$\mathbf{A} = \begin{bmatrix} 1 & 0 \\ 0 & -1 \end{bmatrix}$$

SOLUTION

The eigenvalues of $\mathbf{A}$ are $\lambda_1 = 1$ and $\lambda_2 = -1$ corresponding to $\lambda_1 = 1$, the eigenvectors are the non-zero solutions of the equation

$$\begin{bmatrix} 1 & 0 \\ 0 & -1 \end{bmatrix}\begin{bmatrix} x \\ y \end{bmatrix} = \begin{bmatrix} x \\ y \end{bmatrix}$$

which reduces to the system of equations

$$x = x$$
$$-y = y$$

A solution of this system is of the form $\begin{bmatrix} k \\ 0 \end{bmatrix}$, where k is a real number. So the eigenspace corresponding to $\lambda_1 = 1$, is the set of vectors of the form $\begin{bmatrix} k \\ 0 \end{bmatrix}$, where k is a real number.

The eigenvectors associated with $\lambda_2 = -1$ are given by the equation

$$\begin{bmatrix} 1 & 0 \\ 0 & -1 \end{bmatrix}\begin{bmatrix} x \\ y \end{bmatrix} = -\begin{bmatrix} x \\ y \end{bmatrix}$$

which gives

$$x = -x$$
$$-y = -y$$

A solution of this system of equations is $x = 0$, $y = k$ where k is a real number. So the eigenspace corresponding to $\lambda_2 = -1$ is the set of vectors of the form $\begin{bmatrix} 0 \\ k \end{bmatrix}$ where k is a real number.

Thus the invariant subspaces of $\mathbf{R}^2$ under a reflection about x-axis are the one dimensional vector spaces containing the sets of vectors of the form $\begin{bmatrix} k \\ 0 \end{bmatrix}$ and $\begin{bmatrix} 0 \\ k \end{bmatrix}$ respectively. Geometrically, they represent x- and y-axes respectively.

EXAMPLE 4.20 Find a basis for each eigenspace of the matrix $\mathbf{A} = \begin{bmatrix} 2 & 1 & 1 \\ 2 & 3 & 2 \\ 3 & 3 & 4 \end{bmatrix}$,

$\mathbf{B} = \begin{bmatrix} 2 & -1 & 1 \\ 0 & 3 & 1 \\ 2 & 1 & 3 \end{bmatrix}$, and $\mathbf{C} = \begin{bmatrix} 1 & 1 & 0 \\ 0 & 1 & 0 \\ 1 & 2 & 1 \end{bmatrix}$.

SOLUTION

The characteristic equation of $\mathbf{A}$ is given by

$$|\mathbf{A} - \lambda \mathrm{I}| = 0$$

which gives

$$(\lambda - 1)(\lambda - 1)(\lambda - 7) = 0$$

That is, $\lambda = 1, 1, 7$.

So, the two distinct eigenvalues of **A** are 1 and 7. When $\lambda = 7$, the system $(\mathbf{A} - 7\mathbf{I})\mathbf{X} = \mathbf{O}$ becomes

$$\begin{aligned} -5x + y + z &= 0 \\ 2x - 4y + 2z &= 0 \\ 3x + 3y - 3z &= 0 \end{aligned}$$

The solution of this system is given by

$$x = k, y = 2k, z = 3k \text{ where } k \text{ is a real number.}$$

So, the eigenvectors corresponding to $\lambda = 7$ are the vectors of the form $k\begin{bmatrix} 1 \\ 2 \\ 3 \end{bmatrix}$ where $k \neq 0$. The eigenspace corresponding to $\lambda = 7$ is the set of all vectors of the form $k\begin{bmatrix} 1 \\ 2 \\ 3 \end{bmatrix}$ where k is a real number. So the set $\left\{\begin{bmatrix} 1 \\ 2 \\ 3 \end{bmatrix}\right\}$ acts as a basis for this eigenspace and the dimensions of the eigenspace is 1.

For $\lambda = 1$, the system $(\mathbf{A} - \mathbf{I})\mathbf{X} = \mathbf{O}$ becomes

$$\begin{aligned} x + y + z &= 0 \\ 2x + 2y + 2z &= 0 \\ 3x + 3y + 3z &= 0 \end{aligned}$$

which gives a single equation

$$x + y + z = 0$$

To solve this equation we take $x = p$ and $y = q$, where p and q are arbitrary, and then take $z = -p - q$. Thus every eigenvector corresponding to $\lambda = 1$ has the form

$$\begin{bmatrix} p \\ q \\ -p - q \end{bmatrix} = p\begin{bmatrix} 1 \\ 0 \\ -1 \end{bmatrix} + q\begin{bmatrix} 0 \\ 1 \\ -1 \end{bmatrix} \tag{1}$$

where $p \neq 0$ and $q \neq 0$. The eigenspace corresponding to $\lambda = 1$ is the set of all vectors $\begin{bmatrix} p \\ q \\ -p - q \end{bmatrix}$ where $p, q \in \mathbf{R}$. From (1) it is clear that every vector of such form is a linear combination of the two linearly independent vectors $\begin{bmatrix} 1 \\ 0 \\ -1 \end{bmatrix}$ and $\begin{bmatrix} 0 \\ 1 \\ -1 \end{bmatrix}$. This means $\left\{\begin{bmatrix} 1 \\ 0 \\ -1 \end{bmatrix}, \begin{bmatrix} 0 \\ 1 \\ -1 \end{bmatrix}\right\}$ is a basis of this eigenspace. Hence dimension of the eigenspace corresponding to $\lambda = 1$ is 2.

REMARK: In a similar way, one may proceed for **B** and **C**. Note that the matrix

$\mathbf{B} = \begin{bmatrix} 2 & -1 & 1 \\ 0 & 3 & -1 \\ 2 & 1 & 3 \end{bmatrix}$ also admits two distinct eigenvalues $\lambda = 2$ and $\lambda = 4$.

The eigenvalue $\lambda = 2$ is repeated twice but the eigenspace corresponding to $\lambda = 2$ is of dimension one.

The matrix $\mathbf{C} = \begin{bmatrix} 1 & 1 & 0 \\ 0 & 1 & 0 \\ 1 & 2 & 1 \end{bmatrix}$ has three distinct eigenvalues $\lambda = 0, 1, 2$ and the corresponding eigenspaces are one-dimensional each. Verify these results for the matrices **B** and **C** yourself.

EXERCISE 4.2

1. Let **A** be an $n \times n$ matrix. Prove that the set of all eigenvectors corresponding to an eigenvalue λ of **A** forms a subspace of $\mathbf{R}^n$. What is this subspace called?
2. Find the eigenspaces of the matrices:

   (a) $\mathbf{A} = \begin{bmatrix} -3 & -3 \\ 2 & 2 \end{bmatrix}$ **(b)** $\mathbf{B} = \begin{bmatrix} 5 & 0 \\ 2 & 1 \end{bmatrix}$

   (c) $\mathbf{C} = \begin{bmatrix} 5 & 4 \\ 1 & 2 \end{bmatrix}$ **(d)** $\mathbf{D} = \begin{bmatrix} 1 & 2 \\ 2 & 1 \end{bmatrix}$

   (e) $\mathbf{E} = \begin{bmatrix} 1 & 0 \\ 0 & 1 \end{bmatrix}$ **(f)** $\mathbf{F} = \begin{bmatrix} 1 & 1 \\ 1 & 1 \end{bmatrix}$
3. Find the invariant vector subspaces (eigenspaces) under the following transformations of the plane

   (a) shear parallel to x-axis represented by $\begin{pmatrix} 1 & 2 \\ 0 & 1 \end{pmatrix}$

   (b) shear parallel to y-axis represented by $\begin{pmatrix} 1 & 0 \\ 3 & 1 \end{pmatrix}$

   (c) reflection about the line $y = x$ represented by $\begin{pmatrix} 0 & 1 \\ 1 & 0 \end{pmatrix}$

   (d) projection onto the x-axis represented by $\begin{pmatrix} 1 & 0 \\ 0 & 0 \end{pmatrix}$

   (e) projection onto the line $y = x$ represented by $\begin{pmatrix} 1 & 0 \\ 1 & 0 \end{pmatrix}$

   (f) dilation of factor 2 represented by $\begin{pmatrix} 2 & 0 \\ 0 & 2 \end{pmatrix}$
4. Prove that, in general, no invariant subspaces exists under a rotation of the xy- plane (that is, $\mathbf{R}^2$) about the origin. [Hint. The invariant subspaces exist only when θ is an integral multiple of π.]

5. Find a basis and dimension for each eigenspace of the following matrices:

(a) $\begin{bmatrix} 2 & 1 & 1 \\ 2 & 3 & 4 \\ -1 & -1 & -2 \end{bmatrix}$ **(b)** $\begin{bmatrix} 5 & 4 & 2 \\ 4 & 5 & 2 \\ 2 & 2 & 2 \end{bmatrix}$

(c) $\begin{bmatrix} 2 & -1 & 0 \\ -1 & 2 & 0 \\ 0 & 0 & 3 \end{bmatrix}$ **(d)** $\begin{bmatrix} 0 & 0 & -2 \\ 1 & 2 & 1 \\ 1 & 0 & 3 \end{bmatrix}$

4.3 DIAGONALIZATION

We begin this section by stating the eigenvector problem: Given an $n \times n$ matrix **A**, does there exist a basis for $\mathbf{R}^n$ consisting of eigenvectors of **A**? If such a basis exists, then many numerical computations involving **A** can be simplified. Moreover, such basis have numerous applications in various branches of mathematics.

Our main aim in this section will be to discuss the following equivalent form of the eigenvector problem, called the matrix diagonalization problem and try to establish their equivalence.

The Diagonalization Problem: Given an $n \times n$ matrix **A**, does there exist an invertible matrix **P** such that $P^{-1}AP$ is a diagonal matrix?

The above problem suggests the following definitions.

? **DEFINITION:** A square matrix **A** is called 'diagonalizable' if there exists an invertible matrix **P** such that $\mathbf{P}^{-1}\mathbf{AP}$ is a diagonal matrix. The matrix **P** is said to 'diagonalize' **A**.

? **DEFINITION:** A matrix **B** is said to be similar to a matrix **A** if there exists an invertible matrix **P** such that $\mathbf{B} = \mathbf{P}^{-1}\mathbf{AP}$.

The two definitions together imply the following:

A square matrix **A** is diagonalizable if it is similar to a diagonal matrix.

It must be noted that not every square matrix can be diagonalized. In other words, the statement that every square matrix is similar to a diagonal matrix is not true. Our aim therefore is to find the condition under which a nonsingular matrix **P** exists such that $\mathbf{P}^{-1}\mathbf{AP}$ is diagonal. Moreover if such a matrix **P** exists, then how does one find it. Before answering this, we state the following two important results.

THEOREM 1

A square matrix **A** is invertible if and only if the columns of **A** form a linearly independent set.

THEOREM 2

If the eigenvalues of a matrix are distinct, then the associated eigenvectors are linearly independent.

We prove Theorem 2 for a 2×2 matrix having two distinct eigenvalues.

Let **A** be a 2×2 matrix and λ_1, λ_2 be the two distinct eigenvalues of **A**. Let $\mathbf{X}_1$ and $\mathbf{X}_2$ be the eigenvectors corresponding to the eigenvalues λ_1 and λ_2 respectively. We have to show that $\{\mathbf{X}_1, \mathbf{X}_2\}$ is linearly independent.

Consider $$c_1\mathbf{X}_1 + c_2\mathbf{X}_2 = \mathbf{O} \tag{1}$$

Premultiplying both sides of (1) by **A**.

$$c_1\mathbf{A}\mathbf{X}_1 + c_2\mathbf{A}\mathbf{X}_2 = \mathbf{O} \tag{2}$$

Since $\mathbf{X}_1$ is an eigenvector of **A** corresponding to the eigenvalue λ_1, we have

$$\mathbf{A}\mathbf{X}_1 = \lambda_1\mathbf{X}_1.$$

Similarly $$\mathbf{A}\mathbf{X}_2 = \lambda_2\mathbf{X}_2$$

$\therefore$ (2) becomes

$$c_1\lambda_1\mathbf{X}_1 + c_2\lambda_2\mathbf{X}_2 = \mathbf{O} \tag{3}$$

Multiplying equation (1) by λ_1 and subtracting from (3), we get

$$c_2(\lambda_2 - \lambda_1)\mathbf{X}_2 = \mathbf{O}$$

Since $\mathbf{X}_2$ is an eigenvector, it cannot be zero.

$$\therefore \quad c_2(\lambda_2 - \lambda_1) = 0$$

But λ_1 and λ_2 are distinct, $\therefore \quad c_2 = 0.$ Similarly $c_1 = 0.$

$\therefore$ $\mathbf{X}_1$ and $\mathbf{X}_2$ are linearly independent vectors. Proceeding on the same lines, we can prove that if $\mathbf{X}_1, \mathbf{X}_2, \ldots, \mathbf{X}_n$ are eigenvectors corresponding to the distinct eigenvalues $\lambda_1, \lambda_2, \ldots, \lambda_n$ of an $n \times n$ matrix **A**, then $\{\mathbf{X}_1, \mathbf{X}_2, \ldots, \mathbf{X}_n\}$ is linearly independent.

The following theorem now answers the diagonalization problem.

THEOREM 3

An $n \times n$ matrix **A** is diagonalizable if and only if **A** has n linearly indepednent eigenvectors.

PROOF

We first suppose that **A** has n linearly independent eigenvectors $\mathbf{X}_1, \mathbf{X}_2, \ldots, \mathbf{X}_n$. Form a matrix **P** whose columns are these eigenvectors, that is, $\mathbf{P} = [\mathbf{X}_1\ \mathbf{X}_2 \cdots \mathbf{X}_n]$. Now,

$$\mathbf{AP} = \mathbf{A}[\mathbf{X}_1\ \mathbf{X}_2 \cdots \mathbf{X}_n] = [\mathbf{AX}_1\ \mathbf{AX}_2 \cdots \mathbf{AX}_n] \tag{1}$$

Let $\lambda_1, \lambda_2, \ldots, \lambda_n$ be the eigenvalues associatied with the eigenvectors $X_1, X_2, \ldots, X_n$.

Then,

$$\mathbf{AX}_1 = \lambda_1\mathbf{X}_1, \mathbf{AX}_2 = \lambda_2\mathbf{X}_2, \ldots, \mathbf{AX}_n = \lambda_n\mathbf{X}_n.$$

$$\therefore \quad \mathbf{AP} = [\lambda_1\mathbf{X}_1\ \lambda_2\mathbf{X}_2 \cdots \lambda_n\mathbf{X}_n] \tag{2}$$

Let $\mathbf{D}$ be the diagonal matrix whose diagonal elements are $\lambda_1, \lambda_2, \ldots, \lambda_n$.

or
$$\mathbf{D} = \begin{bmatrix} \lambda_1 & 0 & \cdots & 0 \\ 0 & \lambda_2 & \cdots & 0 \\ \vdots & \vdots & \vdots & \vdots \\ 0 & 0 & \cdots & \lambda_n \end{bmatrix}$$

Then
$$\mathbf{PD} = [\mathbf{X}_1\ \mathbf{X}_2 \cdots \mathbf{X}_n] \begin{bmatrix} \lambda_1 & 0 & \cdots & 0 \\ 0 & \lambda_2 & \cdots & 0 \\ \vdots & \vdots & \vdots & \vdots \\ 0 & 0 & \cdots & \lambda_n \end{bmatrix}$$

$$= [\lambda_1\mathbf{X}_1\ \lambda_2\mathbf{X}_2 \cdots \lambda_n\mathbf{X}_n] \tag{3}$$

From (2) and (3)

$$\mathbf{AP} = \mathbf{PD}$$

By Theorem 1, $\mathbf{P}$ is invertible

$\therefore$
$$\mathbf{P}^{-1}\mathbf{AP} = \mathbf{P}^{-1}\mathbf{PD}$$

or
$$\mathbf{P}^{-1}\mathbf{AP} = \mathbf{D}$$

Conversely, suppose $\mathbf{A}$ is diagonalizable.
Then there exists an invertible matrix $\mathbf{P}$ such that

$$\mathbf{P}^{-1}\mathbf{AP} = \mathbf{D}$$

where $\mathbf{D}$ is a diagonal matrix

$\Rightarrow$
$$\mathbf{AP} = \mathbf{PD} \tag{4}$$

If $\mathbf{X}_1, \mathbf{X}_2, \ldots, \mathbf{X}_n$ are columns of $\mathbf{P}$ and $\lambda_1, \lambda_2, \ldots, \lambda_n$ are diagonal elements of $\mathbf{D}$, then using (1), (3) and (4), we get

$$\mathbf{AX}_1 = \lambda_1\mathbf{X}_1,\ \mathbf{AX}_2 = \lambda_2\mathbf{X}_2, \ldots, \mathbf{AX}_n = \lambda_n\mathbf{X}_n$$

Since, $\mathbf{X}_1, \mathbf{X}_2, \ldots, \mathbf{X}_n$ are columns of the invertible matrix $\mathbf{P}$, they are linearly independent and non-zero. Hence, they are linearly independent eigenvectors of $\mathbf{A}$.

Combining Theorem 2 and Theorem 3, we get

THEOREM 4

Let $\mathbf{A}$ be an $n \times n$ matrix. If $\mathbf{A}$ has n distinct eigenvalues, then $\mathbf{A}$ is diagonalizable.

While Theorem 3 gives a necessary and sufficient condition for a square matrix to be similar to a diagonal matrix, Theorem 4 gives a sufficient but not necessary condition. This does not mean that if a square matrix $\mathbf{A}$ has repeated eigenvalues, it cannot be diagonalized. The identity matrix has same eigenvalues (all equal to 1), but it is already a diagonal matrix. In fact, if the eigenvalues of a square matrix **A** are not distinct, then A may or may not be similar to a diagonal matrix. For example, consider the matrices

$$\mathbf{A} = \begin{bmatrix} 2 & 0 \\ 0 & 2 \end{bmatrix} \quad \text{and} \quad \mathbf{B} = \begin{bmatrix} 2 & 1 \\ 0 & 2 \end{bmatrix}$$

Both matrices have repeated eigenvalues $\lambda_1 = \lambda_2 = 2$. Any non-zero vector of the form $\begin{bmatrix} x \\ y \end{bmatrix}$ is an eigenvector of **A** for λ_1 and λ_2. Hence, it is possible to choose any two linearly independent vectors, such as $\begin{bmatrix} 1 \\ 0 \end{bmatrix}$ and $\begin{bmatrix} 0 \\ 1 \end{bmatrix}$, as eigenvectors of **A** corresponding to λ_1 and λ_2. By Theorem 3, **A** is diagonalizable. In fact, it is already in diagonal form. However, only a vector of the form $\begin{bmatrix} x \\ 0 \end{bmatrix}$, $x \neq 0$ is an eigenvector of **B** for λ_1 and λ_2. Any two vectors of this form are linearly dependent. Since **B** does not admit any set of two linearly independent vectors, **B** cannot be diagonalized.

The proof of Theorem 3 also provides a general procedure for diagonalizing a given $n \times n$ matrix **A**. Following steps describe the method of diagonalizing **A**.

Step 1. Find the eigenvalues as real roots of the characteristic equation $|\mathbf{A} - \lambda\mathbf{I}| = 0$.

Step 2. Find the eigenvectors and basis for the eigenspace corresponding to each eigenvalue.

Step 3. Compute whether there are n linearly independent vectors or not. If not, matrix **A** cannot be diagonalized.

Step 4. If answer to step 3 is yes, then form the matrix **P** whose columns are the n linearly independent vectors.

Step 5. The matrix $\mathbf{P}^{-1}\mathbf{AP}$ will then be a diagonal matrix whose diagonal elements are the eigenvalues of **A** corresponding to the column of **P**.

EXAMPLE 4.21 Diagonalize the matrix $\begin{bmatrix} -4 & -6 \\ 3 & 5 \end{bmatrix}$.

SOLUTION

In Example 4.18 we have proved that this matrix has two eigenvalues $\lambda_1 = 2$ and $\lambda_2 = -1$. $\left\{\begin{bmatrix} -1 \\ 1 \end{bmatrix}\right\}$ is a basis for the eigenspace corresponding to $\lambda_1 = 2$ and $\left\{\begin{bmatrix} -2 \\ 1 \end{bmatrix}\right\}$ is a basis for the eigenspace corresponding to $\lambda_2 = -1$. Since $\begin{bmatrix} -1 \\ 1 \end{bmatrix}$ and $\begin{bmatrix} -2 \\ 1 \end{bmatrix}$ are eigenvectors corresponding to distinct eigenvalues, they are linearly independent. So the matrix can be diagonalized.

Let $\mathbf{P} = \begin{bmatrix} -1 & -2 \\ 1 & 1 \end{bmatrix}$. Then, the matrix $\begin{bmatrix} -4 & -6 \\ 3 & 5 \end{bmatrix}$ is similar to the diagonal matrix $\begin{bmatrix} 2 & 0 \\ 0 & -1 \end{bmatrix}$ whose diagonal elements are the eigenvalues corresponding to the column vectors of **P**. We can verify this by actual computation.

$$\begin{aligned}
\mathbf{P}^{-1}\mathbf{AP} &= \begin{bmatrix} 1 & 2 \\ -1 & -1 \end{bmatrix}\begin{bmatrix} -4 & -6 \\ 3 & 5 \end{bmatrix}\begin{bmatrix} -1 & -2 \\ 1 & 1 \end{bmatrix} \\
&= \begin{bmatrix} 1 & 2 \\ -1 & -1 \end{bmatrix}\begin{bmatrix} -2 & 2 \\ 2 & -1 \end{bmatrix} \\
&= \begin{bmatrix} 2 & 0 \\ 0 & -1 \end{bmatrix}
\end{aligned}$$

REMARKS: It must be noted that the order in which the eigenvectors are taken as columns of the matrix **P** does not matter for a matrix **A** to be diagonalizable. In the above example, we could also have taken **P** to be the matrix $\begin{bmatrix} -2 & -1 \\ 1 & 1 \end{bmatrix}$. **A** is still diagonalizable, but the diagonal matrix is now $\begin{bmatrix} -1 & 0 \\ 0 & 2 \end{bmatrix}$ instead of $\begin{bmatrix} 2 & 0 \\ 0 & -1 \end{bmatrix}$

or
$$\mathbf{P}^{-1}\mathbf{AP} = \begin{bmatrix} -1 & 0 \\ 0 & 2 \end{bmatrix} \quad \text{if} \quad \mathbf{P} = \begin{bmatrix} -2 & -1 \\ 1 & 1 \end{bmatrix}$$

In general we must note that the order in which the eigenvalues are taken as diagonal elements in the diagonal matrix $\mathbf{P}^{-1}\mathbf{AP}$ is the same in which the corresponding eigenvectors are taken as the column vectors of **P**.

EXAMPLE 4.22 If $\mathbf{A} = \begin{bmatrix} 2 & -3 \\ 2 & -5 \end{bmatrix}$ is a 2×2 matrix, find

(i) the eigenvalues of **A**

(ii) the corresponding eigenvectors of **A**

(iii) a non-singular matrix **P** such that $\mathbf{P}^{-1}\mathbf{AP}$ is diagonal.

[DU, B.Sc.(Ph.Sc.), 2006]

SOLUTION

(i) The characteristic equation of **A** is given by

$$|\mathbf{A} - \lambda\mathbf{I}| = 0$$

That is,
$$\begin{vmatrix} 2-\lambda & -3 \\ 2 & -5-\lambda \end{vmatrix} = 0$$

or
$$\lambda^2 + 3\lambda - 4 = 0$$

or
$$(\lambda + 4)(\lambda - 1) = 0$$

$\Rightarrow$
$$\lambda = -4, 1$$

Thus, the eigenvalues of **A** are -4 and 1.

(ii) For $\lambda = -4$, the system of equations $(\mathbf{A} + 4\mathbf{I})\mathbf{X} = \mathbf{O}$ becomes

$$\begin{pmatrix} 6 & -3 \\ 2 & -1 \end{pmatrix}\begin{pmatrix} x \\ y \end{pmatrix} = \begin{pmatrix} 0 \\ 0 \end{pmatrix}$$

That is,
$$6x - 3y = 0$$
$$2x - y = 0$$

which is same as the single equation $2x = y$. A solution is therefore $x = k$, $y = 2k$ for some scalar k. The eigenvectors corresponding to $\lambda = -4$ are therefore vectors of the form $k\begin{bmatrix} 1 \\ 2 \end{bmatrix}$ where $k \neq 0$. For $\lambda = 1$, the system of equations $(\mathbf{A} - \mathbf{I})\mathbf{X} = \mathbf{O}$ becomes

$$\begin{pmatrix} 1 & -3 \\ 2 & -6 \end{pmatrix}\begin{pmatrix} x \\ y \end{pmatrix} = \begin{pmatrix} 0 \\ 0 \end{pmatrix}$$

or
$$x - 3y = 0$$
$$2x - 6y = 0$$

A solution of this system of equations is $x = 3k$, $y = k$, where k is a real number.

$\therefore$ The eigenvectors corresponding to $\lambda = 1$ are the vectors of the form $k\begin{bmatrix} 3 \\ 1 \end{bmatrix}$ where $k \neq 0$.

(iii) Since the 2×2 matrix $\mathbf{A}$ has two distinct eigenvalues, by Theorem 4, $\mathbf{A}$ is diagonalizable.

Let $\mathbf{P} = \begin{bmatrix} 1 & 3 \\ 2 & 1 \end{bmatrix}$.

By Theorem 1, $\mathbf{P}$ is non-singular and invertible. We have

$$\mathbf{P}^{-1} = \frac{1}{5}\begin{bmatrix} 1 & -3 \\ -2 & 1 \end{bmatrix}$$

$\therefore$
$$\begin{aligned} \mathbf{P}^{-1}\mathbf{A}\mathbf{P} &= \frac{1}{5}\begin{bmatrix} 1 & -3 \\ -2 & 1 \end{bmatrix}\begin{bmatrix} 2 & -3 \\ 2 & -5 \end{bmatrix}\begin{bmatrix} 1 & 3 \\ 2 & 1 \end{bmatrix} \\ &= \frac{1}{5}\begin{bmatrix} 1 & -3 \\ -2 & 1 \end{bmatrix}\begin{bmatrix} -4 & 3 \\ -8 & 1 \end{bmatrix} \\ &= \frac{1}{5}\begin{bmatrix} 20 & 0 \\ 0 & -5 \end{bmatrix} \\ &= \begin{bmatrix} 4 & 0 \\ 0 & -1 \end{bmatrix} \end{aligned}$$

which is a diagonal matrix.

EXAMPLE 4.23 Let $\mathrm{T} : \mathbf{R}^2 \to \mathbf{R}^2$ be the reflection about the line $y = x$. Can the standard matrix of T be similar to a diagonal matrix?

SOLUTION

We know that the standard matrix $\mathbf{A}$ of T is given by

$$\mathbf{A} = \begin{bmatrix} 0 & 1 \\ 1 & 0 \end{bmatrix}.$$

Its characteristic equation is $\lambda^2 - 1 = 0$ and the eigenvalues are $\lambda_1 = 1$ and $\lambda_2 = -1$. Since the two eigenvalues of $\mathbf{A}$ are distinct, $\mathbf{A}$ is diagonalizable. The eigenspace corresponding to $\lambda = 1$ has basis $\left\{\begin{bmatrix} 1 \\ 1 \end{bmatrix}\right\}$ and the eigenspace corresponding to $\lambda = -1$ has basis $\left\{\begin{bmatrix} 1 \\ -1 \end{bmatrix}\right\}$.

Let $\mathbf{P} = \begin{bmatrix} 1 & 1 \\ 1 & -1 \end{bmatrix}$. Thus $\mathbf{P}$ is a non-singular matrix such that $\mathbf{P}^{-1}\mathbf{A}\mathbf{P}$ is diagonal.

$$\begin{aligned} \mathbf{P}^{-1}\mathbf{A}\mathbf{P} &= \begin{bmatrix} \frac{1}{2} & \frac{1}{2} \\ \frac{1}{2} & -\frac{1}{2} \end{bmatrix}\begin{bmatrix} 0 & 1 \\ 1 & 0 \end{bmatrix}\begin{bmatrix} 1 & 1 \\ 1 & -1 \end{bmatrix} \\ &= \begin{bmatrix} 1 & 0 \\ 0 & -1 \end{bmatrix} \end{aligned}$$

EXAMPLE 4.24 Diagonalize, if possible, the matrix

$$\mathbf{A} = \begin{bmatrix} 8 & -12 & 5 \\ 15 & -25 & 11 \\ 24 & -42 & 19 \end{bmatrix}$$

SOLUTION

The reader may verify that the characteristic equation of **A** is

$$\lambda^3 - 2\lambda^2 - \lambda + 2 = 0$$

or

$$(\lambda - 1)(\lambda + 1)(\lambda - 2) = 0$$

So, the eigenvalues of A are $\lambda_1 = 1$, $\lambda_2 = -1$ and $\lambda_3 = 2$.

Since **A** has three distinct eigenvalues, **A** can be diagonalized.

The eigenvectors of **A** are the solutions of the system

$$(\mathbf{A} - \lambda \mathbf{I})\mathbf{X} = \mathbf{O}$$

or

$$\begin{bmatrix} 8-\lambda & -12 & 5 \\ 15 & -25-\lambda & 11 \\ 24 & -42 & 19-\lambda \end{bmatrix} \begin{bmatrix} x \\ y \\ z \end{bmatrix} = \begin{bmatrix} 0 \\ 0 \\ 0 \end{bmatrix} \qquad (1)$$

For $\lambda_1 = 1$, the system (1) becomes

$$\begin{aligned} 7x - 12y + 5z &= 0 \\ 15x - 26y + 11z &= 0 \\ 24x - 42y + 18z &= 0 \end{aligned}$$

Solving them we get, $x = y = z$.

The eigenvectors of **A** corresponding to $\lambda_1 = 1$ are the vectors $k\begin{bmatrix} 1 \\ 1 \\ 1 \end{bmatrix}$ where $k \neq 0$.

For $\lambda_2 = -1$, system (1) can be written as

$$\begin{aligned} 9x - 12y + 5z &= 0 \\ 15x - 24y + 11z &= 0 \\ 24x - 42y + 20z &= 0 \end{aligned}$$

A solution of this system is $x = k$, $y = 2k$ and $z = 3k$.

$\therefore$ $k\begin{bmatrix} 1 \\ 2 \\ 3 \end{bmatrix}$ when $k \neq 0$ are the eigenvectors of **A** corresponding to $\lambda_2 = -1$.

For $\lambda_3 = 2$, the system (1) becomes

$$\begin{aligned} 6x - 12y + 5z &= 0 \\ 15x - 27y + 11z &= 0 \\ 24x - 42y + 17z &= 0 \end{aligned}$$

which gives

$$x = k,\ y = 3k \text{ and } z = 6k \text{ as solution.}$$

$\therefore$ a basis for the eigenspace corresponding to $\lambda_3 = 2$ is $\left\{\begin{bmatrix} 1 \\ 3 \\ 6 \end{bmatrix}\right\}$.

It is clear that $\begin{bmatrix}1\\1\\1\end{bmatrix}, \begin{bmatrix}1\\2\\3\end{bmatrix}$ and $\begin{bmatrix}1\\3\\6\end{bmatrix}$ are three linearly independent vectors of $\mathbf{R}^3$.

Let $\mathbf{P} = \begin{bmatrix}1 & 1 & 1\\1 & 2 & 3\\1 & 3 & 6\end{bmatrix}$.

Then **P** is non-singular and $\mathbf{P}^{-1}\mathbf{AP}$ is diagonal.

$$\mathbf{P}^{-1}\mathbf{AP} = \begin{bmatrix}1 & 1 & 1\\1 & 2 & 3\\1 & 3 & 6\end{bmatrix}^{-1}\begin{bmatrix}8 & -12 & 5\\15 & -25 & 11\\24 & -42 & 19\end{bmatrix}\begin{bmatrix}1 & 1 & 1\\1 & 2 & 3\\1 & 3 & 6\end{bmatrix} = \begin{bmatrix}1 & 0 & 0\\0 & -1 & 0\\0 & 0 & 2\end{bmatrix}$$

EXAMPLE 4.25 Consider the following three matrices discussed in Example 4.20.

$$\mathbf{A} = \begin{bmatrix}2 & 1 & 1\\2 & 3 & 2\\3 & 3 & 4\end{bmatrix}, \quad \mathbf{B} = \begin{bmatrix}2 & -1 & 1\\0 & 3 & -1\\2 & 1 & 3\end{bmatrix}, \quad \mathbf{C} = \begin{bmatrix}1 & 1 & 0\\0 & 1 & 0\\1 & 2 & 1\end{bmatrix}.$$

The matrix **C** has distinct eigenvalues $\lambda_1 = 0, \lambda_2 = 1$ and $\lambda_3 = 2$. Therefore, by Theorem 4, **C** can be diagonalized to get the diagonal matrix $\begin{bmatrix}0 & 0 & 0\\0 & 1 & 0\\0 & 0 & 2\end{bmatrix}$.

We now consider the matrix **A**.

The results obtained for **A** in Example 4.20 is summarized below.

Eigenvalue	**Eigenvectors**	**Dimension of Eigenspace**
$\lambda_1 = 7$	$k\begin{bmatrix}1\\2\\3\end{bmatrix}, k \neq 0$	1
$\lambda_2 = \lambda_3 = 1$	$p\begin{bmatrix}1\\0\\-1\end{bmatrix} + q\begin{bmatrix}0\\1\\-1\end{bmatrix}; p, q \neq 0$	2

Since **A** has only two distinct eigenvalues (eigenvalue 1 is repeated twice), Theorem 4 cannot be used to check whether **A** is diagonalizable or not. Now $\left\{\begin{bmatrix}1\\2\\3\end{bmatrix}\right\}$ acts as a basis for the one-dimensional eigenspace of **A** while $\left\{\begin{bmatrix}1\\0\\-1\end{bmatrix}, \begin{bmatrix}0\\1\\1\end{bmatrix}\right\}$ is a basis for the two-dimensional eigenspace of **A**. The three basis vectors $\begin{bmatrix}1\\2\\3\end{bmatrix}, \begin{bmatrix}1\\0\\-1\end{bmatrix}, \begin{bmatrix}0\\1\\-1\end{bmatrix}$ can be easily shown to be linearly independent.

Therefore, now by Theorem 3, **A** can be diagonalized. We can take

$$\mathbf{P} = \begin{bmatrix} 1 & 1 & 0 \\ 2 & 0 & 1 \\ 3 & -1 & -1 \end{bmatrix} \text{ and get } \mathbf{P}^{-1}\mathbf{AP} = \begin{bmatrix} 7 & 0 & 0 \\ 0 & 1 & 0 \\ 0 & 0 & 1 \end{bmatrix}.$$

Lastly, let us consider the case of matrix **B**. Matrix **B** also has repeated eigenvalues. In fact, we can verify the following for matrix **B**.

Eigenvalue	Eigenvectors	Eigenspace
$\lambda_1 = \lambda_2 = 2$	$k\begin{bmatrix} -1 \\ 1 \\ 1 \end{bmatrix}, k \neq 0$	one-dimensional
$\lambda_3 = 4$	$k\begin{bmatrix} 1 \\ -1 \\ 1 \end{bmatrix}, k \neq 0$	one-dimensioanl

$\left\{\begin{bmatrix} -1 \\ 1 \\ 1 \end{bmatrix}, \begin{bmatrix} 1 \\ -1 \\ 1 \end{bmatrix}\right\}$ is linearly independent but any set of three eigenvectors is linearly dependent.

Thus, by Theorem 3, **A** cannot be diagonalized.

REMARK: The three matrices **A**, **B**, **C** in the above example lead us to a general rule for checking diagonalizibility of a matrix discussed below:

Theorem 4 says that if an $n \times n$ matrix has n distinct roots, then it is diagonalizable. Matrix **A** in Example 4.23 and matrix **C** in Example 4.24 satisfy this criterion. But what if a matrix has repeated eigenvalues? We have already discussed that such a matrix may or may not be diagonalized. Asobserved in the above example, we have following general rule for such cases:

An $n \times n$ matrix **A** is diagonalizable if and only if the sum of the dimensions of the distinct eigenspaces equals n.

It is clear from this rule that if an eigenvalue λ of a matrix **A** is repeated k times and the eigenspace corresponding to λ is of dimension less than k, then **A** cannot be diagonalized.

In Example 4.24, **C** has three distinct eigenspaces of dimension one each and **C** is diagonalizable. The matrices **A** and **B** both have repeated eigenvalues. **A** has two eigenspaces corresponding to $\lambda = 1$ and $\lambda = 7$ of dimensions 2 and 1 respectively so that the sum of dimensions is 3. Hence, **A** is diagonalizable. On the other hand, **B** has only two eigenspaces of dimension one each so that the sum of dimensions is not three and therefore **B** is not diagonalizable.

EXERCISE 4.3

1. Let **A** be a 2×2 matrix whose eigenvalues are 3 and 4, and associatied eigenvectors are $\begin{bmatrix} -1 \\ 1 \end{bmatrix}$ and $\begin{bmatrix} 2 \\ 1 \end{bmatrix}$ respectively. Without computation, find a diagonal matrix **D** and a non-singular matrix **P** such that $\mathbf{P}^{-1}\mathbf{AP} = \mathbf{D}$.

2. **A** is a 3×3 matrix with two eigenvalues. Each eigenspace is one-dimensional. Is **A** diagonalizable? Why?

3. Let **A** be a 4×4 matrix with eigenvalues 5, 3 and -2 and suppose that the eigenspace for $\lambda = 3$ is two-dimensional. Do we have enough information to determine if **A** is diagonalizable?

4. For the following 2×2 matrices **A**, find a non-singular matrix **P** and a diagonal matrix **D** such that $\mathbf{P}^{-1}\mathbf{AP} = \mathbf{D}$.

   (a) $\begin{bmatrix} 1 & 1 \\ -2 & 4 \end{bmatrix}$ **(b)** $\begin{bmatrix} 1 & -1 \\ -1 & 1 \end{bmatrix}$ **(c)** $\begin{bmatrix} 1 & -2 \\ 0 & 3 \end{bmatrix}$ **(d)** $\begin{bmatrix} -\frac{1}{\sqrt{2}} & \frac{1}{\sqrt{2}} \\ \frac{1}{\sqrt{2}} & \frac{1}{\sqrt{2}} \end{bmatrix}$

5. Determine whether the following matrices are diagonalizable. If possible, also diagonalize the matrix.

   (a) $\begin{bmatrix} 3 & -1 \\ 4 & -1 \end{bmatrix}$ **(b)** $\begin{bmatrix} 4 & -3 \\ 2 & -1 \end{bmatrix}$ **(c)** $\begin{bmatrix} 4 & -1 \\ 1 & 2 \end{bmatrix}$ **(d)** $\begin{bmatrix} 2 & 1 \\ 0 & 2 \end{bmatrix}$

6. For each of the following matrices, find a non-singular matrix **P** such that $\mathbf{P}^{-1}\mathbf{AP}$ is diagonal.

   (a) $\mathbf{A} = \begin{bmatrix} 0 & 0 & -2 \\ 1 & 2 & 1 \\ 1 & 0 & 3 \end{bmatrix}$ **(b)** $\begin{bmatrix} 5 & -6 & -6 \\ -1 & 4 & 2 \\ 3 & -6 & -4 \end{bmatrix}$

   (c) $\mathbf{A} = \begin{bmatrix} -3 & 0 & 6 \\ 0 & 3 & 6 \\ 6 & 6 & 0 \end{bmatrix}$ **(d)** $\mathbf{A} = \begin{bmatrix} 0 & 0 & 0 \\ 0 & 1 & 0 \\ 1 & 0 & 1 \end{bmatrix}$

7. Diagonalize the following matrices, if possible

   (a) $\begin{bmatrix} 0 & 0 & 1 \\ 0 & 1 & 2 \\ 0 & 0 & 1 \end{bmatrix}$ **(b)** $\begin{bmatrix} 1 & 3 & 3 \\ -3 & -5 & -3 \\ 3 & 3 & 1 \end{bmatrix}$

   (c) $\begin{bmatrix} 2 & 4 & 3 \\ -4 & -6 & -3 \\ 3 & 3 & 1 \end{bmatrix}$ **(d)** $\begin{bmatrix} 0 & 1 & 0 \\ 0 & 0 & 1 \\ 4 & -17 & 8 \end{bmatrix}$

8. Let **A** be a 2×2 matrix such that $\mathbf{A}e_1 = e_2$ and $\mathbf{A}e_2 = \mathbf{O}$ where e_1, e_2 are standard basis vectors of $\mathbf{R}^2$. Find the eigenvalues and the corresponding eigenvectors of **A**. Is the matrix **A** diagonalizable?

9. Let $\mathrm{T}: \mathbf{R}^2 \rightarrow \mathbf{R}^2$ be the following transformations. Find whether the standard matrix representation of T is diagonalizable.

   (a) T is a reflection about x-axis.

   (b) T is a dilation of factor 2.

   (c) T is anticlockwise rotation through $\frac{\pi}{2}$.

10. Let $T: \mathbf{R}^2 \rightarrow \mathbf{R}^2$ be reflection about the line l that passes through origin and makes an angle θ with x-axis. We have found the standard matrix representation $\mathbf{A}$ of T in Example 4.27. Find whether $\mathbf{A}$ is diagonalizable. If so, also find $\mathbf{P}$ and $\mathbf{P}^{-1}\mathbf{AP}$.

11. Show that every 2×2 matrix $\mathbf{A}$ of the form $\mathbf{A} = \begin{bmatrix} a & b \\ b & c \end{bmatrix}$ is diagonalizable [See Problem 13, Exercise 4.1].

12. Let $\mathbf{A} = \begin{bmatrix} a & b \\ c & d \end{bmatrix}$. Find the necessary and sufficient conditions for $\mathbf{A}$ to be diagonalizable.

13. Show that if $\mathbf{A}$ is diagonalizable, then

(a) $\mathbf{A}^{\mathrm{T}}$ is diagonalizable

(b) $\mathbf{A}^k$ is diagonalizable, where k is a positive integer.

14. Recall the definition of similar matrices given in the text.

(a) Prove that $\mathbf{A}$ is similar to $\mathbf{A}$.

(b) If $\mathbf{B}$ is similar to $\mathbf{A}$, then show that $\mathbf{A}$ is similar to $\mathbf{B}$.

(c) Prove that the matrices $\begin{bmatrix} 1 & 1 \\ 0 & 1 \end{bmatrix}$ and $\begin{bmatrix} 1 & 0 \\ 0 & 1 \end{bmatrix}$ have the same eigenvalues but are not similar.

PART II

CHAPTER FIVE

Sequences

■ 5.1 WHAT IS A SEQUENCE?

Informally, a sequence is an ordered list of things. In everyday usage, the word 'sequence' is used to suggest a succession of things or events arranged in some order such as chronological order, size order and so on. It is quite common in many disciplines of applied sciences like physics, biology, medicine and such others in which a variable value or quantity is measured at fixed time intervals and the results are recorded in the form of a list. For example, the pulse rate of a patient, measured on an hourly basis in an observation period of eight hours, is as follows:

Time	7 AM	8 AM	9 AM	10 AM	11 AM	12 Noon	1 PM	2 PM	3 PM
Pulse per min.	82	81	80	79	76	77	76	76	75

The maximum and minimum temperatures of Delhi in the week containing the second equinox day of the year 2005 are as follows:

Day	20 Sept.	21 Sept.	22 Sept.	23 Sept.	24 Sept.	25 Sept.	26 Sept.
Max Temp. (°C)	34.6	36.0	33.8	34.1	27.6	29.8	32.5
Min Temp. (°C)	25.0	26.0	26.6	26.1	23.4	21.5	23.3

In this case, the collections

82, 81, 80, 79, 76, 77, 76, 76, 75;

34.6, 36.0, 33.8, 34.1, 27.6, 29.8, 32.5; and

25.0, 26.0, 26.6, 26.1, 23.4, 21.5, 23.3

are examples of finite sequences. It is clear that we are able to work, mostly with finitely many observations, but theoretically, the observed process could last infinitely. Therefore, we can talk of infinite sequences. The nth member of such a sequence may be denoted by a_n. That is, if for every positive integer n, we are given a real number, which we write as a_n, then the ordered set

$$a_1, a_2, a_3, \ldots, a_n, \ldots$$

also written as $\{a_n\}$, is said to define an infinite sequence. In fact, by a sequence, we shall mean an infinite sequence only, unless stated specially. The precise definition of a sequence is as follows:

> **? DEFINITION:** A 'sequence' $\{a_n\}$ is a function $a : \mathrm{D} \to \mathbf{R}$, where $a(n)$ is denoted by a_n, D is the set of integers greater than or equal to some integer n_0, and **R** is the set of real numbers. Usually $n_0 = 1$ so that D is the set of positive integers. However in some cases, $n_0 = 0$ or some other integer. Unless stated otherwise, we shall take $n_0 = 1$.

EXAMPLE 5.1

(i) The sequence $\{a_n\}$, where $a_n = \frac{1}{n}$. The terms of this sequence are $1, \frac{1}{2}, \frac{1}{3}, \frac{1}{4}, \ldots$ and so on.

(ii) The sequence $\left\{\frac{1}{2^n}\right\}$. We can also write this sequences as $\frac{1}{2}, \frac{1}{2^2}, \frac{1}{2^3}, \frac{1}{2^4}, \ldots$

(iii) The constant sequence $\{a_n\}$, $a_n = c$, where c is a fixed real number.

(iv) The sequence $\{(-1)^n\}$, that is, the sequence $-1, 1, -1, 1, \ldots$

(v) The sequence arising out of decimal representation of a number. For example, the sequence 0.3, 0.33, 0.333, 0.3333, . . . arises from the decimal representation of 1/3 upto 1st, 2nd, 3rd place and so on.

Similarly,

$$1.4, 1.41, 1.414, 1.412, 1.41421, 1.414213, 1.4142135, \ldots$$

is the sequence obtained from the decimal representation of $\sqrt{2}$. The number 2/5 generates the constant sequence 0.4, 0.4, 0.4, ...

(vi) The sequence $\{a_n\}$, where $a_n = \frac{2n}{n^2 - 3n + 2}$. Here the formula makes sense only for $n \geq 3$. Thus, $n_0 = 3$ in this case.

(vii) The sequence $\{\mathrm{A}_n\}$ arising out of the amounts compounded annually of rupees, P, at the rate of r% per annum over n years. Here $\mathrm{A}_n = \mathrm{P}\left(1 + \frac{r}{100}\right)^n$. Thus, the sequence is given by

$$\mathrm{P}\left(1 + \frac{r}{100}\right), \mathrm{P}\left(1 + \frac{r}{100}\right)^2, \mathrm{P}\left(1 + \frac{r}{100}\right)^3, \ldots$$

or by the recursion formula $\mathrm{A}_{n+1} = \mathrm{A}_n\left(1 + \frac{r}{100}\right)$, $\mathrm{A}_1 = \mathrm{P}\left(1 + \frac{r}{100}\right)$.

(viii) The sequence $\{a_n\}$, where a_n is defined by the formula: $a_{2n-1} = 1$, $a_{2n} = 2^n$. The sequence is then given by 1, 2, 1, 4, 1, 8, 1, 16, . . .

(ix) In the nuclear fission reaction, when a slow moving neutron is bombarded on the enriched uranium U_{235} one nucleus splits resulting in the release of energy, one krypton nucleus, one barium nucleus and three more neutrons. These three neutrons then make three other U_{235} nuclei split, repeating the same in each nucleus. If it takes t seconds for a neutron for bombardment after its release then how many bombardments will take place after n seconds from the initial bombardment? It gives us the sequence $3, 3^2, 3^3, \ldots, 3^n, \ldots$

One word of caution here. For a given sequence $\{a_n\}$, we can consider the set of values $A = \{a_n : n \in D\}$. It is important to note that $\{a_n\}$ and A are not the same. The set A is simply the range of the function $a : D \to \mathbf{R}$, whereas $\{a_n\}$ is the function itself. In case of the sequence $\{(-1)^n\}$, we have,

$$\{(-1)^n\} = -1, 1, -1, 1, \ldots; \text{ whereas here } A = \{-1, 1\}.$$

EXERCISE 5.1

1. In each part, find a formula for the general term of the sequence starting with $n = 1$.

(a) $1, -\frac{1}{3}, \frac{1}{9}, -\frac{1}{27}, \ldots$ **(b)** $\frac{1}{2}, \frac{3}{4}, \frac{5}{6}, \frac{7}{8}, \ldots$ **(c)** $\frac{1}{\sqrt{e}}, \frac{4}{3\sqrt{e}}, \frac{9}{4\sqrt{e}}, \frac{16}{5\sqrt{e}}, \ldots$

(d) $0, 3, 8, 15, 24, \ldots$ **(e)** $2, -6, 10, -14, 18, \ldots$ **(f)** $1, 0, 1, 0, 1, \ldots$

(g) $2, 4, 2, 4, 2, 4, \ldots$ **(h)** $0, \frac{3}{2}, \frac{2}{3}, \frac{5}{4}, \frac{4}{5}, \frac{7}{6}, \frac{6}{7}, \ldots$ **(i)** $0, \frac{3}{\sqrt{\pi}}, \frac{8}{\sqrt[3]{\pi}}, \frac{15}{\sqrt[4]{\pi}}, \ldots$

(j) $2, \frac{5}{2}, \frac{10}{3}, \frac{17}{4}, \ldots$

2. In each part, find the two formulae for the general term of the seqeunce, one starting with $n = 1$ and the other with $n = 0$.

(a) $1, -r, r^2, -r^3, \ldots$ **(b)** $r, -r^2, r^3, -r^4, \ldots$

5.2 RECURSION FORMULA FOR SEQUENCES

Sometimes a sequence is constructed by using a recursion formula. A recursion formula is one where succeeding terms are calculated using the previous terms. The following examples will make it clear.

EXAMPLE 5.2

(i) $a_1 = 1, a_{n+1} = \sqrt{2a_n}$. Here the sequence is $1, \sqrt{2}, \sqrt{2\sqrt{2}}, \sqrt{2\sqrt{2\sqrt{2}}}, \ldots$.

(ii) For the sequence $\sqrt{6}, \sqrt{6+\sqrt{6}}, \sqrt{6+\sqrt{6+\sqrt{6}}}, \ldots$, the recursion formula is $a_1 = \sqrt{6}$, $a_{n+1} = \sqrt{6 + a_n}$.

(iii) The 'mechanic's rule' for approximating square roots states that $\sqrt{a} \approx x_{n+1}$, where

$$x_{n+1} = \frac{1}{2}\left(x_n + \frac{a}{x_n}\right), n = 1, 2, 3, \ldots$$

and x_1 is any positive approximation to $\sqrt{a}$.

The Table 5.1 shows the first five terms in an application of the mechanic's rule to approximate $\sqrt{2}$.

Table 5.1

n	$x_1 = 1$ $\quad x_{n+1} = \frac{1}{2}\left(x_n + \frac{2}{x_n}\right)$	Decimal representation
1	$x_1 = 1$ (starting value)	1.00000000000
2	$x_2 = \frac{1}{2}\left(1 + \frac{2}{1}\right) = \frac{3}{2}$	1.50000000000
3	$x_3 = \frac{1}{2}\left[\frac{3}{2} + \frac{2}{3/2}\right] = \frac{17}{12}$	1.41666666667

(Continued)

Table 5.1 (Contd.)

4	$x_4 = \frac{1}{2}\left[\frac{17}{12} + \frac{2}{17/12}\right] = \frac{577}{408}$	1.41421568627
5	$x_5 = \frac{1}{2}\left[\frac{577}{408} + \frac{2}{577/408}\right] = \frac{665857}{470832}$	1.41421356237
6	$x_6 = \frac{1}{2}\left[\frac{665857}{470832} + \frac{2}{665857/470832}\right] = \frac{886731088897}{627013566048}$	1.41421356237

(iv) Many calculators compute reciprocals using the approximation $\frac{1}{a} \approx x_{n+1}$, where

$$x_{n+1} = x_n(2 - ax_n), \qquad n = 1, 2, 3, \ldots$$

and x_1 is an initial approximation to $\frac{1}{a}$. This formula makes it possible to perform division using multiplications and subtractions, which is a faster procedure than dividing directly.

5.2.1 Fibonacci Sequence

One of the most well-known types of sequences to be expressed using recursion formula is the Fibonacci sequence, developed by Leonardo Fibonacci, also known as 'Leonardo of Pisa' (A.D. 1170–1250). We discuss below the Fibonacci sequence arising out of the breeding habit of rabbits.

EXAMPLE 5.3 (Breeding habit of rabbits). Suppose that rabbits live forever and that every month each adult pair gives birth to a new pair which becomes adult within 2 months' time and produces thereafter one pair of rabbits every month. If there is one pair of newborn rabbits now, how many pairs of rabbits will be there in the nth month?

SOLUTION

If F_n denotes the number of pairs of rabbits in the nth month, then we get

$$F_1 = 1,\ F_2 = 1,\ F_n = F_{n-1} + F_{n-2}$$

In fact, this can be explained in the following way: during the first month, a pair of newborn rabbits is present. During the second month, this pair becomes adult. Next month, this pair produces one new pair so that one adult pair and one young pair are present in the third month and so on.

Let us write Y_n for the total number of young pairs at the end of the nth month, A_n for the total number of adult pairs and F_n for the total number of pairs. Then, we have $F_n = A_n + Y_n$ for $n \geq 1$ and for $n \geq 2$, $A_n = A_{n-1} + Y_{n-1}$ and $Y_n = A_{n-1}$. Thus, we have Table 5.2:

Table 5.2

Month	n	1	2	3	4	5	...	n
Young pairs	Y_n	1	0	1	1	2	...	A_{n-1}
Adult pairs	A_n	0	1	1	2	3	...	$Y_{n-1} + A_{n-1}$
Total	F_n	1	1	2	3	5	...	$Y_n + A_n$

Combining these results, we find that

$$A_n = F_{n-1} \quad \text{for} \quad n \geq 2 \qquad \text{and} \qquad Y_n = A_{n-1} = F_{n-2} \quad \text{for} \quad n \geq 3.$$

Therefore, for $n \geq 3$, $F_n = A_n + Y_n = F_{n-1} + F_{n-2}$. Thus, the conditions, $F_n = F_{n-1} + F_{n-2}$, $F_1 = 1, F_2 = 1$ define the famous Fibonacci sequence $\{F_n\}$ and Fibonacci numbers F_n.

EXAMPLE 5.4 Compute F_n for $n = 1, 2, 3, \ldots, 15$.

SOLUTION

Simply use the fact that $F_1 = 1, F_2 = 1, F_n = F_{n-1} + F_{n-2}$ *(Hint: $F_{15} = 610$).*

EXAMPLE 5.5 For a Fibonacci sequence $\{F_n\}$, prove that

(i) $\sum_{k=1}^{n} F_k = F_{n+2} - 1$;

(ii) $\sum_{k=1}^{n} F_k^2 = F_n F_{n+1}$;

(iii) $F_{n+1} F_{n-1} - F_n^2 = (-1)^n$ for $n \geq 2$;

(iv) $F_n = \dfrac{\alpha^n - \beta^n}{\alpha - \beta}$, where α, β are roots of the quadratic equation $x^2 - x - 1 = 0$;

(v) $\alpha^{n-2} \leq F_n \leq \alpha^{n-1}$ for all n.

SOLUTION

(i) For a Fibonacci sequence, we have

$$F_k = F_{k+2} - F_{k+1} \quad \text{for all } k \geq 1. \text{ Therefore,}$$

$$\begin{aligned} F_1 + F_2 + F_3 + \cdots + F_{n-2} + F_n &= (F_3 - F_2) + (F_4 - F_3) + \cdots + (F_{n+2} - F_{n+1}) \\ &= F_{n+2} - F_2 \\ &= F_{n+2} - 1. \end{aligned}$$

(ii) We prove this result by using the principle of mathematical induction. For $n = 1$, L.H.S. $= \sum_{k=1}^{1} F_k^2 = F_1^2 = 1$, R.H.S. $= F_1F_2 = 1$. Thus the result is true for $n = 1$. Assume that it is true for $n = m$. That is

$$\sum_{k=1}^{m} F_k^2 = F_m F_{m+1} \tag{1}$$

Adding F_{m+1}^2 to both sides of (1), we get

$$\begin{aligned} \sum_{k=1}^{m} F_k^2 + F_{m+1}^2 &= F_m F_{m+1} + F_{m+1}^2 \\ \Rightarrow \qquad \sum_{k=1}^{m+1} F_k^2 &= F_{m+1}(F_m + F_{m+1}) \\ &= F_{m+1} F_{m+2}. \end{aligned}$$

Thus, the result is true for $n = m + 1$ also. Hence, by the principle of mathematical induction, it is true for all values of n.

(iii) This result can also be proved by using the principle of mathematical induction. For $n = 2$, L.H.S. $= F_3F_1 - F_2^2 = 2 \cdot 1 - 1^2 = 1$ and R.H.S. $= 1$. So assume that it is true for $n = m$. That is

$$F_{m+1} F_{m-1} - F_m^2 = (-1)^m \tag{2}$$

Now,

$$\begin{aligned}F_{m+2}F_m - F_{m+1}^2 &= (F_{m+1} + F_m)F_m - F_{m+1}^2 \\ &= F_{m+1}F_m + F_m^2 - F_{m+1}^2 \\ &= F_{m+1}[F_m - F_{m+1}] + F_m^2 \\ &= -F_{m+1}F_{m-1} + F_m^2 \\ &= -[F_{m+1}F_{m-1} - F_m^2] \\ &= -(-1)^m = (-1)^{m+1}\end{aligned}$$

Thus the result is true for $n = m + 1$. Hence it is true for all $n \geq 2$.

(iv) Since α, β are roots of the quadratic equation $x^2 - x - 1 = 0$, we have

$$\alpha^2 = \alpha + 1 \quad \text{and} \quad \beta^2 = \beta + 1 \tag{3}$$

We shall show again by the principle of mathematical induction that

$$F_n = \frac{\alpha^n - \beta^n}{\alpha - \beta} \tag{4}$$

For $n = 1$, L.H.S. = R.H.S. For $n = 2$, R.H.S. $= \dfrac{(\alpha + 1) - (\beta + 1)}{\alpha - \beta} = 1 =$ L.H.S.

Assume that (4) is true for $m - 1$ and m for some $m \geq 2$. Then, we have

$$\begin{aligned}F_{m+1} &= F_m + F_{m-1} \\ &= \frac{\alpha^m - \beta^m}{\alpha - \beta} + \frac{\alpha^{m-1} - \beta^{m-1}}{\alpha - \beta} \\ &= \frac{\alpha^{m-1}(\alpha + 1) - \beta^{m-1}(\beta + 1)}{\alpha - \beta} \\ &= \frac{\alpha^{m-1} \cdot \alpha^2 - \beta^{m-1} \cdot \beta^2}{\alpha - \beta} \quad \text{(by (3))} \\ &= \frac{\alpha^{m+1} - \beta^{m+1}}{\alpha - \beta}\end{aligned}$$

Hence, the result follows.

(v) Here $\alpha = \dfrac{1 + \sqrt{5}}{2}$, $\quad F_1 = F_2 = 1$. Hence for $n = 1, n = 2$,

$$\alpha^{n-2} \leq F_n \leq \alpha^{n-1} \tag{5}$$

Assume that (5) holds for $n = m$, and $n = m + 1$. Then,

$$\alpha^{m-2} \leq F \leq \alpha^{m-1}$$

and
$$\alpha^{m-1} \leq F_{m+1} \leq \alpha^m$$

Adding the inequalities, we get

$$\begin{aligned}&\alpha^{m-2} + \alpha^{m-1} \leq F_m + F_{m+1} \leq \alpha^{m-1}(1 + \alpha) \\ \Rightarrow\quad &\alpha^{m-2}(1 + \alpha) \leq F_{m+2} \leq \alpha^{m-1}(1 + \alpha) \qquad (\because\ F_m + F_{m+1} = F_{m+2}) \\ \Rightarrow\quad &\alpha^m \leq F_{m+2} \leq \alpha^{m+1} \qquad (\because\ 1 + \alpha = \alpha^2)\end{aligned}$$

Thus (5) holds for $n = m + 2$. Hence, by the principle of mathematical induction (5) holds for all n.

EXAMPLE 5.6 (Arrangement of leaves). Fibonacci numbers are also observed when arrangement of leaves (phyllotaxis) is studied. We consider the case where leaves around a stem follow a helical pattern. Proceeding upwards, we mark consecutive leaves by L_1, L_2, L_3 and so on. Leaf L_2 will be found standing at a certain angle away from L_1 around the stem and at a certain distance along the stem. In Fig. 5.1, a special case is depicted where the angle is 144°.

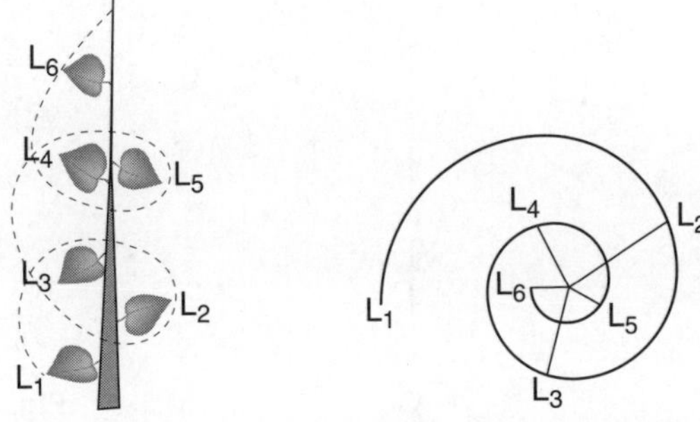

Fig. 5.1

Helical arrangement of leaves on a stem. In the figure, it is assumed that the same pattern with five leaves is repeated after two full windings of the helix. This is the case in roses, some willows, and cherries.

Leaf L_3 is displaced from L_2 in much the same way as L_2 from L_1. We assume the same pattern for all the leaves that follow. In our example, it takes five angles of 144° to arrive at a leaf that has the same orientation as L_1 since $5 \times 144° = 720° = 2 \times 360°$. Thus, we find a periodicity with a period consisting of two windings and five leaves.

In general, we may introduce two numbers:

(a) m = number of complete turns or windings in a period;

(b) n = number of leaves in a period.

In the above case $m = 2$, $n = 5$.

Actual counting on numerous plants, has shown that m and n most frequently take values such as 1, 2, 3, 5, 8, 13, 21, 34, . . . that is, numbers from the Fibonacci sequence. There are cases such as $m = 1$, $n = 2$ in the two row leaves of several bulbous plants as well as the horizontal twigs of elm.

$m = 2$, $n = 5$: very frequent, in willows, roses, stone fruit trees, tulsi

$m = 3$, $n = 8$: in cabbage, asters, and hawkweed

$m = 8$, $n = 21$: scales of spruce and fir cones

Biologists have tried to explain the peculiar prevalence of Fibonacci numbers in phyllotaxis. Symmetry may play a major role because symmetry maintains the mechanical equilibrium of a stem, gives the leaves the best exposure to light and supports regular flow of nutrients. However, science is still far away from a satisfactory explanation.

EXAMPLE 5.7 (Branching habit of trees). Assume that a tree has branches that grow indefinitely and no branch produces a new branch at the end of the first year. Every branch produces a new branch at the end of the second year and each succeeding year of life. At the end of the 1st year (or the beginning of the 2nd year) there is only one branch OP, at the end of the 2nd year (or the beginning of the 3rd year) this branch grows to OPQ and starts a new branch at Q. Effectively, there is only

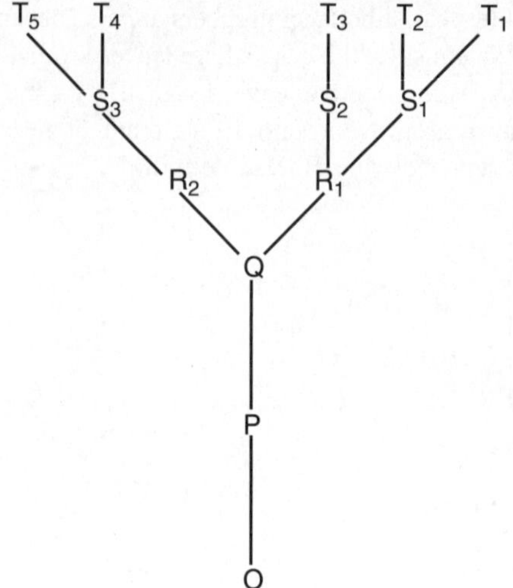

Fig. 5.2

one branch at this stage also. At the end of the 3rd year the original branch grows to $OPQR_1$ and the new branch at Q grows to QR_2. Thus, there are two branches now with tips R_1 and R_2. At this point of time, QR_2 is only one year old and produces no new branch but $OPQR_1$ is three years old and produces a new branch at its tip R_1. At the end of the 4th year, the original branch grows to $OPQR_1S$, the branch QR_2 grows to QR_2S_3 and the tip at R_1 grows to a new branch R_1S_2. Thus, there are three branches now namely, $OPQR_1S_1$, QR_2S_3 and the newest branch R_1S_2 with tips at S_1, S_2 and S_3. At this point of time, the three branches proceed in their original course but a new growth starts at the tips S_1 and S_3. Consequently, at the end of the 5th year, there are five branches namely, $OPQR_1S_1T_1$ $QR_2S_3T_5$, $R_1S_2T_3$, S_1T_2 and S_3T_4 with tips at T_1, T_2, T_3, T_4 and T_5. Table 5.3 makes the picture clear.

Table 5.3

At the end of year	Branches	Number of Branches
1	OP	1
2	OPQ	1
3	$OPQR_1$, QR_2	2
4	$OPQR_1S_1$, QR_2S_3, R_1S_2	3
5	$OPQR_1S_1T_1$, $QR_2S_3T_5$, $R_1S_2T_3$, S_1T_2, $S_3 T_4$	5

At the end of the 6th year, in addition to these five branches, there will be three new branches out of T_1, T_3 and T_5, giving a total of eight branches.

Observe that the numbers at the end of successive years are 1, 1, 2, 3, 5, 8, . . . These numbers are consecutive terms of the Fibonacci sequence.

5.2.2 Tower of Hanoi Game (Tower of Brahma)

Another example where recursion formula is used to define the general term of a sequence is the Tower of Hanoi game. It consists of n circular rings of varying diameters and three pegs on which the rings fit. Initially the rings are placed on the first (left-most) peg in order of their diameters, that is, the largest ring at the bottom covered by successively smaller rings.

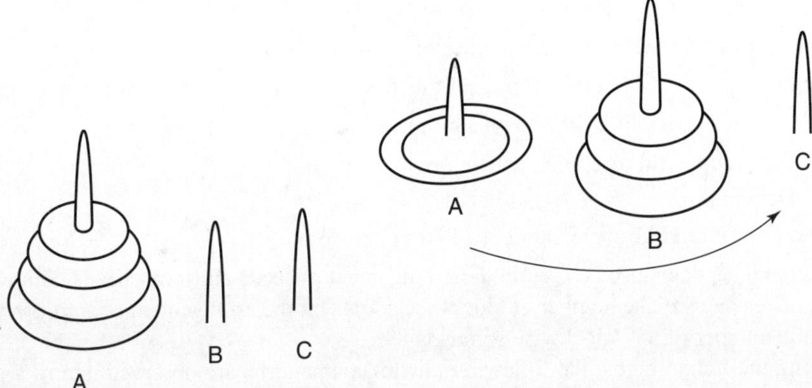

Fig. 5.3

By transferring the rings among the pegs, one seeks to achieve a similarly tapered pile on the third (right-most) peg. The complication is that each time a ring is transferred to a new peg, the transferred ring must be smaller than any of the rings already piled on this new peg; equivalently, at every stage there must be a tapered pile (or no pile) on each peg.

The problem is to find out a recursion formula (or in other words, recurrence relation) for the minimum number of moves required to play the Tower of Hanoi game with n rings.

We observe that if four rings are on peg A (in order of their sizes, of course) and we want to move them to peg C, we must first play the three ring Tower of Hanoi game to get the first three rings from peg A to peg B (via peg C) then move the fourth remaining ring from peg A to peg C, and then again play the three ring game from peg B to peg C (via peg A). (Of course, to move the three rings from A to B, we must play first the two ring game from A to C via B and from C to B via A and so on).

Thus to move the nth ring from A to C, $n-1$ smaller rings must be moved from A to B and later from B to C.

If a_n is the number of moves needed to transfer a tapered pile of n rings from peg A to peg C, then $a_n = a_{n-1} + 1 + a_{n-1} = 2a_n + 1$. The initial condition is $a_1 = 1$ and so $a_2 = 2a_1 + 1 = 3$, $a_3 = 2a_2 + 1 = 7$, $a_4 = 2a_3 + 1 = 15$ and so on.

EXERCISE 5.2

1. In the following, the first term or two, alongwith a recursion formula for the remaining terms of a sequence are provided. Write down the first five terms of the sequence

   (a) $a_1 = 2,\ a_{n+1} = (-1)^{n+1}\dfrac{a_n}{2}$ **(b)** $a_1 = -2,\ a_{n+1} = n\dfrac{a_n}{n+1}$

   (c) $a_1 = 2,\ a_2 = -1,\ a_{n+2} = \dfrac{a_{n+1}}{a_n}$ **(d)** $a_1 = 2,\ a_{n+1} = a_n - \dfrac{1}{2^n}$

   (e) $a_1 = -1,\ a_{n+1} = \dfrac{a_n}{n+1}$

2. Define a Fibonacci sequence $\{F_n\}$. Compute F_{10}, F_{13}, F_{16}.
3. Describe the occurrence of Fibonacci numbers in the nature in
   (i) branching habit of trees
   (ii) arrangement of leaves (phyllotaxis)
   (iii) breeding habit of rabbits

(iv) petals of flowers

(v) plant spirals (such as pine cones, pineapples, sunflowers)

[Hint: In many plants, the number of petals of its flowers is a Fibonacci number. Some examples

3 petals: Lily, iris (often lilies have 6 petals from 2 sets of 3 petals)

5 petals: Buttercup, wild rose, pink and so on.

8 petals: Delphinilims

13 petals: Corn marigold, some varieties of daisy and so on.]

Fibonacci numbers can also be seen in the arrangement of seeds on flower heads. Poppy seeds have 13 ridges on top. The number of clockwise spirals on the arrangement of sunflower seeds is 55 while number of anti-clockwise spirals is 89.

Among the Indian herbs, the following phyllotaxis patterns are observed: (from a project carried out by the physical science students of Rajdhani College, Delhi University)

Tulsi: 2, 5 (No. of turns, No. of leaves)

Neem: 1, 2

Pudina: 1, 2

Tamarind: 1, 2

4. Explain the Tower of Hanoi game with

(i) 3 rings **(ii)** 4 rings **(iii)** n rings

5. For a Fibonacci sequence $\{F_n\}$, show that

(a) $F_1 + F_3 + F_5 + \cdots + F_{2n-1} = F_{2n}$

(b) $F_2 + F_4 + F_6 + \cdots + F_{2n} = F_{2n+1} - 1$

[Hint: Add the equalities $F_1 = F_2$, $F_3 = F_4 - F_2$, $F_5 = F_6 - F_4$ and so on. For the proof of part (b), use the result 5.5(i) of the solved example]

5.3 DIFFERENCE EQUATION

A *difference equation* associated with a sequence $\{x_n\}$ is an equation that involves two or more arbitrary terms of the sequence. The following are examples of difference equations:

$$x_n = x_{n-1}; \qquad x_{k+2} = x_k; \qquad x_n + n^2 x_{n+1} = x_{n-3};$$
$$x_{n+1} = (x_{n+1})^n; \qquad x_{n+1} = ax_n + b; \qquad x_{n+1} = x_n^{-1};$$
$$x_{n+1} = x_n + 2x_n^3.$$

The following are not examples of difference equation.

$$x_n = 3 \text{ (only one term)}; \qquad x_5 = x_6 - x_2 \text{ (no arbitrary terms)}$$

Each of the following represents the same difference equation:

$$\Delta x_n = 2x_n; \qquad x_{n+1} = 3x_n; \qquad x_n = 3x_{n-1}; \qquad x_k = 3x_{k-1};$$

where Δx_n stands for $x_{n+1} - x_n$.

It is the relationship between the terms that characterizes a difference equation. Difference equations are the essential basic mathematical components of all discrete mathematical models. The fundamental concept of a difference equation, namely, the change in quantity between generation (such as $\Delta x_n = x_{n+1} - x_n$), translates naturally into derivative, which is the fundamental concept of differential calculus. When one attempts to apply calculus to real problems that require real solutions frequently, the only way to obtain a solution is to utilize a computer. This requires translating the calculus concepts into discrete step-by-step procedures that are described by difference equations. Some examples of difference equation are given below.

EXAMPLE 5.8 Suppose that female unicorns live for exactly four years. Let u_n be the number of female unicorns at the end of the nth year. The number just born in the nth year is $u_n - u_{n-1}$ and they die in $(n+4)^{\text{th}}$ year after bearing offsprings in $(n+2)^{\text{th}}$ and $(n+3)^{\text{th}}$ years. Thus,

$$\begin{aligned} u_n &= u_{n-1} + (u_{n-2} - u_{n-3}) + (u_{n-3} - u_{n-4}) - (u_{n-4} - u_{n-5}) \\ &= u_{n-1} + u_{n-2} - 2u_{n-4} + u_{n-5}. \end{aligned}$$

EXAMPLE 5.9 We list several common difference equations for growth mechanisms, where x_n indicates the population size at generation n and $\Delta x_n = x_{n+1} - x_n$ is the change in the population size from the nth to the $(n+1)^{\text{th}}$ generation

(a) Constant growth: $\Delta x_n = \text{c}$, where c is some constant

(b) Proportional growth: $\Delta x_n = \text{c}x_n$

(c) Geometric growth: $\Delta x_n = \text{c}r^n$, where r is the common ratio between two successive generations

(d) Logarithmic growth: $\Delta x_n = \log n$

(e) Delayed growth: $\Delta x_n = f(x_{n-t})$, where f is a function to be specified and t is the lag-time or delay-time.

(f) Logistic growth: $\Delta x_n = \text{c}x_n(x_\text{E} - x_n)$, where x_E is an equilibrium population.

(g) Stochastic or probabilistic growth: $\Delta x_n = px_n$, where p is the probability of each individual of the population 'giving birth' to a new individual.

EXAMPLE 5.10 A new park is designed to have a buffalo herd. Due to normal mortality, the herd size is expected to decrease by 5 percent each year. The initial herd size is 85 buffaloes and each year 38 more buffaloes are added to the herd. Express the herd size as a sequence, using difference equation. What is the population of buffaloes in the 3rd generation?

SOLUTION

Let x_n denote the herd size at generation n and $x_0 = 85$ denote the initial herd size. Based upon the above information, the change Δx_n in the herd size will be same at each generation x_n.

$$\Delta x_n = -0.05x_n + 38 \qquad \text{(5\% death + new stock)}$$

Also, $$\Delta x_n = x_{n+1} - x_n$$

Thus, $$x_{n+1} = 0.95x_n + 38,$$

which is the required difference equation from this formula. Now we can evaluate the herd size at generations $n = 1, 2, 3, \ldots$.

$$\begin{aligned} x_1 &= 0.95x_0 + 38 = 118.75 \\ x_2 &= 0.95x_1 + 38 \cong 150.81 \\ x_3 &= 0.95x_2 + 38 \cong 181.27 \end{aligned}$$

and so on. Thus, the herd size in the 3rd generation is approximately 181.

A solution of the difference equation $x_{n+1} = f(n, x_n)$ is a sequence $\{x_n\}$ such that for each choice of $n = 1, 2, 3, \ldots$ the relation $x_{n+1} = f(n, x_n)$ is true.

The sequence $\{x_n\}$ where $x_n = \frac{3}{2^n}$ is a solution of the difference equation $x_{n+1} = \frac{1}{2}x_n$. This may be verified by showing that x_{n+1} equals $\frac{1}{2}x_n$:

$$x_{n+1} = \frac{3}{2^{n+1}} = \frac{3}{2^n} \cdot \frac{1}{2} = \frac{1}{2}x_n.$$

The solution of a difference equation need not be unique. Consider the following example.

EXAMPLE 5.11

(a) The difference equation $x_{n+1} = x_n + 2$ has the solution $\{3, 5, 7, 9, \ldots\}$ and also $\{-4, -2, 0, 2, 4, \ldots\}$.

(b) The difference equation $x_{n+1} \cdot x_n = 1$ has the solution $x_n = 1$ for all n, and also the solution

$$x_n = \begin{cases} \frac{1}{2}, & \text{if } n \text{ is odd;} \\ 2, & \text{if } n \text{ is even.} \end{cases}$$

(c) The difference equation $\Delta x_n = 0$ has as a solution, the constant sequence $\{x_n\}$, where $x_n = $ k, a constant. Then each choice of k results in a different solution.

EXERCISE 5.3

1. A milkman supplies milk in a colony. The number of new customers added every month is directly proportional to the number of existing customers. Express this as a difference equation of sequence $\{x_n\}$, where x_n denotes the number of customers at the end of the nth month. If the milkman started with 5 customers initially and after one month he had total seven customers, find the approximate number of his customers after 6 months.
2. In a school, 500 children are not vaccinated against smallpox. Trend analysis of an outbreak of smallpox has shown that the number of new cases of smallpox reported in the school per week was proportional to (i) the number of children already affected and (ii) the number of unaffected children amongst the non-vaccinated children. Express this as a difference equation of the sequence $\{x_n\}$, x_n denoting the number of affected children after n weeks. Initially 2 children were affected and 3 more were reported affected in the first week of the outbreak. How many children would be affected after (i) 3 weeks (ii) 5 weeks.
3. A factory can produce 50 units more of a product every month than that produced the previous month. Express it as a difference equation. If in the first month it produced 100 units, in which month will it produce 2000 units? What will be the total number of products by then?
4. The growth rate of a company is reported to be 10% every year. Express it as a difference equation of $\{x_n\}$, where x_n is the asset of the company at the end of the nth year. In how many years, will the company be above the double of its assets?

5.4 TYPES OF SEQUENCES

5.4.1. Subsequences

Another way of constructing a sequence is by picking up terms of an existing sequence. The sequence so obtained is called a subsequence of the already existing sequence.

> **? DEFINITION:** Let $\{a_n\}$ be a sequence and $n_1, n_2, n_3, \ldots$ be an increasing sequence of natural numbers. We say that the sequence $\{b_k\}$, where $b_k = a_{nk}$, is a 'subsequence' of the sequence $\{a_n\}$. We also denote this subsequence by $\{a_{nk}\}$.

EXAMPLE 5.12

(i) Let $a_n = (-1)^n$ and $b_k = a_{2k}$. Then the sequence $\{b_k\} = 1, 1, 1, \ldots$ is a subsequence of $\{a_n\} = -1, 1, -1, 1, -1, \ldots$

(ii) Let $a_n = \dfrac{n!}{n^n}$ and $b_k = a_{2k+1}$. Then $\{b_k\} = \left\{\dfrac{(2k+1)!}{(2k+1)^{2k+1}}\right\}$ is a subsequence of $\{a_n\}$.

5.4.2 Other Important Types of Sequences

Depending on their characteristics, sequences can be divided into different types or categories. The following definition describes some important types of sequences:

> **? DEFINITION:** Let $\{a_n\}$ be a sequence. We say that the sequence $\{a_n\}$ is
> (i) *bounded above*, if there exists a real number K such that $a_n \leq K$ for all n;
> (ii) *bounded below*, if there exists a real number k such that $k \leq a_n$ for all n;
> (iii) *bounded*, if it is both bounded above and bounded below. In other words, if there exists some real number M such that $|a_n| \leq M$ for all n;
> (iv) *increasing*, if $a_{n+1} \geq a_n$ for all n;
> (v) *decreasing*, if $a_{n+1} \leq a_n$ for all n;
> (vi) *monotonic*, if it is either increasing or decreasing;
> (vii) *strictly increasing*, if $a_{n+1} > a_n$ for all n;
> (viii) *strictly decreasing*, if $a_{n+1} < a_n$ for all n;
> (ix) *strictly monotonic*, if it is either strictly increasing or strictly decreasing.

EXAMPLE 5.13 Give an example of a sequence $\{a_n\}$ which is unbounded but contains a bounded subsequence.

SOLUTION

The sequence given by $\{a_n\} = \{n^{(-1)^n}\}$ has its members $1, 2, \frac{1}{3}, 4, \frac{1}{5}, \ldots$ and is unbounded. But its subsequence $\{a_{2k-1}\} = 1, \frac{1}{3}, \frac{1}{5}, \ldots$ is bounded.

EXAMPLE 5.14 Verify which of the following sequences are monotonic and bounded:

(a) $\left\{1 - \frac{1}{n}\right\}$ (b) $\{n - 2^n\}$ (c) $\left\{\frac{n!}{n^n}\right\}$ (d) $\left\{\frac{n}{2^n}\right\}$ (e) $\left\{n^{(-1)^n}\right\}$

(f) $\left\{\frac{n^2 + 2n + 1}{n^2 - 3}\right\}$ (g) $\left\{\frac{2^n}{n!}\right\}$ (h) $\left\{\frac{n}{n + 1}\right\}$ (i) $\{n e^{-n}\}$

SOLUTION

(a) $a_{n+1} - a_n = \left(1 - \frac{1}{n + 1}\right) - \left(1 - \frac{1}{n}\right) = \frac{1}{n} - \frac{1}{n + 1} = \frac{1}{n(n + 1)} > 0$

$\therefore$ $\{a_n\}$ is an increasing sequence. Also, $0 < a_n < 1$ for all n. Therefore $\{a_n\}$ is bounded.

(b) $$\begin{aligned} a_{n-1} - a_n &= ((n + 1) - 2^{n+1}) - (n - 2^n) \\ &= 1 - 2^{n+1} + 2^n \\ &= 1 - 2^n(2 - 1) \\ &= 1 - 2^n < 0 \qquad \text{for all } n. \end{aligned}$$

$\therefore$ $\{a_n\}$ is a decreasing sequence.

$\{a_n\}$ is bounded above but not bounded below.

(c) The terms of the sequence are $1, \frac{1}{2}, \frac{2}{9}, \frac{3}{32}, \ldots$

Here $\frac{a_{n+1}}{a_n} = \frac{(n + 1)!}{(n + 1)^{n+1}} \cdot \frac{n^n}{n!} = \left(\frac{n}{n + 1}\right)^n < 1$

Hence, $a_{n+1} < a_n$ for all n.

Therefore, $\{a_n\}$, where $a_n = \frac{n!}{n^n}$ is a decreasing (in fact, strictly decreasing) sequence. It is bounded also, as $0 < a_n \le 1$ for all n.

(d) Here $\frac{b_{n+1}}{b_n} = \frac{n + 1}{2^{n+1}} \Big/ \frac{n}{2^n} = \frac{1}{2}\left(1 + \frac{1}{n}\right) \le 1.$

Therefore, $\{b_n\}$ is a decreasing sequence. It is bounded as well, since $0 < b_n \le \frac{1}{2}$ for all n.

(e) Here, the terms of the sequence are $1, 2, \frac{1}{3}, 4, \frac{1}{5}, 6, \ldots$. It is clear that this sequence is neither increasing nor decreasing, nor bounded. However, it is bounded below as $0 < c_n$ for all n.

(f) Here $d_n = \frac{n^2 + 2n + 1}{n^2 - 3}$; the terms of the sequence are $-2, 9, \frac{8}{3}, \ldots$. Thus, the sequence is not monotonic. We write

$$d_n = \frac{n^2 + 2n + 1}{n^2 - 3} = 1 + \frac{2n + 4}{n^2 - 3} = 1 + c_n,$$

where $c_n = \frac{2n + 4}{n^2 - 3}$. It can be shown that for $n \ge 2$,

$$\frac{c_n}{c_{n+1}} > 1 \Leftrightarrow \frac{2n + 4}{n^2 - 3} \times \frac{(n + 1)^2 - 3}{2(n + 1) + 4} > 1$$

$\Leftrightarrow$ $n^2 + 2n + 5 > 0$, which is true for all $n \ge 2$. (In fact, for $n = 1$, the L.H.S. of the inequality becomes negative). Then $\{c_n\}$ is strictly decreasing for all $n \ge 2$. Therefore $\{d_n\}$ is strictly decreasing for $n \geq 2$. Moreover, $\{d_n\}$ is bounded, as we have $-2 \leq d_n \leq 9$ for all $n \geq 1$.

(g) Here $x_n = \frac{2^n}{n!}$, so that $\frac{x_n}{x_{n+1}} = \frac{2^n}{n!} \times \frac{(n+1)!}{2^{n+1}} = \frac{n+1}{2} \geq 1$ for all $n \geq 1$. Thus, sequence $\{x_n\}$ is decreasing. Moreover $0 < x_n \leq 2$. Thus, $\{x_n\}$ is bounded too.

(h) We employ an alternative method, the differentiation technique, to prove monotonicity of the sequence.

Let $f(x) = \frac{x}{x+1}$, so that $f(a_n) = a_n$ for all n.

Now, $$f'(x) = \frac{(x+1) - x}{(x+1)^2} = \frac{1}{(x+1)^2} > 0.$$

$\therefore f$ is increasing for all x. In particular, $f(n+1) > f(n)$ for all n.

$\therefore a_{n+1} > a_n$ for all n.

Hence $\{a_n\}$ is an increasing sequence. Also, $0 < \frac{n}{n+1} < 1$.

Thus, $\{a_n\}$ is bounded too.

(i) Let $f(x) = xe^{-x}$ so that $f(n) = a_n$ for all n.

Now, $f'(x) = e^{-x} - xe^{-x} = e^{-x}(1-x) \leq 0$ for all $x \geq 1$.

$\therefore$ $f(x)$ is decreasing for all $x \geq 1$. In particular,

$f(n+1) \leq f(n)$ for all n.

$\therefore$ $a_{n+1} \leq a_n$ for all n.

Hence $\{a_n\}$ is a decreasing sequence.

Also, $0 \leq a_n \leq a_1$ for all n; that is, $0 \leq a_n \leq \frac{1}{2}$ for all n. Therefore $\{a_n\}$ is bounded.

EXAMPLE 5.15 Give example of a sequence $\{a_n\}$ which is

(a) bounded but not monotonic;

(b) increasing with $a_n \neq 0$ for all $n \in \mathbf{N}$, but $b_n = \frac{1}{a_n}$ is not monotonic.

SOLUTION

(a) The sequences $\{(-1)^n\}, \left\{1 + \frac{(-1)^n}{n}\right\}$ are bounded but not monotonic.

(b) The sequence $\{2n - 9\}$ is increasing and $a_n \neq 0$ but the sequence $\{b_n\} = \left\{\frac{1}{a_n}\right\} = \left\{-\frac{1}{7}, -\frac{1}{5}, -\frac{1}{3}, -1, 1, \frac{1}{3}, \ldots\right\}$ is not monotonic.

EXAMPLE 5.16 Let $\{a_n\}$ be a sequence; define the sequence $\{b_n\}$ by $b_n = \frac{a_1 + a_2 + \cdots + a_n}{n}$, $n = 1, 2, 3, \ldots$. Show that if $\{a_n\}$ is bounded (respectively monotonic) then $\{b_n\}$ is also bounded (respectively monotonic).

SOLUTION

Assume that $\{a_n\}$ is bounded. Then there exists a real number $M > 0$ such that $|a_n| < M$ for every $n \in \mathbf{N}$.

Now,

$$|b_n| = \left|\frac{a_1 + a_2 + \cdots + a_n}{n}\right| \leq \frac{|a_1| + |a_2| + \cdots + |a_n|}{n} < \frac{nM}{n} = M$$

for all n; thus $\{b_n\}$ is bounded.

Further, we have

$$\begin{aligned} b_{n+1} - b_n &= \frac{a_1 + a_2 + \cdots + a_n + a_{n+1}}{n+1} - \frac{a_1 + a_2 + \cdots + a_n}{n} \\ &= \frac{n(a_1 + a_2 + \cdots + a_n + a_{n+1}) - (n+1)(a_1 + a_2 + \cdots + a_n)}{n(n+1)} \\ &= \frac{na_{n+1} - (a_1 + a_2 + \cdots + a_n)}{n(n+1)} \\ &= \frac{(a_{n+1} - a_1) + (a_{n+1} - a_2) + \cdots + (a_{n+1} - a_n)}{n(n+1)} \end{aligned}$$

This shows that the monotonicity of $\{b_n\}$ follows that of $\{a_n\}$.

EXERCISE 5.4

In Questions 1–10, determine, if the sequence $\{a_n\}$ is bounded above, bounded below or bounded.

1. $a_n = \dfrac{3n+1}{n+4}$

2. $a_n = \dfrac{2^n 5^n}{n!}$

3. $a_n = 4 - \dfrac{2}{n} - \dfrac{1}{2^n}$

4. $a_n = (-1)^n$

5. $a_n = 1 + \dfrac{(-1)^n}{n}$

6. $a_n = 1 + \dfrac{(-1)^n}{n}$

7. $a_n = \left(1 + \dfrac{1}{n}\right)^n$

8. $a_n = 1 + \dfrac{1}{2} + \cdots + \dfrac{1}{n}$

9. $a_n = 1 + \dfrac{1}{2} + \dfrac{1}{2^2} + \cdots + \dfrac{1}{2^{n-1}}$

10. $a_n =$ The nth prime

In Questions 11–25, show that the given sequence is strictly increasing or strictly decreasing.

11. $\left\{\dfrac{1}{n}\right\}$

12. $\left\{\dfrac{n}{3n+1}\right\}$

13. $\left\{\dfrac{n}{5n-1}\right\}$

14. $\{n - 3^n\}$

15. $\{n - n^2\}$

16. $\left\{\dfrac{3^n}{1+3^n}\right\}$

17. $\{2ne^{-n}\}$

18. $\left\{\dfrac{10^n}{(2n)!}\right\}$

19. $\left\{\dfrac{n^n}{n!}\right\}$ **20.** $\left\{\dfrac{5^n}{2^{(n^2)}}\right\}$ **21.** $\left\{\dfrac{1}{n} + \log n\right\}$ **22.** $\left\{\dfrac{\log(n+3)}{n+3}\right\}$

23. $\{\tan^{-1} n\}$ **24.** $a_1 = 1, a_{n+1} - 2a_n - 4$ **25.** $a_1 = 2, a_{n+1} = 3a_n - 2$

■ 5.5 CONVERGENT SEQUENCES

5.5.1 Convergence of Sequences

Consider the graphs of the following sequences:

$\left\{\dfrac{1}{n}\right\}, \{(-1)^n\}, \{n\}, \left\{\dfrac{(-1)^n}{n}\right\}$, represented by the diagrams (a), (b), (c) and (d) respectively.

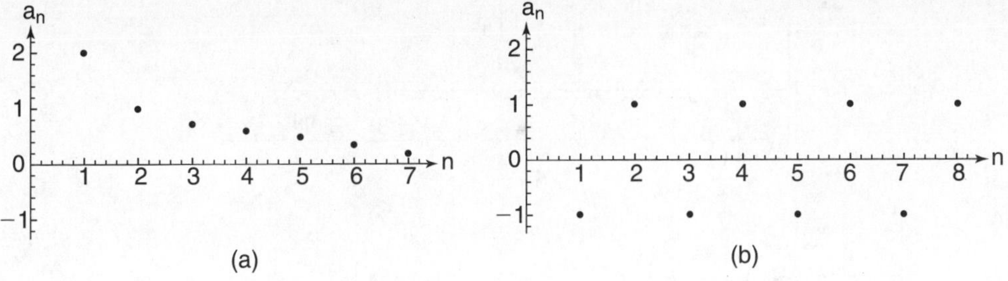

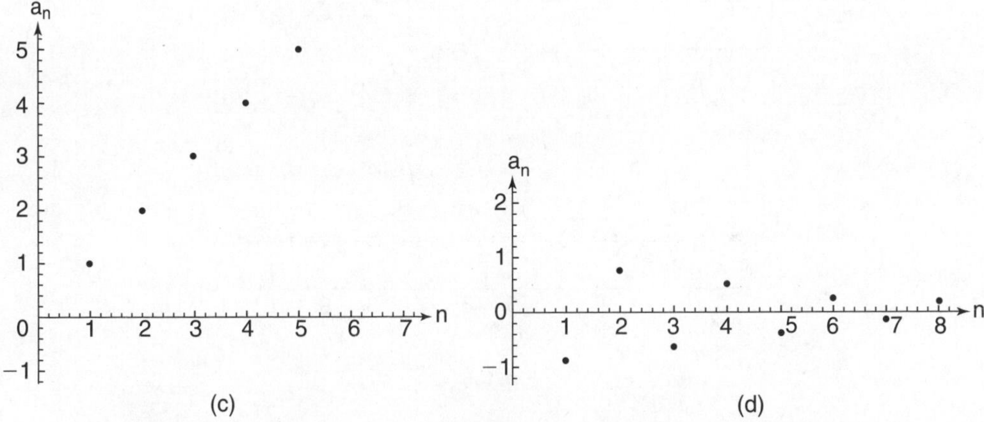

Fig. 5.4

In (a) and (d), as n increases, the terms (points) gradually move towards zero. In (b), the terms (points) oscillate between 1 and -1, in fact, they are alternatively -1 and 1. In (c), a_n increases as n increases. Thus, only in the first and the fourth cases, the terms of the sequences are moving towards a fixed finite number, namely zero. In other words, we can say, the terms of the sequences $\left\{\dfrac{1}{n}\right\}$ and $\left\{\dfrac{(-1)^n}{n}\right\}$ finally 'rest' around zero; however in case of $\{(-1)^n\}$ and $\{n\}$, no such unique 'resting point' is available.

The sequence of the first kind are of much interest to us. Let us observe few more such examples with a closer look.

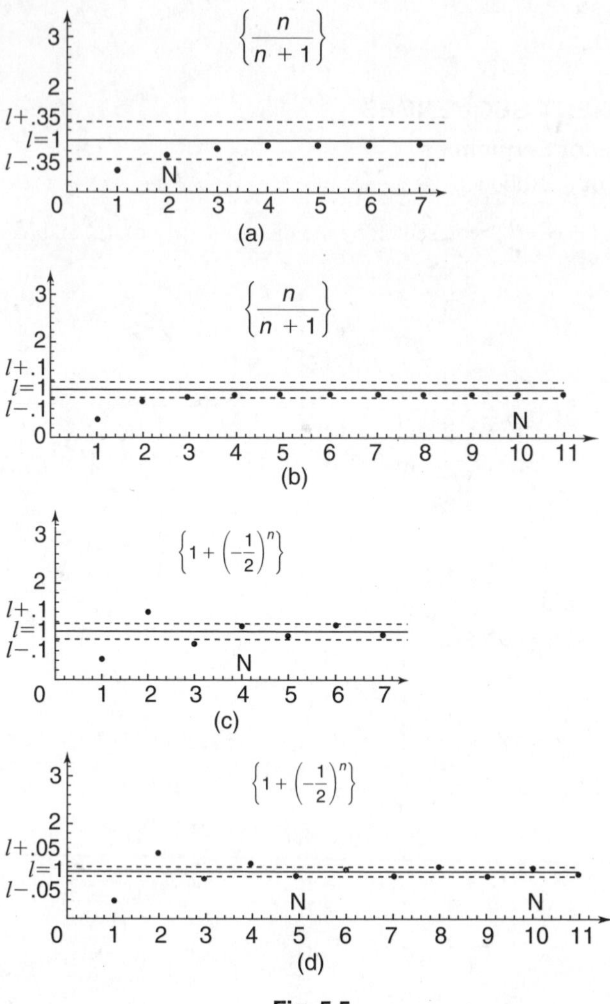

Fig. 5.5

In sequences like $\left\{\frac{n}{n+1}\right\}, \left\{1+\left(-\frac{1}{2}\right)^n\right\}$ and so on, we observe the following facts:

(i) As n increases, a_n's tend to one fixed number, say l.

(ii) If we lay one strip around the line representing l, all a_n's except the first few, that is, all a_n's for $n \geq \mathrm{N}$ (say) lie within that strip.

(iii) Thinner the strip, bigger is the value of N.

Now, we are very close to defining a convergent sequence. In fact, if such a number l, as observed above, exists, we call l the limit of the sequence $\{a_n\}$. Such a sequence is termed as a convergent sequence.

? DEFINITION: A sequence $\{a_n\}$ is said to converge to l, if given $\varepsilon > 0$, there exists a positive integer N such that

$$|a_n - l| < \varepsilon \qquad \text{for all } n \geq N$$

We express it by $\lim_{n \to \infty} a_n = l$ or by $a_n \to l$ as $n \to \infty$.

We also have the following definition.

? DEFINITION: We say that a sequence $\{a_n\}$ diverges to ∞ (respectively to $-\infty$) if for every real number $K > 0$, there exists a positive integer N such that $a_n > K$ (respectively $a_n < -K$) for all $n \geq N$.

We shall use notations $\lim_{n \to \infty} a_n = \infty$ and $\lim_{n \to \infty} a_n = -\infty$ to denote that $\{a_n\}$ diverges to ∞ and $\{a_n\}$ diverges to $-\infty$ respectively.

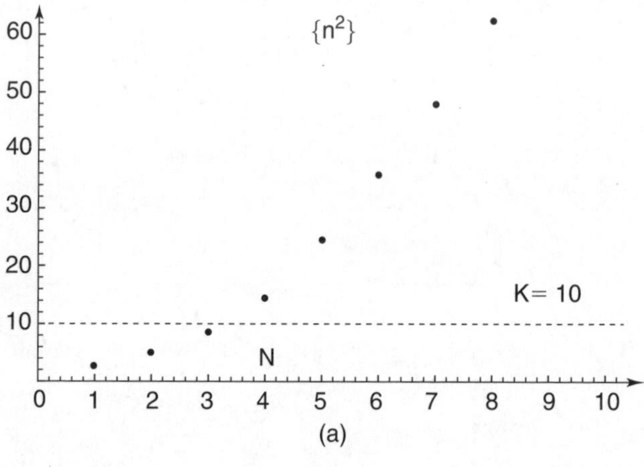

(a)

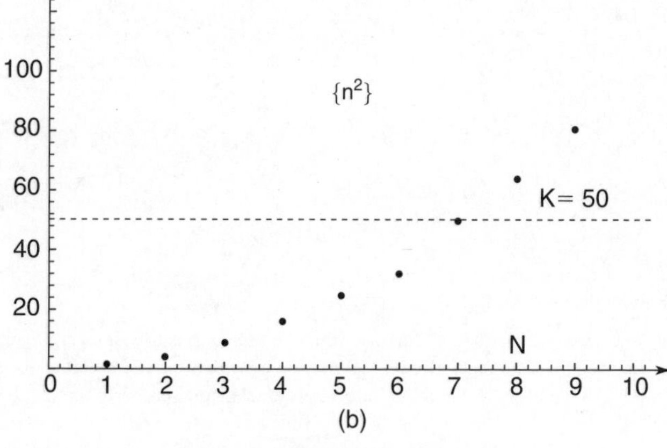

(b)

Fig. 5.6

A sequence which does not converge to any finite limit is said to be divergent. Thus, a sequence of the type $\{(-1)^n\}$ or $\{(-1)^n \cdot n\}$ which oscillates finitely or infinitely are examples of divergent sequences. Thus, apart from the convergent sequences, all the others are divergent sequences.

REMARK: We have provided the definition of convergence. When will a sequence be not convergent?
$\{a_n\}$ diverges $\Leftrightarrow$ For any real number l, there exists $\varepsilon > 0$ such that for every positive integer N, there exists $n \geq N$ with $|a_n - l| \geq \varepsilon$.

We illustrate the definition of convergence with the help of few examples.

EXAMPLE 5.17

(i) For the sequence $\left\{\frac{1}{n}\right\}$, let 0 be the number to be put to test for its limit (why ?). Let $\varepsilon > 0$ be any real number. Now

$$\left|\frac{1}{n} - 0\right| < \varepsilon \Leftrightarrow n > \frac{1}{\varepsilon}$$

Let N be any integer greater than $\frac{1}{\varepsilon}$ (We can always find such an integer, using Archemedian Property[1]). Then, we have,

$$\left|\frac{1}{n} - 0\right| < \varepsilon \qquad \forall n \geq \mathrm{N}$$

Hence, $\lim\limits_{n\to\infty} \frac{1}{n} = 0$.

REMARK: In the inequality $\left|\frac{1}{n} - 0\right| < \varepsilon$, if we take $\varepsilon = .25$, then N can be any integer greater than 4. If $\varepsilon = .001$, then N can be any integer greater than 1000, and so on.

(ii) $\left\{\frac{n}{n+1}\right\}$; The number to be tested here is 1 (why ?). Let $\varepsilon > 0$ be any real number. Now

$$\left|\frac{n}{n+1} - 1\right| < \varepsilon \Leftrightarrow n + 1 > \frac{1}{\varepsilon}$$

[1]**Archemedian Property.** If $x > 0$ and y is any real number, there exists a positive integer n such that $nx > y$. Geometrically, it means that any line segment, no matter how long, may be covered by a finite number of line segments of a given length, no matter how small. Archemedes realized that this was a fundamental property of straight lines and stated it as one of the axioms in geometry.

Let N be any positive integer greater than $\frac{1}{\varepsilon} - 1$. Then, we have

$$\left|\frac{n}{n+1} - 1\right| < \varepsilon \text{ for all } n \geq \text{N}.$$

Therefore, $\frac{n}{n+1} \to 1$ as $n \to \infty$.

REMARK: We could have also taken N as an integer greater than $\frac{1}{\varepsilon}$. Then for any $n \geq$ N, we would get, $n + 1 > \frac{1}{\varepsilon}$, serving our propose.

(iii) $\left\{1 + \left(-\frac{1}{2}\right)^n\right\}$; Again here, we test 1 for possible limit. Let $\varepsilon > 0$ be any real number. Now

$$\left|1 + \left(-\frac{1}{2}\right)^n - 1\right| < \varepsilon \iff n > \frac{\log 1/\varepsilon}{\log 2}.$$

Therefore, by choosing N to be a positive integer greater than $\frac{\log 1/\varepsilon}{\log 2}$, we get

$$\left|1 + \left(-\frac{1}{2}\right)^n - 1\right| < \varepsilon \qquad \text{for all } n \geq \text{N}$$

Hence,
$$\lim_{n\to\infty}\left[1 + \left(-\frac{1}{2}\right)^n\right] = 1.$$

(iv) Let $a_n = n^2$. The terms of this sequence are 1, 4, 9, 16, ... Let $\text{K} > 0$ be any real number. Then $a_n > \text{K}$ for all $n \geq \text{N}$, where N is any positive integer greater than $\sqrt{\text{K}}$. Therefore, the sequence $\{n^2\}$ diverges to ∞.

(v) Let $a_n = (-1)^n$. The terms of this sequence are $-1, 1, -1, 1, -1, \ldots$ Thus, the sequence is oscillating between -1 and 1, instead of approaching any fixed value, as n increases. Therefore $\{a_n\}$ is a divergent sequence.

(vi) $\left\{\frac{(-1)^n}{n}\right\}$. Let $\varepsilon > 0$. Now,

$$\left|\frac{(-1)^n}{n} - 0\right| < \varepsilon \iff n > \frac{1}{\varepsilon}.$$

Hence, as in the first example, $\lim_{n\to\infty} \frac{(-1)^n}{n} = 0$.

(vii) $\left\{\frac{1}{n^2}\right\}$ Let $\varepsilon > 0$ be any real number. Now

$$\left|\frac{1}{n^2} - 0\right| < \varepsilon \iff n > \frac{1}{\sqrt{\varepsilon}}$$

Choose any positive integer $N > \frac{1}{\sqrt{\varepsilon}}$. Then, we have

$$\left|\frac{1}{n^2} - 0\right| < \varepsilon \text{ for all } n \geq N.$$

Hence,
$$\frac{1}{n^2} \to 0 \text{ as } n \to \infty.$$

> **REMARK:** We could have also chosen N as an integer greater than $\frac{1}{\varepsilon}$. Then for any $n \geq N$, $n^2 > \frac{1}{\varepsilon}$ so that $\left|\frac{1}{n^2} - 0\right| < \varepsilon$. Thus, there may be several ways to select N.

(viii) $\left\{\frac{\sin n}{n}\right\}$.

[DU, B.Sc. (Ph.Sc) 2008]

Let $\varepsilon > 0$ be any real number. We observe that

$$\frac{1}{n} < \varepsilon \Leftrightarrow n > \frac{1}{\varepsilon} \text{ and } \left|\frac{\sin n}{n}\right| \leq \frac{1}{n}$$

Let N be any integer greater than $\frac{1}{\varepsilon}$. Then, we have

$$\left|\frac{\sin n}{n} - 0\right| < \varepsilon \text{ for all } n \geq N.$$

Hence,
$$\lim_{n\to\infty} \frac{\sin n}{n} = 0.$$

(ix) $\{x^n\}$ where $|x| < 1$. Let $\alpha = \frac{1 - |x|}{|x|}$. Then $\alpha > 0$ and $|x| = \frac{1}{1 + \alpha}$, so that $|x|^n = \frac{1}{(1 + \alpha)^n}$. Now,

$$\begin{aligned}(1 + \alpha)^n &= 1 + n\alpha + \frac{n(n-1)}{2}\alpha^2 + \cdots + \alpha^n \\ &> 1 + n\alpha;\end{aligned}$$

Thus, $|x|^n = \frac{1}{(1 + \alpha)^n} < \frac{1}{1 + n\alpha}$. Therefore, $|x|^n < \varepsilon$ whenever $\frac{1}{1 + n\alpha} < \varepsilon$; that is, whenever $n > \frac{1 - \varepsilon}{\alpha\varepsilon}$. We choose a positive integer N satisfying $N > \frac{1 - \varepsilon}{\alpha\varepsilon}$. Then, we have

$$|x^n - 0| = |x^n| = |x|^n < \varepsilon \text{ for all } n \geq N.$$

Hence,
$$\lim_{n\to\infty} x^n = 0.$$

5.5.2 Algebra of Convergent Sequences

Below we list out some important properties of convergent sequences. These properties may be used to simplify computations of limits so that it is not necessary to use the definition of convergence every time we evaluate a limit.

THEOREM

(i) The limit of a convergent sequence is unique.

(ii) If $a_n = c$ for each n, then $\lim_{n\to\infty} a_n = c$.

(iii) Let $\{a_n\}$ and $\{b_n\}$ be two sequences with $\lim_{n\to\infty} a_n = \alpha$ and $\lim_{n\to\infty} b_n = \beta$, then

(a) $\lim_{n\to\infty} (a_n + b_n) = \alpha + \beta$;

(b) $\lim_{n\to\infty} (a_n - b_n) = \alpha - \beta$;

(c) $\lim_{n\to\infty} (a_n b_n) = \alpha\beta$;

(d) If $b_n \neq 0$ and $\beta \neq 0$, then $\lim_{n\to\infty} \frac{a_n}{b_n} = \frac{\alpha}{\beta}$;

(e) $\lim_{n\to\infty} (\gamma a_n) = \gamma\alpha$, for any fixed real number γ.

We illustrate the above results with the help of some examples.

EXAMPLE 5.17 In each of the following, determine whether the sequence converges or diverges. If it converges find the limit.

(a) $\left\{\frac{\sin n}{n} \cdot \frac{n}{3n+1}\right\}$ (b) $\left\{\frac{1}{n^2} + \frac{(-1)^{n+1}}{n}\right\}$ (c) $\left\{(-1)^{n+1} \cdot \frac{1}{n}\right\}$

(d) $\{8 - 3n\}$ (e) $\left\{\frac{n}{3n+1}\right\}$ (f) $\left\{(-1)^{n+1} \cdot \frac{n}{3n+1}\right\}$ (g) $\left\{\frac{3 - 7n^4}{n^4 + 1}\right\}$

SOLUTION

(a) We have $\left|\frac{\sin n}{n}\right| \leq \frac{1}{n}$.

For given $\varepsilon > 0$, $\left|\frac{\sin n}{n} - 0\right| < \varepsilon$, whenever $\frac{1}{n} < \varepsilon$.

Therefore, by letting N be any integer greater than $\frac{1}{\varepsilon}$, we find

$$\left|\frac{\sin n}{n} - 0\right| < \varepsilon \qquad \text{for all } n \geq N.$$

Hence,
$$\lim_{n\to\infty} \frac{\sin n}{n} = 0.$$

Again,
$$\lim_{n\to\infty} \frac{n}{3n+1} = \lim_{n\to\infty} \frac{1}{3 + 1/n}$$

$$= \frac{\lim_{n\to\infty} 1}{\lim_{n\to\infty}\left(3 + \frac{1}{n}\right)}$$

$$= \frac{1}{\lim_{n\to\infty} 3 + \lim_{n\to\infty} \frac{1}{n}} = \frac{1}{3+0} = \frac{1}{3}.$$

Therefore, $\lim_{n\to\infty}\left(\frac{\sin n}{n}\cdot\frac{n}{3n+1}\right) = \lim_{n\to\infty}\frac{\sin n}{n}\cdot\lim_{n\to\infty}\frac{n}{3n+1} = 0\cdot\frac{1}{3} = 0$

(b) $\lim_{n\to\infty}\left(\frac{1}{n^2}+\frac{(-1)^{n+1}}{n}\right) = \lim_{n\to\infty}\frac{1}{n^2} + \lim_{n\to\infty}\frac{(-1)^n}{n} = 0 + 0\text{ (how?)} = 0.$

(c) Since $\lim_{n\to\infty}\frac{1}{n} = 0$, the product $(-1)^{n+1}\cdot\frac{1}{n}$ oscillates between positive and negative values, with the odd-numbered terms approaching 0 through positive values and even-numbered terms approaching 0 through negative values. Then $\lim_{n\to\infty}(-1)^{n+1}\cdot\frac{1}{n} = 0$, that is, the sequence converges to zero.

Alternatively, given $\varepsilon > 0$,

$$\left|(-1)^{n+1}\cdot\frac{1}{n} - 0\right| = \frac{1}{n} < \varepsilon \qquad \text{for all } n \geq \text{N}$$

where N is any positive integer greater than $\frac{1}{\varepsilon}$.

Hence, $\lim_{n\to\infty}(-1)^{n+1}\cdot\frac{1}{n} = 0.$

(d) The terms of the sequence are $5, 2, -1, -4, -7, -10, \ldots$.

Thus, the terms are steadily getting smaller as n increases.

Let $\text{K} > 0$ be any real number.

Then $8 - 3n < -\text{K}$ for all $n \geq \text{N}$,

where N is any positive integer $\geq \frac{8+k}{3}$.

Hence, $\lim_{n\to\infty}(8 - 3n) = -\infty$.

Thus the sequence $\{8 - 3n\}$ diverges to $-\infty$.

(e) Discussed in (a) above.

(f) The terms of the sequence are same as of (e), except for its sign. Thus, the terms oscillate between positive and negative values, with the odd numbered terms being identical to those in (e) and even numbered terms being the negatives of those in (e). Thus, the odd-numbered terms in the sequence approach 1/3 and the even-numbered terms approach −1/3. Therefore, the sequence has no limit, it diverges.

(g) $\lim_{n\to\infty}\frac{3-7n^4}{n^4+4} = \lim_{n\to\infty}\frac{\frac{3}{n^4}-7}{1+\frac{4}{n^4}} = \frac{\lim_{n\to\infty}\left(\frac{3}{n^4}-7\right)}{\lim_{n\to\infty}\left(1+\frac{4}{n^4}\right)} = \frac{0-7}{1+0} = -7.$

5.5.3 More on Convergence of Sequences

In this section, we provide some useful results and their applications, related to convergence of sequences. A useful result in this regard is the Sandwich theorem.

▸ THEOREM ◂

The Sandwich Theorem for sequences Let $\{a_n\}, \{b_n\}$ and $\{c_n\}$ be sequences of real numbers such that $a_n \le b_n \le c_n$ for all $n \ge N$, for some positive integer N and $\lim_{n\to\infty} a_n = \alpha = \lim_{n\to\infty} c_n$. Then $\lim_{n\to\infty} b_n = \alpha$. (This property is also known as the squeeze property of sequences).

An immediate consequence of this result is the following corollary:

Corollary If $\lim_{n\to\infty} |a_n| = 0$, then $\lim_{n\to\infty} a_n = 0$.

It follows from the fact that $\quad -|a_n| \le a_n \le |a_n|$.

EXAMPLE 5.18 Examine the convergence of the following sequences:

(a) $\left\{\dfrac{\cos n}{n}\right\}$ (b) $\left\{\dfrac{1}{2^n}\right\}$ (c) $\left\{(-1)^n \cdot \dfrac{1}{n}\right\}$

SOLUTION

(a) Since $0 \le \left|\dfrac{\cos n}{n}\right| \le \dfrac{1}{n}$ and $\dfrac{1}{n} \to 0$ as $n \to \infty$.

Hence by Sandwich Theorem, with $a_n = 0$ for all n,

$$\lim_{n\to\infty}\left|\frac{\cos n}{n}\right| = 0.$$

Consequently, $\lim_{n\to\infty} \dfrac{\cos n}{n} = 0$ (by Corollary).

(b) Since, $0 \le \dfrac{1}{2^n} \le \dfrac{1}{n}$, using Sandwich Theorem,

we get, $$\lim_{n\to\infty} \frac{1}{2^n} = 0.$$

(c) Here, $0 \le \left|(-1)^n \dfrac{1}{n}\right| \le \dfrac{1}{n}$, hence as in (a),

$$\lim_{n\to\infty} (-1)^n \cdot \frac{1}{n} = 0.$$

EXAMPLE 5.19 Discuss the convergence of the sequences

(a) $\left\{\dfrac{2n}{n^2+1}\sin(3n+1)\right\}$ (b) $\left\{\dfrac{1+2+\cdots+n}{n^3+1}\cos n!\right\}$

SOLUTION

(a) $$0 \le \left|\frac{2n}{n^2+1}\sin(3n+1)\right| \le \frac{2n}{n^2+1} \cdot 1$$

$$= \frac{2}{n+\dfrac{1}{n}} \le \frac{2}{n}$$

Now, $\lim_{n\to\infty}\frac{2}{n} = 2\lim_{n\to\infty}\frac{1}{n} = 2\times 0 = 0.$

Hence, by the Sandwich Theorem, $\lim_{n\to\infty}\left|\frac{2n}{n^2+1}\sin(3n+1)\right| = 0$

and by the corollary, $\lim_{n\to\infty}\frac{2n}{n^2+1}\sin(3n+1) = 0.$

(b) Here
$$a_n = \frac{1+2+\ldots+n}{n^3+1}\cos n! = \frac{n(n+1)}{2(n+1)(n^2-n+1)}\cos n! = \frac{1}{2\left(n-1+\frac{1}{n}\right)}\cos n! \le \frac{1}{n-1}$$

Thus, $0 \le |a_n| \le \frac{1}{n-1}$.

Hence, by the Sandwich Theorem $\lim_{n\to\infty}|a_n| = 0$ and therefore, by the corollary $\lim_{n\to\infty} a_n = 0$.

Earlier we have studied the bounded sequences. A bounded sequence need not be convergent as is evident from $\{(-1)^n\}$. However, we have the following results:

THEOREM

(i) Every convergent sequence is bounded.
(ii) A bounded monotonic sequence is convergent. In particular, an increasing sequence which is bounded above is convergent. Similarly, a decreasing sequence which is bounded below is convergent.

EXAMPLE 5.20 Discuss the convergence of the following sequences

(i) $\frac{5^n}{n!}$ (ii) $\frac{x^n}{n!}$, (x is fixed, real) (iii) $\frac{n!}{n^n}$ (iv) $\left(1+\frac{1}{n}\right)^n$

SOLUTION

(i) Here $a_n = \frac{5^n}{n!}, \frac{a_{n+1}}{a_n} = \frac{5^{n+1}}{(n+1)!}\times\frac{n!}{5^n} = \frac{5}{n+1} < 1$ for $n \ge 5$.

Therefore, $a_{n+1} < a_n$ for all $n \ge 5$.

Thus, $\{a_n\}$ is a decreasing sequence for $n \ge 5$ and is bounded below by 0.

Hence, $\{a_n\}$ is convergent. Let $\lim_{n\to\infty} a_n = l$.

Now,
$$a_{n+1} = \frac{5}{n+1}\cdot\frac{5^n}{n!} = \frac{5}{n+1}a_n.$$

Taking limit on both $\lim_{n\to\infty} a_{n+1} = \lim_{n\to\infty} \frac{5}{n+1} a_n = \lim_{n\to\infty} \frac{5}{n+1} \cdot \lim_{n\to\infty} a_n$

$\Rightarrow \quad l = 0 \cdot l$

Hence, $\lim_{n\to\infty} \frac{5^n}{n!} = 0.$

> **REMARK:** The convergence or divergence of a sequence does not depend on the behaviour of its initial terms, that is, finitely many terms in the beginning, but on how the terms behave finally, to say, for $n \geq N$ where N is any finite positive integer.

(ii) $a_n = \frac{x^n}{n!}$; Here same argument works as in (i) as x is a fixed real number.

Therefore, $\lim_{n\to\infty} \frac{x^n}{n!} = 0.$

(iii) $a_n = \frac{n!}{n^n}$; $a_{n+1} = \frac{(n+1)!}{(n+1)^{n+1}} = \frac{n!(n+1)}{(n+1)^{n+1}} = \frac{n!}{(n+1)^n}.$

Hence, $a_n > a_{n+1}$. Thus, $\{a_n\}$ is a monotonic decreasing sequence which is bounded below (as $a_n > 0$ for all n).

Therefore, $\{a_n\}$ is convergent.

Now, $a_n = \frac{n!}{n^n} = \frac{1 \cdot 2 \cdot 3 \ldots n}{n \cdot n \cdot n \ldots n} \leq \frac{1}{n} \cdot \frac{n}{n} \cdot \frac{n}{n} \cdots \frac{n}{n}$

Thus, we have, $0 \leq a_n \leq \frac{1}{n}$; therefore by Sandwich Theorem

$$\lim_{n\to\infty} a_n = 0.$$

(iv) Applying binomial theorem, we get

$$a_n = \left(1 + \frac{1}{n}\right)^n = 1 + n \cdot \frac{1}{n} + \frac{n(n-1)}{2!} \cdot \frac{1}{n^2} + \frac{n(n-1)(n-2)}{3!} \cdot \frac{1}{n^3} + \cdots + \frac{n(n-1) \cdot 2 \cdot 1}{n!} \cdot \frac{1}{n^n}$$

$$= 1 + 1 + \frac{1}{2!}\left(1 - \frac{1}{n}\right) + \frac{1}{3!}\left(1 - \frac{1}{n}\right)\left(1 - \frac{2}{n}\right) + \cdots + \frac{1}{n!}\left(1 - \frac{1}{n}\right)\left(1 - \frac{2}{n}\right) \cdots \left(1 - \frac{n-1}{n}\right)$$

$$\leq 1 + 1 + \frac{1}{2!} + \frac{1}{3!} + \cdots + \frac{1}{n!}$$

$$\leq 1 + 1 + \frac{1}{2} + \frac{1}{2^2} + \cdots + \frac{1}{2^{n-1}}$$

$$= 1 + \frac{1 - 1/2^n}{1 - 1/2}$$

$$= 1 + 2\left(1 - \frac{1}{2^n}\right) \leq 3.$$

Also, $a_{n+1} = \left(1 + \frac{1}{n+1}\right)^{n+1}$

$$= 1 + (n+1)\cdot\frac{1}{n+1} + \frac{(n+1)\cdot n}{2!}\cdot\frac{1}{(n+1)^2} + \frac{(n+1)n(n-1)}{3!}\cdot\frac{1}{(n+1)^3} + \cdots$$

$$= 1 + 1 + \frac{1}{2!}\left(1 - \frac{1}{n+1}\right) + \frac{1}{3!}\left(1 - \frac{1}{n+1}\right)\left(1 - \frac{2}{n+1}\right) + \cdots$$

$$\geq 1 + 1 + \frac{1}{2!}\left(1 - \frac{1}{n}\right) + \frac{1}{3!}\left(1 - \frac{1}{n}\right)\left(1 - \frac{2}{n}\right) + \cdots = a_n$$

Therefore, $\{a_n\}$ is an increasing sequence which is bounded above. Hence, by the above theorem $\{a_n\}$ is convergent.[2] Moreover, $2 \leq a_n \leq 3$ for all n. This leads to the fact that $2 \leq \lim_{n\to\infty} a_n \leq 3$. We denote this limit by the letter 'e'. That is, $\lim_{n\to\infty}\left(1 + \frac{1}{n}\right)^n = e$.

REMARK: In the above example, we found that

$$\left(1 + \frac{1}{n}\right)^n = 1 + 1 + \frac{1}{2!}\left(1 - \frac{1}{n}\right) + \frac{1}{3!}\left(1 - \frac{1}{n}\right)\left(1 - \frac{2}{n}\right) + \cdots + \frac{1}{n!}\left(1 - \frac{1}{n}\right)\left(1 - \frac{2}{n}\right)\cdots\left(1 - \frac{n-1}{n}\right)$$

Taking limit on both the sides, as $n \to \infty$

$$\lim_{n\to\infty}\left(1 + \frac{1}{n}\right)^n = 1 + 1 + \frac{1}{2!} + \frac{1}{3!} + \cdots$$

Thus we get,

$$e = 1 + 1 + \frac{1}{2!} + \frac{1}{3!} + \frac{1}{4!} + \cdots$$

This expression can be used to define e, which is an irrational number.

EXAMPLE 5.21 Does the sequence given by the recursion formula

$$a_1 = 1, \qquad a_{n+1} = \sqrt{2a_n}$$

converge? If so, find its limit.

SOLUTION

The given sequence is $1, \sqrt{2}, \sqrt{2\sqrt{2}}, \sqrt{2\sqrt{2\sqrt{2}}}\ldots$ We show that $\{a_n\}$ is bounded above and increasing.
Here,

$$a_1 < 2, \qquad a_2 < 2.$$

So let us assume that

$$a_n < 2.$$

[2]For an alternative proof of (iv), see (v) on page 180.

$$\Rightarrow \qquad 2a_n < 4$$

$$\Rightarrow \qquad \sqrt{2a_n} < 2$$

$$\Rightarrow \qquad a_{n+1} < 2$$

Hence, by the principle of mathematical induction $a_n < 2$ for all n.
To, show that $\{a_n\}$ is increasing, we observe that $a_2 > a_1$, and assume that $a_n > a_{n-1}$.

$$\Rightarrow \qquad 2a_n > 2a_{n-1}$$

$$\Rightarrow \qquad \sqrt{2a_n} > \sqrt{2a_{n-1}}$$

$$\Rightarrow \qquad a_{n+1} > a_n$$

Hence, by the principle of mathematical induction, $a_{n+1} > a_n$ for all n.
Thus $\{a_n\}$ is an increasing sequence which is bounded above.
Hence $\{a_n\}$ is convergent.
Let $\lim_{n\to\infty} a_n = l$.
Now, $a_{n+1} = \sqrt{2a_n}$.
Taking limit on both the sides.

$$\lim_{n\to\infty} a_{n+1} = \lim_{n\to\infty} \sqrt{2a_n}$$

$$l = \sqrt{2l}$$

$$\Rightarrow \qquad l^2 - 2l = 0$$

$$\Rightarrow \qquad l = 2 \text{ or } l = 0$$

$$\Rightarrow \qquad l = 2 \qquad \text{as } a_1 = 1 \text{ and rest of the terms are greater than } a_1.$$

Below we provide a useful result which enables us to use L' Hospital's rule[3] to evaluate the limits of certain sequences.

THEOREM

Suppose that $f(x)$ is a function defined for all $x \geq$ N and $\{a_n\}$ is a sequence of real numbers such that $a_n = f(n)$ for all $x \geq$ N, where N is some positive integers. Then,

$$\lim_{x\to\infty} f(x) = l \Rightarrow \lim_{n\to\infty} a_n = l$$

EXAMPLE 5.22 Find the limits of the following

(i) $\left\{\dfrac{\log n}{n}\right\}$ (ii) $\left\{\dfrac{3^n}{5n}\right\}$ (iii) $\left\{\dfrac{n^2-4}{n+2}\right\}$ (iv) $\left\{\dfrac{\log n}{n^c}\right\}, c > 0$ (v) $\left\{\dfrac{n}{e^n}\right\}$

[3]**L' Hospital's Rule.** For two functions f and g, suppose that $f(a) = g(a) = 0$, and that $f'(a)$ and $g'(a)$ exist and that $g'(a) \neq 0$. Then

$$\lim_{x\to a} \frac{f(x)}{g(x)} = \frac{f'(a)}{g'(a)}.$$

The rule also applies to quotients that lead to indeterminate form $\dfrac{\infty}{\infty}$.

SOLUTION

(i) The function $f(x) = \dfrac{\log x}{x}$ is defined for all $x \geq 1$ and agrees with the given sequence at positive integers.

Therefore, by the above theorem, $\lim\limits_{n\to\infty} \dfrac{\log n}{n}$ will equal $\lim\limits_{x\to\infty} \dfrac{\log x}{x}$, if the latter exists.

Applying L' Hospital's rule we get

$$\lim_{x\to\infty} \frac{\log x}{x}\left(\frac{\infty}{\infty}\right) = \lim_{x\to\infty} \frac{1/x}{1} = \frac{0}{1} = 0$$

Hence, $\lim\limits_{n\to\infty} \dfrac{\log n}{n} = 0.$

(ii) By L' Hospital's rule, $\lim\limits_{x\to\infty} \dfrac{3^x}{5x}\left(\dfrac{\infty}{\infty}\right) = \lim\limits_{x\to\infty} \dfrac{3^x \log 3}{5} = \infty.$

Hence, $\lim\limits_{n\to\infty} \dfrac{3^n}{5n} = \infty$; that is, $\left\{\dfrac{3^n}{5n}\right\}$ diverges to ∞.

(iii) $\lim\limits_{x\to\infty} \dfrac{x^2 - 4}{x + 2} = \lim\limits_{x\to\infty} \dfrac{2x}{1} = \infty$ (L' Hospital's rule).

Therefore, $\lim\limits_{n\to\infty} \dfrac{n^2 - 4}{n + 2} = \infty.$

(iv) $\lim\limits_{x\to\infty} \dfrac{\log x}{x^c}\left(\dfrac{\infty}{\infty}\right) = \lim\limits_{x\to\infty} \dfrac{1/x}{c x^{c-1}} = \lim\limits_{x\to\infty} \dfrac{1}{c x^c} = 0$ (L' Hospital's rule).

Hence[4], $\lim\limits_{n\to\infty} \dfrac{\log n}{n^c} = 0$

(v) $\lim\limits_{x\to\infty} \dfrac{x}{e^x} = \lim\limits_{x\to\infty} \dfrac{1}{e^x}$ (L' Hospital's rule)

$= 0$

Hence, $\lim\limits_{n\to\infty} \dfrac{n}{e^n} = 0.$

Here is one more result where continuity of a function is used for evaluation of limit.

THEOREM

If $a_n \to l$ and f is a function that is continuous at l and defined at all the a_n's, then $f(a_n) \to f(l)$.

EXAMPLE 5.23 Evaluate the following limits:

(i) $\lim\limits_{n\to\infty} \sqrt{(n-1)/n}$ (ii) $\lim\limits_{n\to\infty} 5^{1/n}$ (iii) $\lim\limits_{n\to\infty} \sqrt[n]{n}$

(iv) $\lim\limits_{n\to\infty} x^{1/n}$ $(x > 0)$ (v) $\lim\limits_{n\to\infty} \left(1 + \dfrac{x}{n}\right)^n$ (any x).

[4]It shows that log n increases more slowly than any positive power of n.

SOLUTION

(i) We know that $a_n = \dfrac{n-1}{n} = 1 - \dfrac{1}{n} \rightarrow 1.$

Therefore, $f(a_n) \rightarrow f(1)$, where f is any function is defined for every a_n and is continuous at $x = 1$.

We choose $f(x) = \sqrt{x}$. Then f satisfies the required conditions.

Therefore, $f(a_n) \rightarrow f(1)$. That $\sqrt{\dfrac{n-1}{n}} \rightarrow \sqrt{1}$.

$$\therefore \qquad \lim_{n\to\infty} \sqrt{\frac{n-1}{n}} = 1$$

(ii) We know that $a_n = \dfrac{1}{n} \rightarrow 0$. Therefore, $f(a_n) \rightarrow f(0)$, where f is any function such that f is defined for every a_n and is continuous at $x = 0$.

We take $f(x) = 5^x$, which satisfies the required conditions.

Hence, $f\left(\dfrac{1}{n}\right) \rightarrow f(0)$. That is $5^{1/n} \rightarrow 5^0$.

$$\therefore \qquad \lim_{n\to\infty} 5^{1/n} = 5 = 1\,.$$

(iii) $\sqrt[n]{n} = n^{1/n} = e^{\log n^{1/n}} = e^{\frac{1}{n}\log n}$

Let
$$a_n = \frac{\log n}{n}.$$

Now,

$$\lim_{n\to\infty} \frac{\log x}{x} \left(\frac{\infty}{\infty}\right)$$

$$= \lim_{n\to\infty} \frac{1}{x} \text{ (L' Hospital's rule)}$$

$$= 0$$

Therefore,

$$\lim_{n\to\infty} \frac{\log n}{n} = 0$$

$$\Rightarrow \qquad \lim_{n\to\infty} f\left(\frac{\log n}{n}\right) = \lim_{n\to\infty} f(a_n) = f(0)$$

where f is any function which is defined for all a_n and is continuous at $x = 0$.

We choose $f(x) = e^x$. Then f satisfies the required conditions.

$$\therefore \qquad \lim_{n\to\infty} e^{\frac{\log n}{n}} = e^0 = 1$$

$$\Rightarrow \qquad \lim_{n\to\infty} n^{1/n} = 1$$

(iv) Let $a_n = \log x^{1/n}$; then $a_n = \dfrac{\log x}{n} \to 0$ as $n \to \infty$ (since x is fixed).

Hence, $f(a_n) \to f(0)$, where f is any function which is defined for every a_n and is continuous at $x = 0$. Let $f(y) = e^y$. Then f is continuous at 0 and is defined at every a_n.

Hence, by the above theorem, we have, $f(a_n) \to f(0)$. That is $x^{1/n} \to 1$.

Then $\lim_{n\to\infty} x^{1/n} = 1$.

(v) Let $a_n = \log\left(1 + \dfrac{x}{n}\right)^n = n\log\left(1 + \dfrac{x}{n}\right) = \dfrac{\log\left(1 + \dfrac{x}{n}\right)}{1/n}$

$$
\begin{aligned}
\lim_{n\to\infty} a_n &= \lim_{n\to\infty} \frac{\log\left(1 + \frac{x}{n}\right)}{1/n} \quad \left(\frac{0}{0}\right) \\
&= \lim_{n\to\infty} \frac{\frac{1}{1 + \frac{x}{n}}\left(-\frac{x}{n^2}\right)}{-1/n^2} \qquad \text{(L' Hospital's rule, } x \text{ fixed)} \\
&= x
\end{aligned}
$$

Hence, $f(a_n) \to f(x)$, where f is any function which is defined for every a_n and is continuous at x.

Let $f(x) = e^x$; then f is continuous at x and is defined for every a_n. Hence, by the above theorem

$$f(a_n) \to f(x) \qquad \text{as } n \to \infty.$$

Therefore,

$$\lim_{n\to\infty}\left(1 + \frac{x}{n}\right)^n = e^x.$$

REMARK: If $x = 1$ in the above result, then we get $\lim_{n\to\infty}\left(1 + \dfrac{1}{n}\right)^n = e$. Thus (v) provides an alternative proof for the result $\lim_{n\to\infty}\left(1 + \dfrac{1}{n}\right)^n = e$.

5.5.4. Some Important Limits

From the examples provided so far, we list below few results on limits, which can be used for evaluating the limits of other similar sequences:

(i) $\lim_{n\to\infty} \dfrac{\log n}{n} = 0$ (ii) $\lim_{n\to\infty} \sqrt[n]{n} = 1$

(iii) $\lim_{n\to\infty} \sqrt[n]{x} = 1 \ (x > 0)$ (iv) $\lim_{n\to\infty} x^n = 0 \ (|x| < 1)$

(v) $\lim_{n\to\infty}\left(1 + \dfrac{x}{n}\right)^n = e^x$ (any x) (vi) $\lim_{n\to\infty} \dfrac{x^n}{n!} = 0$ (any x).

EXERCISE 5.5

In Questions 1–3, find out a positive integer N which satisfies the given inequalities.

1. $\left|\dfrac{1}{n+1} - 0\right| < \varepsilon$ for all $n \geq$ N, where $\varepsilon = .25, .01, .004, .0254$

2. $\left|\dfrac{1}{2^n} - 0\right| < \varepsilon$ for all $n \geq$ N, where $\varepsilon = 2, .9, .033, .0008$

3. $\left|\dfrac{\sin n}{n} - 0\right| < \varepsilon$ for all $n \geq$ N, where $\varepsilon = .2, .05, .001, .0175$

4. Use the definition of limit to show that:

(a) $\lim\limits_{n\to\infty} \dfrac{2n+3}{n+1} = 2$ **(b)** $\lim\limits_{n\to\infty} \dfrac{(-1)^n}{n^3+1} = 0$

(c) $\lim\limits_{n\to\infty} 2^{\sqrt{n}} = \infty$ **(d)** $\lim\limits_{n\to\infty} \dfrac{\sin(n^2)}{n+1} = 0$ **(e)** $\lim\limits_{n\to\infty} (100 - 3^n) = -\infty$

5. Compute the following limits:

(a) $\lim\limits_{n\to\infty} \dfrac{n}{n^2+1}\sin(2n+1)$ **(b)** $\lim\limits_{n\to\infty} \dfrac{1+2+3+\cdots+n}{n^3+4}\cos(2n)$

(c) $\lim\limits_{n\to\infty} \dfrac{1+4+7+\cdots+(3n-2)}{n^2+1}$ **(d)** $\lim\limits_{n\to\infty} \dfrac{1^2+2^2+3^2+\cdots+n^2}{6n^3-n^2+3n+4}$

(e) $\lim\limits_{n\to\infty} \left(\dfrac{1+2+\cdots+n}{n+1} - \dfrac{n}{2}\right)$ **(f)** $\lim\limits_{n\to\infty} \left[\sqrt{n^4+n^2} - \sqrt{n^4-n^2}\right]$

(g) $\lim\limits_{n\to\infty} \dfrac{1+\frac{1}{2}+\frac{1}{4}+\cdots+\frac{1}{2^n}}{1+\frac{1}{3}+\frac{1}{4}+\cdots+\frac{1}{3^n}}$ **(h)** $\lim\limits_{n\to\infty} \dfrac{\sqrt{n^2+4}-n}{\sqrt{n^2+3}-n}$

(i) $\lim\limits_{n\to\infty} \dfrac{(n+3)!+(n+2)!}{(n+3)!-(n+2)!}$ **(j)** $\lim\limits_{n\to\infty} \dfrac{7\cdot 3^{n+1}+3\cdot 4^n}{5\cdot 2^n+4^{n+2}}$

Find out which of the sequence $\{a_n\}$ in questions 6–39 converge and which ones diverge. Find the limit of each convergent sequence.

6. $a_1 = 1, a_{n+1} = 3a_n - 2$ **7.** $a_1 = 4 + (.3)^n$

8. $a_n = \dfrac{2n+(-1)^n}{n}$ **9.** $a_n = 2 + (-1)^n$

10. $a_n = \dfrac{2n+3}{1-2\sqrt{n}}$ **11.** $a_n = \left(4 - \dfrac{1}{2^n}\right)\left(3 + \dfrac{1}{2^n}\right)$

12. $a_n = \sin\left(\dfrac{\pi}{2} + \dfrac{1}{n}\right)$ **13.** $a_n = n\pi\cos\left(n\dfrac{\pi}{2}\right)$

14. $a_n = n\pi\cos(n\pi)$ **15.** $a_n = \dfrac{\cos^2 n}{3^n}$

16. $a_n = (1.5)^{1/n^2}$ **17.** $a_n = \dfrac{\log n}{n^{1/n}}$

18. $a_n = \dfrac{\log(n+2)}{\sqrt{n+2}}$

19. $a_n = \left(1 - \dfrac{2}{n}\right)^n$

20. $a_n = \sqrt[n]{5n}$

21. $a_n = \left(\dfrac{4}{n}\right)^{3/n}$

22. $a_n = \sqrt[n]{3^n n}$

23. $a_n = \dfrac{1}{n}\displaystyle\int_1^n \dfrac{1}{x}dx$

24. $a_n = \sqrt[n]{3^{2n+4}}$

25. $a_n = \dfrac{n!}{10^{2n}}$

26. $a_n = \left(\dfrac{1}{n}\right)^{1/(\log n)}$

27. $a_n = \dfrac{10^n}{n!}$

28. $a_n = \log\left(1 + \dfrac{1}{n}\right)^n$

29. $a_n = \left(\dfrac{4n+1}{4n-1}\right)^n$

30. $a_n = \tanh n$

31. $a = \dfrac{2^n \cdot 3^n}{5^{-n} \cdot n!}$

32. $a_n = \tan^{-1}(n^2)$

33. $a_n = \dfrac{2n^2}{3n-1}\sin\dfrac{1}{n}$

34. $a_n = \left(1 - \dfrac{1}{n^2}\right)^n$

35. $a_n = \left(\dfrac{x^n}{3n+1}\right)^{1/n}, x > 0$

36. $a_n = \dfrac{(\log n)^3}{n}$

37. $a_1 = 1, a_{n+1} = \sqrt{3a_n}$

38. $a_1 = 1, a_{n+1} = \sqrt{6a_n}$

39. $a_n = \displaystyle\int_1^n \dfrac{1}{x^p}dx, p > 1$

CHAPTER SIX

Functions and Their Graphs

■ 6.1 INTRODUCTION

Many scientific laws are discovered by collecting, organizing and analysing experimental data. Graphs play a major role in studying these data.

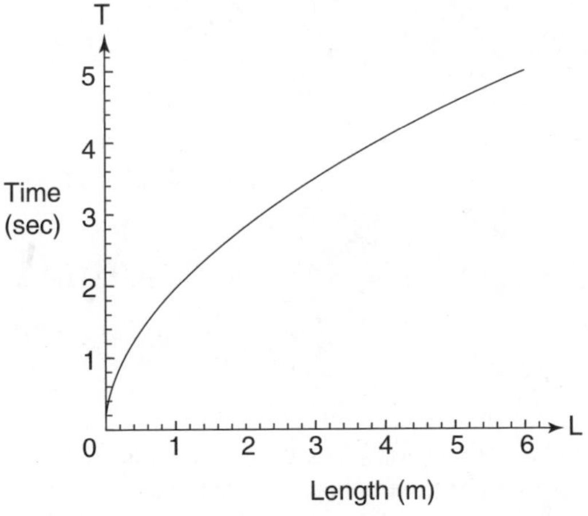

Fig. 6.1 The L-T graph for a simple pendulum reveals the relationship $L \propto T^2$.

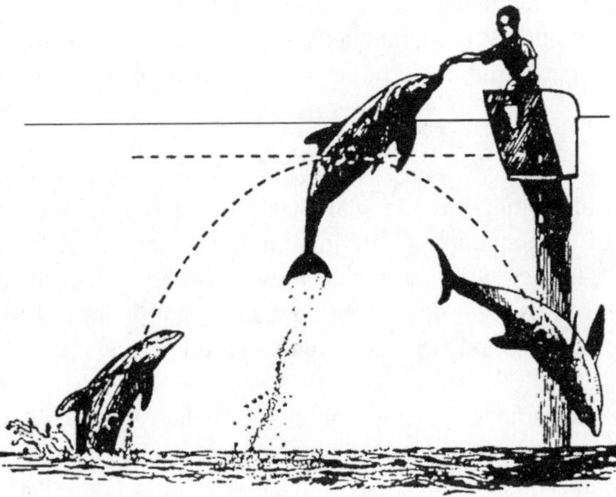

Fig. 6.2 The centre of gravity describes a parabola no matter how the animal acts during the motion.

From Fig. 6.1, it becomes clear that the time period of a simple pendulum is proportional to the square-root of the length of the pendulum. Fig. 6.2 confirms Galileo's Law which states that the centre of gravity is subject to the same acceleration for all bodies, provided air resistance is negligible.

However, graphs are susceptible to wrong interpretation too. Look at the following data (Table 6.1) and its two possible graphs given in Fig. 6.3.

Table 6.1 Average world yield of major grains measured in quintals per hectare

Year	Yield	Year	Yield	Year	Yield.
1960	13.9	1965	14.7	1970	17.3
1961	13.4	1966	16.2	1971	18.8
1962	14.3	1967	16.2	1972	18.5
1963	13.9	1968	16.7	1973	19.4
1964	14.5	1969	16.9		

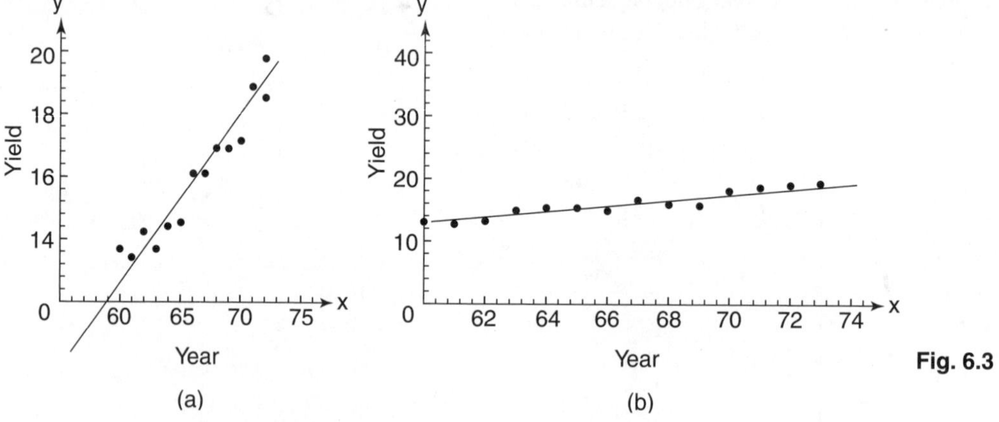

Fig. 6.3

Here the difference in visual impact is due to the scale-factor. While the first gives an idea that the yield is increasing at a sufficiently high rate, the second one presents a poor yield rate. The difference has occurred due to the change in the amount that a unit length represents along the axes in the two graphs.

Thus careful objectivity needs to be maintained while interpreting a graph. In this chapter, we discuss about the graphs of some commonly used functions such as polynomial, trigonometric, inverse trigonometric, exponential, logarithmic, hyperbolic functions and so on which occur frequently in scientific data analysis.

6.1.1 Graph of a Function

Mathematically speaking, the set of all points in the *xy*-plane whose co-ordinates satisfy an equation in x and y is called the graph of that equation. Graph may however be plotted from physical data also, as seen in the above cases. In fact, tables, graphs and equations provide three methods – numerical, visual and algebraic respectively – for describing how one quantity depends on another.

Recall that a function f is a rule or correspondence that assigns a unique value $f(x)$ to each input value x.

If f is a real-valued function of a real variable x, then the graph of f in the *xy*-plane is defined to be the graph of the equation $y = f(x)$. For example, the graph of the function $f(x) = x$ is the graph of the equation $y = x$. Fig. 6.4 shows the graphs of some functions which may be already familiar to you.

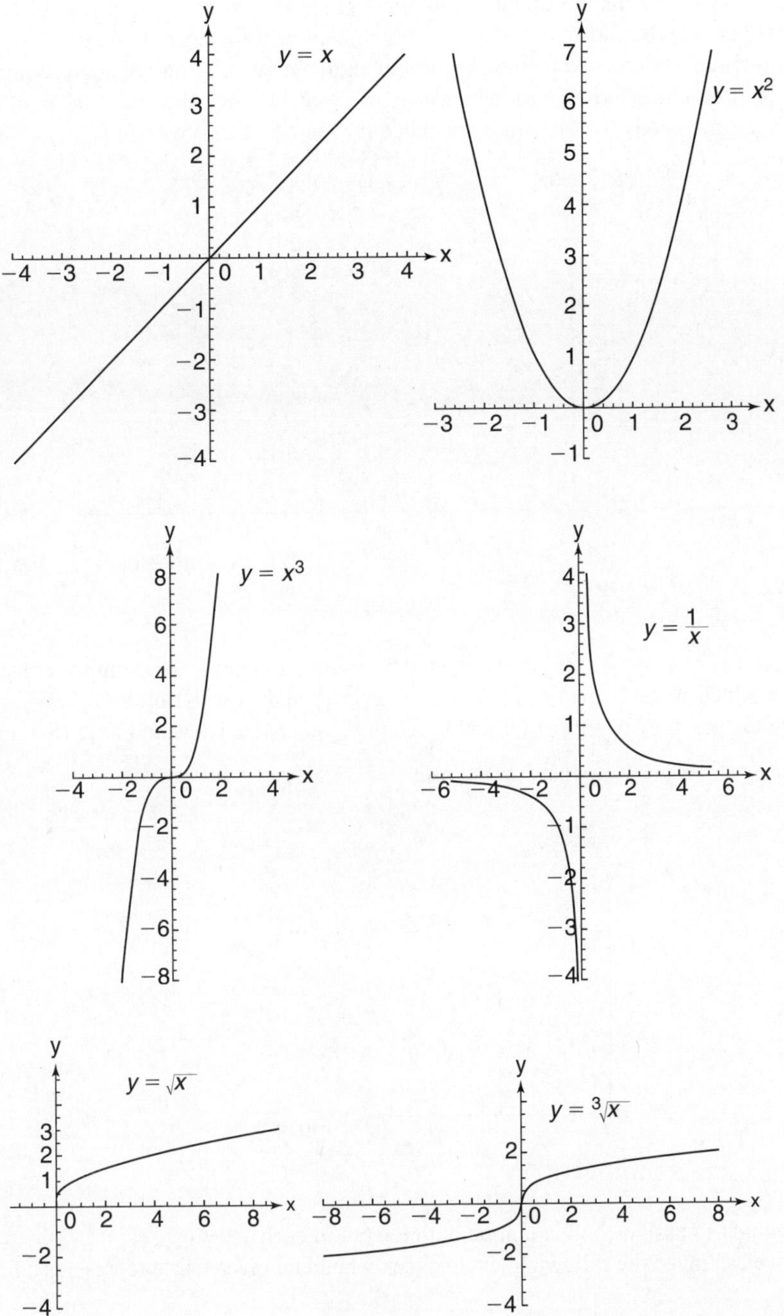

Fig. 6.4

Graphs provide visual information about a function. The points on the graph are of the form $(x, f(x))$. Hence, each y-coordinate is the value of the function at the corresponding x-coordinate. The domain and range of f can be obtained by projecting the graph of f onto the x-axis and y-axis respectively (Fig. 6.5). The values of x for which $f(x) = 0$ are the x-coordinates of the points where the graph intersects the x-axis. These values are called the 'zeroes' of f, 'roots' of $f(x) = 0$, or the 'x-intercepts' of $y = f(x)$.

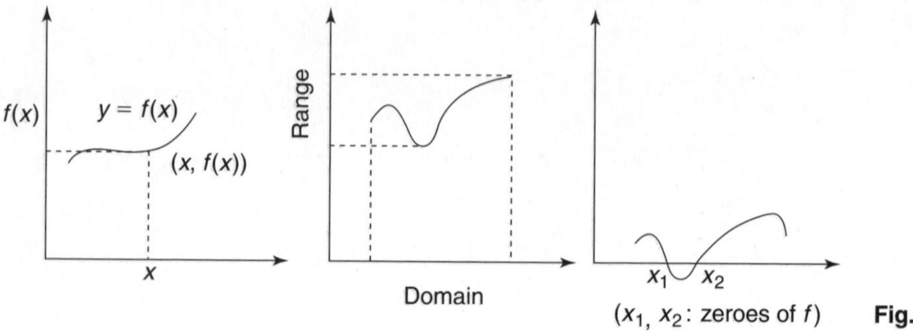

Fig. 6.5

Vertical line test Every graph on xy-plane may not represent a function. For example, consider the following curve which is cut at two distinct points (x_1, y_1) and (x_1, y_2) by a vertical line. This curve cannot be the graph of $y = f(x)$ for any function f; otherwise we would have (See Fig. 6.5)

$$f(x_1) = y_1 \quad \text{and} \quad f(x_1) = y_2$$

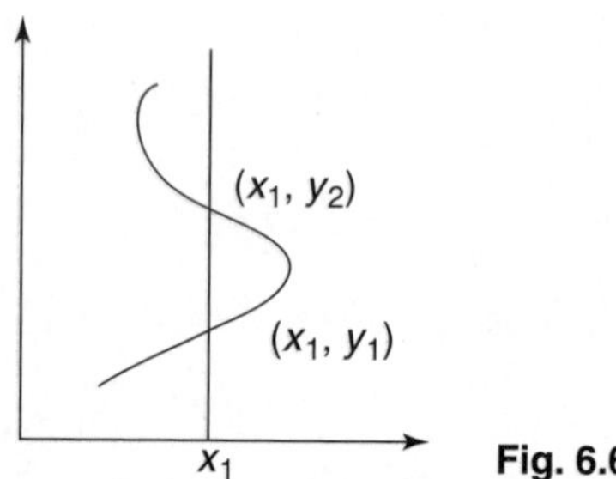

Fig. 6.6

which is impossible as f assigns only a unique output $f(x)$ to each x.

In general, we can make the following test to check whether a given curve represents a function or not.

A curve in the xy-plane is the graph of some function f if and only if no vertical line intersects the curve in more than one point.

EXAMPLE 6.1 The graph of the equation $x^2 + y^2 = 16$ is a circle of radius 4, centered at the origin, as shown in Fig. 6.7. From the vertical line test, it is clear that it does not represent any function. However, its lower half and upper half represent the functions f_1 and f_2 respectively, where $f_1(x) = -\sqrt{4 - x^2}$ and $f_2(x) = \sqrt{4 - x^2}$.

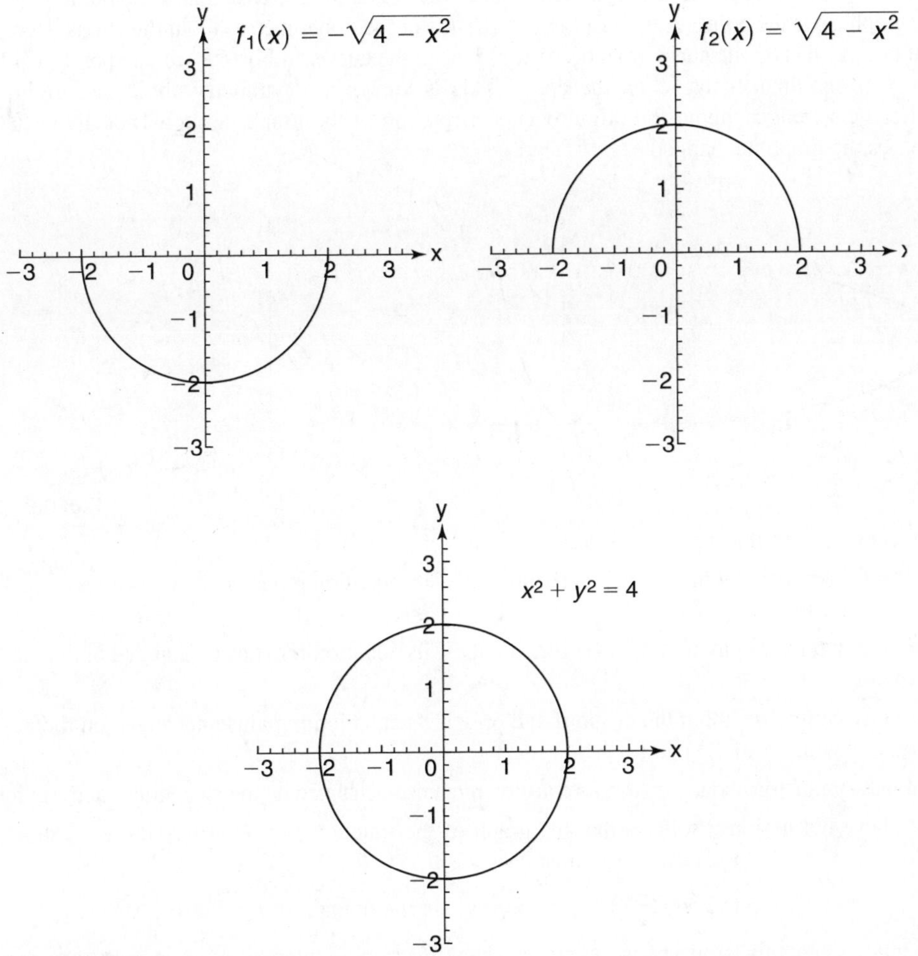

Fig. 6.7

Horizontal line test A function is one-to-one if and only if any horizontal line intersects graph at the most in one point. For example, $f(x) = e^x,\ x \in (-\infty, \infty)$ (See Fig. 6.8).

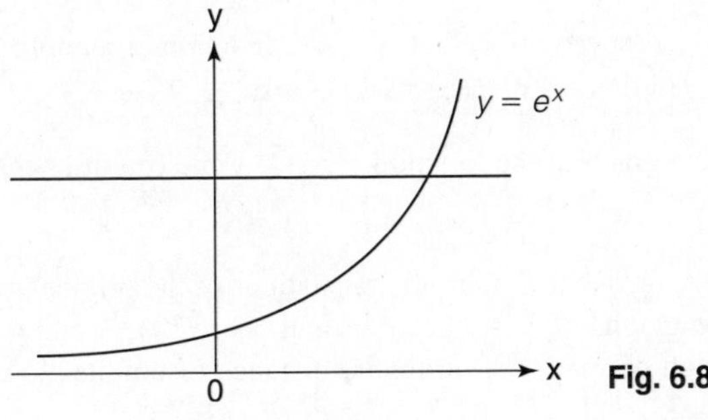

Fig. 6.8

Symmetry test Look at the following curves. In Fig. 6.9 (a), the x-axis and in Fig. 6.9 (b), the y-axis are respectively acting as mirrors for the curve. In Fig. 6.9 (c), the curve in the third quadrant may be realized as image of the curve in the first quadrant through consecutive reflections on the axes. In Fig. 6.9 (a), the curve is symmetric about the x-axis in the sense that if the point (x, y) is on the graph, so is the point $(x, -y)$. In Fig. 6.9 (b), the curve is symmetric about the y-axis. Here, the point $(-x, y)$ lies on the curve whenever (x, y) lies on the curve. In Fig. 6.9 (c), the point (x, y) and $(-x, -y)$ simultaneously lie on the curve. This is known as 'symmetry about the origin'. Geometrically, symmetry about the origin occurs if rotation of the graph through 180° about the origin leaves the graph unchanged.

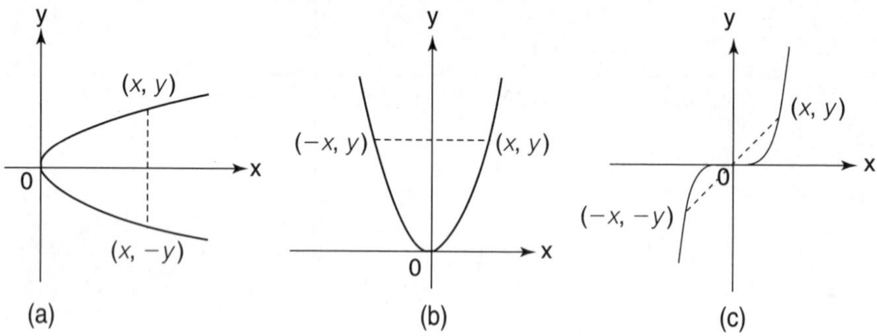

Fig. 6.9

We can summarise our discussion as follows:

(a) A curve is symmetric about the x-axis if and only if its equation remains unchanged on replacing y with $-y$ in it.

(b) A curve is symmetric about the y-axis if and only if its equation remains unchanged on replacing x with $-x$ in it.

(c) A curve is symmetric about the origin if and only if its equation remains unchanged on replacing x and y with $-x$ and $-y$ in it.

(d) A curve is symmetric about $y = x$ if its equation remains unchanged on interchanging x and y in it.

From the above discussion it follows that the graph of the function $y = f(x)$ is symmetric about y-axis if $y = f(x) = f(-x)$. That is, we must have

$$f(x) = f(-x) \qquad \text{for every } x \text{ in the domain of } f.$$

A function with this property is called an even function. Examples of even functions are $f(x) = 2x^2 + x^4$, $f(x) = \cos x$, $f(x) = 2x^2 - 4x^6 + 7x^8 + 9$ etc.

On the other hand, the graph of the function $y = f(x)$ is symmetric about the origin if and only if $y = f(x)$ and $-y = f(-x)$ are the same equation. That is, we must have

$$f(-x) = -f(x) \qquad \text{for every } x \text{ in the domain of } f.$$

A function with this property is called an odd function. Examples of odd functions are $f(x) = x^3 - 3x^5 + 7$, $f(x) = \sin x$, $f(x) = 2x^3 + 4x^5 - 9$ etc.

Translation Based on the graph of the function $y = f(x)$ one can draw the graphs of

$$y = f(x) + \text{c}, \qquad y = f(x) - \text{c}, \qquad y = f(x + \text{c}), \qquad y = f(x - \text{c})$$

where c is a positive constant, as the graphs are translations of the original graph of $y = f(x)$ along the x-axis or y-axis. The graphs of $y = f(x) + \text{c}$ and $y = f(x) - \text{c}$ are obtained by translating (that is, shifting) the graph of $y = f(x)$ through a distance of c units along y-axis in the positive (upward) and negative (downward) directions respectively. On the other hand, the graphs of $y = f(x + c)$ and $y = f(x - c)$ are translations of that of $y = f(x)$ along the x-axis, through a distance of c units, in the negative (left) and positive (right) direction respectively.

We can make Table 6.2.

Table 6.2

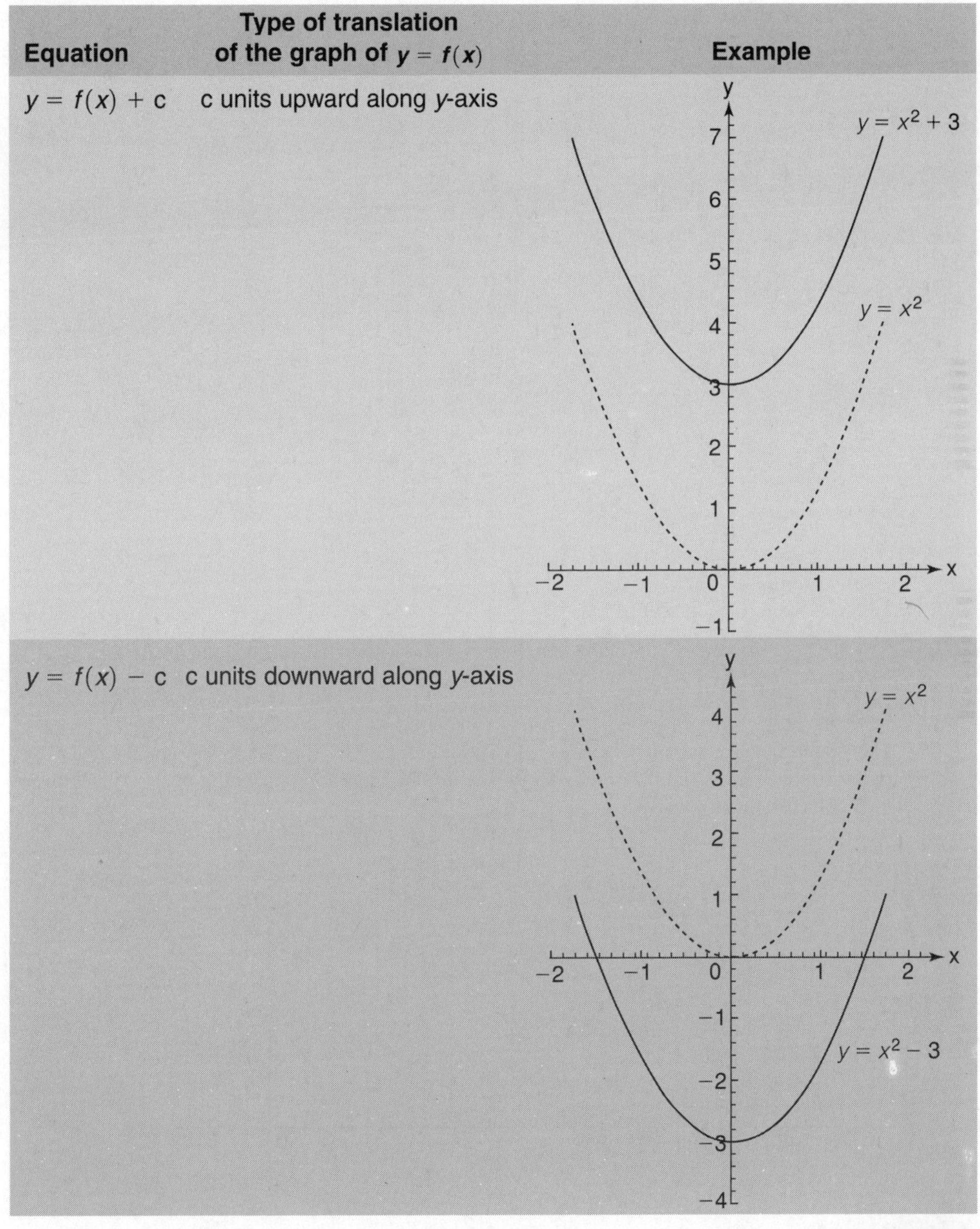

Equation	Type of translation of the graph of $y = f(x)$	Example
$y = f(x) + c$	c units upward along y-axis	
$y = f(x) - c$	c units downward along y-axis	

(Continued)

Table 6.2 (Contd.)

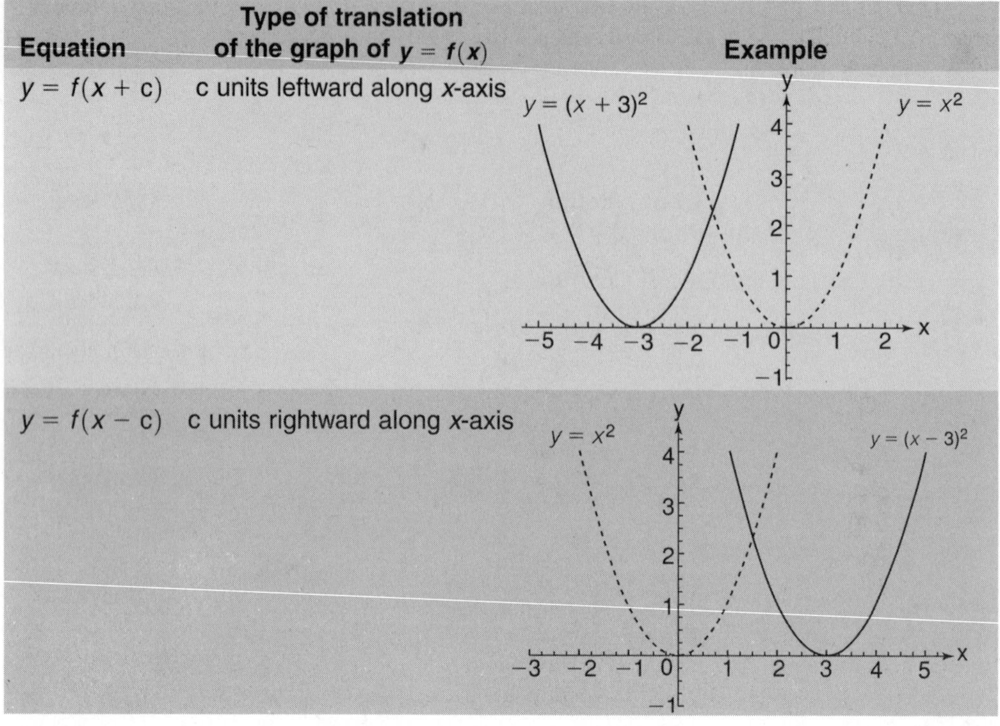

Equation	Type of translation of the graph of $y = f(x)$	Example
$y = f(x + c)$	c units leftward along x-axis	$y = (x + 3)^2$; $y = x^2$
$y = f(x - c)$	c units rightward along x-axis	$y = x^2$; $y = (x - 3)^2$

EXAMPLE 6.2 Sketch the graph of

(a) $y = \sqrt{x - 3}$ (b) $y = \sqrt{x + 3}$

SOLUTION

The graph of the equation $y = \sqrt{x - 3}$ can be obtained by translating the graph of $y = \sqrt{x}$ rightward by 3 units, and the graph of $y = \sqrt{x + 3}$ by translating the graph of $y = \sqrt{x}$ leftward by 3 units (Fig. 6.10).

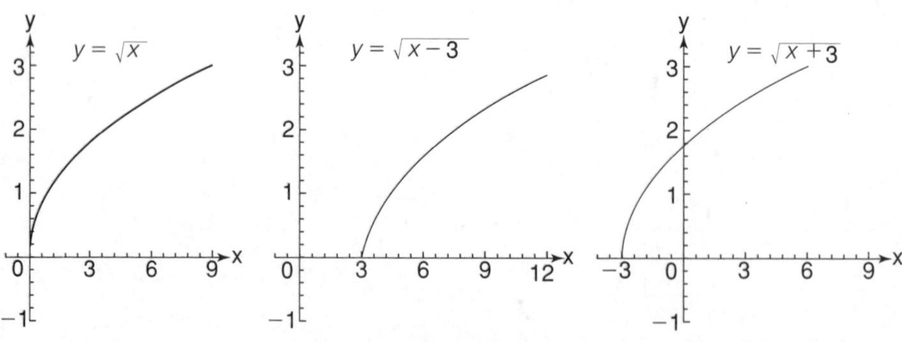

Fig. 6.10

EXAMPLE 6.3 Sketch the graph of $y = |x - 2| + 1$

SOLUTION

The graph can be obtained by two translations: first translate the graph of $y = |x|$ rightward by 2 units to obtain the graph of $y = |x - 2|$, then translate this graph upward by 1 unit to obtain the graph of $y = |x - 2| + 1$.

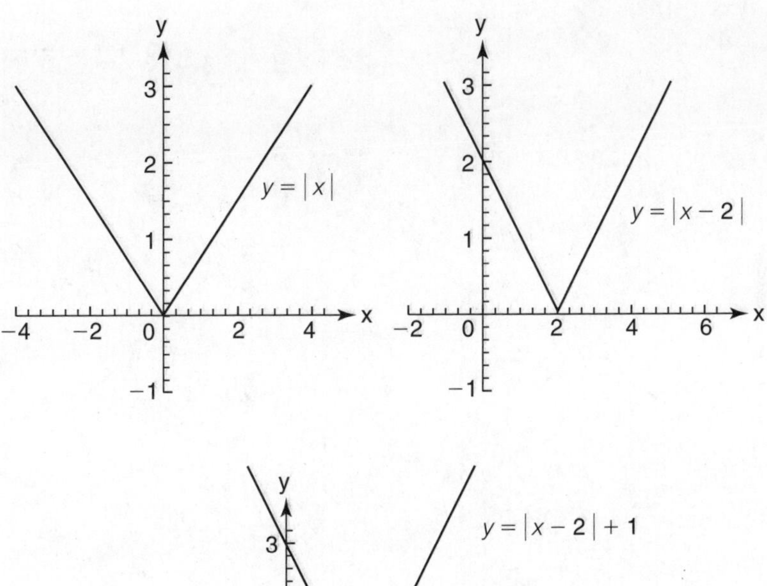

Fig. 6.11

Reflection The graphs of $y = f(-x)$ and $y = -f(x)$ can be obtained from the graph of $y = f(x)$, by taking its reflection (image) about the y-axis and x-axis respectively. If we take its reflection about $y = x$, we get the graph of the inverse function $y = f^{-1}(x)$. Look at Table 6.3.

Table 6.3

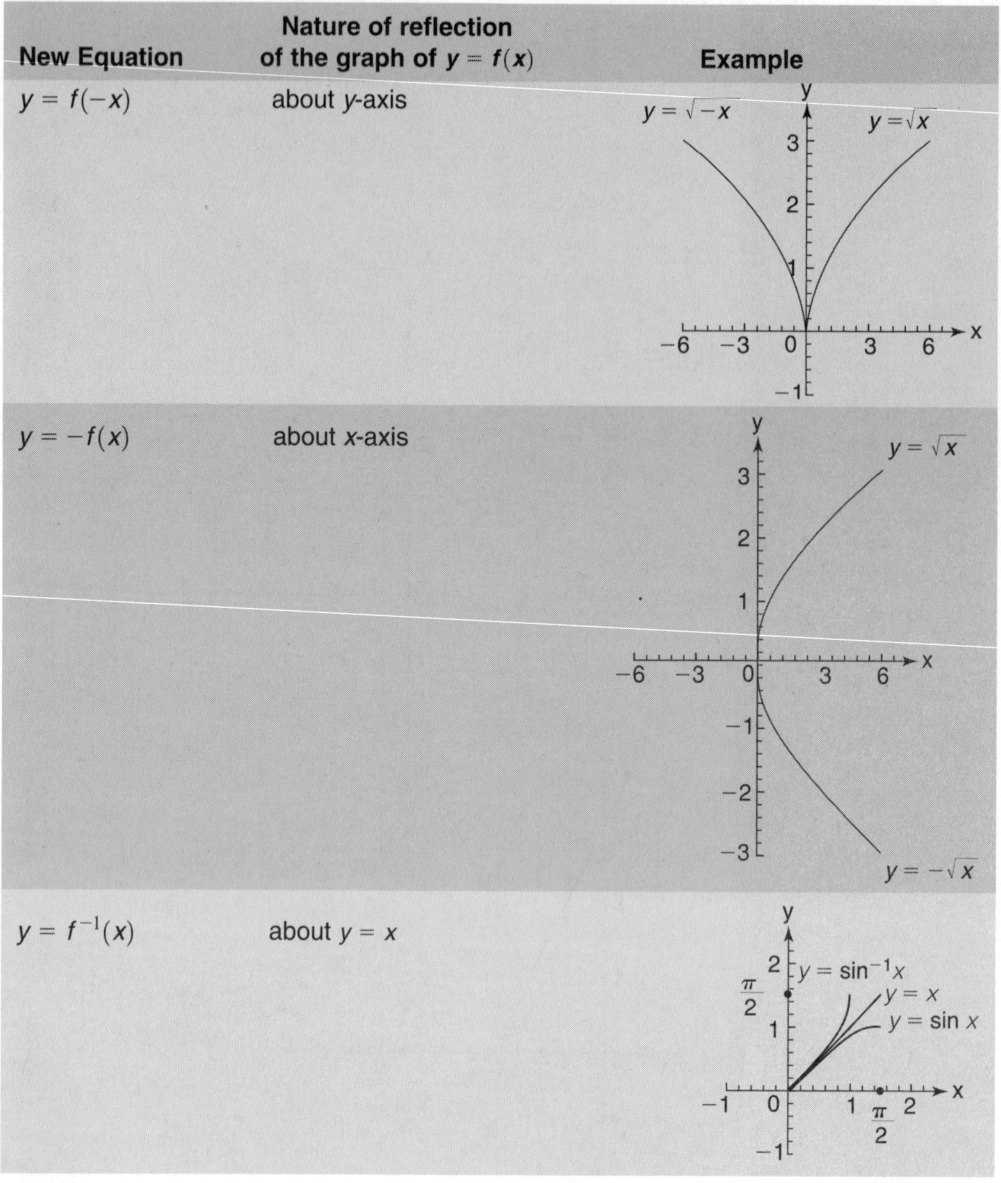

New Equation	Nature of reflection of the graph of $y = f(x)$	Example
$y = f(-x)$	about y-axis	$y = \sqrt{-x}$, $y = \sqrt{x}$
$y = -f(x)$	about x-axis	$y = \sqrt{x}$, $y = -\sqrt{x}$
$y = f^{-1}(x)$	about $y = x$	$y = \sin^{-1}x$, $y = x$, $y = \sin x$

EXAMPLE 6.4 Sketch the graph of $y = \sqrt[3]{2 - x}$.

SOLUTION

The graph can be obtained by a reflection and a translation; first reflect the graph of $y = \sqrt[3]{x}$ about the y-axis to obtain the graph of $y = \sqrt[3]{-x}$, then translate this graph rightward by 2 units to obtain the graph of the equation $y = \sqrt[3]{-(x - 2)} = \sqrt[3]{2 - x}$.

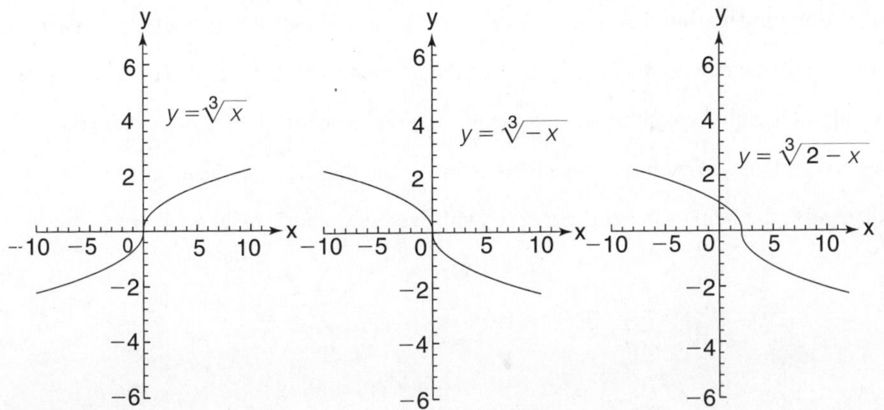

Fig. 6.12

EXAMPLE 6.5 Sketch the graph of $y = 4 - |x - 2|$.

SOLUTION

The graph can be obtained by a reflection and two translations: first translate the graph of $y = |x|$ rightward by 2 units to obtain the graph of $y = |x - 2|$; then reflect this graph about the x-axis to obtain the graph of $y = -|x - 2|$; and then translate this graph upward by 4 units to obtain the graph of the equation $y = -|x - 2| + 4 = 4 - |x - 2|$.

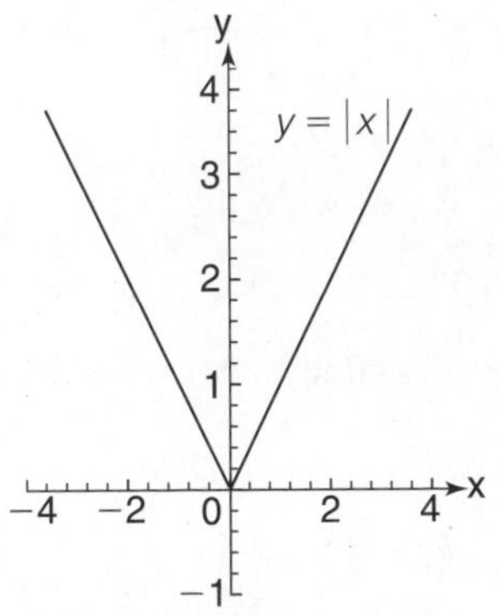

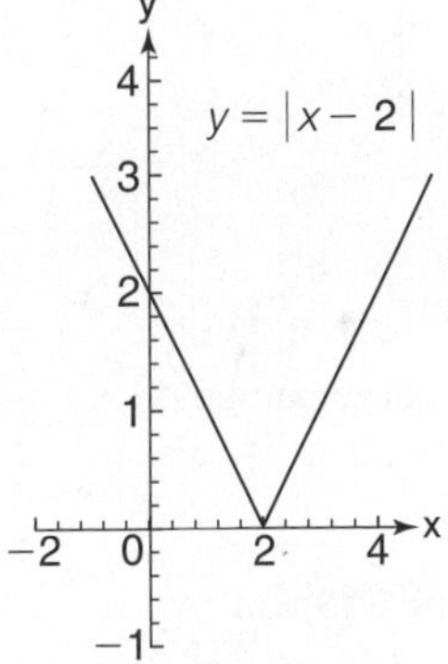

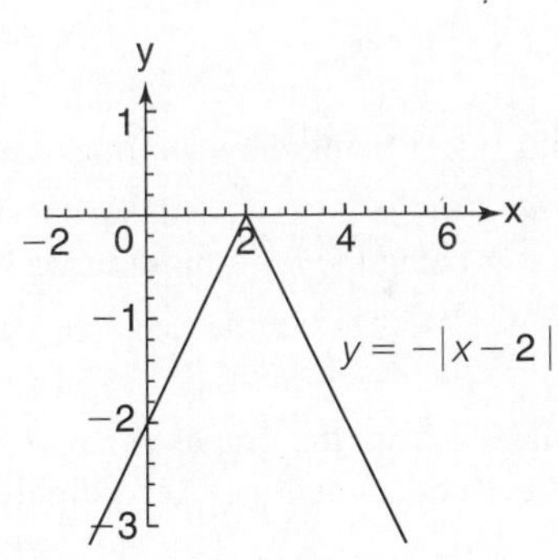

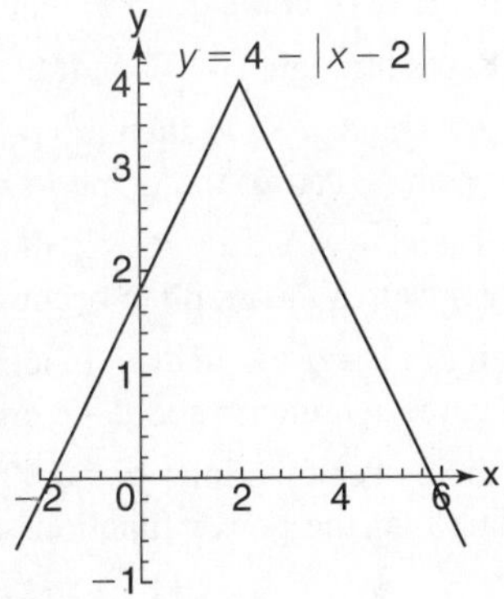

Fig. 6.13

Compression and magnification For a function $y = f(x)$ and a constant c, consider the function $y = cf(x)$. If c $= -1$, the new graph is reflection of the former about the x-axis. For $0 < c < 1$, $c > 1$ the new graph is called compression and magnification respectively of $y = f(x)$ by the factor c. Compression and magnification change the size but retain the shape of the original curve. We can explain them with the help of the curve $y = x^2$ with factors $\frac{1}{2}$ and 2. (Fig. 6.14)

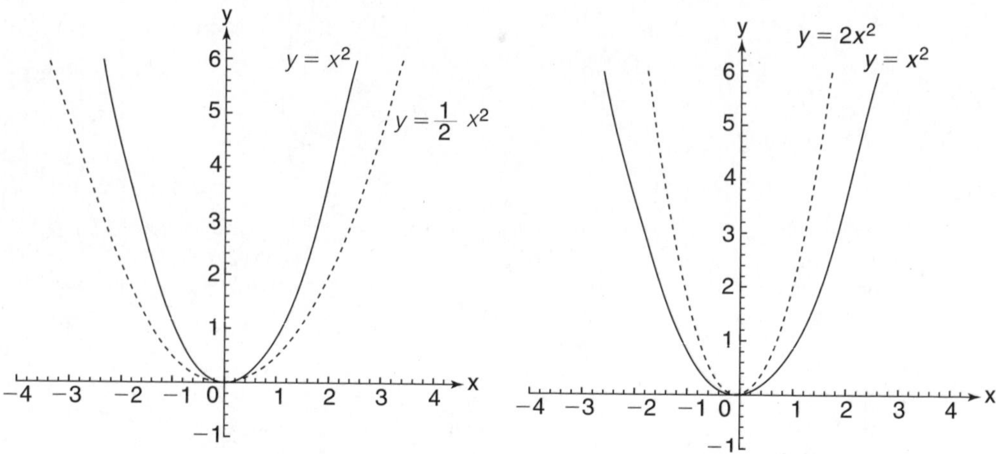

Fig. 6.14

If c is negative, the curve can be drawn by applying compression/magnification first, and reflection later.

6.1.2 Some Useful Graphs

The graph of the power function $f(x) = x^n$ The graph of the equation $y = x^n$ of the power function $f(x) = x^n$ for x in the interval $(0, \infty)$ has the property that, as the variable x increases, so does the variable y. Further we observe that for $n = 1, 2, 3, \ldots$

(i) $y = x^n$ passes through $(0, 0)$ and $(1, 1)$

(ii) If $0 < x < 1$ and $n < m$ then $x^n > x^m$; thus in $(0, 1)$ $x > x^2 > x^3 > x^4$ and so on. Thus increasing n causes the graph to become flatter with $y = x$ being the uppermost curve.

(iii) For $x > 1$ and $n < m$, $x^n < x^m$; thus in $(1, \infty)$, $x < x^2 < x^3 < x^4$ and so on. Thus increasing n causes the graph to become steeper with $y = x$ being the lowermost curve.

(iv) If n is even $f(x) = x^n$ is an even function and the graph of $y = x^n$ is symmetric about y-axis; if n is odd, it is symmetric about the origin.

The graph of the power functions $y = x^n$ for $x < 0$ are obtained from the graphs for $x > 0$ by utilizing the fact that the power functions are either 'even' or 'odd' functions. Accordingly we have Fig. 6.15.

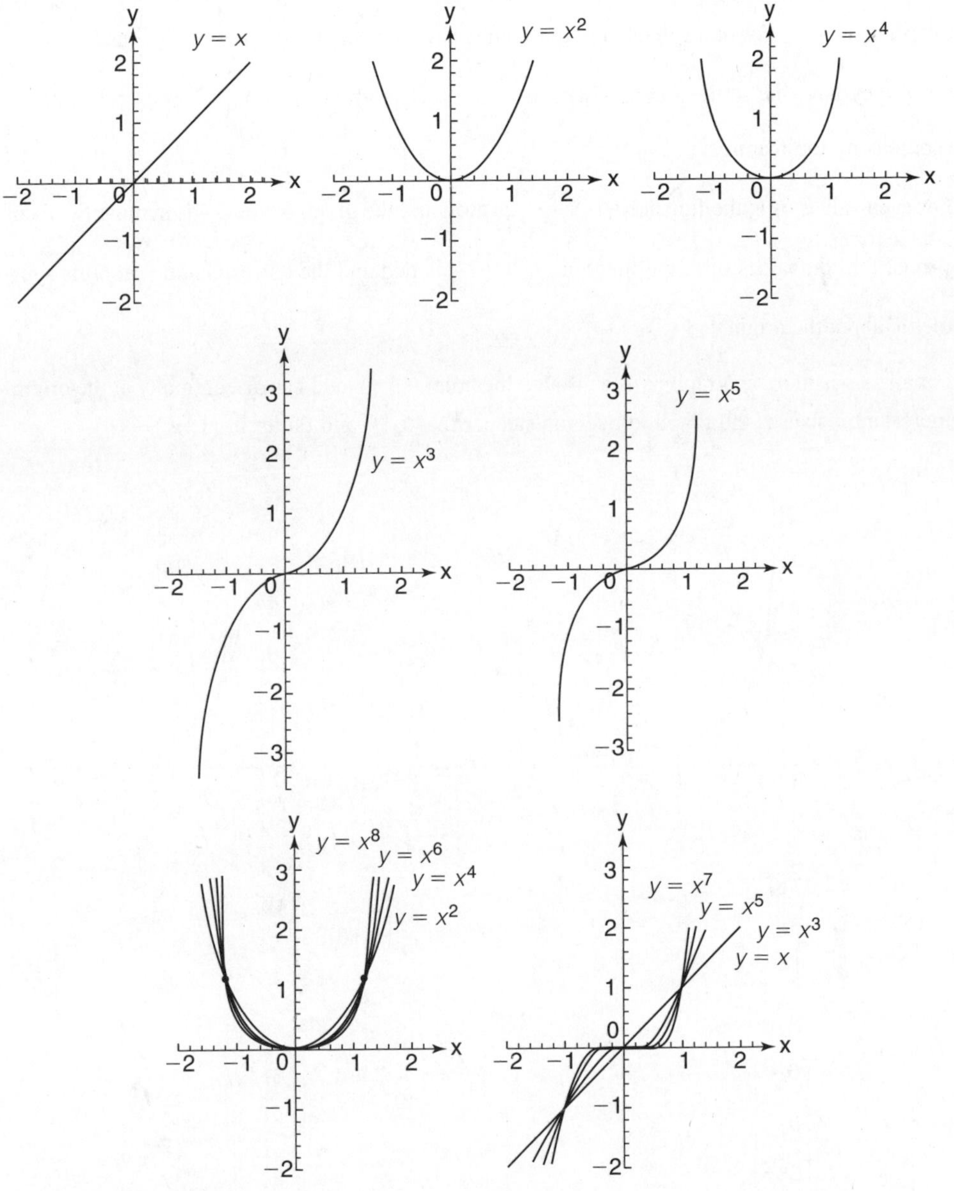

Fig. 6.15

Examples of power function

(i) It is experimental fact that muscle strength is approximately proportional to the muscle volume provided we compare muscles of the same shape. Again, the volume of a body of any shape is proportional to the cube of any of its linear dimension (length, height and so on) Hence

$$\text{muscle strength} \propto l^3$$

(ii) It is found that animals, in running away from predators or in searching for food, consume energies that are approximately proportional to the fourth power of their linear dimension.

The graph of $y = \frac{1}{x^n}$ For odd values of n, the graphs have the general shape as $y = \frac{1}{x}$ and for even value of n they have the same general shape as $y = \frac{1}{x^2}$. The graphs of $y = \frac{1}{x^n}$, $n = 1, 2, 3, \ldots$ share some common features:

(i) For even values of n, the function $f(x) = \frac{1}{x^n}$ is even and the graph of $y = \frac{1}{x^n}$ is symmetric about y-axis; for odd values of n, the function $f(x) = \frac{1}{x^n}$ is odd and the corresponding graph is symmetric about the origin.

(ii) For all values of n, the graph passes through the point $(1, 1)$ and has discontinuity at the origin.

(iii) Increasing n causes the graph to become steeper in $(0, 1)$ and flatter in $(1, \infty)$

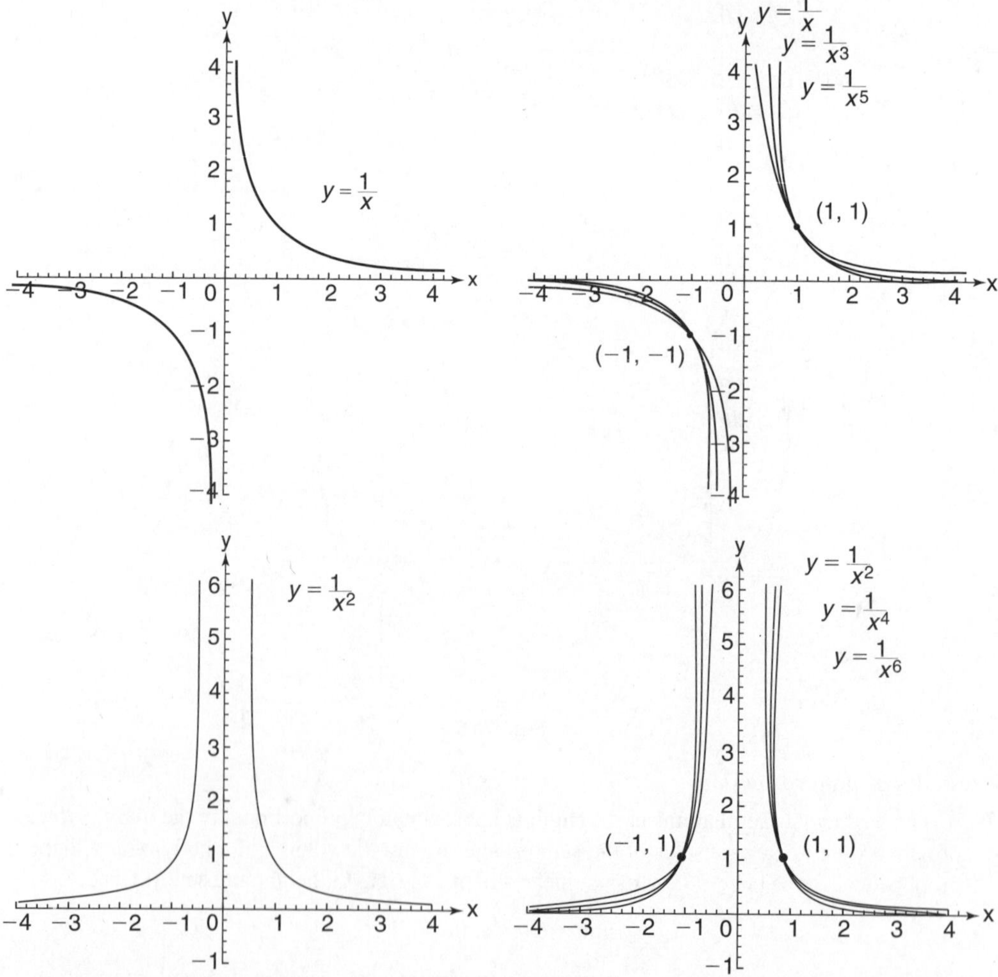

Fig. 6.16

The graph of $y = x^{1/n}$ The graph of $y = \sqrt{x}$ (that is, $n = 2$) is the upper half of the parabola $x = y^2$. It is defined only for non-negative x. However $y = \sqrt[3]{x}$ is defined for the entire x-axis and is symmetric about the origin. See Fig. 6.17.

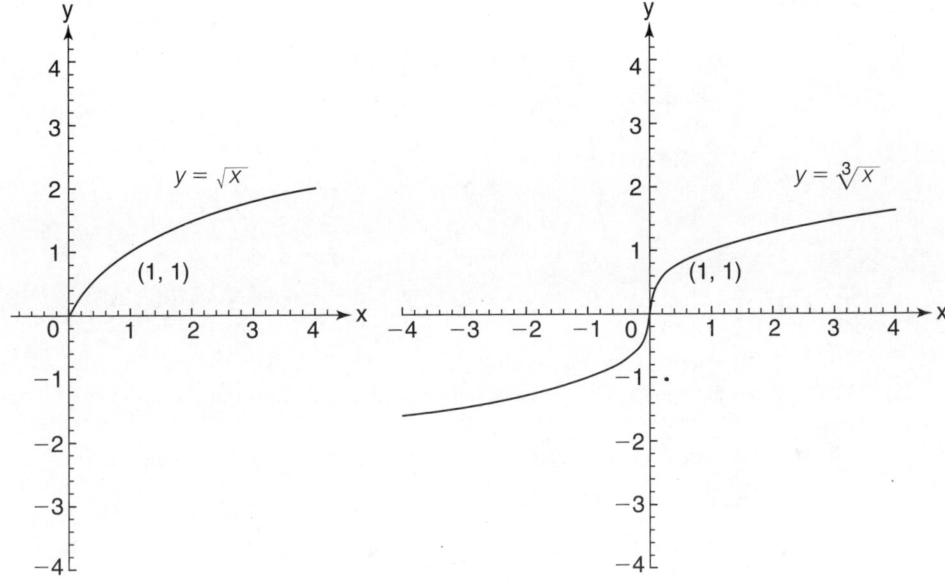

Fig. 6.17

For even values of n the graphs of $y = \sqrt[n]{x}$ have the same general shape as that of $y = \sqrt{x}$; for odd values of n, they have the same general shape as of $y = \sqrt[3]{x}$.

REMARK: Students should remember the graphs of $y = x^2, x^3, \frac{1}{x}, \frac{1}{x^2}, \sqrt{x}, \sqrt[3]{x}$ and others in order to draw the graphs of the types $y = x^n, \frac{1}{x^n}, x^{1/n}$ respectively.

EXAMPLE 6.6 Draw the graphs of

(a) $y = x^5$ (b) $y = \frac{1}{x^3}$ (c) $y = \sqrt[4]{x}$

SOLUTION

(a) The graph of $y = x^5$ has the following characteristics:

(i) It is defined for all real values of x.

(ii) It passes through $(0, 0)$ and $(1, 1)$.

(iii) It is symmetric about the origin as $f(x) = x^5$ is an odd function.

(iv) As x increases, y also increases.

(v) It follows the pattern of $y = x^3$.

(vi) Some points on the graph are $(0, 0)$, $(1, 1)$, $(2, 32)$, $(3, 243)$.

A rough sketch of the curve is shown in Fig. 6.18.

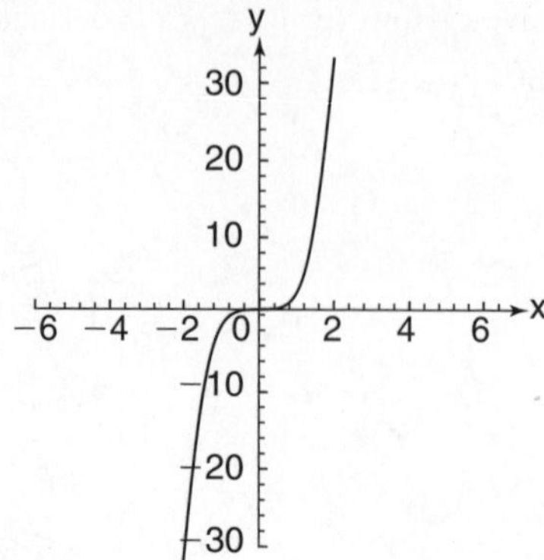

Fig. 6.18

(b) The graph of $y = \dfrac{1}{x^3}$ has the following characteristics:

(i) It is defined for all real x except 0.

(ii) It passes through $(1, 1)$. In the open interval $(0, 1)$, it moves from ∞ to 1 and in $(1, \infty)$, it moves from 1 to 0.

(iii) It is symmetric about the origin as the function $f(x) = \dfrac{1}{x^3}$ is an odd function.

(iv) It follows the pattern of $y = \dfrac{1}{x}$.

(v) Some points on the graph are $(-5, 8)$, $(1, 1)$, $(2, .125)$, $(3, .036)$.

A rough sketch of the graph is shown in Fig. 6.19.

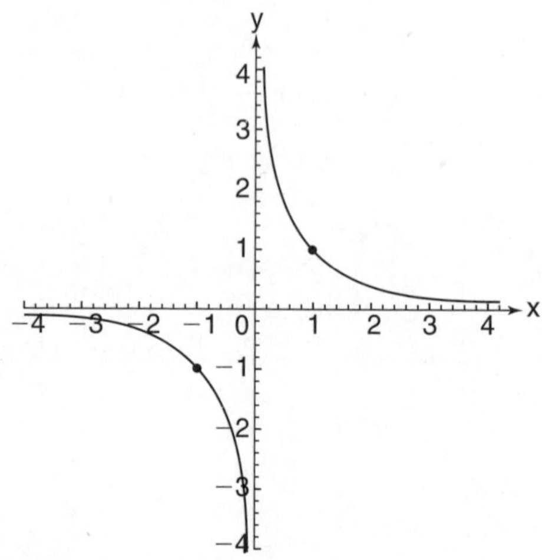

Fig. 6.19

(c) The graph of $y = x^{1/4}$ has the following characteristics:

(i) It is defined for non-negative real values of x.

(ii) It passes through $(0, 0)$ and $(1, 1)$.

(iii) As x increases, y also increases.

(iv) It follows the pattern of $y = \sqrt{x}$.

(v) Some points on the graph are $(0, 0)$, $(1, 1)$, $(4, 1.41)$, $(16, 2)$.

A rough sketch of the graph is shown in Fig. 6.20.

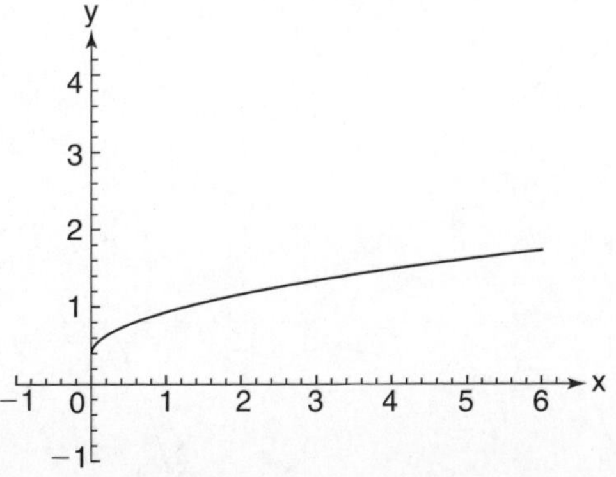

Fig. 6.20

Polynomial functions A polynomial function is the sum of constant multiples of power functions. A polynomial function f of degree n is expressed either as

$$f(x) = a_0 + a_1x + \cdots + a_nx^n$$

or
$$y = a_0 + a_1x + \cdots + a_nx^n$$

In case $n = 0$ or $n = 1$, it reduces to trivial cases where graphs are straight lines parallel to x-axis or with slope a_1 respectively.

Second degree polynomials give rise to parabolic curves. We illustrate with the help of Example 6.7.

EXAMPLE 6.7 Draw the graphs of (a) $y = \frac{1}{2}x^2 - 3x + \frac{11}{2}$ (b) $y = -2x^2 - 4x - 4$

SOLUTION

We write the equations as $y = \frac{1}{2}(x - 3)^2 + 1 \qquad y = -2(x + 1)^2 - 2$
Accordingly, the graphs may be obtained as shown in Fig. 6.21.

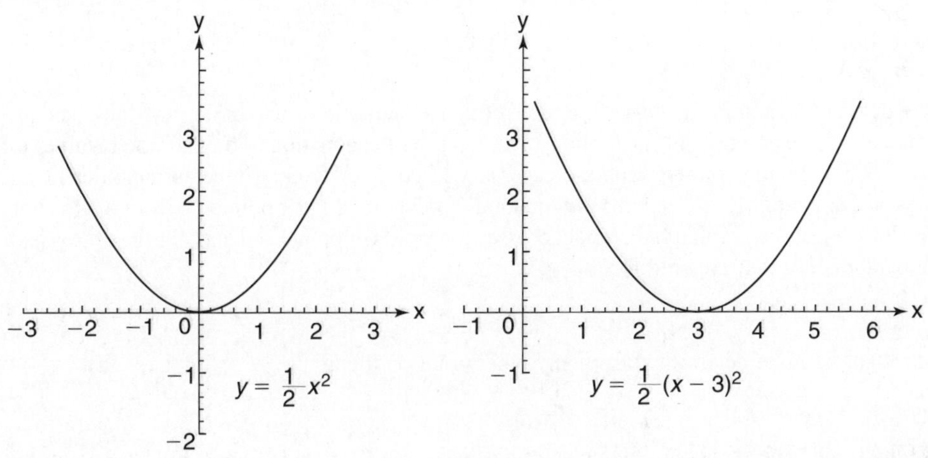

Fig. 6.21 (*continued*)

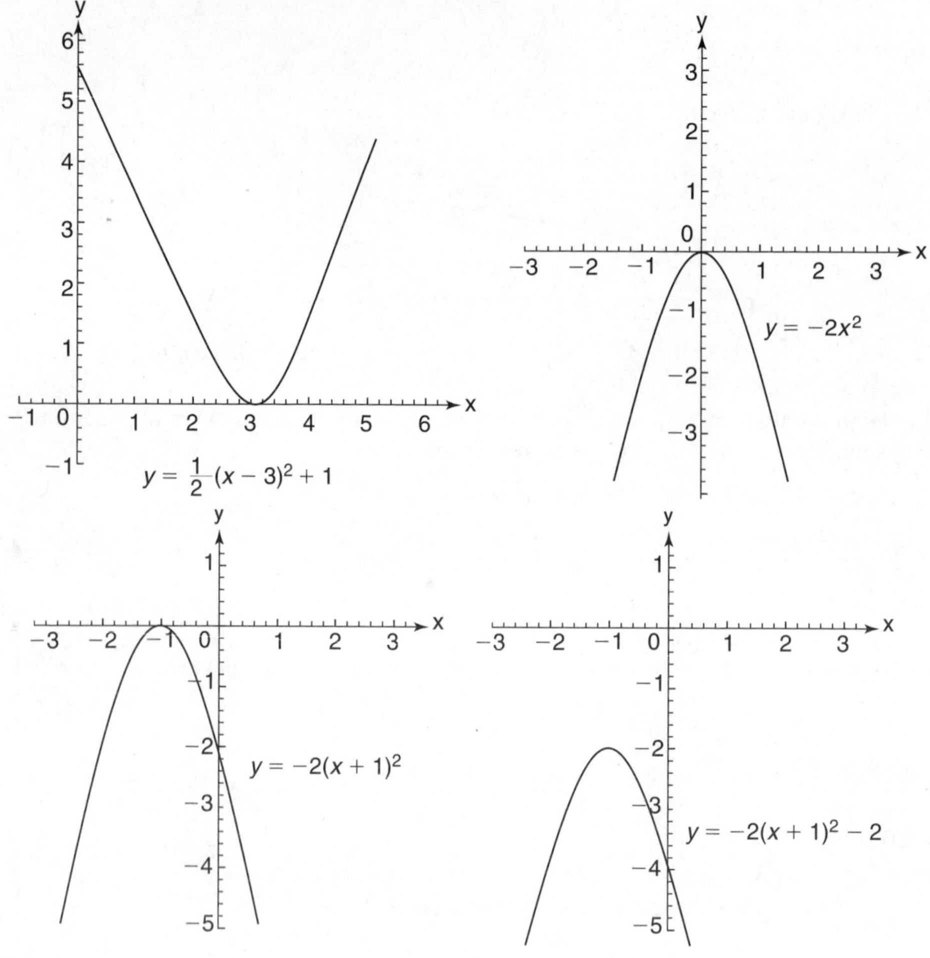

Fig. 6.21

Second degree polynomial functions are referred to as quadratic or parabolic functions. The graph of a second degree polynomial function is a parabola. As the equation of a quadratic involves three parameters, three points on its graph are necessary to completely determine a quadratic. The general determination of the coefficients of a quadratic polynomial from points on its graph is complicated. However, if the graph of a parabola crosses the x-axis at points x_1 and x_2, then the equation of the parabola has the 'factored form'

$$y = k(x - x_1)(x - x_2).$$

The constant k can be determined from any other point on the graph.

EXAMPLE 6.8 The parabola passing through the points (2, 0), (3, 0) and (4, 1) is given by an equation of the form

$$y = k(x - 2)(x - 3)$$

The constant k is determined by substituting the point (4, 1) and solving

$$1 = k(4 - 2)(4 - 3) = 2k$$

or
$$k = \tfrac{1}{2}.$$

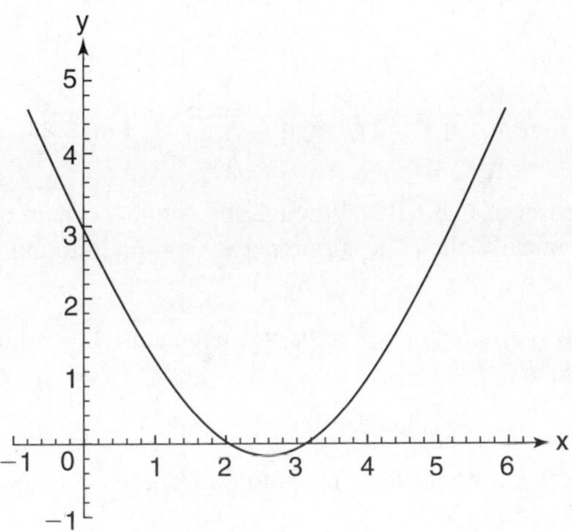

Fig. 6.22

Application Consider a unidirectional bimolecular chemical reaction in which two compounds A and B combine to form a new compond X. Symbolically, this is expressed as

$$A + B \rightarrow X$$

This is the simplest type of second-order chemical reaction, that is a reaction in which the rate of formation of the compound X is given by a quadratic function. Using standard chemistry notation, let the initial concentration of the compounds A and B be denoted by $[A]_0$ and $[B]_0$ respectively, and let the concentration of X be denoted by $[X]$. The function R, which gives the rate of formation of X, has the form

$$R([X]) = k([A]_0 - [X])([B]_0 - [X]) \qquad (1)$$

In this equation k is a proportionality constant and the terms $([A]_0 - [X])$ and $([B]_0 - [X])$ represent the concentration of A and B compounds in the reaction mixture when the concentration of X has reached the level $[X]$.

EXAMPLE 6.9 Assume $k = 2$, $[A]_0 = 0.5$, $[B]_0 = 0.6$. Then equation (1) becomes

$$R([X]) = 2(0.5 - [X])\,(0.6 - [X]).$$

Completing the square, we can express the equation as

$$R([X]) = 2([X] - 0.55)^2 + (-0.005).$$

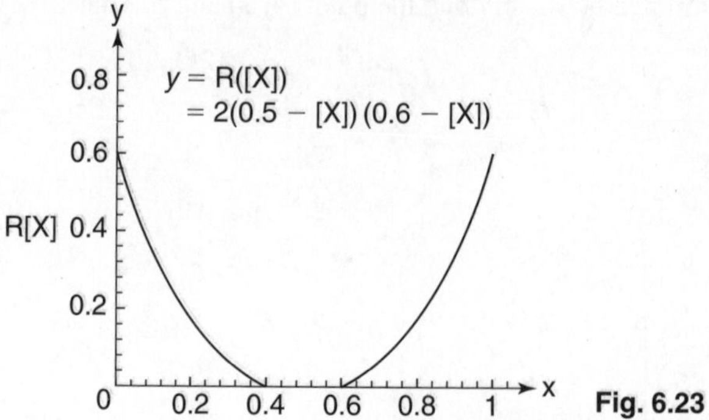

Fig. 6.23

The graph of this equation is given in Fig. 6.23. Although the natural domain of the function R is the set of all real numbers, chemically, the concentration $[X]$ is limited to numbers between zero and one.

Exponential function For any real number $a > 0$, the exponential function with base a is defined on $(-\infty, \infty)$ by the equation

$$y = a^x, \quad \text{or equivalently} \quad y = \exp_a(x)$$

For $a > 1$, a^x increases gradually from 0 to ∞ as x-moves from $-\infty$ to ∞. The curve passes through the point (0, 1) as shown in Fig. 6.24.

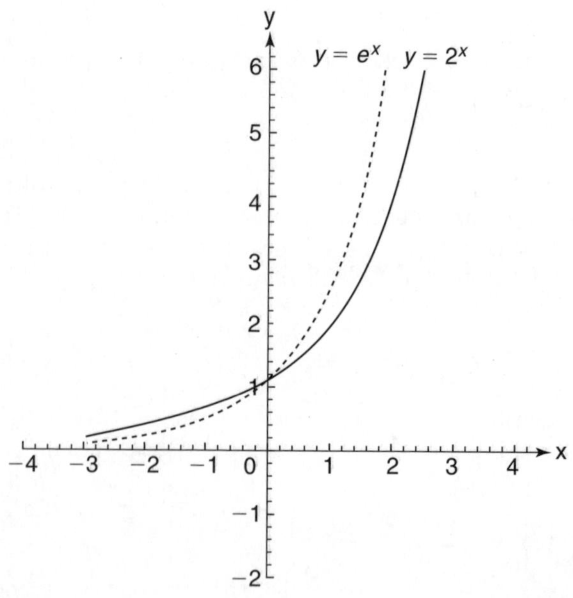

Fig. 6.24

If $0 < a < 1$, then $\dfrac{1}{a} > 1$and $y = a^x = \left(\dfrac{1}{a}\right)^{-x}$ can be realized as reflection of the graph of the function $f(x) = \left(\dfrac{1}{\text{a}}\right)^x$ about the y-axis.

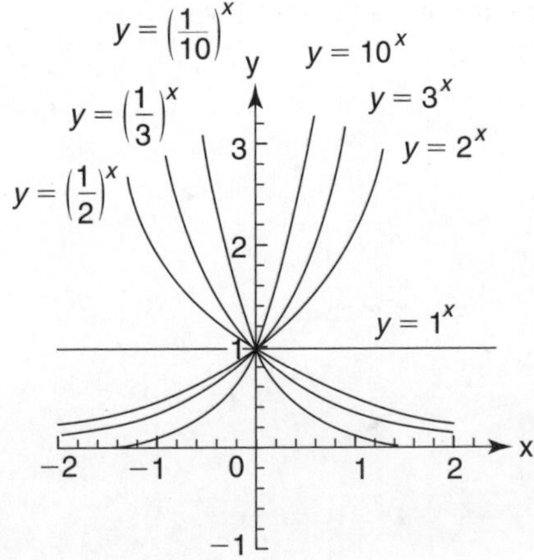

Fig. 6.25

The exponential functions which are most frequently encountered are those with base 2, base 10 and base e. Exponential processes are characterized by the way they change with time. Many biological and other natural quantities change instantaneously over a period of time. Suppose a quantity y (velocity, temperature, electric current, amount of bacteria and so on) increases or decreases at a rate that is, at any given time t, proportional to the amount present. If, when $t = 0$, the amount of the quantity y is y_0, then we can find y as a function of t by solving the following initial value problem.

$$\frac{dy}{dt} = ky; \text{ initial condition } y = y_0 \text{ when } t = 0.$$

If y is increasing, then k is positive; if y is decreasing k is negative. Solving this differential equation, we get

$$y = y_0 e^{kt}$$

The number k is called the rate constant of the equation.

Detailed discussions about the occurrence of these types of curves are provided in the next chapter.

EXAMPLE 6.10 Draw the graphs of

(a) $y = e^{-x} + 1$ (b) $y = e^{|x|}$ (c) $y = \left(\frac{1}{x}\right)^x$

SOLUTION

(a) (i) The graph may be obtained from $y = e^x$ by first applying reflection about y-axis and then upward shifting of one unit.

(ii) It is defined for all real x; y moves from ∞ to 1 as x moves from $-\infty$ to ∞.

(iii) It passes through $(0, 2)$, $(-1, 3.73)$, $(1, 1.3)$ (approx.)

A rough sketch of the graph may be as in Fig. 6.26:

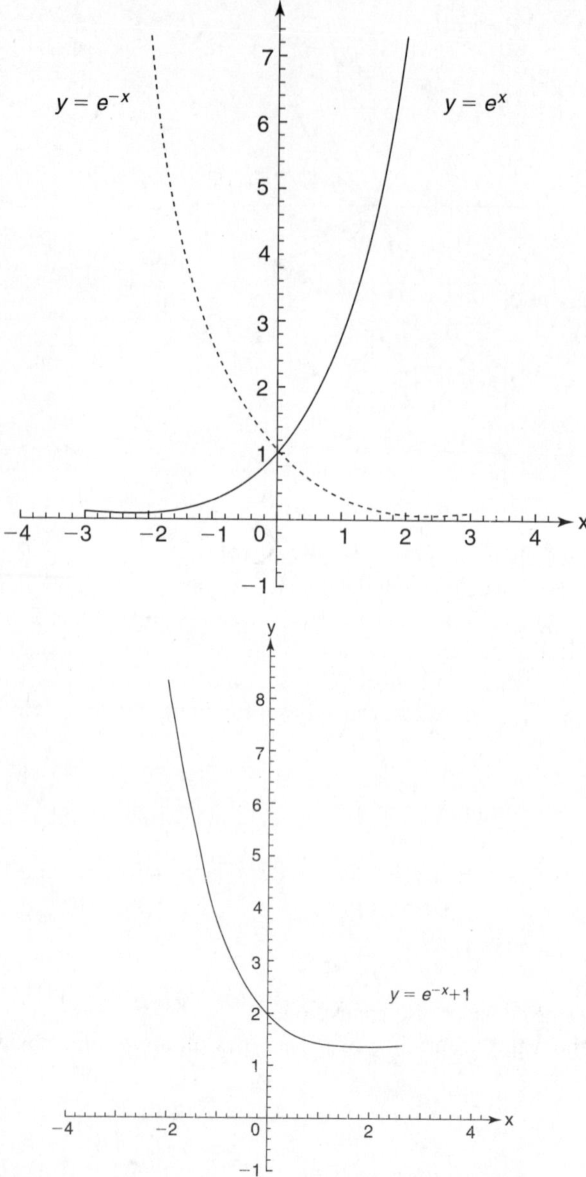

Fig. 6.26

(b) The function $f(x) = e^{|x|}$ can be rewritten as follows:

$$f(x) = e^x, \quad x \geq 0$$
$$= e^{-x}, \quad x < 0$$

Therefore, the graph of $y = e^{|x|}$ consists of two parts: for $x \geq 0$, $y = e^x$ and for $x < 0$, $y = e^{-x}$. That is, $y = e^x$ for $x \geq 0$ and its reflection about y-axis for $x < 0$.

The graph is shown in Fig. 6.27.

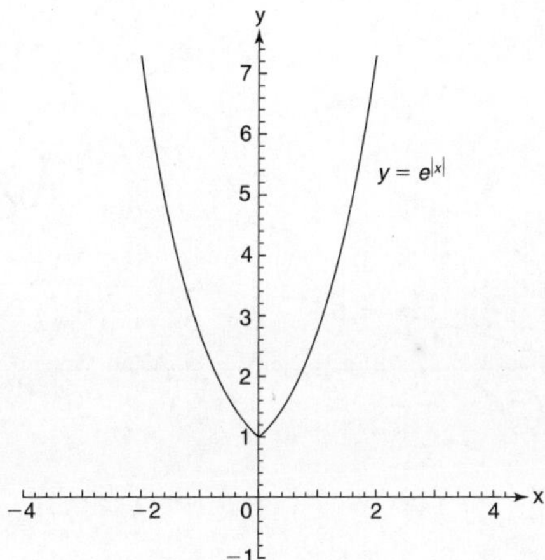

Fig. 6.27

(c) Since $y = (\frac{1}{2})^x = 2^{-x}$, its graph has the following characteristics:

(i) It is reflection of $y = 2^x$ about y-axis.

(ii) It is defined for all real x; as x moves from $-\infty$ to ∞, y moves from ∞ to 0.

(iii) It passes through $(-3, 8)$, $(-2, 4)$,. $(0, 1)$, $(1, .5)$, $(2, .25)$ and so on.

A rough sketch of the graph is given in Fig. 6.28.

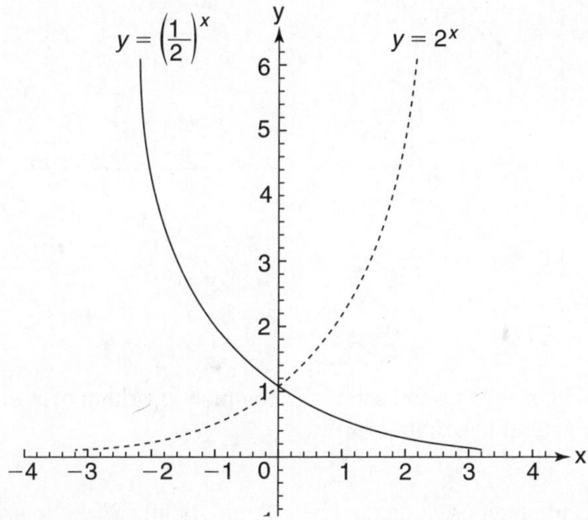

Fig. 6.28

6.1.3 Logarithmic Functions

From the diagram of $y = a^x$, it is clear by horizontal line test that, if a is any positive number other than 1, the function $y = a^x$ is one-to-one. It therefore possesses inverse. The inverse function of $y = a^x$ is the logarithmic function of x with base a, and is denoted by $y = \log_a x$. Its graph can be obtained by reflecting the graph of $y = a^x$ across the line $y = x$. (Fig. 6.29).

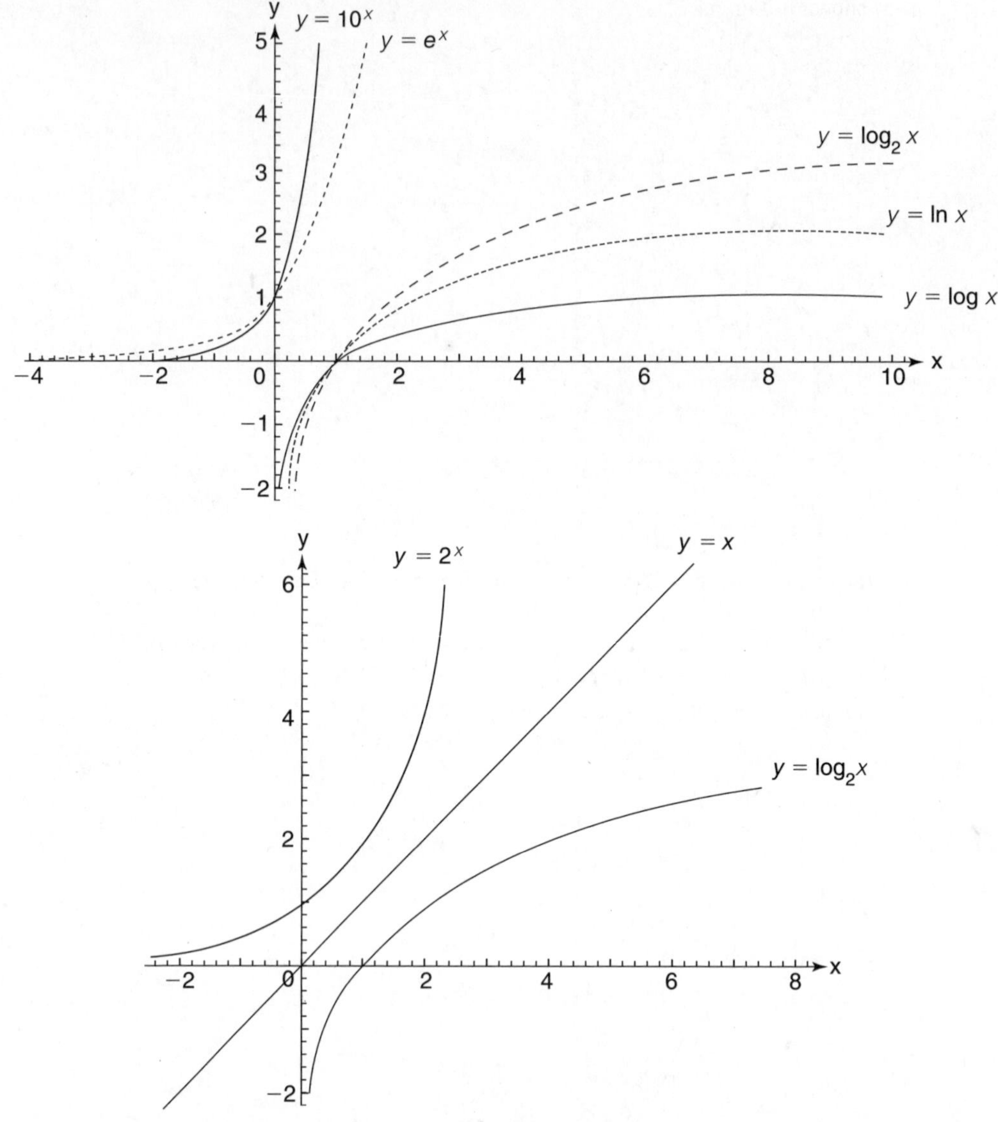

Fig. 6.29

Conventionally $\log_{10} x$ is written as $\log x$ and called the common logarithm of x, and $\log_e x$ is written as $\ln x$ and is called the natural logarithm of x.

EXAMPLE 6.11

(i) Earthquake intensity is often reported on the Logarithmic Richter Scale. Its formula is

$$\mathrm{R} = \log\left(\frac{a}{\mathrm{T}}\right) + B,$$

where R is the magnitude of the intensity, a is the amplitude of the ground motion in microns at the receiving station, T is the period of the seismic wave in seconds, and B is an empirical factor that allows for the weakening of the seismic wave with increasing distance from the epicenter of the earthquake.

(ii) The pH scale for measuring acidity of a solution is a base 10 logarithmic scale. The pH value (hydrogen potential) of the solution is the common logarithm of the reciprocal of the solutions of hydronium ion concentration, $[H_3O^+]$. Thus

$$pH = \log\frac{1}{[H_3O^+]} = -\log[H_3O^+].$$

The hydronium ion concentration is measured in moles per litre. Vinegar has a pH of 3, distilled water a pH of 7, seawater a pH of 8.15 and so on.

(iii) Another example of the use of common logarithm is the decibel scale (db) for measuring loudness. If I is the intensity of sound in watts per square meter, the decibel level of sound is,

$$\text{Sound Level} = 10\log(I \times 10^{12})\ \text{db}$$

Rustle of leaves produces sound level of 10 db, average whisper a level of 20 db, ordinary conversation a level of 65 db and so on whereas threshold of pain is considered to be 170 db.

6.1.4. Trigonometric Functions

All six trigonometric functions are periodic; tan *x* and cot *x* have a period of π, rest have a period of 2π. (A period for a function *f* is the smallest positive number *p* such that $f(x + p) = f(x)$ for all *x*).

Due to periodicity, the graph of these functions repeats itself after an interval equal to their periods along the *x*-axis. Thus we need to know their behaviour only for one period along *x*-axis, say in $[0, 2\pi]$ for $f(x) = \sin x$, cos x and so on. Table 6.4 provides values with a gap of $\pi/6$. The function $f(x) = \sin x$ has a period of 2π.

Table 6.4

x	0	$\pi/6$	$\pi/3$	$\pi/2$	$2\pi/3$	$5\pi/6$	π	$7\pi/6$	$4\pi/3$	$3\pi/2$	$5\pi/3$	$11\pi/6$	2π
sin x	0	.5	.87	1	.87	.5	0	−.5	−.87	−1	−.87	−.5	0

Accordingly, the graph of $f(x) = \sin x$ may be drawn as in Fig. 6.30.

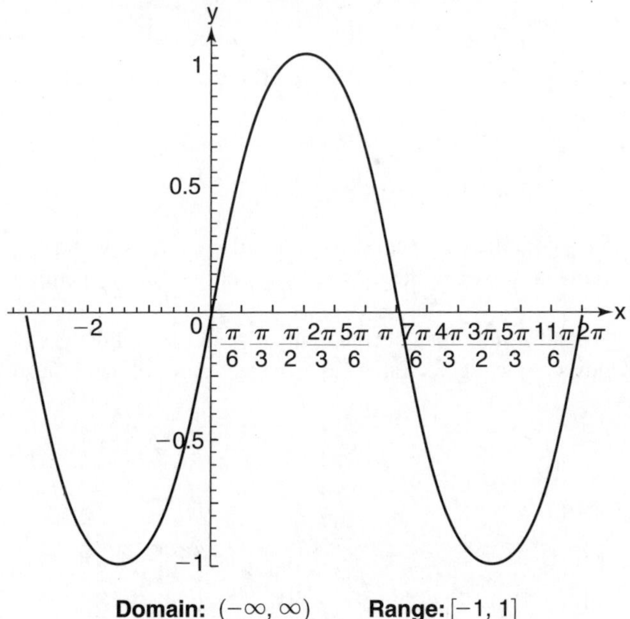

Domain: $(-\infty, \infty)$ **Range:** $[-1, 1]$ **Fig. 6.30**

$y = \cos x$

Using the facts that $\sin(\frac{\pi}{2} - x) = \cos x$ and $\cos(\pi - x) = -\cos x$, Table 6.5 may be obtained.

Table 6.5

x	0	$\pi/6$	$\pi/3$	$\pi/2$	$2\pi/3$	$5\pi/6$	π	$7\pi/6$	$4\pi/3$	$3\pi/2$	$5\pi/3$	$11\pi/6$	2π
$\cos x$	1	.87	.5	0	−.5	−.87	−1	−.87	−.5	0	.5	.87	1

The graph may be drawn as in Fig. 6.31.

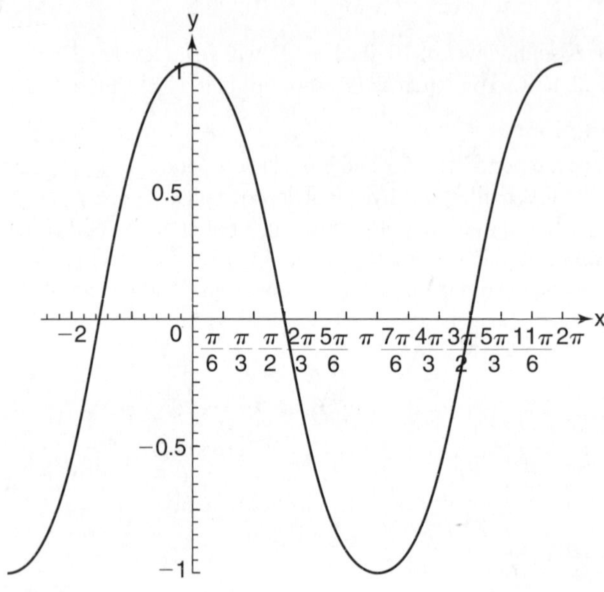

Domain: $(-\infty, \infty)$ **Range:** $[-1, 1]$ **Fig. 6.31**

$y = \tan x$

Since the tangent function is periodic with period π, it is sufficient to draw the graph over an interval of length π. In the remaining part of the x-axis, the graph will be a repetition of the graph in this part.

The function $f(x) = \tan x$ is not defined at $x = \frac{\pi}{2}$. It tends to ∞ there. Since $\tan(-x) = -\tan x$, it tends to $-\infty$ at $-\frac{\pi}{2}$. Table 6.6 gives the values of $\tan x$ at some points in the interval $(-\pi/2, \pi/2)$.

Table 6.6

x	$-\pi/3$	$-\pi/6$	0	$\pi/6$	$\pi/3$
$\tan x$	−1.73	−.58	0	.58	1.73

Accordingly, the graph of $y = \tan x$ may be drawn as in Fig. 6.32.

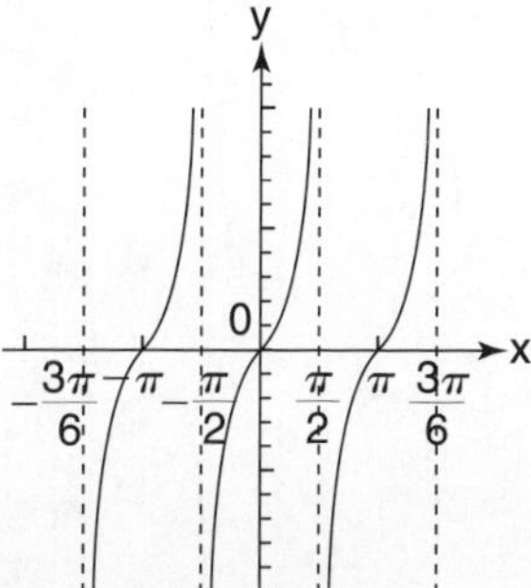

Domain: The set of all real numbers except the odd integral multiples of $\frac{\pi}{2}$. **Range:** $(-\infty, \infty)$

Fig. 6.32

The graph of cot *x*, sec *x*, cosec *x* These three functions are reciprocals of $\tan x$, $\cos x$, $\sin x$ respectively (whereever they are defined). Therefore *y*-value at any point may be obtained from the corresponding values of the above graphs. We provide the graphs of these three functions in Figs. 6.33, 6.34 and 6.35.

$y = \cot x$

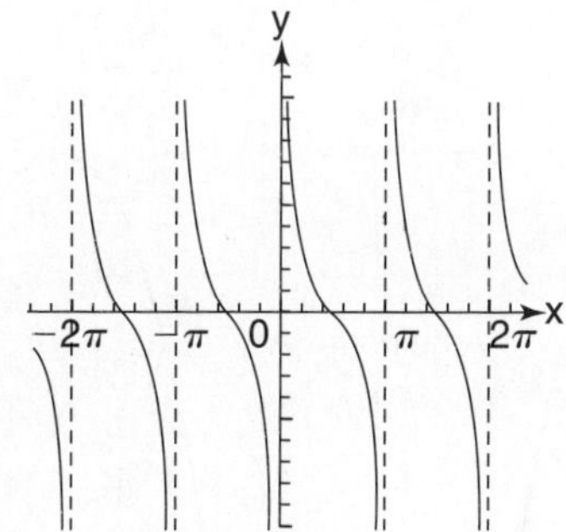

Domain: The whole of **R** except the integral multiples of π **Range:** $(-\infty, \infty)$

Fig. 6.33

$y = \sec x$

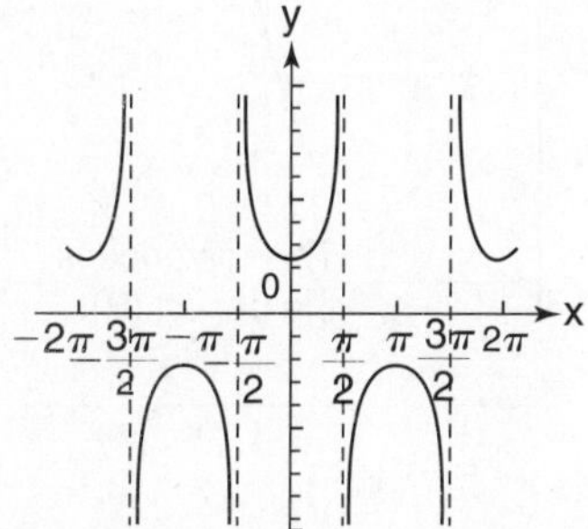

Domain: The whole of **R** except the odd multiples of $\pi/2$ **Range:** The whole of **R** except $(-1, 1)$

Fig. 6.34

$y = \mathbf{cosec}\,x$

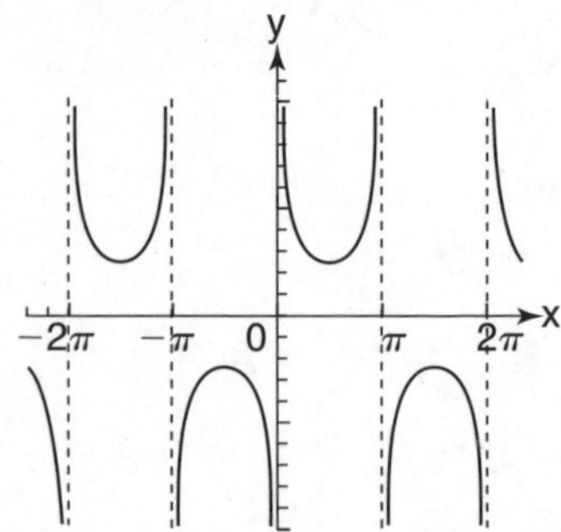

Domain: Whole **R** except the integral multiples of π **Range:** $\mathbf{R} \sim (-1, 1)$

Fig. 6.35

General sine curves A general sine curve is of the form

$$f(x) = \mathrm{A}\sin\left(\frac{2\pi}{\mathrm{B}}(x - \mathrm{C})\right) + \mathrm{D}$$

where |A| is the amplitude, |B| is the period, C is the horizontal shift and D is the vertical shift.

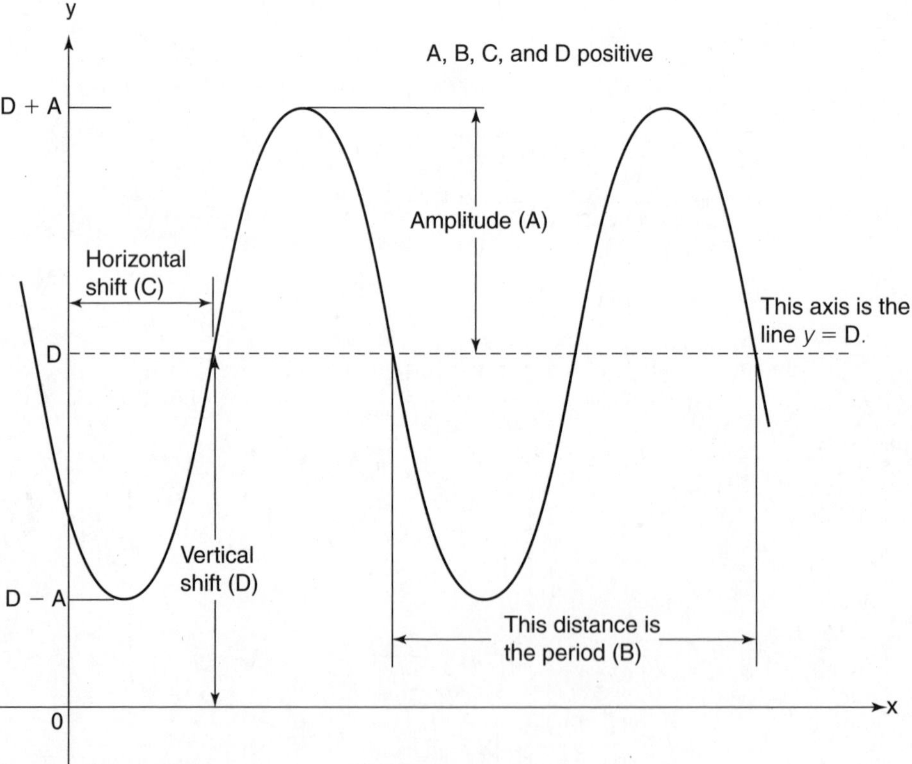

The general sine curve

Fig. 6.36

Some more graphs are provided in Fig. 6.37.

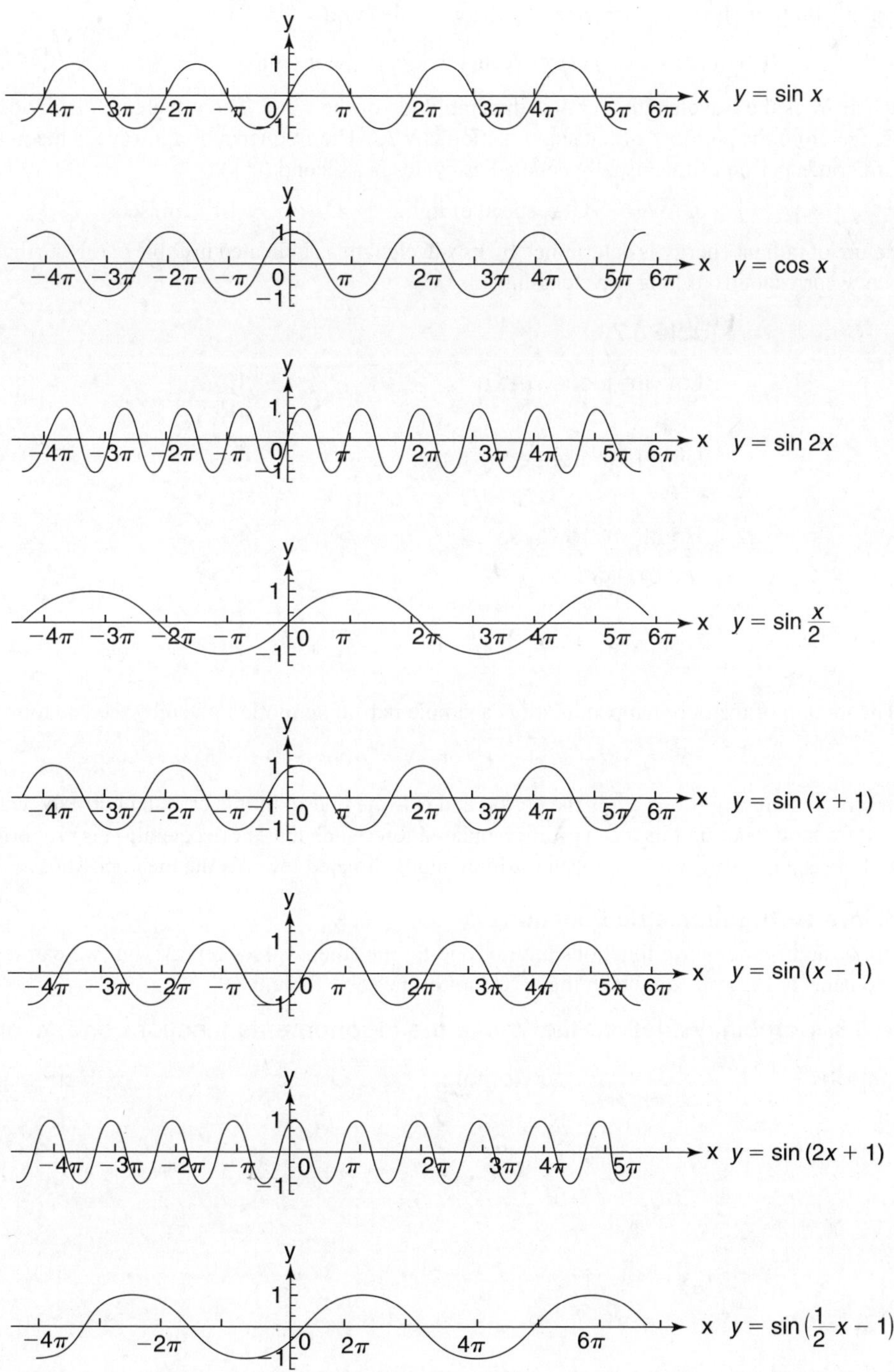

Fig. 6.37

EXAMPLE 6.12

(i) Radiant energy is described by a sinusodial function which describes how the energy is propagated with time. It is thus described by the wave function

$$f(t) = A\sin\left(\frac{2\pi Ct}{W}\right) + \text{constant}$$

where W is the wavelength and A is the amplitude of the wave. The wavelength is divided by C, therefore the period T of the sine function is W/C. The frequency f of a wave is the number of periods per unit time, usually denoted as cycles per second (cps).

$$f = C/W, \qquad C = \text{speed of light}, \qquad C \cong 3 \times 10^{10}\,\text{cm/sec}.$$

The nature of radiant energy is determined by its wavelength as indicated in Table 6.7. Note that low frequency corresponds to long wavelength.

Table 6.7

Low frequency radio	10^4
	10^2
Ultra-high frequency radio	0
	10^{-2}
Radar microwaves	10^{-4}
Visible light	10^{-6}
	10^{-8}
X-rays	10^{-10}

(ii) The motion of the bob in a pendulum is a simple harmonic motion given by the equation

$$x = A\cos(\omega t + \alpha) \quad \text{or} \quad x = a\cos\omega t + b\sin\omega t.$$

where A is the amplitude, $\frac{2\pi}{\omega}$ is the period and α is the phase. In this case the bob executes a to and fro motion around its mean position under a force such that the acceleration is proportional to the displacement from the mean position and is directed towards the mean position.

6.1.5. Inverse Trigonometric Functions

The six basic trigonometric functions are not one-to-one (their values repeat), but we can restrict their domains to intervals on which they are one-to-one (See Table 6.8).

Table 6.8. Domain restrictions that make the trigonometric functions one-to-one

Function	Domain	Range
sin x	$[-\pi/2, \pi/2]$	$[-1, 1]$
cos x	$[0, \pi]$	$[-1, 1]$
tan x	$(-\pi/2, \pi/2)$	$(-\infty, \infty)$
cot x	$(0, \pi)$	$(-\infty, \infty)$
sec x	$[0, \pi/2) \cup (\pi/2, \pi]$	$(-\infty, -1] \cup [1, \infty)$
cosec x	$[-\pi/2, 0) \cup (0, \pi/2]$	$(-\infty, -1] \cup [1, \infty)$

Since these restricted functions are now one-to-one, they have inverses, which we denote by

$$y = \sin^{-1}x \quad \text{or} \quad y = \text{arc}\sin x, \qquad y = \cos^{-1}x \quad \text{or} \quad y = \text{arc}\cos x \text{ and so on.}$$

The graph of the inverse function is obtained by taking the reflection of the original curve about the line $y = x$ as in Fig. 6.38.

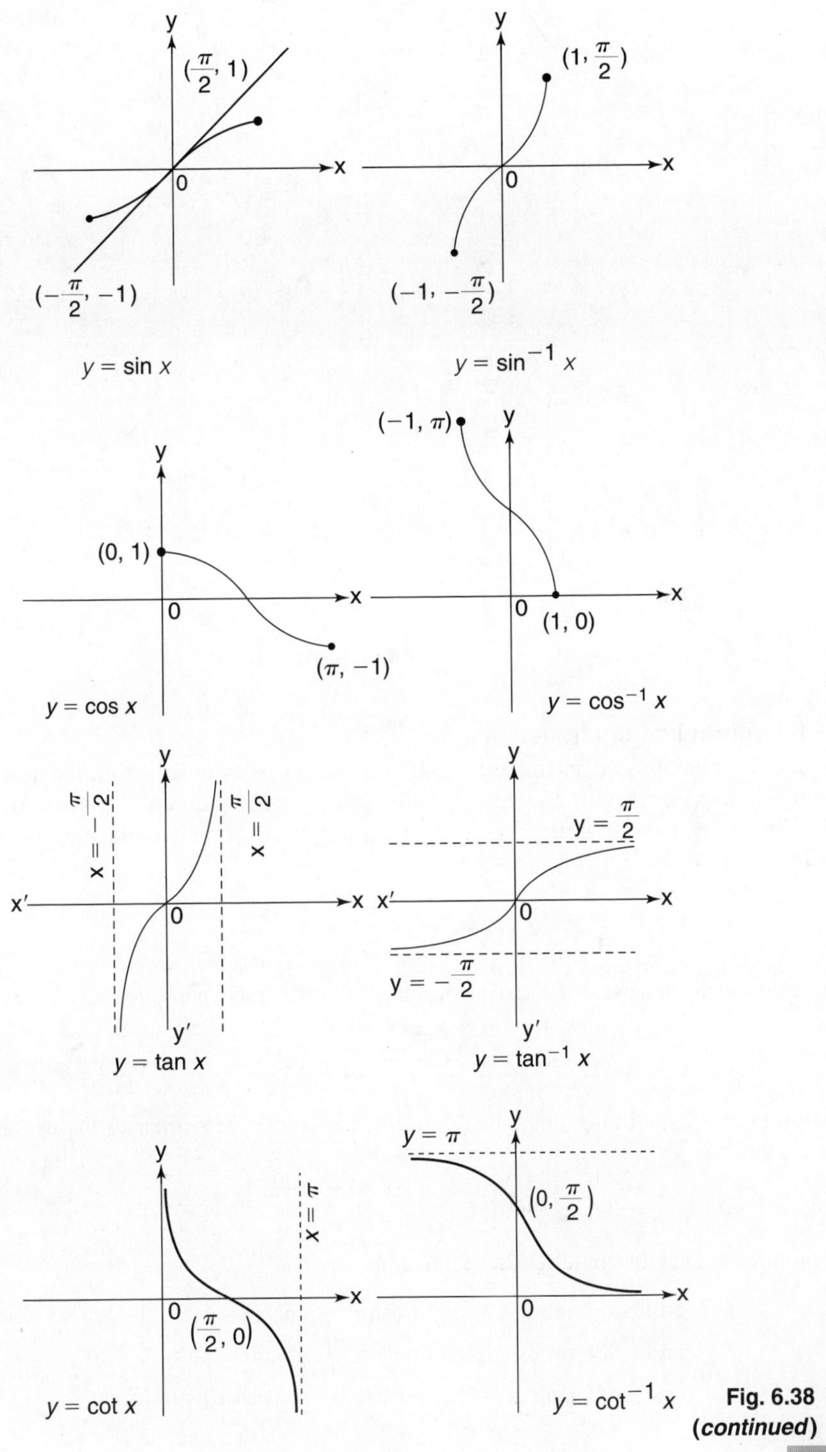

Fig. 6.38
(continued)

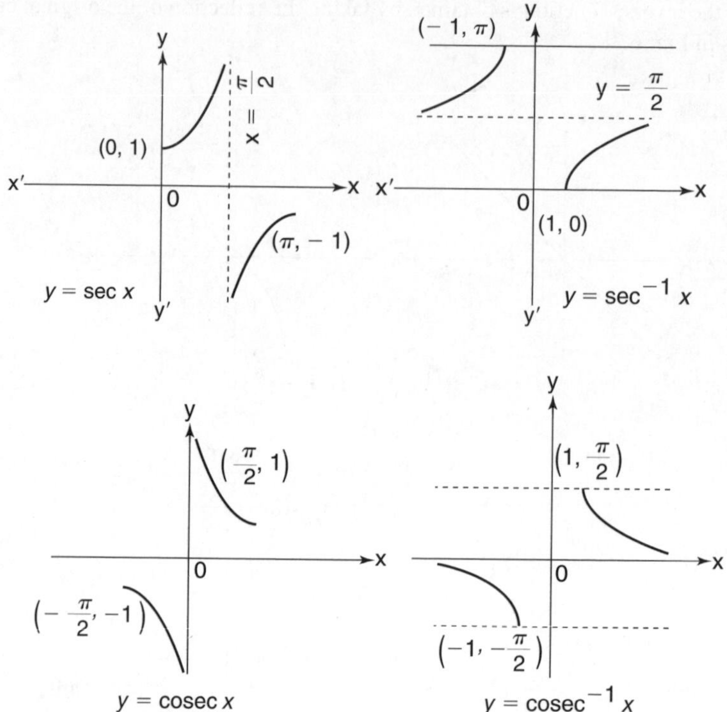

Fig. 6.38

6.1.6 Hyperbolic Functions

Hyperbolic functions occur in the motions of waves in elastic solids, in the shapes of hanging electric power lines and the temperature distributions in metal cooling fins. To define hyperbolic functions, first we write e^x as a sum of one even function and one odd function in the following way:

$$e^x = \frac{e^x + e^{-x}}{2} + \frac{e^x - e^{-x}}{2}$$

The even and the odd parts of e^x define hyperbolic cosine and hyperbolic sine of x respectively. Thus

$$\sinh x = \frac{e^x - e^{-x}}{2}, \qquad \cosh x = \frac{e^x + e^{-x}}{2}.$$

The other four hyperbolic functions are defined in a manner similar to the trigonometric functions. Thus

$$\tanh x = \frac{\sinh x}{\cosh x} = \frac{e^x - e^{-x}}{e^x + e^{-x}} \text{ and so on.}$$

The following identities hold for these functions

$$\cosh x + \sinh x = e^x, \qquad \cosh x - \sinh x = e^{-x},$$
$$\cosh(-x) = \cosh x, \qquad \sinh(-x) = -\sinh x$$
$$\cosh^2 x - \sinh^2 x = 1, \qquad \cosh 2x = \cosh^2 x + \sinh^2 x$$

$$\sinh 2x = 2\sinh x \cosh x, \quad 1 - \operatorname{sech}^2 x = \tanh^2 x, \quad 1 + \operatorname{cosech}^2 x = \coth^2 x$$

$$\sinh(x + y) = \sinh x \cosh y + \cosh x + \sinh y,$$

$$\cosh(x + y) = \cosh x \cosh y + \sinh x \sinh y$$

The graph of the hyperbolic function are shown in Fig. 6.39:

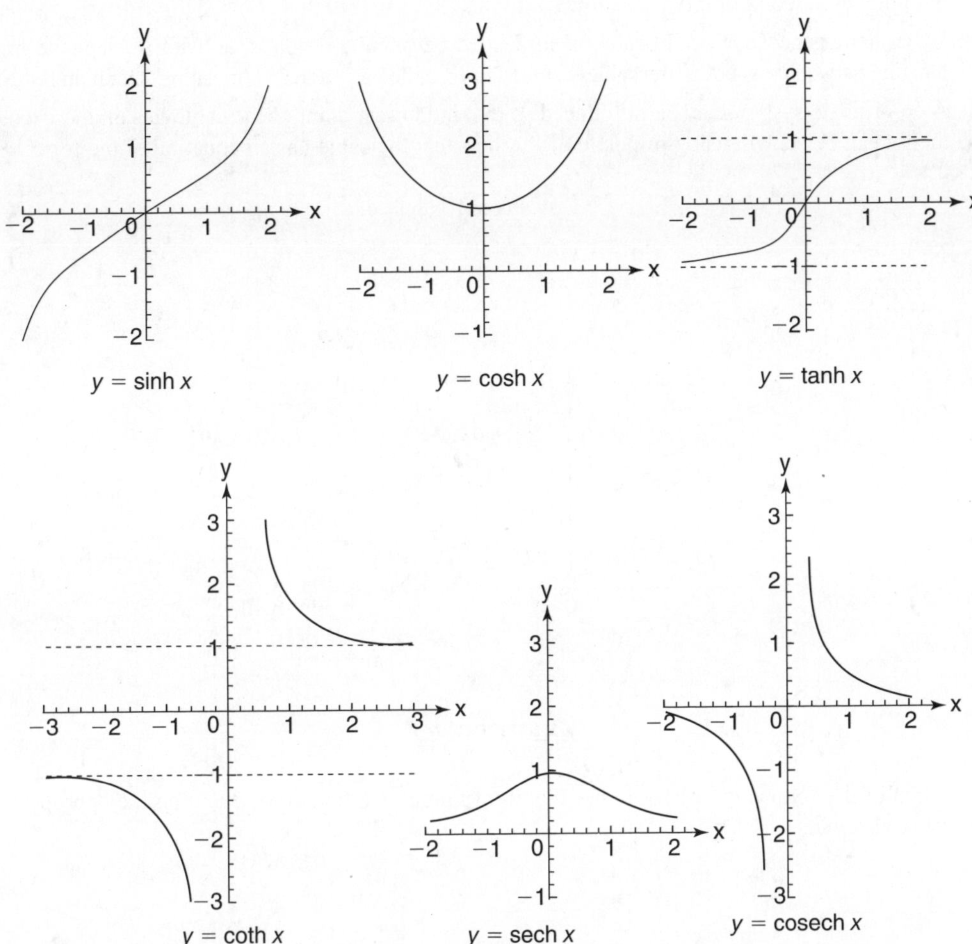

Fig. 6.39

Applications

(i) If a body moves along a straight line under a force such that its acceleration is proportional to the displacement but is directed away from the origin (mean position), then the motion is described by $x = a \cosh \omega t + b \sinh \omega t$, where x is the distance of the particle from the origin at time t.

(ii) In skydiving (or in any kind of free fall from rest) under the action of gravity, the body encounters an air resistance proportional to the square of the velocity. Then at any time t, its motion is given by

$$m\frac{dv}{dt} = mg - kv^2$$

where m is the mass of the body and k is constant that depends on the density of air and on the body's aerodynamic properties. Solving this equation gives

$$v = \sqrt{\frac{mg}{k}} \tanh\left(\sqrt{\frac{gk}{m}}\, t\right)$$

with the initial condition that $v = 0$ when $t = 0$.

(iii) When homogeneous, flexible cable is suspended between two points, as with a telephone line hanging between two poles, the cable forms a curve, called a 'catenary' (meaning 'chain' in Latin). A co-ordinate system may be introduced so that the lowest point of the cable lies on the y-axis, then it can be shown, using principle of Physics, that the cable has an equation of the form

$$y = a \cosh\left(\frac{x}{a}\right) + \text{c}.$$

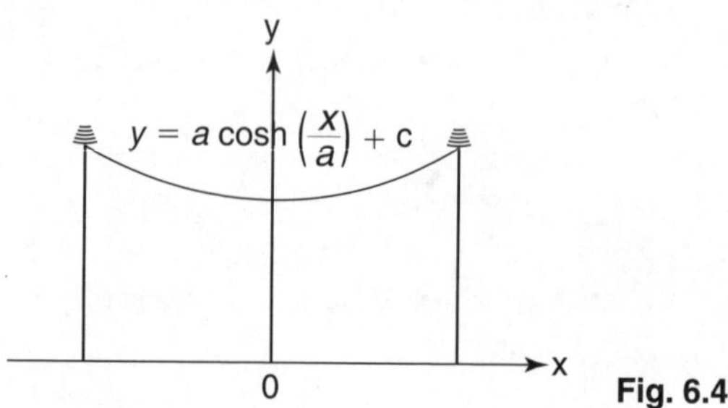

Fig. 6.40

EXERCISE 6.1

1. Which of the following curves do not represent graph of a function ? Mention the appropriate test to check it.

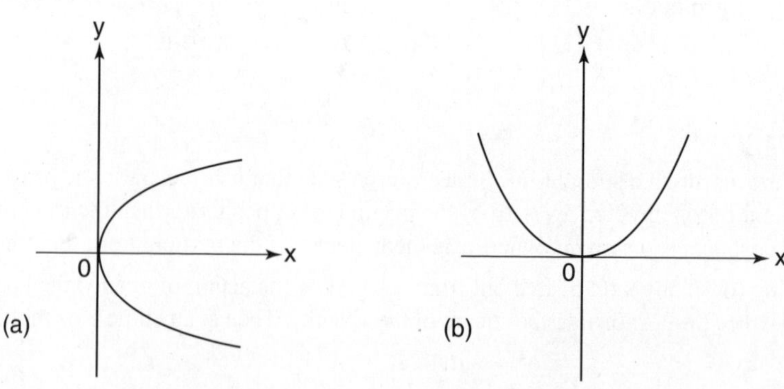

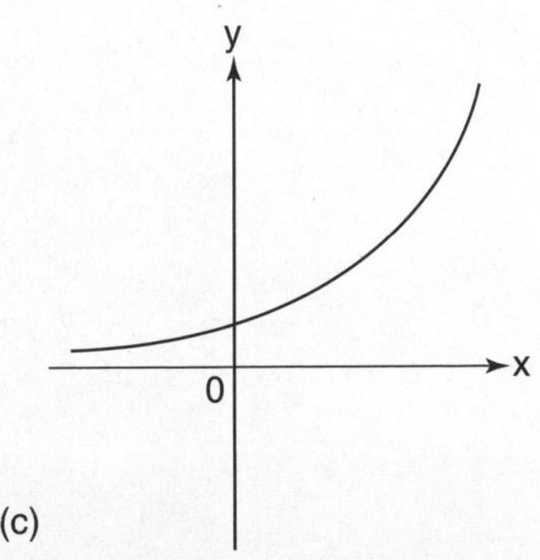

(c)

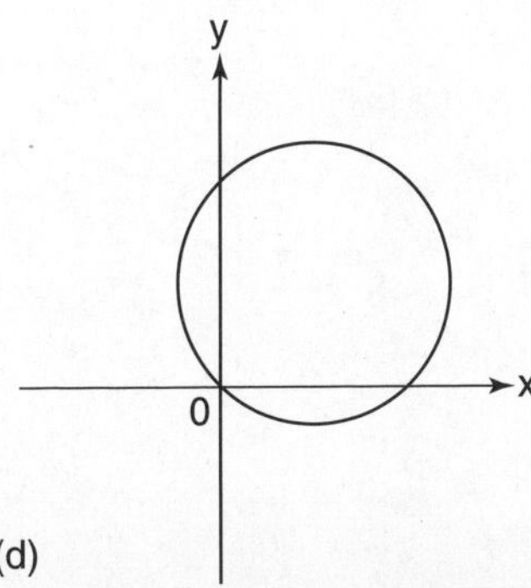

(d)

Where the curve represents the graph of some function, check whether the function is one-to-one, using an appropriate test.

2. Draw the graphs of the following functions

(a) $f(x) = 2x^2 + 3$ **(b)** $f(x) = -\frac{1}{2}x^2 + 5$

(c) $f(x) = (x - 2)^2 + 5$ **(d)** $f(x) = x^2 + 6x + 7$,

starting with the graph of $y = x^2$, showing the effect of different steps used (such as magnification, translation, reflection and so on.)

3. Using the method adopted in Question 2 above, draw the graphs of

(a) (i) $y = 2\sqrt{x - 3}$ **(ii)** $y = -\sqrt{x + 2}$ **(iii)** $y = -\frac{1}{2}\sqrt{x} + 3$

(b) (i) $y = \frac{1}{2}|x - 2| + 4$ **(ii)** $y = 3|x + 2| - 5$ **(iii)** $y = -2|x - 2| + 4$

(c) (i) $y = e^{2x} - 3$ **(ii)** $y = e^{\frac{1}{2}x} + 3$ **(iii)** $y = 3e^{-2x} + 2$

(d) (i) $y = \ln(x + 1)$ **(ii)** $y = -\ln(x - 2)$ **(iii)** $y = 2\ln x + 1$

(e) (i) $y = 2\sin\left(x + \frac{\pi}{2}\right)$ **(ii)** $y = -\sin(x + 1)$ **(iii)** $y = 2\sin(2x + 3)$

(f) (i) $y = \cos\left(x - \frac{\pi}{2}\right)$ **(ii)** $y = 3\cos\left(x + \frac{\pi}{2}\right)$ **(iii)** $y = -\cos(2x + 1)$

(g) (i) $y = \sin^{-1}x + 2$ **(ii)** $y = 3\sin^{-1}x - 2$ **(iii)** $y = \sin^{-1}(2x) + 4$

(h) (i) $y = 2\cos^{-1}x$ **(ii)** $y = -\cos^{-1}(x + 1)$ **(iii)** $y = \cos^{-1}(2x) + 1$

(i) (i) $y = 2x^3 + 1$ **(ii)** $y = (2x)^3 - 4$ **(iii)** $y = -x^3 + 3$

(j) (i) $y = 2^x + 1$ **(ii)** $y = 2^{-x} + 1$ **(iii)** $y = 3 \cdot 2^x - 2$

(k) (i) $y = \sinh(x + 1)$ **(ii)** $y = 2\sinh(x - 2)$ **(iii)** $y = -\sinh(2x)$

(l) (i) $y = 2x^2 + 1$ **(ii)** $y = -x^2 + 3$ **(iii)** $y = -(x + 2)^2 + 3$

(m)(i) $y = \frac{1}{x + 1}$ **(ii)** $y = -\frac{1}{x + 1}$ **(iii)** $y = -\frac{1}{x + 2} + 3$.

CHAPTER SEVEN

Differential Equations in Mathematical Modelling

7.1 INTRODUCTION

A differential equation is an equation involving one or more derivatives of an unknown function. Differential equations occur naturally in many fundamental laws of physical and social sciences which involve rate of change. The order of a differential equation is the order of the highest derivative that it contains. The power of the highest derivative is called the degree of the differential equation. For example,

Differential Equation	Order	Degree
$\frac{dy}{dx} = 5x$	1	1
$\left(\frac{dy}{dx}\right)^2 = -4x + 7$	1	2
$\left(\frac{d^2y}{dx^2}\right)^3 - 6\frac{dy}{dx} + 2y = 0$	2	3
$\frac{d^2y}{dx^2} - \left(\frac{dy}{dx}\right)^3 + 5y = 0$	2	1

In this chapter, we shall discuss about the formation and solution of differential equations arising out of different physical and social situations such as population growth, radioactive decay, cell-division, administration of medicine, heat transfer and so on.

7.1.1 EXPONENTIAL GROWTH MODEL

A very common observation about 'population growth' is that unless constrained by environmental or other limitations, populations (human beings, bacteria, plants and so on) tend to grow at a rate which is proportional to the size of the population. Larger the population, higher is the growth rate.

To convert it into a mathematical problem, let us assume that $x = x(t)$ is the population at time t. Also, let us assume that at time $t = 0$, population is $x = x_0$. At any point of time, the rate of increase of the population with respect to time is $\frac{dx}{dt}$. According to the given situation,

$$\frac{dx}{dt} \propto x \qquad \text{so that}$$

$$\frac{dx}{dt} = \text{k}x;$$

where k is a constant of proportionality. Since the direction of change is positive (as larger the population, higher is the growth), k is positive. Usually it can be determined experimentally. Thus, we are left with

$$\frac{dx}{dt} = \text{k}x, \qquad x(0) = x_0$$

EXAMPLE 7.1 If the population of a country doubles itself in 50 years, in how many years will it treble under the assumption that the rate of increase is proportional to the number of inhabitants?

SOLUTION

Let $x = x(t)$ denote the population when year $= t$ and x_0 be the population when $t = 0$. According to the question,

$$\frac{dx}{dt} \propto x$$

$$\Rightarrow \quad \frac{dx}{dt} = \text{k}x, \qquad \text{where k is a constant of proportionality.}$$

$$\Rightarrow \quad \frac{dx}{x} = \text{k}dt$$

On integrating, we get

$$\ln x = \text{k}t + \ln \text{c}$$

For $t = 0, x = x_0$,
therefore, $\text{c} = x_0$

$$\Rightarrow \quad x = x_0 \text{e}^{\text{k}t}$$

when $t = 50$ years, $x = 2x_0$
so that, $\text{e}^{50\text{k}} = 2$

$$\Rightarrow \quad \text{k} = \frac{\ln 2}{50}$$

Suppose, time required for population to be $3x_0$ is T years.

$$\Rightarrow \quad 3x_0 = x_0 \text{e}^{\text{kT}}$$

$$\Rightarrow \quad \text{kT} = \ln 3$$

$$\Rightarrow \quad \text{T} = \frac{\ln 3}{\text{k}} = \frac{50 \ln 3}{\ln 2} = 79 \text{ years (approx.)}$$

Giving a mathematical shape to a physical situation, as has been done above, is an important topic in mathematics, known as mathematical modelling. In fact, mathematical modelling acts as the interface between mathematics and the physical world around us. A problem involving a differential equation along with some initial condition is also called an initial value problem. By solving the above type of initial value problem, we get $x = x_0 e^{kt}$ as the population at any time t. Due to this, the above type of population growth model (where rate of increase is proportional to the quantity present) is called exponential growth model.

EXAMPLE 7.2

(a) A certain culture of bacteria grows at a rate that is proportional to the number present. It is found that the number doubles in 4 hours. How many may be expected at the end of 24 hours?

(b) Draw an approximate graph of the number of bacteria vs time, if initially there are two bacteria and constant of proportionality is one.

[DU, B.Sc.(Ph. Sc.), 2006]

SOLUTION

(a) Let $x = x(t)$ denote the number of bacteria present at time t hours.
It is given that

$$\frac{dx}{dt} \propto x$$

$$\Rightarrow \quad \frac{dx}{dt} = \mathrm{k}x, \qquad \text{where k is constant of proportionality.}$$

$$\Rightarrow \quad \frac{dx}{x} = \mathrm{k}dt,$$

Integrating, we get

$$\ln x = \mathrm{k}t + \ln \mathrm{C}$$

Let there be x_0 number of bacteria when $t = 0$. This implies that $\mathrm{C} = x_0$.
Hence

$$x = x_0 e^{kt}$$

It is given that $x = 2x_0$ when $t = 4$.

$$\Rightarrow \quad e^{4k} = 2$$

$$\Rightarrow \quad \mathrm{k} = \frac{\ln 2}{4}$$

If X denote the number of bacteria at the end of 24 hours, then

$$\mathrm{X} = x_0 e^{24k}$$

$$\Rightarrow \quad \mathrm{X} = x_0 e^{24\left(\frac{\ln 2}{4}\right)} = 2^6 x_0 = 64x_0$$

∴ Number of bacteria will be 64 times the original number after 24 hours.

(b) The equation obtained is

$$x = x_0 e^{kt}$$

We are given that $x_0 = 2$, $\mathrm{k} = 1$. Thus the required equation is

$$x = 2e^t$$

The required graph is of the type $y = ce^x$, where c is magnifying factor. Replacing x, y and c by t, x and 2 in $y = ce^x$, we get the required graph as given in Fig. 7.1.

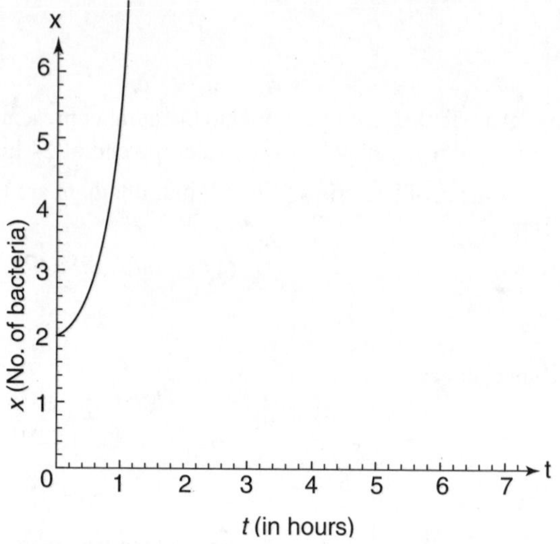

Fig. 7.1
No. of bacteria vs time graph

The process of cell-division also exhibits exponential growth model. Look at the following initial value problem which uses exponential growth model.

EXAMPLE 7.3 A cell of the bacterium E. Coli divides into two cells every 20 minutes when placed in a nutrient culture. Let $x = x(t)$ be the number of cells that are present t minutes after a single cell is placed in the culture. Assume that the growth of bacteria is approximated by a continuous exponential growth model.

(a) Find an initial value problem whose solution is $x(t)$.

(b) Find a formula for $x(t)$.

(c) How many cells are present after 2 hours?

(d) How long does it take for the number of cells to reach 1,000,000?

(e) Draw an approximate x-t graph of the above.

SOLUTION

(a) It is given that the growth model is approximated by a continuous exponential growth model. Therefore,

$$\frac{dx}{dt} \propto x$$

$$\Rightarrow \quad \frac{dx}{dt} = kx, \qquad \text{where k is the constant of proportionality.}$$

The differential equation $\frac{dx}{dt} = kx$ along with the initial condition $x_0 = x(0) = 1$ constitute the initial value problem.

(b) Solving the differential equation, we get

$$\int \frac{dx}{x} = \int k dt + \ln C$$

$$\ln x = kt + \ln C$$

$$\Rightarrow \qquad x = Ce^{kt}$$

From the initial condition that $t = 0$, $x = 1$, we get $C = 1$.
Thus, we have

$$x(t) = e^{kt}$$

It is given that for $t = 20$, $x = 2$. This gives

$$2 = e^{20k}$$

$$\Rightarrow \qquad k = \frac{\ln 2}{20}$$

Thus the required formula for $x(t)$ is

$$x(t) = e^{\left(\frac{\ln 2}{20}\right)t} = 2^{t/20} \qquad (1)$$

(c) After 2 hours (= 120 minutes), number of cells present is given by

$$x(120) = 2^6$$

$$\Rightarrow \qquad x = 64$$

(d) Time required for number of cells to reach 1,000,000 is given by

$$1{,}000{,}000 = 2^{t/20}$$

$$\Rightarrow \qquad \ln 10^6 = \frac{t}{20} \ln 2$$

$$\Rightarrow \qquad t = 20 \times \frac{6 \ln 10}{\ln 2}$$

$$= 399 \text{ minutes (approx.)}$$

$$= 6 \text{ hours } 39 \text{ minutes}$$

(e) The function $x = 2^{t/20}$ has the following characteristics

(i) It is an exponential function of the type $y = a^x$, where $a = 2$.

(ii) $2^{t/20}$ increases gradually from 1 to ∞ as t increases from 0 to ∞.

(iii) The curve passes through the points (0, 1), (20, 2), (40, 4), (60, 8) and so on.

A rough sketch of the graph is as in Fig. 7.2.

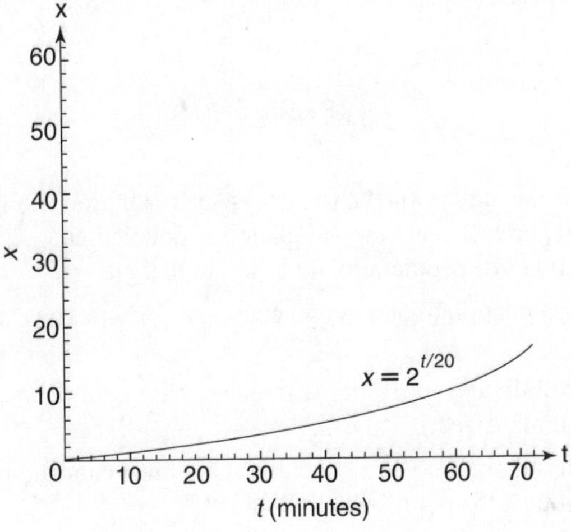

Fig. 7.2

REMARK: In case of exponential growth, the constant k can be written as

$$k = \frac{dx/dt}{x}$$

which states that the growth rate, as a fraction of the entire population, remains constant over time and this constant is k. For this reason, k is called the relative growth rate of the population. It is also expressed as a percentage. Thus, a relative growth rate of 5% time in an exponential growth model means that $k = 0.05$. Similarly, we define relative decay rate in case of exponential decay model.

EXAMPLE 7.4 According UN data, the world population in 1998 was approximately 5.9 billion and growing at a rate of about 1.33% per year. Assuming an exponential growth rate, estimate the world population by the end of the year 2023.

SOLUTION

We assume that the population by the year end of 1998 is 5.9 billion. We take

t = time elapsed from the end of 1998 (in year)

x = world population (in billion)

Then, we have

$$x_0 = x(0) = 5.9$$

Here, following exponential growth model, we get

$$x = x_0 e^{kt}$$

Now, $x_0 = 5.9$, $k = .0133$. Thus

$$x = 5.9\, e^{(.0133)t}$$

By the end of year 2023, $t = 25$.
Hence the required population is

$$x = 5.9\, e^{(.0133)25} \approx 8.2$$

Thus the world population by the end of year 2023 will be approximately 8.2 billion.

EXERCISE 7.1

1. The population of a certain city is known to increase at a rate proportional to the number of people living in the city. If after 2 years the population is doubled and after 3 years the population is 20,000, find the number of people initially living in that city.
2. A town has an initial population of 500 that increases by 15% in 10 years. What will be the population in 30 years?
3. A particular type of cell divides itself into two in every six hours. How many hours will it take to be five times its original size?
4. At time $t = 0$, a bacterial culture has x_0 bacteria. One hour later the population has grown by 25%. How long will it take the population to double?

5. Suppose that an initial population of 10,000 bacteria grows exponentially at a rate of 1% per hour and that $y = y(t)$ is the number of bacteria present t hours later.
   (a) Find an initial-value problem whose solution is $y(t)$,
   (b) Find a formula for $y(t)$;
   (c) How long does it take for the initial population of bacteria to double?
   (d) How long does it take for the population of bacteria to reach 45,000?
6. Suppose that 100 fruit flies are placed in a breeding container that can support at most 5000 flies. Assuming that the population grows exponentially at a rate of 2% per day, how long will it take for the container to reach capacity?
7. In each part, find an exponential growth model $y = y_0 e^{kt}$ that satisfies the stated conditions.
   (a) $y_0 = 2$, doubling time T $= 5$
   (b) $y(0) = 5$, growth rate $= 1.5\%$
   (c) $y(1) = 1$, $y(10) = 100$
   (d) $y(1) = 1$, doubling time T $= 5$
8. A bacteria culture is known to grow at a rate proportional to the number present. After one hour, 1000 bacteria are observed in the culture and after 4 hours, it is 3000. Determine the number of bacteria originally present in the culture.
9. In a certain unicellular bacteria culture, the rate of increase in the number of bacteria is proportional to the number present. If the number of bacteria trebles after 5 hours, how many bacteria will be present in the culture after 10 hours?

■ 7.2 EXPONENTIAL DECAY MODEL

Suppose that the quantity of a certain material decreases at a rate proportional to the amount of material present. Let $x(t)$ denote the quantity of material present at time t. Then we have

$$\frac{dx}{dt} = -kx$$

where $k > 0$ is a constant of proportionality, also called the decay constant. Suppose the initial quantity of the material is $x = x_0$ when $t = 0$. Then, we can obtain a general formula for $x(t)$ by solving the initial-value problem

$$\frac{dx}{dt} = -kx, \qquad x(0) = x_0 \tag{1}$$

On solving (1), we get

$$x = x_0 e^{-kt}$$

which represents the quantity of the material remaining after time t.

The above mathematical model is known as exponential decay model. Usually it is observed in radioactive disintegration, administration of medicines and so on. Uranium, radium and other radio-active substances are known to decay at a rate proportional to the amount present. In such cases, the amount of time required to disintegrate to one-half of its original mass is called the half-life of the material.

When a drug (say, penicillin, aspirin and so on) is administrated to an individual, it enters the blood stream and then is absorbed by the body over time. Medical research has shown that more the amount of drug present in the body, the more rapidly it is absorbed by the body. Over the time, as the amount of drug present in the body decreases, the absorption also becomes slower. Let $x(t)$ be the amount of the drug present in the blood stream at time t. At any point of time t, the rate of change in x with respect to t (here the rate of absorption) is $\frac{dx}{dt}$, so the above observation gives rise to the following differential equation

$$\frac{dx}{dt} = -kx,$$

where k is a positive constant of proportionality that depends on the drug and can be denoted experimentally. The negative sign is required because x decreases with time. This along with the initial dosage of the drug say $x = x_0$, at time $t = 0$ forms the initial-value problem

$$\frac{dx}{dt} = -kx, \qquad x(0) = x_0.$$

EXAMPLE 7.5 Radium is known to decay at a rate proportional to the amount present. If half-life of radium is 1600 years, what percentage of radium will remain in a given sample after 800 years?

SOLUTION

Let $x = x(t)$ denote the amount of radium left after time t (in years). By the given condition

$$\frac{dx}{dt} = -kx, \qquad \text{where k is a constant of proportionality}$$

$$\frac{dx}{x} = -kdt$$

$$x = ce^{-kt}, \qquad \text{c is an arbitrary constant.}$$

Initially, let $x = x_0$ for $t = 0$. Then $c = x_0$

$$x = x_0 e^{-kt}$$

It is given, that when $t = 1600$ years, $x = \frac{x_0}{2}$. This leads to

$$\frac{x_0}{2} = x_0 e^{-1600k}$$

$$k = \frac{1}{1600}\ln 2$$

The amount left after 800 years is given by

$$x = x_0 e^{-\left(\frac{\ln 2}{1600}\right)800} = x_0 e^{-\left(\frac{1}{2}\ln 2\right)} = \frac{x_0}{\sqrt{2}}$$

∴ Percentage of the amount left after 800 years

$$= \frac{x_0}{\sqrt{2}} \times \frac{1}{x_0} \times 100\%$$

$$= 50\sqrt{2}\%$$

$$= 70.5\% \text{ (approx.)}$$

EXAMPLE 7.6 Assume that the rate at which radioactive nuclei decay is proportional to the number of nuclei that are present in a given sample. In a certain sample, 10% of the original number of radioactive nuclei have undergone disintegration in a period of 200 years.

(i) What percentage of the original radioactive nuclei will remain after 1000 years?

(ii) In how many years will only one fourth of the original number remain?

(iii) Draw a graph of the amount remaining against time for this problem, assuming $x_0 = 400$.

SOLUTION

Let $x = x(t)$ denote the amount of the radioactive nuclei left after time t (in years). Then proceeding as in the previous case we get

$$x = x_0 e^{-kt}$$

where x_0 is the original amount of the radioactive nuclei. It is given that $x = 90\%$ of x_0 when $t = 200$ years.

$$\Rightarrow \qquad \frac{9}{10}x_0 = x_0 e^{-200k}$$

$$\Rightarrow \qquad k = -\frac{1}{200}\ln\frac{9}{10}$$

$$\Rightarrow \qquad x = x_0 e^{\frac{1}{200}\left(\ln\frac{9}{10}\right)t} = x_0\left(\frac{9}{10}\right)^{t/200}, \text{ after time } t.$$

(i) For $t = 1000$

$$x = x_0\left(\frac{9}{10}\right)^5 = x_0(.9)^5$$

Percentage of radioactive nuclei left after 1000 years

$$= \frac{x_0(.9)^5}{x_0} \times 100\% = 55.05\% \text{ (approx.)}$$

(ii) Suppose after T years, only 1/4th of the original amount of radioactive nuclei is left. Then,

$$\frac{1}{4}x_0 = x_0\,(.9)^{T/200}$$

$$\Rightarrow \qquad T = 200\frac{\ln(.25)}{\ln .9} = 200\frac{\ln 4}{\ln\left(\frac{10}{9}\right)} = 263 \text{ yeas (approx.)}$$

(iii) $x(t) = 400(.9)^{t/200}$

The graph is magnification of the exponential graph $x = (.9)^{t/200}$, which is of the type $x = a^{t/200}$, $a = .9 < 1$. As t moves 0 to ∞, x moves from 400 to 0. The graph passes through $(0, 400)$, $(100, 120\sqrt{10})$, $(200, 360)$ and so on.

A rough sketch of the graph is given in Fig. 7.3.

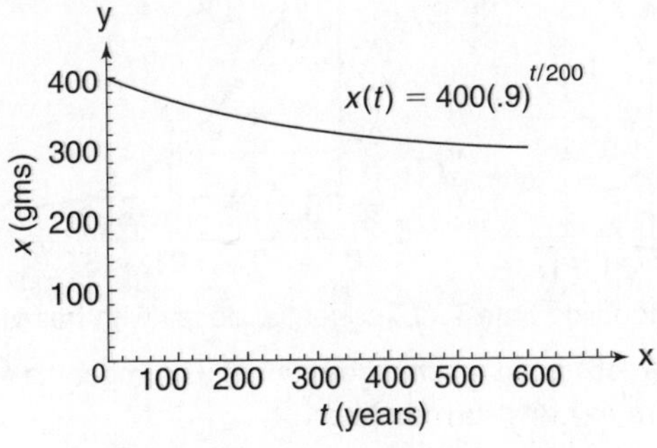

Fig. 7.3

EXAMPLE 7.7 An amount of 10 ml of a medicine is injected into a patient's body. Half the amount of the medicine is absorbed by the patient's body in 10 hours. How long will it take the patient to absorb 70% of the medicine?

SOLUTION

It is an exponential decay problem having the following differential equation and initial condition

$$\frac{dx}{dt} = -kx, \qquad x_0 = x(0) = 10,$$

where $x(t)$ denotes the amount of medicine left to be absorbed.

Solving the differential equation, we get

$$x = ce^{-kt}$$

Initially, $t = 0$, $x = 10$. Thus we get

$$x = 10e^{-kt}$$

For $t = 10$, $x = \frac{x_0}{2} = 5$. Therefore

$$\frac{1}{2} = e^{-k(10)} \Rightarrow 10k = \ln 2 \Rightarrow k = \frac{1}{10}\ln 2$$

If t is the time for absorption of 70% of the medicine, then, we have

$$30\% \text{ of } x_0 = 10e^{-t\left(\frac{\ln 2}{10}\right)}$$

$$\Rightarrow \qquad 3 = 10e^{-t\left(\frac{\ln 2}{10}\right)} \qquad (\because x_0 = 10)$$

$$\Rightarrow \qquad t = \frac{10}{\ln 2}\ln\left(\frac{10}{3}\right) = 17.43 \text{ hours (approx.)}$$

REMARK: If T is the time period for the original population to be double or half respectively in an exponential growth or decay respectively, then population versus time graph in the two cases will appear as in Fig. 7.4.

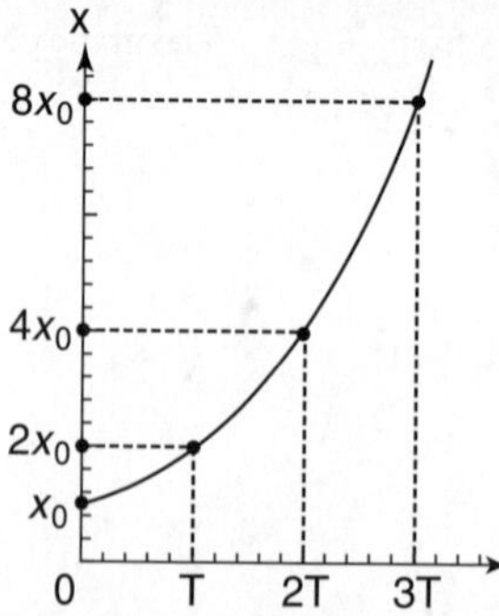

Exponential growth with doubling time T

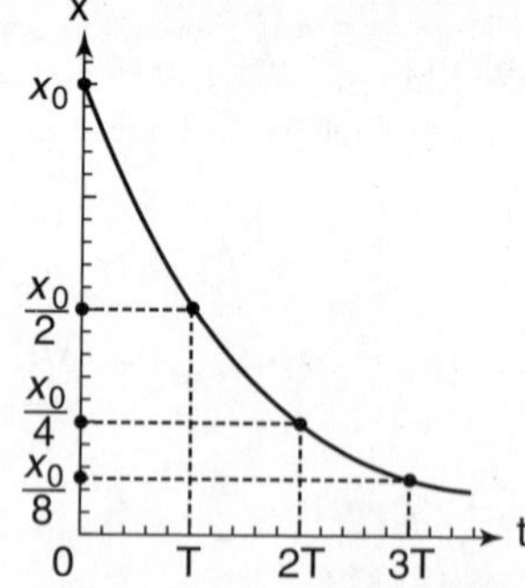

Exponential decay with half-life T

Fig. 7.4

Exponential decay is also observed when a body loses its temperature to the surrounding environment or gains temperature from its surroundings.

Newton's Law of Cooling states that the rate at which the temperature of a cooling object decreases and the rate at which a warming object gains temperature are proportional to the difference between the temperature of the object and the temperature of the surrounding medium. Let $T(t)$ be the temperature of an object at time t placed in a medium with constant temperature T_S. According to Newton's Law of Cooling, we have

$$\frac{dT}{dt} = -k(T - T_S)$$

where $k > 0$ is a constant of proportionality. In case of a cooling body, $T > T_S$, so that the right side is negative and $\frac{dT}{dt}$ thus represents the rate of decrease of temperature of the hot body. In case of a warming body, $T < T_S$, so that right hand side is positive and thus $\frac{dT}{dt}$ represent the rate of increase of temperature of the cold body.

If at $t = 0$, the temperature of the body is T_0, then we get the following initial-value problem

$$\frac{dT}{dt} = -k(T - T_S), \qquad T(0) = T_0.$$

Solving it, we get that in both the cases, the temperature $T(t)$ of the body at any time t is given by

$$T(t) = T_S + (T_0 - T_S)e^{-kt},$$

where k is a positive constant.

EXAMPLE 7.8 A body with initial temperature of 100°C is allowed to cool in air which remains at a constant temperature of 20°C. It is given that after 10 minutes, the body has cooled to 40°C.

(i) Find the temperature of the body after half an hour.

(ii) Draw a graph of temperature vs time.

SOLUTION

Let $T(t)$ denote the instantaneous temperature of the body in degree celcius at time t (in minutes). According to the Newton's Law of Cooling,

$$\frac{dT}{dt} = -k(T - 20)$$

where k is a positive constant of proportionality.

$$\Rightarrow \qquad \frac{dT}{T - 20} = -kdt$$

Integrating, $\ln(T - 20) = -kt + \ln C$, C being the constant of integration

$$\Rightarrow \qquad T - 20 = Ce^{-kt}$$

Initially, $T = 100, t = 0 \Rightarrow C = 80$

$$\Rightarrow \qquad T - 20 = 80e^{-kt}$$

It is given that $T(10) = 40$

$$\Rightarrow \qquad 80e^{-k(10)} = 20$$

$$\Rightarrow \qquad k = \frac{1}{10}\ln 4$$

$$\Rightarrow \qquad T = 20 + 80e^{-\left(\frac{1}{10}\ln 4\right)t}$$

$$\Rightarrow \qquad T = 20 + 80\left(\frac{1}{4}\right)^{t/10}$$

Temperature after half an hour

$$= 20 + 80\left(\frac{1}{4}\right)^{3}$$

$$= 21.25^{\circ} \text{ Celsius}$$

(ii) The required graph of

$$T(t) = 20 + 80\left(\frac{1}{4}\right)^{t/10}$$

may be obtained from the graph of $f(t) = \left(\frac{1}{4}\right)^{t/10}$, first by magnifying it by 80 and then by shifting upward trough 20 units.

The graph of $f(t) = \left(\frac{1}{4}\right)^{t/10}$ is of exponential type, $y = a^x$,

where $a = \frac{1}{4}$, $\quad x = \frac{t}{10}$.

As t increases from 0 to ∞ (t is non negative), $f(t)$ decreases from 1 to 0 so that T(t) decreases from 100 to 20. Some points on the graph of $T(t) = 20 + 80\left(\frac{1}{4}\right)^{t/10}$ are (0, 100), (10, 40), (20, 25), (30, 21.25) and so on. A rough sketch of the graph is given in Fig. 7.5.

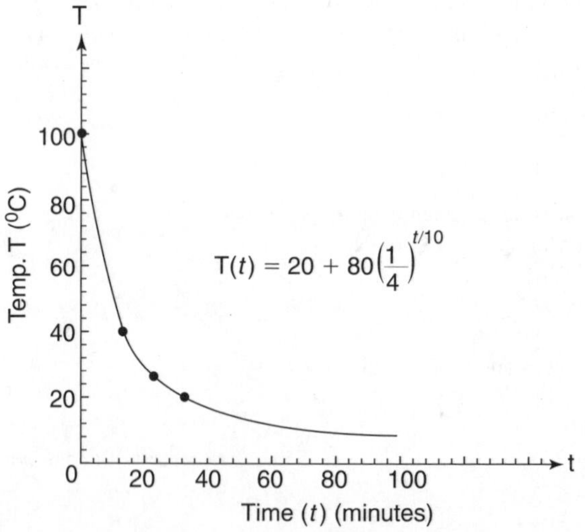

Fig. 7.5

EXAMPLE 7.9 According to Newton's Law of Cooling, the rate at which a substance cools in air is proportional to the difference between the temperature of the substance and that of the surrounding air. If the temperature of the air is 30°C and the substance cools from 100°C to 70°C in 15 minutes, find when the temeprature will be 40°C.

SOLUTION

Let $T = T(t)$ be the temperature of the substance in degree celcius at any time t (in minutes).

According to Newton's Law of Cooling

$$\frac{dT}{dt} = -k(T - 30), \qquad T(0) = 100^{\circ}C$$

$$\Rightarrow \qquad \frac{dT}{T - 30} = -k\,dt$$

Integrating, we get

$$\ln(T - 30) = -kt + \ln C$$

$$\Rightarrow \qquad T - 30 = Ce^{-kt}$$

Initially, $T = 100$, $t = 0 \Rightarrow C = 70$

$$\Rightarrow \qquad T = 30 + 70e^{-kt}$$

It is given that $T(15) = 70$.

$$\therefore \qquad 70 = 30 + 70e^{-15k}$$

$$15k = \ln\left(\frac{7}{4}\right) = 0.56$$

$$\Rightarrow \qquad k = .0373$$

$$\Rightarrow \qquad T = 30 + 70e^{-.0373\,t}$$

If t_0 be the time when temperature will be 40°C, then

$$\Rightarrow \qquad 40 = 30 + 70e^{-(0.0373)t_0}$$

$$\Rightarrow \qquad \log 7 = (0.0373)t_0$$

$$\Rightarrow \qquad t_0 = \frac{\log 7}{0.0373} = 52 \text{ minutes (approx.)}$$

EXERCISES 7.2

1. In each part, find an exponential decay model $y = y_0e^{-kt}$ that satisfies the stated conditions
   (a) $y(1) = 10$, half life $T = 6$
   (b) $y(1) = 100$, $y(10) = 10$
   (c) $y(0) = 20$, decay rate $= 2\%$
   (d) $y_0 = 10$, half life $= 5$
2. A research student wants to determine the half-life of a certain radioactive substance. He finds that in exactly 5 days a ten milligram sample of the substance decays to 3.5 milligrams. Based on this data, what should he infer about the half-life of the substance?
3. Radon-222 is a radioactive gas with a half-life of 3.83 days. The gas is a health hazard because it tends to get trapped in the basement of the house. To prevent entry of the gas, basements may be sealed. Suppose that 5.0×10^7 radon atoms are trapped in a basement at the time it is sealed and $y(t)$ is the number of atoms present in the basement t days later.
   (a) Find an initial value problem whose solution is $y(t)$.
   (b) Find a formula for $y(t)$.
   (c) How many atoms will be present after 30 days?
   (d) How long will it take for 90% of the original quantity of gas to decay?
4. The half-life of radioactive carbon-14 is about 5730 years. Find its decay constant. If 100 gms of radioactive carbon-14 are stored for 100 years, how many grams will be left at the end of that time?
5. In an archaeological excavation, bones of some extinct animals were discovered. It was found that the bones contained around 30% of their original carbon-14. Using the fact that half-life of carbon-14 is about 5730 years, find the age in which the animal was found on earth.

6. A certain radioactive material is known to decay at a rate proportional to the amount present. If initially 1/2 gm of the material was present and 0.1 percent of the original mass has decayed after one week, find the half-life of the material.

7. A dose of 5 ml of a medicine is injected into a patient's body. In one hour 1 ml of the medicine was absorbed by the body. In how much time will 90% of the medicine be absorbed by the body?

8. An asthmatic bout of a patient comes under control when 4 ml of a certain medicine is absorbed by his body. On occurrence of a bout, the patient is given an injection of 10 ml of the medicine. If it took 15 minutes for the patient to absorb 2 ml of the medicine after it was injected, how much time will it take to control the bout?

9. A metal rod at a temperature of 100°F is placed in a room at a constant temperature of 0°F. If after 20 minutes the temperature of the rod comes down to half of its original temperature, find an expression for the temperature of the rod at time t.

10. A body at a temperature 72°F is taken outdoors, where the temperature is 20°F. After 5 minutes, the temperature of the body is 55°F. How long will it take the body to reach a temperature of 43°F?

11. A cup of water at a temperature of 95°C is placed in a room with a constant temperature 21°C.

(a) Assuming that Newton's Law of Cooling applies, set up and solve an initial-value problem whose solution is the temperature of the water t minutes after it is placed in the room.

(b) How many minutes will it take for the water to reach a temperature of 51°C if it cools to 85°C in one minute?

(c) Draw an approximate graph of temperature of the water as a function of time.

7.3 POPULATION GROWTH WITH LOGISTIC GROWTH MODEL

While discussing exponential growth model for population, it was assumed that the population was not constrained by the environment. However, in real life this assumption is usually not valid. In fact, the environmental constraints become influential once the population attains a certain level. It is observed that an ecological system has a limited capacity to support only upto a certain number of individuals of a population of a particular species. Let C be this number. Then C is called the Carrying Capacity of the system. Let $x = x(t)$ be the actual number of individuals of that population in that time t.

If $x > \mathrm{C}$, the population exceeds the capacity of the ecological system and tends to decrease toward C. When $x < \mathrm{C}$, the population is below the capacity of the ecological system and tends to increase towards C. When $x = \mathrm{C}$, the population tends to remain stable. From these observations, we get that

$$\frac{dx}{dt} < 0 \qquad \text{if } \frac{x}{\mathrm{C}} > 1$$

$$\frac{dx}{dt} > 0 \qquad \text{if } \frac{x}{\mathrm{C}} < 1$$

$$\frac{dx}{dt} = 0 \qquad \text{if } \frac{x}{\mathrm{C}} = 1$$

Also, it is observed that when the population is far below the carrying capacity (that is, $\frac{x}{\mathrm{C}} \approx 0$), then the environmental constraints have very little effect and the growth of population exhibits exponential model; that is,

$$\frac{dx}{dt} \approx \mathrm{k}x \qquad \text{if } \frac{x}{\mathrm{C}} \approx 0$$

Thus, the rate of charge of population depends on two opposing factors: x and $C - x$. Accordingly, we get

$$\frac{dx}{dt} \propto x(C - x) \Rightarrow \frac{dx}{dt} \propto Cx\left(1 - \frac{x}{C}\right)$$

$$\Rightarrow \quad \frac{dx}{dt} = kx\left(1 - \frac{x}{C}\right),$$

where k is the positive constant of proportionality. Thus, if k and C are determined experimentally and the population is known at some point of time, say $x(0) = x_0$, then a general formula for the population $x(t)$ can be determined by solving the initial-value problem

$$\frac{dx}{dt} = k\left(1 - \frac{x}{C}\right)x, \qquad x(0) = x_0$$

This differentiation equation is called logistic differential equation and the growth model is called the logistic model or inhibited growth model. It can be shown that the solution of this initial-value problem is

$$x = \frac{x_0 C}{x_0 + (C - x_0)e^{-kt}}$$

EXAMPLE 7.10 Suppose that a population x(t) grows in accordance with the logistic model

$$\frac{dx}{dt} = 10(1 - 0.1x)x$$

(a) What is the carrying capacity?

(b) What is the value of k?

(c) For what value of x is the population growing most rapidly?

(d) Find the solution of this differential equation if it is given that the initial population was $x(0) = 2$.

(e) Draw a graph of this solution.

SOLUTION

Comparing the given equation with the standard differential equation of logistic model

$$\frac{dx}{dt} = k\left(1 - \frac{x}{C}\right)x;$$

we get

(a) Carrying Capacity, C = 10

(b) Value of k = 10

(c) A variable takes maximum value at a point where its first derivative is zero and the second derivative is negative, provided that the variable is twice differentiable at that point. Therefore, the growth rate $\frac{dx}{dt}$ is maximum where

$$\frac{d^2x}{dt^2} = 0 \text{ and } \frac{d^3x}{dt^3} < 0.$$

Now, $$\frac{d^2x}{dt^2} = 10(1 - .2x), \qquad \frac{d^3x}{dt^3} = -2$$

$\therefore$ $$\frac{d^2x}{dt^2} = 0 \Rightarrow x = 5$$

And, $$\left.\frac{d^3x}{dt^3}\right|_{x=5} = -2 < 0.$$

Hence the growth rate is maximum at $x = 5$.

(d) $$\frac{dx}{dt} = 10(1 - 0.1x)x$$

$$\Rightarrow \quad \frac{dx}{x(1 - .1x)} = 10\,dt$$

$$\Rightarrow \quad \frac{dx}{x(10 - x)} = dt$$

$$\Rightarrow \quad \left(\frac{1}{x} + \frac{1}{10 - x}\right)dx = 10dt$$

On integrating, we get

$$\ln x - \ln(10 - x) = 10t + \ln C$$

$$\frac{x}{C(10 - x)} = e^{10t}$$

It is given that $x = 2$ for $t = 0$. This gives $C = \frac{1}{4}$.

Hence the required solution of the given differential equation is

$$x = \frac{10}{1 + 4e^{-10t}}$$

(e) Here $$x = \frac{10}{1 + 4e^{-10t}}$$

For $t = 0$, $x = 2$. As t tends to ∞, x tends to 10.

Also, $$\frac{dx}{dt} = -\frac{10}{(1 + 4e^{-10t})^2} \times (-40\,e^{-10t})$$

$$= \frac{400\,e^{-10t}}{(1 + 4e^{-10t})^2} > 0$$

Therefore, the function $x = x(t)$ is an increasing function.

From (c), we know that $\frac{dx}{dt}$ is maximum at $x = 5$. Thus tangent to the curve has maximum slope at $x = 5$. Also, for $x = 5$, $t = \frac{\ln 2}{5}$. A rough sketch of the graph is as shown in Fig. 7.6.

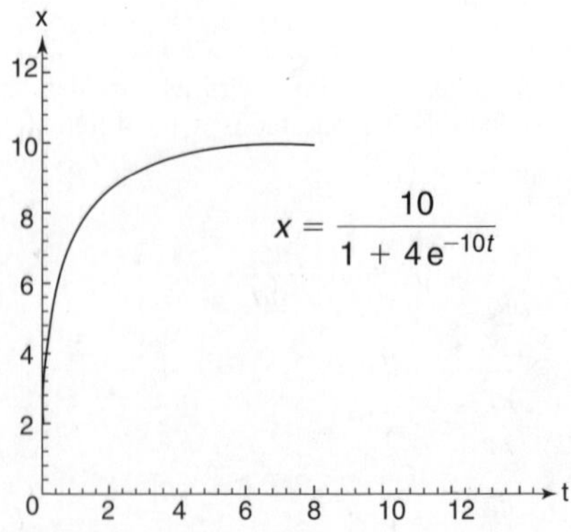

Fig. 7.6

EXAMPLE 7.11 A reserved forest has the capacity to preserve 100 hounds of a particular variety. Initially there were 10 such animals. The number grew to 50 in 5 years. In how many years will this number grow to 90?

SOLUTION

Here carrying capacity is 100. It is a problem of logistic growth. The differential equation of this growth process is

$$\frac{dx}{dt} = \mathrm{k}x(100 - x),$$

where $x = x(t)$ is the number of hounds at time t. Rewriting, we get,

$$\frac{dx}{x(100 - x)} = \mathrm{k}dt$$

$$\Rightarrow \qquad \left(\frac{1}{x} + \frac{1}{100 - x}\right)dx = 100\,\mathrm{k}dt$$

On integrating, we get

$$\ln x - \ln(100 - x) = 100\,\mathrm{k}t + \ln \mathrm{C}$$

$$\Rightarrow \qquad \frac{x}{\mathrm{C}(100 - x)} = \mathrm{e}^{100\,\mathrm{k}t}$$

when $t = 0, x = 10 \Rightarrow C = \dfrac{1}{9}$

$$\Rightarrow \qquad \frac{x}{100 - x} = \frac{1}{9}\mathrm{e}^{100\,\mathrm{k}t}$$

It is given that $x = 50$ when $t = 5$, therefore

$$\mathrm{e}^{500\,\mathrm{k}} = 9$$

$$\Rightarrow \qquad \mathrm{k} = \frac{1}{500}\ln 9$$

Therefore, we have,

$$\frac{x}{100 - x} = \frac{1}{9}\mathrm{e}^{\left(\frac{1}{5}\ln 9\right)t} = 9^{\frac{t}{5} - 1}$$

Suppose, the number of hounds grow to 90 in T years of time. Then, $\dfrac{90}{10} = 9^{\frac{\mathrm{T}}{5} - 1} \Rightarrow \mathrm{T} = 10$ years.

Logistic model is also observed in the pattern in which a disease spreads in a locality. Suppose that there are L individuals living in a particular locality and a disease begins to spread in the locality. At any point of time, the rate at which the disease spreads will depend on how many individuals are already affected and how many are not. With more individuals affected, the opportunity to spread the disease tends to increase, but at the same time there are fewer individuals to catch the disease, so the opportunity to spread the disease tends to decrease. There are two conflicting influences on the rate at which the disease spreads.

Let $x = x(t)$ be the number of individuals to have caught the disease at time t, so that $\mathrm{L} - x$ is the number of individuals who do not have the diseasde at time t. The rate at which the disease spreads is proportional to x as well as to $\mathrm{L} - x$. Thus, we have

$$\frac{dx}{dt} = \mathrm{k}x(\mathrm{L} - x)$$

where k is the positive constant of proportionality and depends on the nature of the disease and the behaviour pattern of the individuals. It can be determined experimentally. If $x = x_0$ is the number of individuals at time $t = 0$, then the above differential equation may be rewritten as an initial value problem as

$$\frac{dx}{dt} = \mathrm{k}\mathrm{L}\left(1 - \frac{x}{\mathrm{L}}\right)x, \qquad x(0) = x_0$$

This represents the logistic model of growth pattern for spreading of diseases.

EXAMPLE 7.12 A university hostel has 1000 students. After the summer holidays, 20 students returned infected with viral fever and after 5 days, 35 students were having the fever.

(a) Using logistic model, set up an initial-value problem where $x = x(t)$ is the number of students, who were having the fever after t days.

(b) Solve the initial value problem and use the given data to find the constant of proportionality.

(c) Draw a rough graph which shows how the disease spread over the first 2-week period.

SOLUTION

(a) Assume that $x = x(t)$ denotes the number of infected students after t days. It is given that $x(0) = 20$, $x(5) = 35$. The carrying capacity is $\mathrm{C} = 1000$. With the logistic model, the differential equation for the spread of the disease is

$$\frac{dx}{dt} = \mathrm{k}x(1000 - x)$$

where k is the constant of proportionality.

This differential equation, along with the initial condition $x(0) = 20$ forms the required initial value problem.

(b)

$$\frac{dx}{dt} = \mathrm{k}x(1000 - x)$$

$$\Rightarrow \qquad \frac{dx}{x(1000 - x)} = \mathrm{k}dt$$

$$\Rightarrow \qquad \left(\frac{1}{x} + \frac{1}{1000 - x}\right)dx = 1000\mathrm{k}dt$$

Integrating

$$\ln x - \ln(1000 - x) = 1000\,\mathrm{k}t + \ln \mathrm{C}$$

$$\Rightarrow \qquad \ln\frac{x}{\mathrm{C}(1000 - x)} = 1000\,\mathrm{k}t$$

$$\Rightarrow \qquad \frac{x}{1000 - x} = \mathrm{C}e^{1000\,\mathrm{k}t}$$

Initially $t = 0,\ x = 20 \Rightarrow \mathrm{C} = \dfrac{20}{980} = \dfrac{1}{49}$

$$\Rightarrow \qquad \frac{x}{1000 - x} = \frac{1}{49}e^{1000\,\mathrm{k}t}$$

It is given that for $t = 5,\ x = 35$

$$\Rightarrow \qquad \frac{35}{965} = \frac{1}{49}e^{5000\,\mathrm{k}}$$

$$\Rightarrow \qquad e^{5000k} = \frac{343}{193}$$

$$\Rightarrow \qquad k = \frac{1}{5000}\log\left(\frac{343}{193}\right) = .000115$$

$\therefore$ We have
$$\frac{x}{1000 - x} = \frac{1}{49}e^{.115t}$$

$$\Rightarrow \qquad x = \frac{1000}{1 + 49e^{-.115t}}$$

(c) A table of x and t is shown in Fig.7.7 (calculated through a calculating device) for number of persons getting infected over the first two weeks.

t	0	1	2	3	4	5	6	7	8	9	10	11	12	13	14
$x(t)$	20	22	25	28	31	35	39	44	49	54	61	67	75	83	93

A graph of $t - x$ can be drawn as shown in Fig. 7.7.

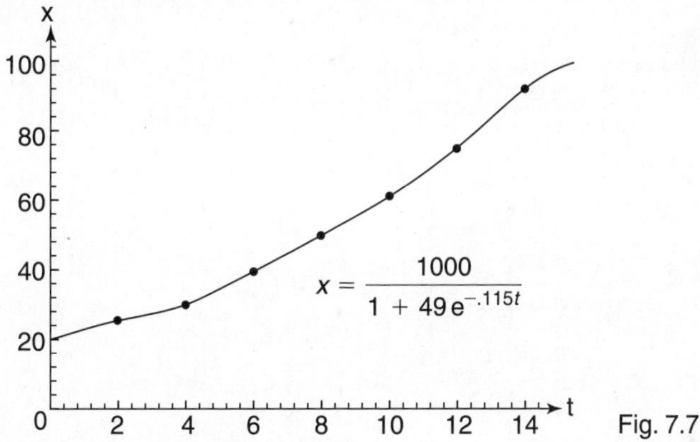

Fig. 7.7

In business and industries there are many examples of growth in which the rate of growth is slow at first, becomes faster and then slows again as a limit is reached. Logistic model is ideal for explaining this kind of growth.

EXAMPLE 7.13 A company has introduced a new model of TV It is predicted that the market will saturate at 200,000 TVs and that the total sales will be governed by the equation

$$\frac{dS}{dt} = \frac{1}{4}S(2 - S)$$

where S is the total sale in millions of TVs and t is measured in months. If 1000 TVs are sold off at the time of launching, find as expression for total sale S at any time t. Draw an approximate graph for the same.

SOLUTION

We have the initial value problem:

$$\frac{dS}{dt} = \frac{1}{4}S(2 - S),$$ $S(0) = 1000$ TVs $= .001$ (in million), as carrying capacity is 2 which stands for 2 million

$$\Rightarrow \qquad \frac{dS}{S(2-S)} = \frac{1}{4}dt$$

$$\Rightarrow \qquad \left(\frac{1}{S} + \frac{1}{2-S}\right)dS = \frac{1}{2}dt$$

On integrating

$$\ln S - \ln(2-S) = \frac{1}{2}t + \ln C$$

$$\Rightarrow \qquad \frac{S}{C(2-S)} = e^{\frac{1}{2}t}$$

$$\Rightarrow \qquad \frac{S}{2-S} = Ce^{\frac{1}{2}t}$$

$$S(0) = .001 \Rightarrow C = \frac{1}{1999}$$

$$\Rightarrow \qquad S = \frac{2e^{\frac{1}{2}t}}{1999 + e^{\frac{1}{2}t}} = \frac{2}{1 + 1999\,e^{-.5t}}$$

Its graph passes through the points (0, .001), (2, .0027), (10, .1382), (20, 1.8336) and so on. As t tends to ∞, S tends to 2.

A graph of the above function may be drawn as in Fig. 7.8.

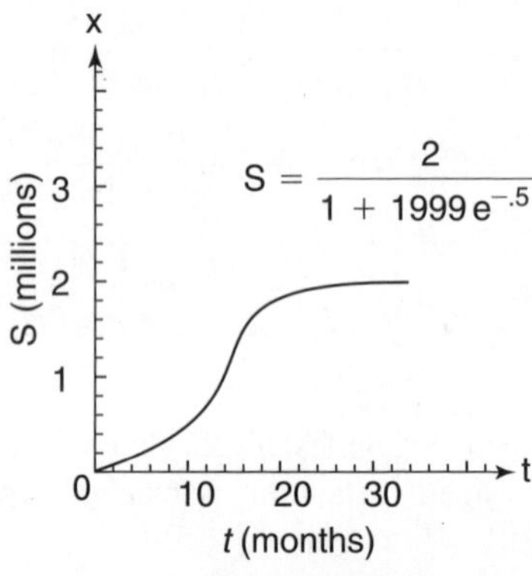

Fig. 7.8

EXERCISE 7.3

1. Suppose that the growth of a population $x = x(t)$ is given by the logistic equation

$$x(t) = \frac{60}{5 + 7e^{-t}}$$

(a) What is the population at time $t = 0$?

(b) What is the carrying capacity C?

(c) What is the constant k?

(d) When does the population reach half of the carrying capacity ?

(e) Find an initial-value problem whose solution is $x(t)$

(f) Draw an approximate x-t graph.

2. Suppose that a population $x(t)$ grows in accordance with the logistic model

$$\frac{dx}{dt} = 50x - 0.001x^2$$

(a) What is the carrying capacity?

(b) What is the value of k?

(c) For what value of x is the population growing most rapidly?

3. In a school, there are 1000 students and all are likely to get infected with eye-flu virus. Initially 20 students got infected and within 2 weeks, 100 students got infected with the disease. In how much time would the majority of students be infected by the eye-flue virus? It is given that the disease spreads with logistic growth model.

4. In a reserve forest, there is capacity to maintain 500 elephants. Initially, there were 20 elephants. Within 2 years time, the number rose to 30. Assuming logistic growth model, find out how much time it will take for the population to grow to 200.

CHAPTER EIGHT

Successive Differentiation

8.1 HIGHER ORDER DERIVATIVES

The derivative of a derivable function *f* is itself a function, which also may be derivable. Thus, we may talk of the first derivative of *f*, the second derivative of *f* and so on, which are usually denoted by f', f'', . . . and so on. In general, the nth derivative of *f*, denoted by f^n is defined as

$$f^n(x) = \frac{d}{dx}[f^{n-1}(x)], \qquad n \geq 2, 3, 4 \ldots$$

In particular, $f^1(x) = f'(x) = \dfrac{d}{dx}[f(x)],\ f^2(x) = f''(x) = \dfrac{d}{dx}\left[\dfrac{d}{dx}(f(x))\right] = \dfrac{d^2}{dx^2}(f(x)).$

If we write, $y = f(x)$, then $\dfrac{d^n y}{dx^n}$, y_n or $f^n(x)$ are all used to denote the nth derivative of *f*. The symbols

$$f^n(\text{a}),\ \left[\frac{d^n y}{dx^n}\right]_{x=\text{a}},\ y_n(\text{a}),\ \left.\frac{d^n y}{dx^n}\right|_{x=\text{a}}$$

all denote the value of the nth derivative of $y = f(x)$ at $x = \text{a}$. The process of finding out the nth derivative $(n \geq 2)$ of a function is called 'successive differentiation'.

EXAMPLE 8.1 If $x = \text{a}(\cos\theta + \theta\sin\theta)$, $y = \text{a}(\sin\theta - \theta\cos\theta)$, find $\dfrac{d^2y}{dx^2}$ at $\theta = \dfrac{\pi}{3}$.

SOLUTION

We have,

$$\frac{dx}{d\theta} = \text{a}(-\sin\theta + \sin\theta + \theta\cos\theta) = \text{a}\theta\cos\theta$$

$$\frac{dy}{d\theta} = \text{a}(\cos\theta - \cos\theta + \theta\sin\theta) = \text{a}\theta\sin\theta$$

$$\frac{dy}{dx} = \frac{dy}{d\theta}\cdot\frac{d\theta}{dx} = \frac{dy}{d\theta} \div \frac{dx}{d\theta} = \tan\theta$$

$$\Rightarrow \quad \frac{d^2y}{dx^2} = \frac{d}{dx}(\tan\theta) = \frac{d}{d\theta}(\tan\theta)\frac{d\theta}{dx} = \sec^2\theta\cdot\frac{1}{\text{a}\theta\cos\theta} = \frac{\sec^3\theta}{\text{a}\theta}$$

$$\therefore \quad \left.\frac{d^2y}{dx^2}\right|_{\theta=\frac{\pi}{3}} = \frac{\sec^3\pi/3}{\text{a}\,\pi/3} = \frac{24}{\text{a}\pi}$$

REMARK: The result holds good for all those values of θ for which $\frac{dx}{d\theta} \neq 0$.

EXAMPLE 8.2 If $y = \sin(\sin x)$, prove that

$$\frac{d^2y}{dx^2} + \tan x \frac{dy}{dx} + y\cos^2 x = 0.$$

SOLUTION

We have,

$$\frac{dy}{dx} = \frac{d}{dx}\Big(\sin(\sin x)\Big) = \cos(\sin x) \cdot \cos x$$

$$\frac{d^2y}{dx^2} = -\sin(\sin x) \cdot \cos^2 x - \cos(\sin x) \cdot \sin x$$

$$\Rightarrow \quad \frac{d^2y}{dx^2} + y\cos^2 x = -\cos(\sin x) \cdot \sin x$$

$$\Rightarrow \quad \frac{d^2y}{dx^2} + y\cos^2 x + \cos(\sin x) \cdot \cos x \cdot \tan x = 0$$

$$\Rightarrow \quad \frac{d^2y}{dx^2} + \tan x \cdot \frac{dy}{dx} + y\cos^2 x = 0$$

REMARK: The result holds good for all those values of x for which $\cos x \neq 0$.

EXAMPLE 8.3 If $x = \sin t$, $y = \sin pt$, prove that

$$(1 - x^2)\frac{d^2y}{dx^2} - x\frac{dy}{dx} + p^2 y = 0.$$

SOLUTION

We have, $\frac{dx}{dt} = \cos t$, $\quad \frac{dy}{dx} = p\cos pt$

$$\therefore \quad \frac{dy}{dx} = \frac{dy}{dt}\Big/\frac{dx}{dt} = \frac{p\cos pt}{\cos t}$$

Now, $$\frac{d^2y}{dx^2} = \frac{d}{dt}\left(\frac{dy}{dx}\right) \cdot \frac{dt}{dx}$$

$$= \frac{d}{dt}\left(\frac{p\cos pt}{\cos t}\right) \cdot \frac{1}{\cos t}$$

$$= \frac{-p^2 \sin pt \cos t + p\cos pt \sin t}{\cos^2 t} \cdot \frac{1}{\cos t}$$

$$\therefore \qquad \cos^2 t \frac{d^2y}{dx^2} = -p^2 \sin pt + p \frac{\cos pt \cdot \sin t}{\cos t}$$

$$\Rightarrow \qquad (1 - \sin^2 t)\frac{d^2y}{dx^2} = -p^2 \sin pt + p \frac{\cos pt}{\cos t} \cdot \sin t$$

$$\Rightarrow \qquad (1 - x^2)\frac{d^2y}{dx^2} = -p^2 y + x\frac{dy}{dx}$$

$$\Rightarrow \qquad (1 - x^2)\frac{d^2y}{dx^2} - x\frac{dy}{dx} + p^2 y = 0$$

EXAMPLE 8.4 If $p^2 = a^2 \cos^2\theta + b^2 \sin^2\theta$, prove that $p + \dfrac{d^2p}{d\theta^2} = \dfrac{a^2b^2}{p^3}$.

(DU, B.Sc.(H), 1991)

SOLUTION

Differentiating $p^2 = a^2 \cos^2\theta + b^2 \sin^2\theta$, with respect to θ, we get

$$2p\frac{dp}{d\theta} = -2a^2 \cos\theta \sin\theta + 2b^2 \sin\theta \cos\theta$$

$$2p\frac{dp}{d\theta} = (b^2 - a^2)\sin 2\theta \qquad (1)$$

Differentiating (1) with respect to θ, we get

$$2p\frac{d^2p}{d\theta^2} + 2\left(\frac{dp}{d\theta}\right)^2 = 2(b^2 - a^2)\cos 2\theta \qquad (2)$$

$$\Rightarrow \qquad 4p^3\frac{d^2p}{d\theta^2} + 4p^2\left(\frac{dp}{d\theta}\right)^2 = 4(b^2 - a^2)p^2\cos 2\theta \qquad (3)$$

Substituting the value of $2p\dfrac{dp}{d\theta}$ from (1) in (3), we get

$$4p^3\frac{d^2p}{d\theta^2} + (b^2 - a^2)^2\sin^2 2\theta = 4(b^2 - a^2)p^2\cos 2\theta$$

$$\Rightarrow \qquad 4p^3\frac{d^2p}{d\theta^2} = 4(b^2 - a^2)p^2\cos 2\theta - (b^2 - a^2)^2\sin^2 2\theta$$

$$\Rightarrow \qquad 4p^3\frac{d^2p}{d\theta^2} = 4(b^2 - a^2)p^2\cos 2\theta - (b^2 - a^2)^2 4\sin^2\theta\cos^2\theta$$

$$\Rightarrow \qquad p^3\frac{d^2p}{d\theta^2} = (b^2 - a^2)p^2\cos 2\theta - (b^2 - a^2)^2\sin^2\theta\cos^2\theta$$

$$\Rightarrow \qquad p^4 + p^3\frac{d^2p}{d\theta^2} = p^4 + (b^2 - a^2)p^2\cos 2\theta - (b^2 - a^2)^2\sin^2\theta\cos^2\theta$$

$$= (a^2\cos^2\theta + b^2\sin^2\theta)^2 + (b^2 - a^2)(a^2\cos^2\theta + b^2\sin^2\theta)$$

$$\times (\cos^2\theta - \sin^2\theta) - (b^2 - a^2)^2 \sin^2\theta\cos^2\theta$$

$$= (a^4 + a^2b^2 - a^4)\cos^4\theta + (b^4 - b^4 + a^2b^2)\sin^4\theta$$

$$+ (2a^2b^2 - 2a^2b^2 + a^4 + b^4 - b^4 - a^4 + 2b^2a^2)\sin^2\theta\cos^2\theta$$

$$= a^2b^2\cos^4\theta + a^2b^2\sin^4\theta + 2a^2b^2\sin^2\theta\cos^2\theta$$

$$= a^2b^2(\cos^2\theta + \sin^2\theta)^2$$

$$= a^2b^2$$

$$\therefore \qquad p + \frac{d^2p}{d\theta^2} = \frac{a^2b^2}{p^3}.$$

EXAMPLE 8.5 Change the independent variable to z in the equation

$$\frac{d^2y}{dx^2} + \cot x\frac{dy}{dx} + 4y\,\mathrm{cosec}^2 x = 0$$

by means of the transformation $z = \log\tan\frac{x}{2}$.

SOLUTION

$$z = \log\tan\frac{x}{2}, \quad \frac{dz}{dx} = \frac{1}{\tan\frac{x}{2}}\cdot\sec^2\frac{x}{2}\cdot\frac{1}{2} = \mathrm{cosec}\,x$$

$$\frac{dy}{dx} = \frac{dy}{dz}\cdot\frac{dz}{dx}$$

$$= \frac{dy}{dz}\mathrm{cosec}\,x$$

$$\frac{d^2y}{dx^2} = \frac{d}{dz}\left(\frac{dy}{dz}\cdot\mathrm{cosec}\,x\right)\cdot\frac{dz}{dx}$$

$$= \mathrm{cosec}\,x\cdot\left[\frac{d^2y}{dz^2}\cdot\mathrm{cosec}\,x + \frac{dy}{dz}\cdot\frac{d}{dz}(\mathrm{cosec}\,x)\right]$$

$$= \mathrm{cosec}^2x\frac{d^2y}{dz^2} + \mathrm{cosec}\,x\cdot\frac{dy}{dz}\cdot\frac{d}{dx}(\mathrm{cosec}\,x)\cdot\frac{dx}{dz}$$

$$= \mathrm{cosec}^2x\cdot\frac{d^2y}{dz^2} + \mathrm{cosec}\,x\cdot\frac{dy}{dz}(-\mathrm{cosec}\,x\cot x)\cdot\frac{1}{\mathrm{cosec}\,x}$$

$$= \mathrm{cosec}^2x\cdot\frac{d^2y}{dz^2} - \mathrm{cosec}\,x\cot x\frac{dy}{dz}$$

$$\therefore \quad \frac{d^2y}{dx^2} + \cot x\cdot\frac{dy}{dx} + 4y\,\mathrm{cosec}^2x$$

$$= \mathrm{cosec}^2x\cdot\frac{d^2y}{dz^2} - \mathrm{cosec}\,x\cdot\cot x\frac{dy}{dz} + \cot x\cdot\frac{dy}{dz}\cdot\mathrm{cosec}\,x + 4y\,\mathrm{cosec}^2x$$

$$= \operatorname{cosec}^2 x\left[\frac{d^2y}{dz^2} + 4y\right]$$

$$\therefore \quad \frac{d^2y}{dx^2} + \cot x\frac{dy}{dx} + 4y\operatorname{cosec}^2 x = 0$$

$$\Rightarrow \quad \operatorname{cosec}^2 x\left[\frac{d^2y}{dz^2} + 4y\right] = 0$$

$\Rightarrow \quad \dfrac{d^2y}{dz^2} + 4y = 0$, which is the required equation with z as the independent variable.

EXAMPLE 8.6 If $y = \left(\frac{1}{x}\right)^x$, show that $y_2(1) = 0$.

SOLUTION

We have, $y = \left(\frac{1}{x}\right)^x$

Taking logarithm on both sides

$$\ln y = x\ln\left(\frac{1}{x}\right)$$

$$= x(\ln 1 - \ln x)$$

$$\Rightarrow \quad \ln y = -x\ln x$$

Differentiating with respect to x, we get

$$\frac{1}{y}\frac{dy}{dx} = -x\cdot\frac{1}{x} - \ln x$$

$$\Rightarrow \quad \frac{dy}{dx} = -y(1 + \ln x)$$

Differentiating again, we get

$$\frac{d^2y}{dx^2} = -y\left(\frac{1}{x}\right) - \frac{dy}{dx}(1 + \ln x)$$

$$= -\frac{y}{x} + y(1 + \ln x)^2$$

$$= \left(\frac{1}{x}\right)^x\left[-\frac{1}{x} + (1 + \ln x)^2\right]$$

$$\therefore \quad y_2(1) = -1 + (1 + 0)^2$$

$$= 0$$

EXAMPLE 8.7 If $\sin(x + y) = py$, where p is a constant, prove that

$$\frac{d^2y}{dx^2} = -y\left(1 + \frac{dy}{dx}\right)^3.$$

SOLUTION

$$\sin(x+y) = py$$

Differentiating both sides with respect to x,

$$\cos(x+y)(1+y_1) = py_1$$

$$\Rightarrow \quad \cos(x+y) + y_1\big(\cos(x+y) - p\big) = 0$$

Differentiating again,

$$-\sin(x+y)(1+y_1) + y_2\big(\cos(x+y) - p\big) + y_1\big(-\sin(x+y)(1+y_1)\big) = 0$$

$$\Rightarrow \quad -py(1+y_1) + y_2\left(\frac{py_1}{1+y_1} - p\right) + y_1\big(-py(1+y_1)\big) = 0$$

$$\Rightarrow \quad y(1+y_1)^2(1+y_1) + y_2 = 0$$

$$\Rightarrow \quad y_2 = -y(1+y_1)^3$$

$$\Rightarrow \quad \frac{d^2y}{dx^2} = -y\left(1+\frac{dy}{dx}\right)^3$$

EXERCISE 8.1

1. If $y = \mathrm{A}\cos nx + \mathrm{B}\sin nx$, prove that $\dfrac{d^2y}{dx^2} + n^2y = 0$.

2. If $y = \mathrm{A}e^{mx} + \mathrm{B}e^{-mx}$, prove that $\dfrac{d^2y}{dx^2} = m^2y$.

3. If $y = \dfrac{a+bx}{c+dx}$, then show that $2y_1y_3 = 3y_2^2$.

4. If $y = x\ln[(ax)^{-1} + a^{-1}]$, prove that $x(x+1)\dfrac{d^2y}{dx^2} + x\dfrac{dy}{dx} = y - 1$.

5. If $y = (\mathrm{A} + \mathrm{B}t)e^{nt}$, show that $\dfrac{d^2y}{dt^2} - 2n\dfrac{dy}{dt} + n^2y = 0$.

6. Change the independent variable to θ in the equation

$$\frac{d^2y}{dx^2} + \frac{2x}{1+x^2}\frac{dy}{dx} + \frac{y}{(1+x^2)^2} = 0,$$

by means of the transformation $x = \tan\theta$.

7. If $y = \tan^{-1}x$, prove that $(1+x^2)\dfrac{d^2y}{dx^2} + 2x\dfrac{dy}{dx} = 0$.

8. If $y = \mathrm{A}e^{px} + \mathrm{B}e^{qx}$, prove that $\dfrac{d^2y}{dx^2} - (p+q)\dfrac{dy}{dx} + pqy = 0$. *(DU, B.Sc. (H), 1994).*

9. If $x = 2\cos t - \cos 2t$, $y = 2\sin t - \sin 2t$, find the value of $\dfrac{d^2y}{dx^2}$ at $t = \dfrac{\pi}{2}$.
[DU, B.Sc.(H), 1989].

10. If $x = a\sin^3 t$, $y = b\cos^3 t$, find $\dfrac{d^2y}{dx^2}$.

11. If $ax^2 + 2hxy + by^2 = 1$, prove that $\dfrac{d^2y}{dx^2} = \dfrac{h^2 - ab}{(hx + by)^3}$.

12. If $x^3 + y^3 - 3axy = 0$, show that $\dfrac{d^2y}{dx^2} = -\dfrac{2a^3xy}{(y^2 - ax)^3}$.

8.2 THE *n*th DERIVATIVES OF SOME IMPORTANT FUNCTIONS

1. Let $y = (ax + b)^m$

Then
$$y_1 = ma(ax + b)^{m-1}$$
$$y_2 = m(m - 1)a^2(ax + b)^{m-2}$$
$$y_3 = m(m - 1)(m - 2)a^3(ax + b)^{m-3},$$

so that in general,

$$y_n = m(m - 1)(m - 2)\ldots(m - n + 1)a^n(ax + b)^{m-n}$$

In case, m is a positive integer, y_n can be written as $\dfrac{m!}{(m - n)!}a^n(ax + b)^{m-n}$. So that, the mth derivative of $(ax + b)^m$ is constant, that is, $m!a^m$, and $(m + 1)^{\text{th}}$ derivative is zero.

Corollary 1. For $m = -1$, we get

$$y_n = (-1)(-2)\cdots(-n)a^n(ax + b)^{-1-n}$$

Thus,
$$\frac{d^n}{dx^n}\left(\frac{1}{ax + b}\right) = \frac{(-1)^n n!a^n}{(ax + b)^{n+1}}$$

Corollary 2. Let $y = \ln(ax + b)$. Then $y_1 = \dfrac{a}{ax + b}$.

$$\therefore \quad y_n = (y_1)_{n-1} = \frac{(-1)^{n-1}(n - 1)!}{(ax + b)^n} \cdot a^n$$

2. Let $y = e^{ax}$. Then

$$y_1 = ae^{ax},\ y_2 = a^2e^{ax},\ y_3 = a^3e^{ax} \text{ and so on}$$

Thus, $y_n = a^ne^{ax}$

3. Let $y = \sin(ax + b)$. Then

$$y_1 = a\cos(ax + b) = a\sin\left(ax + b + \frac{\pi}{2}\right)$$

$$y_2 = a^2\cos\left(ax + b + \frac{\pi}{2}\right) = a^2\sin\left(ax + b + \frac{2\pi}{2}\right)$$

$$y_3 = a^3\cos\left(ax + b + \frac{2\pi}{2}\right) = a^3\sin\left(ax + b + \frac{3\pi}{2}\right), \text{ and so on}$$

Hence, $y_n = a^n\sin\left(ax + b + \dfrac{n\pi}{2}\right)$.

4. Let $y = \cos(ax + b)$. Then proceeding as above, we get

$$y_n = a^n \cos\left(ax + b + \frac{n\pi}{2}\right)$$

5. Let $y = e^{ax}\sin(bx + c)$. Then

$$\begin{aligned} y_1 &= ae^{ax}\sin(bx + c) + e^{ax} \cdot b\cos(bx + c) \\ &= e^{ax}[a\sin(bx + c) + b\cos(bx + c)] \end{aligned}$$

Now, on substituting $a = r\cos\theta$, $b = r\sin\theta$, we get

$$\begin{aligned} y_1 &= re^{ax}[\sin(bx + c)\cos\theta + \cos(bx + c)\sin\theta] \\ &= re^{ax}\sin(bx + c + \theta) \end{aligned}$$

Similarly, $y_2 = r^2e^{ax}\sin(bx + c + 2\theta)$ and so on.
Hence, in general,

$$y_n = r^n e^{ax}\sin(bx + c + n\theta), \text{ where}$$

$$r = \sqrt{a^2 + b^2}, \qquad \theta = \tan^{-1}\left(\frac{b}{a}\right).$$

6. Let $y = e^{ax}\cos(bx + c)$.
Then, proceedings as above,

$$y_n = r^n e^{ax}\cos(bx + c + n\theta),$$

where $r = \sqrt{a^2 + b^2}$, $\theta = \tan^{-1}\left(\frac{b}{a}\right)$.

REMARK: In order to determine the nth derivative of a rational function, we have to decompose it into partial fractions. Sometimes it also becomes necessary to apply De Moivre's Theorem, which states that

$$(\cos\theta \pm i\sin\theta)^n = \cos n\theta \pm i\sin n\theta,$$

where n is any integer, positive or negative and $i = \sqrt{-1}$.

EXAMPLE 8.8 Find the nth derivative of $y = \dfrac{x}{1 + 3x + 2x^2}$

SOLUTION

We have, $y = \dfrac{x}{1 + 3x + 2x^2} = \dfrac{x}{(2x + 1)(x + 1)}$

Let $\dfrac{x}{(2x + 1)(x + 1)} = \dfrac{A}{2x + 1} + \dfrac{B}{x + 1}$

$\Rightarrow \quad x = A(x + 1) + B(2x + 1)$

Putting $x = -1, -1/2$ respectively, we get, $B = +1$, $A = -1$

Thus,

$$y = -\frac{1}{2x+1} + \frac{1}{x+1}$$

$$\therefore \qquad y_n = -\frac{(-1)^n n! 2^n}{(2x+1)^{n+1}} + \frac{(-1)^n n!}{(x+1)^{n+1}}$$

$$= (-1)^n n! \left[\frac{1}{(x+1)^{n+1}} - \frac{2^n}{(2x+1)^{n+1}}\right].$$

EXAMPLE 8.9 Find the nth derivative of $y = \dfrac{2x}{x^2 + a^2}$

SOLUTION

We have,

$$y = \frac{2x}{x^2+a^2} = \frac{2x}{(x+ai)(x-ai)}$$

Let

$$\frac{2x}{(x+ai)(x-ai)} = \frac{A}{x+ai} + \frac{B}{x-ai}$$

$$\Rightarrow \qquad 2x = A(x-ai) + B(x+ai)$$

Comparing the coefficient of the like terms on both sides, we get

$$A + B = 2, \qquad (A - B)ai = 0$$

$$\Rightarrow \qquad A = 1, \qquad B = 1$$

Thus,

$$y_n = \frac{(-1)^n n!}{(x+ai)^{n+1}} + \frac{(-1)^n n!}{(x-ai)^{n+1}}$$

$$= (-1)^n n! \left[\frac{1}{(x+ai)^{n+1}} + \frac{1}{(x-ai)^{n+1}}\right]$$

To remove 'i' from the expression, we substitute

$$x = r\cos\theta, \qquad a = r\sin\theta$$

Then

$$\frac{1}{(x+ai)^{n+1}} = \frac{1}{r^{n+1}}(\cos\theta + i\sin\theta)^{-(n+1)}$$

$$= \frac{1}{r^{n+1}}\big(\cos(n+1)\theta - i\sin(n+1)\theta\big)$$

And

$$\frac{1}{(x-ai)^{n+1}} = \frac{1}{r^{n+1}}(\cos\theta - i\sin\theta)^{-(n+1)}$$

$$= \frac{1}{r^{n+1}}\big(\cos(n+1)\theta + i\sin(n+1)\theta\big)$$

Therefore,

$$y_n = \frac{(-1)^n n!}{r^{n+1}} \cdot 2\cos(n+1)\theta,$$

where

$$r = \sqrt{x^2+a^2}, \qquad \theta = \tan^{-1}(a/x)$$

EXAMPLE 8.10 Find $\dfrac{d^n y}{dx^n}$ where (i) $y = \cos^4 x$ (ii) $y = e^{ax}\cos^2 x \sin x$

SOLUTION

(i) We have $\cos^2 x = \dfrac{1 + \cos 2x}{2}$

$$\Rightarrow \quad \cos^4 x = \left(\frac{1 + \cos 2x}{2}\right)^2$$

$$= \frac{1}{4} + \frac{\cos 2x}{2} + \frac{\cos^2 2x}{4}$$

$$= \frac{1}{4} + \frac{1}{2}\cos 2x + \frac{1}{8}(1 + \cos 4x)$$

$$= \frac{3}{8} + \frac{1}{2}\cos 2x + \frac{1}{8}\cos 4x$$

$$\therefore \quad y_n = \frac{1}{2}\cdot 2^n \cos\left(2x + \frac{n\pi}{2}\right) + \frac{1}{8}\cdot 4^n \cos\left(4x + \frac{n\pi}{2}\right)$$

(ii) $$\cos^2 x \sin x = \frac{1}{2}(1 + \cos 2x)\sin x$$

$$= \frac{1}{2}\sin x + \frac{1}{4}\cdot 2\sin x \cos 2x$$

$$= \frac{1}{2}\sin x + \frac{1}{4}(\sin 3x - \sin x)$$

$$= \frac{1}{4}(\sin x + \sin 3x)$$

Hence,

$$y_n = \frac{d^n}{dx^n}(e^{ax}\cos^2 x \sin x)$$

$$= \frac{1}{4}\frac{d^n}{dx^n}(e^{ax}\sin x) + \frac{1}{4}\frac{d^n}{dx^n}(e^{ax}\sin 3x)$$

$$= \frac{1}{4}(a^2 + 1)^{n/2} e^{ax}\sin(x + n\tan^{-1}1/a) + \frac{1}{4}(a^2 + 9)^{n/2} e^{ax}\sin\left(3x + \tan^{-1}\frac{3}{a}\right)$$

EXAMPLE 8.11 If $y = \dfrac{1}{1 + x + x^2 + x^3}$, show that

$$y_n = \frac{1}{2}(-1)^n n! \sin^{n+1}\theta\,[\sin(n + 1)\theta - \cos(n + 1)\theta + (\sin\theta + \cos\theta)^{-(n+1)}]$$

SOLUTION

$$y = \frac{1}{1 + x + x^2 + x^3}$$

$$= \frac{1}{(1 + x)(1 + x^2)}$$

Let $$\frac{1}{(1+x)(1+x^2)} = \frac{A}{1+x} + \frac{Bx+C}{(x^2+1)}$$

$$\Rightarrow \quad 1 = A(x^2+1) + (Bx+C)(1+x)$$

Putting $x = -1, i, -i$, we get

$$A = \frac{1}{2}, \quad Bi + C = \frac{1}{1+i}, \quad -Bi + C = \frac{1}{1-i}$$

$$\Rightarrow \quad A = \frac{1}{2}, \quad B = -\frac{1}{2}, \quad C = \frac{1}{2}$$

$$\Rightarrow \quad y = \frac{1}{2}\left[\frac{1}{x+1} + \frac{-x+1}{x^2+1}\right] = \frac{1}{2}\left[\frac{1}{x+1} - \frac{x-1}{(x+i)(x-i)}\right]$$

Let $$\frac{x-1}{(x+i)(x-i)} = \frac{C}{x+i} + \frac{D}{x-i}$$

$$\Rightarrow \quad x - 1 = C(x-i) + D(x+i)$$

Putting $x = i, -i, \quad C = \frac{1}{2}(1-i), \quad D = \frac{1}{2}(1+i)$

$$\Rightarrow \quad y = \frac{1}{2}\left(\frac{1}{x+1}\right) - \frac{1}{4}\left(\frac{1-i}{x+i}\right) - \frac{1}{4}\left(\frac{1+i}{x-i}\right)$$

$$\Rightarrow \quad y_n = \frac{1}{2}\frac{(-1)^n n!}{(x+1)^{n+1}} - \frac{1}{4}(1-i)\frac{(-1)^n n!}{(x+i)^{n+1}} - \frac{1}{4}(1+i)\frac{(-1)^n n!}{(x-i)^{n+1}}$$

We take $x = \text{r}\cos\theta, \ 1 = \text{r}\sin\theta$, so that

$$\theta = \cot^{-1}x, \quad \text{r} = \sqrt{1+x^2}$$

$$\therefore \quad y_n = \frac{1}{2}(-1)^n n!\Big[(\sin\theta + \cos\theta)^{-(n+1)} - \frac{1}{2}(1-i)\{\cos(n+1)\theta - i\sin(n+1)\theta\}$$

$$- \frac{1}{2}(1+i)\{\cos(n+1)\theta + i\sin(n+1)\theta\}\Big]\text{r}^{-(n+1)}$$

$$= \frac{1}{2}(-1)^n n![(\sin\theta + \cos\theta)^{-(n+1)} - \cos(n+1)\theta + \sin(n+1)\theta]\text{r}^{-(n+1)}$$

$$= \frac{1}{2}(-1)^n n![(\sin\theta + \cos\theta)^{-(n+1)} - \cos(n+1)\theta + \sin(n+1)\theta](1+\cot^2\theta)^{-\frac{(n+1)}{2}}$$

$(\because x = \cot\theta)$

$$= \frac{1}{2}(-1)^n n! \sin^{(n+1)}\theta\left[\sin(n+1)\theta - \cos(n+1)\theta + (\sin\theta + \cos\theta)^{-(n+1)}\right]$$

$$= \frac{1}{2}(-1)^n n! \sin^{(n+1)}\theta[\sin(n+1)\theta - \cos(n+1)\theta + (\sin\theta + \cos\theta)^{-(n+1)}]$$

EXAMPLE 8.12 If $y = \tan^{-1}x$, show that

$$y_n = (-1)^{n-1}(n-1)!\sin n\left(\frac{\pi}{2} - y\right)\sin^n\left(\frac{\pi}{2} - y\right)$$

[DU, B.Sc.(H), 1996, 1990]

SOLUTION

$$y = \tan^{-1}x$$

$$y = \frac{1}{1+x^2} = \frac{1}{(x+i)(x-i)}$$

$$= \frac{1}{2i}\left[\frac{1}{x-i} - \frac{1}{x+i}\right] \quad \text{(By partial fractionating)}$$

$\therefore$ Differentiating $(n-1)$ times

$$y_n = \frac{1}{2i}(-1)^{n-1}(n-1)!\left[\frac{1}{(x-i)^n} - \frac{1}{(x+i)^n}\right]$$

$$= \frac{1}{2i}(-1)^{n-1}(n-1)!\left[\frac{1}{(\tan y - i)^n} - \frac{1}{(\tan y + i)^n}\right]$$

$$= \frac{1}{2i}(-1)^{n-1}(n-1)!\left[\frac{\cos^n y}{(\sin y - i\cos y)^n} - \frac{\cos^n y}{(\sin y + i\cos y)^n}\right]$$

$$= \frac{1}{2i}(-1)^{n-1}(n-1)!\cos^n y\left[\left(\cos\left(\frac{\pi}{2}-y\right) - i\sin\left(\frac{\pi}{2}-y\right)\right)^{-n} - \left(\cos\left(\frac{\pi}{2}-y\right) + i\sin\left(\frac{\pi}{2}-y\right)\right)^{-n}\right]$$

$$= \frac{1}{2i}(-1)^{n-1}(n-1)!\sin^n\left(\frac{\pi}{2}-y\right)\left[\cos n\left(\frac{\pi}{2}-y\right) + i\sin n\left(\frac{\pi}{2}-y\right) - \cos n\left(\frac{\pi}{2}-y\right) + i\sin n\left(\frac{\pi}{2}-y\right)\right]$$

(*By DeMoiver's Theorem*)

$$= (-1)^{n-1}(n-1)!\sin^n\left(\frac{\pi}{2}-y\right)\sin n\left(\frac{\pi}{2}-y\right)$$

EXERCISE 8.2

1. Find the nth derivative of

(i) $\dfrac{x+1}{x^2-4}$ **(ii)** $\dfrac{4x}{(x-1)^2(x+1)}$ **(iii)** $\dfrac{1}{(x^2+1)}$ **(iv)** $\dfrac{x}{1+3x+2x^2}$ **(v)** $\dfrac{1}{x^4-a^4}$

(vi) $\tan^{-1}x$ **(vii)** $\tan^{-1}\dfrac{1+x}{1-x}$ [Hint : $y = \dfrac{\pi}{4} + \theta$ where $\theta = \tan^{-1}x$]

(viii) $\tan^{-1}\dfrac{2x}{1-x^2}$ *[DU, B.Sc.(G),1991)]*

(ix) $\tan^{-1}\left(\dfrac{x}{a}\right)$ *[DU, B.Sc. (G), 2001]*

(x) $\tan^{-1}\left(\dfrac{x\sin\alpha}{1-x\cos\alpha}\right)$ *[DU, B.Sc.(H), 1987]*

[Hint : $y_1 = \dfrac{a}{z^2+a^2}$, where $z = x - \cos\alpha$, $a = \sin\alpha$, now proceed as in (viii)]

2. If $y = x \ln \dfrac{x-1}{x+1}$, prove that

$$\frac{d^n y}{dx^n} = (-1)^n (n-2)! \left[\frac{x-n}{(x-1)^n} - \frac{x+n}{(x+1)^n} \right]$$

3. If $y = x(x+1)\ln (x+1)^3$, prove that

$$\frac{d^n y}{dx^n} = \frac{3(-1)^{n-1}(n-3)!\,(2x+n)}{(x+1)^{n-1}}$$ *[DU, B.Sc.(H), 1991, 1988]*

4. If $y = \dfrac{1}{x^2+x+1}$, prove that $\dfrac{d^n y}{dx^n} = \dfrac{2(-1)^n n! \, \sin(n+1)\theta}{\sqrt{3}\, r^{n+1}}$, where $r^2 = x^2+x+1, \theta = \cot^{-1}\left(\dfrac{2x+1}{\sqrt{3}}\right)$

8.3 LEIBNITZ'S THEOREM

The following theorem, known as Leibnitz's Theorem, is used to find higher order derivatives for the product of two functions.

THEOREM

If U and V are two functions of x possessing derivatives of the nth order, then

$$(UV)_n = U_nV + {}^nC_1U_{n-1}V_1 + {}^nC_2U_{n-2}V_2 + \cdots + {}^nC_rU_{n-r}V_r + \cdots + {}^nC_nUV_n$$

PROOF

We prove this theorem by the principle of mathematical induction.

For $n = 1$

$$\begin{aligned}(UV)_1 &= \frac{d}{dx}(UV) = U\frac{dV}{dx} + V\frac{dU}{dx}\\ &= UV_1 + VU_1\\ &= U_1V + UV_1\end{aligned}$$

Hence the theorem holds for $n = 1$. We assume that it holds for $n = k$. Then,

$$\begin{aligned}(UV)_k = U_kV &+ {}^kC_1U_{k-1}V_1 + {}^kC_2U_{k-2}V_2 + \cdots + {}^kC_{r-1}U_{k-r+1}V_{r-1}\\ &+ {}^kC_rU_{k-r}V_r + \cdots + {}^kC_kUV_k\end{aligned}$$

Differentiating both sides with respect to x, we get

$$\begin{aligned}(UV)_{k+1} &= (U_{k+1}V + U_kV_1) + {}^kC_1(U_kV_1 + U_{k-1}V_2) + \cdots + {}^kC_{r-1}(U_{k-r+1}V_r\\ &\quad + U_{k-r+2}V_{r-1}) + {}^kC_r(U_{k-r}V_{r+1} + U_{k-r+1}V_r) + \cdots + {}^kC_kUV_{k+1}\\ &= U_{k+1}V + (1 + {}^kC_1)U_kV_1 + ({}^kC_1 + {}^kC_2)U_{k-1}V_2 + \cdots\\ &\quad + ({}^kC_{r-1} + {}^kC_r)U_{k-r+1}V_k + \cdots + {}^kC_kUV_{k+1}\end{aligned}$$

We know that

$${}^kC_{r-1} + {}^kC_r = {}^{k+1}C_r,\ 1 + {}^kC_1 = 1 + k = {}^{k+1}C_1,\ {}^kC_k = 1 = {}^{k+1}C_{k+1}$$

$$\begin{aligned}\therefore\ (UV)_{k+1} &= U_{k+1}V + {}^{k+1}C_1U_kV_1 + {}^kC_2U_{k-1}V_2 + \cdots + {}^{k+1}C_rU_{k-r+1}V_r + \cdots\\ &\quad + {}^{k+1}C_{k+1}UV_{k+1}\end{aligned}$$

Thus the theorem is true for $n = k + 1$ also.
Hence, by the principle of mathematical induction, it is true for all values of n.
This completes the proof.

EXAMPLE 8.13 Find $(x^2 e^x \cos x)_n$.

SOLUTION

To find the nth derivative, we look upon $e^x \cos x$ as the first factor and x^2 as the second and apply Leibnitz's Theorem.

$$\therefore \quad (x^2 e^x \cos x)_n = (e^x \cos x)_n x^2 + {}^nC_1 (e^x \cos x)_{n-1} \cdot 2x + {}^nC_2 (e^x \cos x)_{n-2} \cdot 2$$

$$= 2^{n/2} e^x \cos(x + n \tan^{-1} 1) x^2 + n \cdot 2^{(n-1)/2} e^x \cos\left(x + (n-1)\tan^{-1} 1\right) 2x$$

$$+ \frac{n(n-1)}{2} \cdot 2^{(n-2)/2} e^x \cos\left(x + (n-2)\tan^{-1} 1\right) 2$$

$$= 2^{\frac{1}{2}(n-2)} \cdot e^x \left[2x^2 \cos\left(x + \frac{n\pi}{4}\right) + 2^{3/2} \cdot nx \cdot \cos\left(x + \overline{n-1}\frac{\pi}{4}\right) \right.$$

$$\left. + n(n-1) \cos\left(x + \overline{n-2}\frac{\pi}{4}\right) \right]$$

EXAMPLE 8.14 If $y = a \cos(\ln x) + b \sin(\ln x)$, show that

$$x^2 y_{n+2} + (2n+1) x y_{n+1} + (n^2 + 1) y_n = 0.$$

[DU, B.Sc.(G) 2002, B.A.(P), 2000]

SOLUTION

Differentiating $y = a \cos(\ln x) + b \sin(\ln x)$, we get

$$y_1 = \frac{-a \sin(\ln x)}{x} + \frac{b \cos(\ln x)}{x}$$

$$\Rightarrow \quad x y_1 = -a \sin(\ln x) + b \cos(\ln x)$$

Differentiating again,

$$x y_2 + y_1 = -\frac{a \cos(\ln x)}{x} - \frac{b \sin(\ln x)}{x}$$

$$\Rightarrow \quad x(x y_2 + y_1) = -\left[a \cos(\ln x) + b \sin(\ln x)\right]$$

$$\Rightarrow \quad x^2 y_2 + x y_1 + y = 0$$

Differentiating n times using Leibnitz's theorem, we obtain

$$(x^2 y_2)_n + (x y_1)_n + y_n = 0$$

$$\Rightarrow \quad [x^2 y_{n+2} + {}^nC_1 \cdot 2x \cdot y_{n+1} + {}^nC_2 \cdot 2 \cdot y_n] + [x y_{n+1} + {}^nC_1 y_n] + y_n = 0$$

$$\Rightarrow \quad x^2 y_{n+2} + (2n+1) x y_{n+1} + (n^2 + 1) y_n = 0$$

EXAMPLE 8.15 If $y = e^{m \sin^{-1} x}$, show that

$(1 - x^2) y_{n+2} - (2n+1) x y_{n+1} - (n^2 + m^2) y_n = 0$. Hence, find the value of $y_n(0)$.

[DU, B.Sc.(G), 1998]

SOLUTION

Let $$y_1 = e^{m\sin^{-1}x} \tag{i}$$

then $$y_1 = e^{m\sin^{-1}x} \cdot \frac{m}{\sqrt{1-x^2}} \tag{ii}$$

$$\Rightarrow \quad (1-x^2)y_1^2 = m^2y^2$$

Differentiating, we get

$$(1-x^2)\,2y_1y_2 - 2xy_1^2 = 2m^2yy_1$$

Dividing both the sides by $2y_1$, we get

$$(1-x^2)y_2 - xy_1 = m^2y \tag{iii}$$

Differentiating n times using, Leibnitz's theorem, we get

$$[(1-x^2)\,y_{n+2} + {}^nC_1(-2x)y_{n+1} + {}^nC_2(-2)y_n] - [xy_{n+1} + ny_n] = m^2y_n$$

$$\Rightarrow \quad (1-x^2)y_{n+2} - (2n+1)xy_{n+1} - (n^2+m^2)y_n = 0$$

Putting $x = 0$ in the above expression, we get

$$y_{n+2}(0) = (n^2+m^2)y_n(0) \tag{iv}$$

From (i), (ii), and (iii), we get

$$y(0) = 1, \quad y_1(0) = m, \quad y_2(0) = m^2.$$

Putting $n = 1, 2, 3, 4$ and so on in (iv), we get

$$y_3(0) = (1^2+m^2)\,y_1(0) = m(1^2+m^2)$$

$$y_4(0) = (2^2+m^2)\,y_2(0) = m^2(2^2+m^2);$$

$$y_5(0) = (3^2+m^2)\,y_3(0) = m(1^2+m^2)(3^2+m^2);$$

$$y_6(0) = (4^2+m^2)\,y_4(0) = m^2(2^2+m^2)(4^2+m^2); \text{ and so on.}$$

In general

$$y_n(0) = \begin{cases} m^2(2^2+m^2)(4^2+m^2)\cdots\cdots[(n-2)^2+m^2], & \text{when } n \text{ is even} \\ m(1^2+m^2)(3^2+m^2)\cdots\cdots[(n-2)^2+m^2], & \text{when } n \text{ is odd} \end{cases}$$

EXAMPLE 8.16 If $y = \sin(m\sin^{-1}x)$, prove that

$$(1-x^2)\,y_{n+2} = (2n+1)\,xy_{n+1} + (n^2-m^2)y_n.$$

[DU, B.Sc.(Ph. Sc.), 2006, B.A.(P), 2001, B.Sc.(G), 1993]

SOLUTION

We have, $$y = \sin(m\sin^{-1}x)$$

$$\Rightarrow \quad y_1 = \cos(m\sin^{-1}x) \cdot \frac{m}{\sqrt{1-x^2}}$$

$$\Rightarrow \quad (1-x^2)y_1^2 = m^2\cos^2(m\sin^{-1}x)$$

$$\Rightarrow \quad (1-x^2)y_1^2 = m^2(1-\sin^2(m\sin^{-1}x))$$

$$\Rightarrow \quad (1-x^2)y_1^2 = m^2(1-y^2)$$

$$\Rightarrow \quad (1-x^2)y_1^2 + m^2y^2 = m^2$$

Differentiating, we get

$$(1 - x^2)2y_1 y_2 + y_1^2(-2x) + m^2 2y y_1 = 0$$

Dividing by $2y_1$, we get

$$(1 - x^2)y_2 - xy_1 + m^2 y = 0$$

Differentiating n times by Leibnitz's theorem, we get

$$\left[y_{n+2}(1 - x^2) + {}^nC_1 y_{n+1}(-2x) + {}^nC_2 y_n \cdot (-2)\right] - \left[xy_{n+1} + {}^nC_1 \cdot 1 \cdot y_n\right] + m^2 y_n = 0$$

$$\Rightarrow \quad (1 - x^2)y_{n+2} - (2n + 1)xy_{n+1} + (m^2 - n^2)y_n = 0$$

EXAMPLE 8.17 If $f(x) = \tan x$, prove that

$$f^n(0) - {}^nC_2 f^{n-2}(0) + {}^nC_4 f^{n-4}(0) - \cdots = \sin\frac{n\pi}{2}$$

SOLUTION

$$f(x) = \frac{\sin x}{\cos x}$$

$$\Rightarrow \quad f(x)\cos x = \sin x$$

Differentiating n times using Leibnitz's theorem, we get

$$f^n(x) \cdot \cos x + {}^nC_1 f^{n-1}(x)(-\sin x) + {}^nC_2 f^{n-2}(x)(-\cos x) + {}^nC_3 f^{n-3}(\sin x) + {}^nC_4 f^{n-4}(x) \cdot (\cos x) + \cdots = \sin\left(x + \frac{n\pi}{2}\right)$$

Putting $x = 0$ in the above equation, we get

$$f^n(0) - {}^nC_2 f^{n-2}(0) + {}^nC_4 f^{n-4}(0) - \cdots = \sin\frac{n\pi}{2}.$$

EXAMPLE 8.18 If $y = (\sin^{-1}x)^2$, prove that

$$(1 - x^2)\frac{d^2y}{dx^2} - x\frac{dy}{dx} - 2 = 0.$$

[DU, B.A.(P), 2005, 2003]

Hence show that

$$(1 - x^2)y_{n+2} - (2n + 1)xy_{n+1} - n^2 y_n = 0.$$

SOLUTION

Here, $y = (\sin^{-1}x)^2$

$$\frac{dy}{dx} = 2\sin^{-1}x \cdot \frac{1}{\sqrt{1 - x^2}}$$

Squaring both the sides, we get

$$(1 - x^2)\left(\frac{dy}{dx}\right)^2 = 4(\sin^{-1}x)^2 = 4y$$

Differentiating again, we get

$$(1 - x^2)2\frac{dy}{dx}\cdot\frac{d^2y}{dx^2} + (-2x)\left(\frac{dy}{dx}\right)^2 = 4\frac{dy}{dx}$$

Dividing by $2\frac{dy}{dx}$,

$$(1 - x^2)\frac{d^2y}{dx^2} - x\frac{dy}{dx} = 2$$

$$\Rightarrow \quad (1 - x^2)\frac{d^2y}{dx^2} - x\frac{dy}{dx} - 2 = 0$$

Differentiating n times by Leibnitz's theorem, we get

$$\left[(1 - x^2)y_{n+2} + {}^nC_1(-2x)y_{n+1} + {}^nC_2(-2)y_n\right] - \left[xy_{n+1} + {}^nC_1\cdot 1\cdot y_n\right] = 0$$

$$\Rightarrow \quad (1 - x^2)y_{n+2} - (2n + 1)xy_{n+1} - n^2y_n = 0.$$

EXAMPLE 8.19 $\cos^{-1}\left(\frac{y}{b}\right) = \ln\left(\frac{x}{n}\right)^n$, prove that

$$x^2y_{n+2} + (2n + 1)xy_{n+1} + 2n^2y_n = 0.$$

SOLUTION

We have, $\cos^{-1}\left(\frac{y}{b}\right) = \ln\left(\frac{x}{n}\right)^n = n(\log x - \log n)$

Differentiating , we get

$$\frac{-1}{\sqrt{1 - (\frac{y}{b})^2}}\cdot\frac{1}{b}\cdot\frac{dy}{dx} = n\cdot\frac{1}{x}$$

$$\Rightarrow \quad x^2y_1^2 = n^2(b^2 - y^2)$$

Differentiating again, we get

$$2xy_1^2 + 2x^2y_1y_2 = -2n^2yy_1$$

$$\Rightarrow \quad xy_1 + x^2y_2 = -n^2y$$

$$\Rightarrow \quad x^2y_2 + xy_1 + n^2y = 0$$

Differentiating n-times by using Leibnitz's theorem, we get,

$$[x^2y_{n+2} + {}^nC_1(2x)y_{n+1} + {}^nC_2 2y_n] + [xy_{n+1} + {}^nC_1\cdot y_n] + n^2y_n = 0$$

$$\Rightarrow \quad [x^2y_{n+2} + 2nxy_{n+1} + (n^2 - n)y_n] + [xy_{n+1} + ny_n] + n^2y_n = 0$$

$$\Rightarrow \quad x^2y_{n+2} + (2n + 1)xy_{n+1} + 2n^2y_n = 0$$

EXAMPLE 8.20 If $y = \left[x + \sqrt{1 + x^2}\right]^m$, show that

$$(1 + x^2)y_{n+2} + (2n + 1)xy_{n+1} + (n^2 - m^2)y_n = 0$$

Also, find $y_n(0)$ *[DU, B.Sc.(H), 1991]*

SOLUTION

Here,
$$y = \left[x + \sqrt{1 + x^2}\right]^m \tag{I}$$

$\therefore$
$$y_1 = m\left[x + \sqrt{1 + x^2}\right]^{m-1} \cdot \left[1 + \frac{1}{2} \times \frac{2x}{\sqrt{1 + x^2}}\right]$$
$$= m\left[x + \sqrt{1 + x^2}\right]^{m-1} \cdot \left[1 + \frac{x}{\sqrt{1 + x^2}}\right]$$
$\Rightarrow$
$$\sqrt{1 + x^2}y_1 = m\left[x + \sqrt{1 + x^2}\right]^m = my$$

Squaring both sides
$$(1 + x^2)y_1^2 = m^2y^2 \tag{II}$$

Differentiating this equation with respect to x,
$$(1 + x^2)2y_1y_2 + 2xy_1^2 = 2m^2yy_1$$
$$(1 + x^2)y_2 + xy_1 = m^2y \tag{III}$$

Now, differentiating the above equation n times and using Leibnitz's rule, we get
$$[(1 + x^2)y_{n+2} + {}^nC_1(2x)y_{n+1} + {}^nC_2(2)y_n] + [xy_{n+1} + {}^nC_1 \cdot y_n] = m^2y_n$$
$$\Rightarrow \quad (1 + x^2)y_{n+2} + (2nx + x)y_{n+1} + (n^2 - n + n - m^2)y_n = 0$$
$$\Rightarrow \quad (1 + x^2)y_{n+2} + (2n + 1)xy_{n+1} + (n^2 - m^2)y_n = 0 \tag{IV}$$

Putting $x = 0$ in (I), (II), (III), we get
$$y(0) = 1, \qquad y_1(0) = m, \qquad y^2(0) = m^2$$

Putting $x = 0$ in (IV), we get
$$y_{n+2}(0) = (m^2 - n^2)y_n(0) \tag{V}$$

Now, putting $n = 1, 2, 3, 4, \ldots$ in (V), we get
$$y_3(0) = (m^2 - 1^2)y_1(0) = (m^2 - 1^2)m$$
$$y_4(0) = (m^2 - 2^2)y_2(0) = (m^2 - 2^2)m^2$$
$$y_5(0) = (m^2 - 3^2)y_3(0) = (m^2 - 3^2)(m^2 - 1^2)m$$
$$y_6(0) = (m^2 - 4^2)y_4(0) = (m^2 - 4^2)(m^2 - 2^2)m^2 \text{ and so on.}$$

In general, $y_n(0) = m^2(m^2 - 2^2)(m^2 - 4^2)\cdots(m^2 - (n - 2)^2)$, if n is even
$= m(m^2 - 1^2)(m^2 - 3^2)\cdots(m^2 - (n - 2)^2)$, if n is odd.

EXERCISE 8.3

1. If $y = e^{m\cos^{-1}x}$, show that $(1 - x^2)y_{n+2} - (2n + 1)xy_{n+1} - (n^2 + m^2)y_n = 0$
2. If $y = \cos(m\sin^{-1}x)$, show that $(1 - x^2)y_{n+2} - (2n + 1)xy_{n+1} + (m^2 - n^2)y_n = 0$
3. If $y = \ln(x + \sqrt{1 + x^2})$, prove that $(1 + x^2)y_{n+2} + (2n + 1)xy_{n+1} + n^2y_n = 0$
4. If $y = \sin^{-1}x$, then show that $(1 - x^2)y_{n+2} - (2n + 1)xy_{n+1} - n^2y_n = 0$.
5. If $y = \tan^{-1}x$, show that $(1 + x^2)y_{n+2} + 2(n + 1)xy_{n+1} + n(n + 1)y_n = 0$
6. State Leibnitz's Theorem. Find the nth derivative of $x \ln x$.

[DU, B.Sc.(H) Chem,1994]

7. If $y = \dfrac{\sin^{-1}x}{\sqrt{1-x^2}}$, show that $(1 - x^2)y_{n+2} - (2n + 3)xy_{n+1} - (n + 1)^2 y_n = 0$.

[DU, B.Sc.(Ph.Sc), 2008] , *[DU, B.Sc.(H), 1994]*

8. If $y = (x^2 - 1)^x$, then prove that $(x^2 - 1)y_{n+2} + 2xy_{n+1} - n(n + 1)^2 y_n = 0$.

[DU, BA(P), 1994, 1992]

9. If $y = x^2 e^x$, then show that $y_n = \dfrac{1}{2}n(n - 1)y_2 - n(n - 2)y_1 + \dfrac{1}{2}(n - 1(n - 2)y$.

10. After solving the question given in Q1, 2, 3, 4, 5 find $y_n(0)$ in each case.

11. If $y = x^2 \sin x$, prove that

$$\frac{d^n y}{dx^n} = (x^2 - n^2 + n)\sin\left(x + \frac{n\pi}{2}\right) - 2nx\cos\left(x + \frac{n\pi}{2}\right).$$

CHAPTER NINE

Polynomial Approximation of Functions

9.1 TAYLOR POLYNOMIAL AND MACLAURIN POLYNOMIAL

If a function f is differentiable at $x = a$, then the equation of the tangent line to the graph of the function through $P(a, f(a))$ is $y = f(a) + f'(a)(x - a)$. Since this line closely approximates the graph of f at values of x near a, it follows that

$$f(x) \approx f(a) + f'(a)(x - a),$$

provided that x is close to a. The expression $f(a) + f'(a)(x - a)$ is called the linear approximation of f at a. It is of the form

$$P(x) = c_1 + c_2(x - a),$$

satisfying $P(a) = f(a)$, $P'(a) = f'(a)$.

However, if the curve has a bend at a, then a linear approximation of f at this point will not be much closer to the actual curve. In that case, we may look for a quadratic or higher approximation of f at a. A quadratic approximation of f at a is a quadratic expression of the form

$$P(x) = c_0 + c_1(x - a) + c_2(x - a)^2$$

satisfying $P(a) = f(a), \quad P'(a) = f'(a), \quad P''(a) = f''(a)$.

Thus, a quadratic approximation of f at a is the expression

$$f(x) \approx f(a) + f'(a)(x - a) + \frac{f''(a)}{2!}(x - a)^2$$

We illustrate the linear and quadratic approximations of f with the help of the Example 1.

EXAMPLE 9.1 Find the linear and quadratic approximation of e^x at $x = 0$ and draw the graph of e^x and of the two approximations together.

SOLUTION

If we let $f(x) = e^x$, then $f'(x) = f''(x) = e^x$.
Hence, $f(0) = f'(0) = f''(0) = 1$.

Thus, the linear and quadratic approximations of e^x at $x = 0$ are $e^x \approx 1 + x$ and $e^x \approx 1 + x + \frac{1}{2}x^2$ respectively.

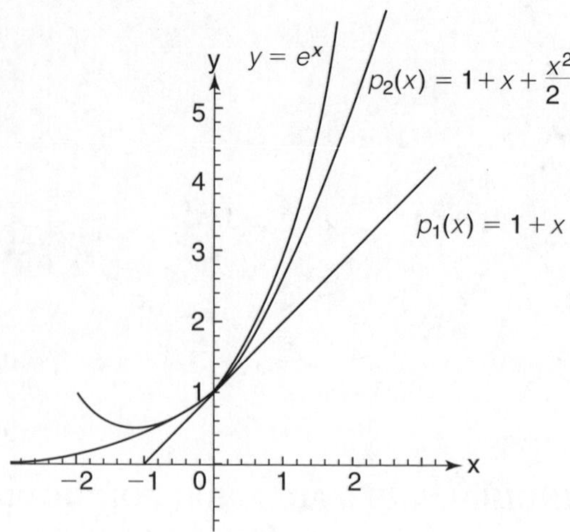

Fig. 9.1

As expected, the quadratic approximation is more accurate than the linear approximation around $x = 0$. Thus, it is natural to think of an approximation of higher order, in order to get more accuracy. Therefore, we go further with the kind of polynomials $P(x)$ discussed above for a given function in the following way:

Let f be a function possessing derivatives upto n^{th} order at $x = a$. We look for an approximation of f at $x = a$ of the form

$$P(x) = c_0 + c_1(x - a) + c_2(x - a)^2 + \cdots + c_n(x - a)^n,$$

under the condition that $P(a) = f(a), P'(a) = f'(a), \ldots, P^n(a) = f^n(a)$.

As calculated earlier, it may be shown that

$$c_0 = f(a), c_1 = f'(a), c_2 = \frac{f''(a)}{2!}, c_3 = \frac{f'''(a)}{3!}, \ldots, c_n = \frac{f^n(a)}{n!}.$$

We give a special name to this polynomial.

? DEFINITION: Let f be a function having derivatives upto n^{th} order at $x = a$. Then the 'n^{th} order Taylor polynomial' of f at $x = a$ is the polynomial

$$P_n(x) = f(a) + f'(a)(x - a) + \frac{f''(a)}{2!}(x - a)^2 + \frac{f'''(a)}{3!}(x - a)^3 + \cdots + \frac{f^n(a)}{n!}(x - a)^n$$

If $a = 0$, then the polynomial $P_n(x)$ is called 'n^{th} order Maclaurin polynomial' of f at $x = a$.

EXAMPLE 9.2 Find the first four Taylor Polynomials for $\ln x$ at $x = 2$.

SOLUTION

Let $f(x) = \ln x$. Then

$$f'(x) = \frac{1}{x}, \; f''(x) = \frac{-1}{x^2}, \; f'''(x) = \frac{2}{x^3}.$$

The first four Taylor polynomials of $f(x)$ at $x = 2$ are

$$P_0(x) = f(2) = \ln 2$$

$$P_1(x) = f(2) + f'(2)(x-2) = \ln 2 + \frac{1}{2}(x-2)$$

$$P_2(x) = f(2) + f'(2)(x-2) + \frac{f''(2)}{2!}(x-2)^2 = \ln(2) + \frac{1}{2}(x-2) - \frac{1}{8}(x-2)^2$$

$$P_3(x) = f(2) + f'(2)(x-2) + \frac{f''(2)}{2!}(x-2)^2 + \frac{f'''(2)}{3!}(x-2)^3$$
$$= \ln 2 + \frac{1}{2}(x-2) - \frac{1}{8}(x-2)^2 + \frac{1}{24}(x-2)^3$$

EXAMPLE 9.3 Find the nth Maclaurin polynomials for

(i) $\sin x$ (ii) $\cos x$ (iii) $\dfrac{1}{1-x}$

SOLUTION

(i) Here $f(x) = \sin x$ has derivatives of all orders and

$$f(x) = \sin x,\ f'(x) = \cos x,\ f''(x) = -\sin x,\ f'''(x) = -\cos x,\ f^{\text{iv}}(x) = \sin x, \ldots$$

Thus, $f(0) = 0, f'(0) = 1, f''(0) = 0, f'''(0) = -1, f^{(\text{iv})}(0) = 0, \ldots$

Therefore, the successive Maclaurin polynomials for $\sin x$ are

$$P_0(x) = f(0) = 0$$

$$P_1(x) = f(0) + f'(0) \cdot x = x$$

$$P_2(x) = f(0) + f'(0) \cdot x + \frac{f''(0)}{2!} \cdot x^2 = x$$

$$P_3(x) = f(0) + f'(0) \cdot x + \frac{f''(0)}{2!} \cdot x^2 + \frac{f'''(0)}{3!} \cdot x^3 = x - \frac{x^3}{3!}$$

$$P_4(x) = f(0) + f'(0) \cdot x + \frac{f''(0)}{2!} \cdot x^2 + \frac{f'''(0)}{3!} \cdot x^3 + \frac{f^{\text{iv}}(0)}{4!} \cdot x^4 = x - \frac{x^3}{3!}$$

$$P_5(0) = f(0) + f'(0) \cdot x + \frac{f''(0)}{2!} \cdot x^2 + \frac{f'''(0)}{3!} \cdot x^3 + \frac{f^{\text{iv}}(0)}{4!} \cdot x^4 + \frac{f^{\text{v}}(0)}{5!} \cdot x^5 = x - \frac{x^3}{3!} + \frac{x^5}{5!}$$

and so on.

Since, in general,

$$f^{2n}(0) = 0,\ f^{2n+1}(0) = (-1)^n; \quad n = 0, 1, 2, \ldots$$

we have, for $n = 0, 1, 2, 3, \ldots$

$$P_{2n+1}(x) = P_{2n+2}(x) = x - \frac{x^3}{3!} + \frac{x^5}{5!} - \frac{x^7}{7!} + \cdots + (-1)^n \frac{x^{2n+1}}{(2n+1)!}$$

(ii) Here $f(x) = \cos x, f'(x) = -\sin x, f''(x) = -\cos x, f'''(x) = \sin x$

$f^4(x) = \cos x$ and so on. In general, $f^{2n}(x) = (-1)^n \cos x$

$f^{2n+1}(x) = (-1)^{n+1} \sin x$. At $x = 0$, $f^{2n}(x) = (-1)^n, f^{2n+1}(0) = 0.$

Proceeding as in part (i), it can be shown that

$$P_0(x) = P_1(x) = 1$$

$$P_2(x) = P_3(x) = 1 - \frac{x^2}{2!}$$

$$P_4(x) = P_5(x) = 1 - \frac{x^2}{2!} + \frac{x^4}{4!} \text{ and so on.}$$

In general, the Maclaurin polynomials for $\cos x$ are given by

$$P_{2n}(x) = P_{2n+1}(x) = 1 - \frac{x^2}{2!} + \frac{x^4}{4!} - \frac{x^6}{6!} + \cdots + (-1)^n \frac{x^{2n}}{(2n)!}, \; n = 0, 1, 2, \ldots$$

The Fig. 9.2 below shows the graphs of approximating polynominals for $\cos x$.

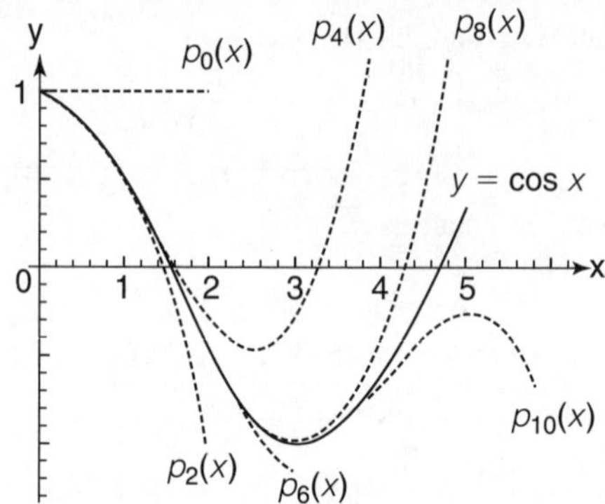

Fig. 9.2

(iii) Let $f(x) = \dfrac{1}{1-x}$. The values of f and its first n derivatives at $x = 0$ are given by

$$f'(x) = \frac{1}{(1-x)^2}, f''(0) = \frac{2}{(1-x)^3}, f'''(x) = \frac{3!}{(1-x)^4}, \ldots, f^n(x) = \frac{n!}{(1-x)^{n+1}}$$

$$f(0) = 1, f'(0) = 1, f''(0) = 2!, f'''(0) = 3!, \ldots, f^n(0) = n!$$

Therefore, the nth Maclaurin Polynomial for $f(x) = \dfrac{1}{1-x}$ is

$$P_n(x) = f(0) + f'(0)x + \frac{f''(0)}{2!}x^2 + \frac{f'''(0)}{3!}x^3 + \cdots + \frac{f^n(0)}{n!}x^n$$

$$= 1 + x + x^2 + x^3 + \cdots + x^n$$

EXAMPLE 9.4 Find the nth Taylor Polynomial for $1/x$ at $x = 1$.

SOLUTION

Let $f(x) = \dfrac{1}{x}$. We have,

$$f'(x) = -\frac{1}{x^2}, f''(x) = \frac{2}{x^3}, f'''(x) = -\frac{3!}{x^4}, \ldots, f^n(x) = (-1)^n \frac{n!}{x^{n+1}}$$

For $x = 1, f(1) = 1, f'(1) = -1, f''(1) = 2, f'''(x) = -3!, \ldots, f^n(1) = (-1)^n n!$.

Therefore, the nth Taylor Polynomial for $f(x) = \frac{1}{x}$ at $x = 1$ is

$$\mathrm{P}_n(x) = f(1) + f'(1)(x-1) + \frac{f''(1)}{2!}(x-1)^2 + \frac{f'''(1)}{3!}(x-1)^3 + \cdots + \frac{f^n(1)}{n!}(x-1)^n$$

$$= 1 - (x-1) + (x-1)^2 - (x-1)^3 + \cdots + (-1)^n(x-1)^n$$

EXERCISE 9.1

In Questions 1–8, find the Taylor Polynomials of orders 0, 1, 2 and 3 generated by f at a.

1. $f(x) = \ln x,\ a = 4$
2. $f(x) = \ln(1 + x),\ a = 0$
3. $f(x) = \frac{1}{x},\ a = 1$
4. $f(x) = \frac{1}{(x+2)},\ a = 0$
5. $f(x) = \sin x,\ a = \frac{\pi}{4}$
6. $f(x) = \cos x,\ a = \frac{\pi}{4}$
7. $f(x) = \sqrt{x},\ a = 3$
8. $f(x) = \sqrt{x + 4},\ a = 2$

In Questions 9–17, find Maclaurin Polynomials of order m generated by f.

9. $f(x) = e^{-x^2},\ m = 3$
10. $f(x) = e^{-2x},\ m = n$
11. $f(x) = \sin 3x,\ m = 2n + 1$
12. $f(x) = \cos 2x,\ m = 2n$
13. $f(x) = \sinh x,\ m = n$ $\left[\text{Hint: } \sinh x = \frac{1}{2}(e^x - e^{-x}),\ \cosh x = \frac{1}{2}(e^x + e^{-x})\right]$
14. $f(x) = \cosh x,\ m = n$
15. $f(x) = \ln(\cos x),\ m = 6$
16. $f(x) = \sin(\sin x),\ m = 3$
17. $f(x) = \ln\left(\frac{\sin x}{x}\right),\ m = 6$

9.2 POLYNOMIAL APPROXIMATIONS OF FUNCTIONS AND ERROR ESTIMATION

In the previous section, we have introduced n^{th} order Taylor Polynomial and Maclaurin Polynomial of a function. We have observed that higher the order, better is the approximation of the function in terms of the polynomial. Here, we introduce the famous Taylor's formula, which provides the exact measurement of the error that occurs in this approximation.

TAYLOR'S THEOREM

Let f be a real-valued function defined over the interval I (closed or open), possessing derivatives upto $(n+1)^{\text{th}}$ order in the interval.
Let $a \in \text{I}$. Then for any $x \in \text{I}$, we have

$$f(x) = f(a) + \frac{f'(a)}{1!}(x-a) + \frac{f''(a)}{2!}(x-a)^2 + \frac{f'''(a)}{3!}(x-a)^2$$

$$+ \cdots + \frac{f^n(a)}{n!}(x-a)^n + \text{R}_n(x) \qquad (1)$$

where $\text{R}_n(x)$, called remainder or error of order n, is an expression with the following different forms:

(i) (Lagrange's form) $\text{R}_n(x) = \dfrac{f^{n+1}(c)}{(n+1)!}(x-a)^{n+1}$, for some c between x and a;

(ii) (Cauchy's form) $\text{R}_n(x) = \dfrac{f^{n+1}(c)}{n!}(x-c)^n(x-a)$, for some c between x and a.

Thus, if a function f satisfies the condition of Taylor's Theorem then we may write,

$$f(x) = \text{P}_n(x) + \text{R}_n(x),$$

where $\text{P}_n(x)$ is the Taylor polynomial of $f(x)$ of order n at $x = a$ and $\text{R}_n(x)$ is the remainder or error of order n. Thus, $f(x) = \text{P}_1(x) + \text{R}_1(x) = \text{P}_2(x) + \text{R}_2(x) = \text{P}_3(x) + \text{R}_3(x) = \cdots$ and so on.

It may be observed that higher the order n, smaller is the magnitude of the error and hence better is the approximation of f in terms of $\text{P}_n(x)$.

EXAMPLE 9.5 Estimate the error for the following approximations:

(a) $\sin x \approx x - \dfrac{x^3}{6}$ for $|x| \le \dfrac{1}{2}$;

(b) $\tan x \approx x + \dfrac{x^3}{3}$ for $|x| \le 0.1$;

(c) $e^x \approx 1 + x + \dfrac{x^2}{2!} + \cdots + \dfrac{x^n}{n!}$ for $0 \le x \le 1$;

(d) $\sqrt{1+x} \approx 1 + \dfrac{x}{2} - \dfrac{x^2}{8}$ for $0 \le x \le 1$.

SOLUTION

(a) Here $f(x) = \sin x$, $\text{I} = \left[-\dfrac{1}{2}, \dfrac{1}{2}\right]$. Since f possesses derivatives of every order in I, by Taylor's Theorem, for any $a, x \in \text{I}$, we have,

$$f(x) = \text{P}_n(x) + \text{R}_n(x), \qquad n = 0, 1, 2, 3, \ldots$$

where $P_n(x) = f(a) + f'(a)(x - a) + \frac{f''(a)}{2!}(x - a)^2 + \cdots + \frac{f^n(a)}{n!}(x - a)^n$

and $R_n(x) = \frac{f^{n+1}(c)}{(n + 1)!}(x - a)^{n+1}$ (Lagrange's form of remainder)

where c is some real number between x and a.

For $a = 0$, we get

$$P_0(x) = 0, \quad R_0(x) = \frac{\cos(c)}{1!}x,$$

$$P_1(x) = x, \quad R_1(x) = -\frac{\sin(c)}{2!}x^2,$$

$$P_2(x) = x, \quad R_2(x) = -\frac{\cos(c)}{3!}x^3,$$

$$P_3(x) = x - \frac{x^3}{6}, \quad R_3(x) = \frac{\sin(c)}{4!}x^4,$$

$$P_4(x) = x - \frac{x^3}{6}, \quad R_4(x) = \frac{\cos(c)}{5!}x^5,$$

where c is a number between 0 and x . Therefore,

$$\begin{aligned} \left|\sin x - \left(x - \frac{x^3}{6}\right)\right| &= |f(x) - P_4(x)| \\ &= |R_4(x)| \\ &= \left|\frac{\cos(c)}{5!}x^5\right| \\ &= \frac{|\cos(c)|}{5!}|x|^5 \leq \frac{1}{5! \cdot 2^5} = \frac{1}{3840} = 0.00026041666\ldots \end{aligned}$$

Thus the maximum possible error in this approximation is .0002604167.

Here, although both $P_3(x) = P_4(x) = x - \frac{x^3}{6!}$, we go for $P_4(x)$ to get a better approximation.

(b) Here $f(x) = \tan x$, $I = [-0.1, 0.1]$. Since f possesses derivatives of every order in I, by Taylors Theorem, for any $a, x \in I$ we have

$$f(x) = P_n(x) + R_n(x), \qquad n = 0, 1, 2, \ldots$$

where $P_n(x) = f(a) + f'(a)(x - a) + \frac{f''(a)}{2!}(x - a)^2 + \cdots + \frac{f^n(a)}{n!}(x - a)^n$

and $R_n(x) = \frac{f^{n+1}(c)}{(n + 1)!}(x - a)^{n+1}$ (Lagrange's form of remainder)

where c is some number between a and x. For $a = 0$, we get

$$P_0(x) = f(0) = 0, \quad R_0(x) = \frac{f'(x)}{1!}x = \frac{1 + \tan^2 c}{1!}x$$

$$P_1(x) = f(0) + \frac{f'(0)}{1!}x = x, \quad R_1(x) = \frac{f''(c)}{2!}x^2 = \frac{2\tan c(1 + \tan^2 c)}{2}x^2$$

$$P_2(x) = f(0) + \frac{f'(0)}{1!}x + \frac{f''(0)}{2!}x^2 = x$$

$$R_2(x) = \frac{f'''(c)}{3!}x^3 = \frac{16\tan c + 40\tan^3 c + 24\tan^5 c}{4!}x^4$$

$$P_3(x) = f(0) + \frac{f'(0)}{1!}x + \frac{f''(0)}{2!}x^2 + \frac{f'''(0)}{3!}x^3 = x + \frac{x^3}{3}$$

$$R_3(x) = \frac{f^4(c)}{4!}x^4 = \frac{16\tan c + 40\tan^3 c + 24\tan^5 c}{4!}x^4$$

$$P_4(x) = f(0) + \frac{f'(0)}{1!}x + \frac{f''(0)}{2!}x^2 + \frac{f'''(0)}{3!}x^3 + \frac{f^4(0)}{4!}x^4 = x + \frac{x^3}{3}$$

$$R_4(x) = \frac{f^5(c)}{5!}x^5 = \frac{136\tan^2 c + 240\tan^4 c + 120\tan^6 c + 16}{5!}x^5$$

where c is in between 0 and x.
Consequently

$$\left|\tan x - \left(x + \frac{x^3}{3}\right)\right| = |f(x) - P_4(x)| = |R_4(x)|$$

$$= \frac{136\tan^2 c + 240\tan^4 c + 120\tan^6 c + 16}{5!}|x|^5$$

Now, $|\tan c| \le \tan(0.1) = \dfrac{\sin(0.1)}{\cos(0.1)} \le \dfrac{0.1}{\cos(0.1)}$

$$\left[\because \frac{d}{dx}(x - \sin x) \ge 0,\ x - \sin x \text{ is an increasing function}\right]$$

and $\cos(0.1) = 1 - 2\sin^2(0.05) \ge 1 - 2\sin(0.05) \ge 1 - 2 \times 0.5 = 0.9$

thus $|\tan c| \le \dfrac{0.1}{0.9} = \dfrac{1}{9}$ and we have

$$\left|\tan x - \left(x + \frac{x^3}{3}\right)\right| \le \frac{136 \times 9^{-2} + 240 \times 9^{-4} + 120 \times 9^{-6} + 16}{5!}\left(\frac{1}{10}\right)^5$$

$$= \frac{12256}{8303765625} \approx .1476318181 \times (10)^{-5}$$

(c) Here, $f(x) = e^x$, $I = [0, 1]$. Since f possesses derivatives of every order in I, proceeding as in above, taking $a = 0$ and using Taylor's Theorem, we get $f(x) = P_n(x) + R_n(x)$, $n = 0, 1, 2, \ldots$

where $P_n(x) = 1 + x + \dfrac{x^2}{2!} + \cdots + \dfrac{x^n}{n!}$, $\quad R_n(x) = \dfrac{e^c}{(n+1)!}x^{n+1}$ (Lagrange's form),

where $0 < c < x \le 1$. Therefore, since $e^c < e^1$, we have

$$\left|e^x - \left(1 + x + \frac{x^2}{2!} + \cdots + \frac{x^n}{n!}\right)\right| = |f(x) - P_n(x)| = |R_n(x)|$$

$$= \frac{e^c}{(n+1)!}x^{n+1} \le \frac{e}{(n+1)!} < \frac{3}{(n+1)!} \qquad [\because e \le 3]$$

For example,

$$|f(x) - P_4(x)| < \frac{3}{5!} = \frac{1}{40}$$

$$|f(x) - P_{10}(x)| < \frac{3}{11!} = \frac{1}{1209600}$$

$$|f(x) - P_{15}(x)| < \frac{3}{16!} = \frac{1}{4358914556000}$$

(d) We have, $f(x) = \sqrt{1 + x}$, $I = [0, 1]$. Hence, proceeding as in above, by applying Taylor's Theorem with $a = 0$, we get $f(x) = P_n(x) + R_n(x)$, $n = 0, 1, 2, \ldots$

$$P_0(x) = f(0) = 1, \qquad R_0(x) = \frac{f'(c)}{1!}x = \frac{x}{2\sqrt{1 + c}} \quad \text{(Lagrange's Form)}$$

$$P_1(x) = f(0) + \frac{f'(0)}{1!}x = 1 + \frac{x}{2}$$

$$R_1(x) = \frac{f''(c)}{2!}x = \frac{-x^2}{8(1 + c)^{3/2}}$$

$$P_2(x) = f(0) + \frac{f'(0)}{1!}x + \frac{f''(0)}{2!}x^2 = 1 + \frac{x}{2} - \frac{x^2}{8}$$

$$R_2(x) = \frac{f''(c)}{3!}x^3 = \frac{x^3}{16(1 + c)^{5/2}}, \quad \text{where } 0 < c < x. \text{ Then}$$

$$\left|\sqrt{1 + x} - \left(1 + \frac{x}{2} - \frac{x^2}{8}\right)\right| = |f(x) - P_2(x)| = |R_2(x)|$$

$$= \left|\frac{x^3}{16(1 + c)^{5/2}}\right| \le \frac{1}{16} = 0.0625$$

EXAMPLE 9.6 Find an approximate value of e correct upto five decimal places.

SOLUTION

The function e^x has derivative of all orders throughout the interval $(-\infty, \infty)$. Using Taylor's formula with Lagrange's form of remainder for $a = 0$, we get

$$e^x = 1 + x + \frac{x^2}{2!} + \cdots + \frac{x^n}{n!} + R_n(x),$$

where $R_n(x) = \frac{e^c}{(n + 1)!}x^{n+1}$ for some c between 0 and x. For $x = 1$, we get

$$e = 1 + 1 + \frac{1}{2!} + \frac{1}{3!} + \cdots + \frac{1}{n!} + R_n(1)$$

with $R_n(1) = \frac{e^c}{(n + 1)!}$ for some c between 0 and 1.

It is known to us that $e < 3$. Also, as $c < 1$, we have, $e^c < e^1 < 3$. Therefore, we have

$$|R_n(1)| < \frac{3}{(n + 1)!}$$

Now our problem is to determine how many terms to include to achieve five decimal place accuracy; that is, to choose n so that the absolute value of the nth remainder $R_n(1)$ satisfies

$$|R_n(1)| < .000005$$

Thus, we can achieve a five decimal place accuracy by choosing n so that

$$\frac{3}{(n+1)!} \leq 0.000005 \quad \text{or} \quad (n+1)! \geq 6{,}00{,}000$$

Since $9! = 3{,}62{,}880$ and $10! = 36{,}28{,}800$, the smallest value of n which satisfies the given condition is $n = 9$. Thus, the value of e upto five decimal place accuracy is

$$\mathrm{e} \approx 1 + 1 + \frac{1}{2!} + \frac{1}{3!} + \cdots + \frac{1}{9!} \approx 2.71828.$$

EXAMPLE 9.7 For what values of x can we replace $\sin x$ by $x - \dfrac{x^3}{3!}$ with an error of magnitude no greater than 3×10^{-4}?

SOLUTION

The function $f(x) = \sin x$ possesses derivatives of all orders in the interval $(-\infty, \infty)$. Therefore, by applying Taylor's formula with Lagrange's form of remainder

$$f(x) = \mathrm{P}_n(x) + \mathrm{R}_n(x)$$

for $n = 4$, $a = 0$, we get

$$\sin x = \sin 0 + (\cos 0) + \frac{(-\sin 0)}{2!}x^2 + \frac{(-\cos 0)}{3!}x^3 + \frac{(\sin 0)}{4!}x^4 + \frac{(\cos \mathrm{c})}{5!}x^5$$

where c is a real number with $0 < \mathrm{c} < x$.

$$\Rightarrow \qquad \sin x = x - \frac{x^3}{3!} + \frac{\cos \mathrm{c}}{5!}x^5$$

According to the given condition, we need to find out the values of x such that

$$\left|\frac{\cos \mathrm{c}}{5!}x^5\right| \leq 3 \times 10^{-4}$$

As $|\cos \mathrm{c}| \leq 1$, we look for x satisfying, $\left|\dfrac{x^5}{5!}\right| \leq 3 \times 10^{-4}$, that is

$$|x|^5 \leq 360 \times 10^{-4},$$

$$|x| \leq (\sqrt[5]{360 \times 10^{-4}}) \approx 0.514, \text{ after rounding off.}$$

REMARK: In the above example, both $n = 3, 4$ give $\mathrm{P}_n(x) = x - \dfrac{x^3}{3!}$. However, larger the value of n, better is the accuracy.

EXAMPLE 9.8 Calculate e with an error of magnitude less than 10^{-6}.

SOLUTION

The function $f(x) = \mathrm{e}^x$ possess derivatives of all orders in the interval $(-\infty, \infty)$. Hence, by applying Taylor's formula with Lagrange's form of remainder for $a = 0$ we get,

$$f(x) = \mathrm{P}(x) + \mathrm{R}_n(x)$$

$$\Rightarrow \qquad \mathrm{e}^x = 1 + x + \frac{1}{2!}x^2 + \frac{1}{3!}x^3 + \cdots + \frac{1}{n!}x^n + \mathrm{R}_n(x)$$

If we take $x = 1$, then

$$e = 1 + 1 + \frac{1}{2!} + \frac{1}{3!} + \cdots + \frac{1}{n!} + R_n(1),$$

where $R_n(1) = \dfrac{e^c}{(n+1)!}$ for some c between 0 and 1. Thus, we have

$$e \approx 1 + 1 + \frac{1}{2!} + \frac{1}{3!} + \cdots + \frac{1}{n!},$$

with an error of magnitude equal to $\left|\dfrac{e^c}{(n+1)!}\right|$. Therefore, we need to find out the value of n for which this error is less than 10^{-6}. As $c < 1$, we have, $e^c < e^1$. Again, for simplicity's sake, we use the fact that $e < 3$. Thus, we look for n satisfying

$$\frac{3}{(n+1)!} \leq 10^{-6}$$

By calculating, we find that $\dfrac{3}{9!} > 10^{-6}$ while $\dfrac{3}{10!} < 10^{-6}$.

Thus, n should be at least 9 so that the error is less than 10^{-6}. Thus, with an error less than 10^{-6}, the value of e is

$$e = 1 + 1 + \frac{1}{2!} + \frac{1}{3!} + \cdots + \frac{1}{9!} \approx 2.718282$$

EXAMPLE 9.9 Use Taylor's formula to approximate the value of $\sqrt[3]{30}$.

SOLUTION

Consider the function $f(x) = x^{1/3}$ in $I = [27, 30]$. f has derivatives of every order in I. Hence by applying Taylor's Theorem with Lagrange's form of remainder, we get, for any $a, x \in I$,

$$f(x) = P_n(x) + R_n(x), \quad \text{where} \quad f(x) = x^{1/3} \text{ and}$$

$$P_n(x) = f(a) + f'(a)(x-a) + \frac{f''(a)}{2!}(x-a)^2 + \cdots + \frac{f^n(a)}{n!}(x-a)^n,$$

$$= a^{1/3} + \left(\frac{1}{3}\right)\cdot a^{-2/3}(x-a) + \frac{1}{3}\left(-\frac{3}{2}\right)a^{-5/3}\cdot\frac{1}{2!}(x-a)^2 + \cdots$$

$$+ \frac{1}{3}\left(-\frac{2}{3}\right)\cdots\left(\frac{1}{3} - n + 1\right)a^{\frac{1}{3}-n}\cdot\frac{1}{n!}\cdot(x-a)^n$$

$$R_n(x) = \frac{f^{n+1}(c)}{(n+1)!}(x-a)^{n+1}, \quad \text{for some c lying between a and } x.$$

$$= \frac{1}{3}\left(-\frac{2}{3}\right)\cdots\left(\frac{1}{3} - n\right)\frac{c^{\frac{1}{3}-n-1}}{(n+1)!}(x-a)^{n+1}$$

For $x = 30$, $a = 27$, we get

$$(30)^{1/3} = P_n(30) + R_n(30)$$

For $n = 1, 2, \ldots$, we get

$$n = 1, \qquad (30)^{1/3} = \left[3 + \frac{1}{3}(27)^{-2/3}\cdot 3\right] + \frac{1}{3}\frac{\left(-\frac{2}{3}\right)(c)^{-5/3}}{2!} \times 3^2$$

$$n = 2, \quad (30)^{1/3} = \left[3+\frac{1}{3}(27)^{-2/3}\cdot 3+\frac{1}{3}\cdot\left(-\frac{2}{3}\right)\cdot(27)^{-5/3}\cdot\frac{3^2}{2!}\right]+\frac{1}{3}\left(-\frac{2}{3}\right)\left(-\frac{5}{3}\right)c^{-8/3}\frac{3^3}{3!}$$

$$n = 3, \quad (30)^{1/3} = \left[3+\frac{1}{3}(27)^{-2/3}\cdot 3+\frac{1}{3}\left(-\frac{2}{3}\right)(27)^{-5/3}\cdot\frac{3^2}{2!}+\frac{1}{3}\left(-\frac{2}{3}\right)\left(-\frac{5}{3}\right)(27)^{-8/3}\frac{3^3}{3!}\right]$$

$$+\frac{1}{3}\left(-\frac{2}{3}\right)\left(-\frac{5}{3}\right)\left(-\frac{8}{3}\right)(c)^{-11/3}\cdot\frac{3^4}{4!}$$

and so on.

Since c is a number lying between 27 and 30,

$c^{-5/3} < (27)^{-5/3}, \quad c^{-8/3} < (27)^{-8/3}, \ldots, \quad c^{\frac{1}{3}-n-1} < 3^{-3n-2}$. Therefore

$$|(30)^{1/3} - P_n(30)| = |R_n(30)| < \left|\frac{\frac{1}{3}\left(-\frac{2}{3}\right)\cdots\left(\frac{1}{3}-n\right)3^{-2n-1}}{(n+1)!}\right| = |E_n(30)| \quad \text{(say)}$$

Thus from the above, we get the following approximate values $\left(P_n(30)\right)$ of $\sqrt[3]{30}$ with the corresponding errors of estimation $\left(E_n(30)\right)$.

n	$P_n(30) \approx \sqrt{30}$	$E_n(30)$
1	$3+\frac{1}{9} = 3.111\ldots$	$\frac{1}{3^5} = .0041152$
2	$3+\frac{1}{9}-\frac{1}{3^5} = 3.1069948$	$\frac{5}{3^9} = .0002540$
3	$3+\frac{1}{9}-\frac{1}{3^5}+\frac{5}{3^9} = 3.1072488$	$\frac{10}{3^{12}} = .0000188$

and so on.

EXAMPLE 9.10 Use Taylor's Theorem to compute an approximate value of $\sqrt{e}$ with an estimate of the corresponding error.

SOLUTION

Since the function $f(x) = e^x$ possesses derivatives of every order in any interval, we may apply Taylor's Theorem with Lagrange's form of remainder with $a = 0$ to get

$$f(x) = P_n(x) + R_n(x)$$

where $\quad f(x) = e^x$

$$P_n(x) = f(0) + \frac{f'(0)}{1!}x + \frac{f''(0)}{2!}x^2 + \cdots + \frac{f^n(0)}{n!}x^n$$

$$= 1 + \frac{x}{1!} + \frac{x^2}{2!} + \cdots + \frac{x^n}{n!}$$

$$R_n(x) = \frac{f^{n+1}(c)}{(n+1)!}x^{n+1} = \frac{e^c}{(n+1)!}x^{n+1}, \qquad \text{for some } 0 < c < x.$$

Now, by taking $x = \frac{1}{2}$, we get

$$\sqrt{e} = P_n\left(\frac{1}{2}\right) + R_n\left(\frac{1}{2}\right)$$

$$\Rightarrow \qquad \left[\sqrt{e} - P_n\left(\frac{1}{2}\right)\right] = \left|R_n\left(\frac{1}{2}\right)\right|$$

$$\Rightarrow \qquad \left|\sqrt{e} - \left(1 + \frac{1}{2} + \frac{1}{2^2 \cdot 2!} + \cdots + \frac{1}{2^n \cdot n!}\right)\right| = \left|\frac{e^c}{2^{n+1}(n+1)!}\right|$$

$$\leq \frac{e^{1/2}}{2^{n+1}(n+1)!} \qquad \left(\because \; 0 < c < \frac{1}{2}\right)$$

$$< \frac{2}{2^{n+1}(n+1)!} = E_n\left(\frac{1}{2}\right) \text{ (say) } (\because e < 4)$$

Thus, the approximate value $P_n\left(\frac{1}{2}\right)$ of $\sqrt{e}$ can be obtained by calculating the value of the expression $1 + \frac{1}{2} + \frac{1}{2^2 \cdot 2!} + \cdots + \frac{1}{2^n \cdot n!}$ for different values of n with the corresponding error estimation $E_n\left(\frac{1}{2}\right) = \frac{1}{2^n(n+1)!}$. We may calculate the following values:

n	$P_n\left(\frac{1}{2}\right)$	$E_n\left(\frac{1}{2}\right)$
1	$\frac{3}{2} = 1.5$	$\frac{1}{4} = .25$
2	$\frac{13}{8} = 1.625$	$\frac{1}{24} = .41666\ldots$
3	$\frac{79}{48} = 1.64583$	$\frac{1}{192} = .005208333\ldots$
4	$\frac{211}{128} = 1.6484375$	$\frac{1}{1920} = .00520883\ldots$

EXERCISE 9.2

1. Show that the value of e, correct to seven decimal places is 2.7182818.
2. For what value of x, can we approximate e^x by $1 + x + \frac{x^2}{2}$, correct to four decimal places?
3. For approximately what values of x, can we replace $\sin x$ by $x - \frac{x^3}{6}$ with an error of magnitude not greater than 5×10^{-4}?
4. Estimate the error in the approximation of $\sinh x = x + \frac{x^3}{3!}$ when $|x| < 0.5$ [Hint: use $R_4(n)$ not $R_3(n)$].

5. If $\cos x$ is replaced by $1 - \frac{x^2}{2}$ and $|x| < 0.2$, what estimate can be made of the error? Is the estimate $1 - \frac{x^2}{2}$ of $\cos x$ an under estimate?

6. Estimate the error in the approximation $\sin x = x$ when $|x| < 10^{-3}$. For which of these values of x is $x < \sin x$?

7. For the approximation $e^x = 1 + x + \frac{x^2}{2}$, estimate the error when $|x| < 0.1$.

8. For what values of x can you replace $\ln(1 + x)$ by x with an error of magnitude not greater than 1% of the value of x?

9. When $0 \le x \le 0.01$, show that e^x may be replaced by $1 + x$ with an error of magnitude not greater than 0.6 % of x.

10. Use Taylor's Theorem with $a = 0$ and $n = 3$ to find the cubic approximation of $f(x) = \frac{1}{1-x}$ at $x = 0$. Give an upper bound for the magnitude of the error in the approximation when $|x| \le 0.1$.

11. Show that for $0 \le x \le .25$, $\cos x$ can be approximated by $1 - \frac{x^2}{2!} + \frac{x^4}{4!}$ with an error not greater than $\frac{(.25)^6}{6!}$.

12. The estimate $\sqrt{1+x} = 1 + \frac{x}{2}$ is used when x is small. Estimate the error when $|x| < 0.01$.

13. Use Taylor's formula to approximate the following values

(i) $\sqrt[5]{250}$ (ii) $\sin 18^\circ$ (iii) $\ln(1.2)$ (iv) $(1.1)^{1.2}$ (v) $\sqrt[3]{e}$

9.3 SERIES EXPANSION OF FUNCTIONS

In the fifth chapter, we have defined a sequence and discussed its convergence. If $\{u_n\}_{n=1}$ is a sequence, then $\sum_{n=1}^{\infty} u_n$ defined by $\sum_{n=1}^{\infty} u_n = u_1 + u_2 + \cdots + u_3 + \cdots$ is called an *infinite series*.

For example, $\sum \frac{1}{n} = 1 + \frac{1}{2} + \frac{1}{3} + \cdots + \frac{1}{n} + \cdots$ is an infinite series.

If we add the first two terms of the series $\Sigma\, u_n$, that is, if we find out $u_1 + u_2$ and then add u_3 to it, and go on adding each term to the sum of the previous terms, we see that, as there is no last term of the series, we will never arrive at the end of the process. Thus, in the ordinary sense, the expression 'the sum of an infinite series' has no meaning. A meaning is assigned to this expression by employing the notion of limit in the following way:

Let S_n denote the sum of the first n terms of a series. That is, $S_n = u_1 + u_2 + \cdots + u_n$; S_n is called the nth *partial sum of* $\Sigma\, u_n$. If $\{S_n\}$ tends to a finite limit (as a sequence) to, say s, as n tends to infinity, then the series $\Sigma\, u_n$ is said to be convergent and s is said to be its *sum*.

EXAMPLE 9.11 Consider the series $\sum_{n=0}^{\infty} \frac{1}{2^n} = 1 + \frac{1}{2} + \frac{1}{2^2} + \frac{1}{2^3} + \cdots + \frac{1}{2^n} + \cdots$.

Here,
$$S_n = \frac{1\left(1 - \frac{1}{2^n}\right)}{1 - \frac{1}{2}} = 2\left(1 - \frac{1}{2^n}\right)$$

$$\lim_{n\to\infty} S_n = \lim_{n\to\infty} 2\left(1 - \frac{1}{2^n}\right) = 2.$$

Hence, the series $\sum_{n=0}^{\infty} \frac{1}{2^n}$ is convergent and 2 is its sum.

EXAMPLE 9.12 Consider the infinite geometric series $1 + r + r^2 + \cdots + r^n + \cdots$

Here, $S_n = \dfrac{1 - r^n}{1 - r}$, $r \neq 1$; $S_n = n$, where $r = 1$.

We examine $\lim_{n\to\infty} S_n$ for different values of r:

If $r = 1$, then $S_n = n$ which tends to ∞ as n tends to ∞.

If $|r| < 1$, then $\lim_{n\to\infty} r_n = 0$, therefore $\lim_{n\to\infty} S_n = \dfrac{1}{1 - r}$;

If $r > 1$, then $\lim_{n\to\infty} r^n = \infty$, thus $\lim_{n\to\infty} S_n = \infty$;

If $r \leq -1$, then $\lim_{n\to\infty} r^n$ and $\lim_{n\to\infty} S_n$ do not exist.

Hence, we see that the infinite geometric series Σr^n converges if and only if $|r| < 1$ and the sum of the series is $\frac{1}{1 - r}$.

9.3.1 Taylor and Maclaurin Series

Here, we further generalize the concept of Taylor Polynomial to obtain an infinite power series from a given function f.

? DEFINITION: Let f be a function having derivatives of all orders throughout some interval containing a as an interior point. Then the 'Taylor series' generated by f at $x = a$ is the power series

$$f(a) + \frac{f'(a)}{1!}(x - a) + \frac{f''(a)}{2!}(x - a)^2 + \cdots + \frac{f^n(a)}{n!}(x - a)^n + \cdots$$

The Taylor series generated by f at $x = 0$ is called the 'Maclaurin series' generated by f. Thus the Maclaurin series generated by f is the series

$$f(0) + \frac{f'(0)}{1!}x + \frac{f''(0)}{2!}x^2 + \cdots + \frac{f^n(0)}{n!}x^n + \cdots$$

EXAMPLE 9.13 Find the Taylor series generated by $f(x) = \dfrac{1}{x}$ at $x = 2$. When does this series converge to $\dfrac{1}{x}$?

SOLUTION

We need to find $f(2)$, $f'(2)$, $f''(2)$, $f'''(2)$, ...

Now, $f(x) = \dfrac{1}{x}$, $f'(x) = -\dfrac{1}{x^2}$, $f''(x) = (-1)^2\dfrac{2!}{x^3}$, $f'''(x) = \dfrac{(-1)^3 3!}{x^4}$, ...

Thus, $f(2) = \frac{1}{2}, f'(2) = -\frac{1}{2^2}, f''(2) = \frac{2!}{2^3}, f'''(2) = -\frac{3!}{2^4}, \cdots$

Therefore, the Taylor series for $f(x) = \frac{1}{x}$ at $x = 2$ is

$$f(2) + f'(2)(x-2) + \frac{f''(2)}{2!}(x-2)^2 + \frac{f'''(2)}{3!}(x-2)^3 + \frac{f^4(2)}{4!}(x-2)^4 + \cdots$$

$$= \frac{1}{2} + \left(-\frac{1}{2^2}\right)(x-2) + \left(\frac{2!}{2^3}\right)\frac{(x-2)^2}{2!} + \left(-\frac{3!}{2^4}\right)\frac{(x-2)^3}{3!} + \left(\frac{4!}{2^5}\right)\frac{(x-2)^4}{4!} + \cdots$$

$$= \frac{1}{2} - \frac{x-2}{2^2} + \frac{(x-2)^2}{2^3} - \frac{(x-2)^3}{2^4} + \frac{(x-2)^4}{2^5} - \cdots$$

$$= \frac{1}{2}\left[1 - \left(\frac{x-2}{2}\right) + \left(\frac{x-2}{2}\right)^2 - \left(\frac{x-2}{2}\right)^3 + \left(\frac{x-2}{2}\right)^4 - \cdots\right]$$

This is a geometric series with common ratio $r = -\frac{x-2}{2}$. It converges for $|r| < 1$, that is, for $0 < x < 4$ and its sum is $\frac{1}{2} \cdot \frac{1}{1 + \left(\frac{x-2}{2}\right)} = \frac{1}{x}$.

EXAMPLE 9.14 Find the Taylor series generated by $f(x) = e^x$ at $x = 0$ and at $x = a$.

SOLUTION

Since

$$f(x) = e^x, f'(x) = e^x, f''(x) = e^x, f'''(x) = e^x, \ldots, f^n(x) = e^x, \ldots$$

we have,

$$f(0) = 1, f'(0) = 1, f''(0) = 1, \ldots, f^n(0) = 1, \ldots$$

$$f(a) = e^a, f'(a) = e^a, f''(a) = e^a, f''(a) = e^a, \ldots, f^n(a) = e^a, \ldots$$

Therefore, the Taylor series generated by f at $x = 0$ is

$$f(0) + f'(0) + \frac{f'(0)}{2!}x^2 + \frac{f'''(0)}{3!}x^3 + \cdots + \frac{f^n(0)}{n!}x^n + \cdots$$

$$= 1 + x + \frac{x^2}{2!} + \frac{x^3}{3!} + \cdots + \frac{x^n}{n!} + \cdots$$

$$= \sum_{n=0}^{\infty} \frac{x^n}{n!}$$

Similarly, the Taylor series of $f(x) = e^x$ at $x = a$ is given by

$$f(a) + f'(a)(x-a) + f''(a)\frac{(x-a)^2}{2!} + f'''(a)\frac{(x-1)^3}{3!} + \cdots + f^n(a)\frac{(x-a)^n}{n!} + \cdots$$

$$= e^a\left[1 + (x-a) + \frac{(x-a)^2}{2!} + \frac{(x-a)^4}{4!} + \cdots + \frac{(x-a)^n}{n!} + \cdots\right].$$

REMARK: (i) By definition, the first series in Example 9.14 is also the Maclaurin series for e^x.

(ii) The series generated by $f(x)$ need not converge to $f(x)$. For example, take the function $f(x)$ defined by

$$f(x) = \begin{cases} 0, & x = 0 \\ e^{-1/x^2}, & x \neq 0 \end{cases}.$$

Here, f has derivative of every order at $x = 0$ and $f^n(0) = 0$ for all n. This means that the Taylor's series generated by f at $x = 0$ is

$$f(0) + f'(0) \cdot x + \frac{f''(0)}{2!} \cdot x^2 + \cdots + \frac{f^n(0)}{n!} \cdot x^n + \cdots$$

$$= 0 + 0 \cdot x + 0 \cdot x^2 + \cdots + 0 \cdot x^n + \cdots = 0$$

Thus, this series converges for every x (its sum is 0) but it converges to $f(x)$ only at $x = 0$.

EXAMPLE 9.15 Find the Taylor series generated by $f(x) = \sin x$ at $x = 0$.

SOLUTION

Here

$$f(x) = \sin x,\ f'(x) = \cos x,\ f''(x) = -\sin x,\ f'''(x) = -\cos x, \ldots,$$
$$f^n(x) = (-1)^n \sin x,\ f^{2n+1}(x) = (-1)^n \cos x$$

At $x = 0$, we have

$f(0) = 0,\ f'(0) = 1,\ f''(0) = 0,\ f'''(0) = -1,\ \ldots,\ f^{2n}(0) = 0,\ f^{2n+1}(0) = (-1)^n, \ldots$

Therefore, the Taylor series generated by $f(x) = \sin x$ at $x = 0$ is

$$f(0) + f'(0).x + \frac{f''(0)}{2!}x^2 + \frac{f'''(0)}{3!}x^3 + \cdots + \frac{f^{2n}(0)}{(2n)!}x^{2n} + \frac{f^{2n+1}(0)}{(2n+1)!}x^{2n+1} + \cdots$$

$$= x - \frac{x^3}{3!} + \frac{x^5}{5!} - \cdots + (-1)^n \frac{x^{2n+1}}{(2n+1)!} + \cdots$$

$$= \sum_{n=0}^{\infty} (-1)^n \frac{x^{2n+1}}{(2n+1)!}$$

Again, as in Example 14, this also is the Maclaurin series for the function $f(x) = \sin x$.

EXERCISE 9.3

In Questions 1–10, find the Taylor series generated by f at $x = a$.

1. $f(x) = 2x^4 + x^3 - 3x + 7, \quad a = 1$
2. $f(x) = x^3 - 2x + 4, \quad a = 2$
3. $f(x) = 3x^4 - 2x^3 + x - 4, \quad a = -1$
4. $f(x) = x^4 + x^2 + 1, \quad a = -2$
5. $f(x) = 3^x, \quad a = -1$

6. $f(x) = e^x, \quad a = 1, 2$

7. $f(x) = e^{-x}, \quad a = \ln 2$

8. $f(x) = \frac{1}{x}, \quad a = -1$

9. $f(x) = \sin \pi x, \quad a = \frac{1}{2}$

10. $f(x) = \ln x, \quad a = 1$

In Questions 11–18, find the Maclaurin series generated by the function *f*.

11. $f(x) = e^{-2x}$

12. $f(x) = \cos \pi x$

13. $f(x) = \sin 3x$

14. $f(x) = \frac{1}{1 + x}$

15. $f(x) = 9\cos(-x)$

16. $f(x) = \cos hx \qquad \left[\text{Hint: } \cos hx = \frac{e^x + e^{-x}}{2}\right]$

17. $f(x) = \sin hx$

18. $f(x) = x \sin x$

19. $f(x) = x^5 - 7x^3 + 2x^2 - 4$

20. $f(x) = (x - 1)^3$

■ 9.4 CONVERGENCE OF TAYLOR SERIES

In the example of $f(x) = \frac{1}{x}$ of the previous section, we have seen that the Taylor series generated by the function $f(x) = \frac{1}{x}$ at $x = 2$ converges to $f(x)$ if x lies in the interval $(0, 4)$. That is, we indeed have,

$$\frac{1}{x} = \frac{1}{2}\left[1 - \left(\frac{x-2}{2}\right) + \left(\frac{x-2}{2}\right)^2 - \left(\frac{x-2}{2}\right)^3 + \left(\frac{x-2}{2}\right)^4 - \cdots\right]$$

provided, $0 < x < 4$.

Here, the question arises: given a function *f*, that has derivatives of all orders at $x = a$, does there exist an open interval containing a, such that $f(x)$ is the sum of its Taylor series at $x = a$ for each x in that interval? In other words, when can we have,

$$f(x) = f(a) + \frac{f'(a)}{1!}(x - a) + \frac{f''(a)}{2!}(x - a)^2 + \cdots$$

The answer to this question may be obtained by invoking Taylor's Theorem provided in Section 9.2. Taylor's Theorem states, that if a function f possesses derivatives upto $(n + 1)$-order in an interval I and $a \in \mathrm{I}$ then, for every $x \in \mathrm{I}$,

$$f(x) = \mathrm{P}_n(x) + \mathrm{R}_n(x)$$

where $\mathrm{P}_n(x)$ is the nth order Taylor polynomial at $x = a$ and $\mathrm{R}_n(x)$ is the remainder of order n.

First let us assume that f possesses derivatives of all orders throughout the interval I (which may be open or closed), so that we may talk of the Taylor series of f at $x = a$. Then the Taylor polynomial $\mathrm{P}_n(x)$ is nothing but S_n $(n = 0, 1, 2, \ldots)$ forming the sequence of partial sums for the Taylor series of f at $x = a$. Since $f(x) = \mathrm{P}_n(x) + \mathrm{R}_n(x)$, in view of the Taylor's Theorem, the

Taylor Series at $x = a$ converges to $f(x)$ if $R_n(x)$ tends to zero as n tends to infinity. Thus, if $R_n(x) \to 0$ as $n \to \infty$ for all x in the interval, we say that the Taylor series generated by f at $x = a$ converges to $f(x)$ and we write,

$$f(x) = \sum_{k=0}^{\infty} \frac{f^k(a)}{k!}(x - a)^k$$

If $0 \in I$, then by taking $a = 0$ in the above case, we say that the Maclaurin series generated by f converges to f and write it as

$$f(x) = \sum_{k=0}^{\infty} \frac{f^k(0)}{k!} x^k$$

EXAMPLE 9.16 Show that the Maclaurin's series for $\cos x$ converges to $\cos x$ for all x; that is

$$\cos x = \sum_{k=0}^{\infty} (-1)^k \frac{x^{2k}}{(2k)!} = 1 - \frac{x^2}{2!} + \frac{x^4}{4!} - \frac{x^6}{6!} + \cdots \qquad (-\infty < x < \infty)$$

SOLUTION

The function $f(x) = \cos x$, possesses derivatives of all orders in the interval $(-\infty, \infty)$ and for all x,

$$f^{n+1}(x) = \pm \cos x \qquad \text{or} \qquad f^{n+1}(x) = \pm \sin x$$

Thus, we have, $|f^{n+1}(x)| \leq 1$. Now, by Taylor's Theorem with $a = 0$ and Lagrange's form of remainder,

$$f(x) = P_n(x) + R_n(x), \qquad \text{where } R_n(x) = \frac{f^{n+1}(c)}{(n + 1)!} x^{n+1}$$

for some c between 0 and x. Therefore, we have

$$0 \leq |R_n(x)| \leq \frac{|x|^{n+1}}{(n + 1)!}.$$

Since $\lim_{n \to \infty} \frac{x^n}{n!} = 0$, it follows that $R_n(x) \to 0$ as $n \to \infty$. Hence, we have

$$f(x) = f(0) + f'(0)x + \frac{f''(0)}{2!}x^2 + \frac{f'''(0)}{3!}x^3 + \frac{f^4(0)}{4!}x^4 + \cdots$$

That is,

$$\cos x = 1 - \frac{x^2}{2!} + \frac{x^4}{4!} - \cdots$$

REMARK: In the same way, we can prove that the Maclaurin series for $\sin x$ converges to $\sin x$ for all x in $(-\infty, \infty)$. That is

$$\sin x = x - \frac{x^3}{3!} + \frac{x^5}{5!} - \cdots$$

EXAMPLE 9.17 Use Maclaurin series for $\sin x$ to approximate $\sin 3^\circ$ to five decimal-place accuracy.

SOLUTION

The function $f(x) = \sin x$ possesses derivatives of all orders, and for all x in $(-\infty, \infty)$, we have

$$f^n(x) = \pm \sin x \qquad \text{or} \qquad f^n(x) = \pm \cos x$$

We take $x = \frac{\pi}{60}$, the radian measure for 3°, as the differentiation formulas for the trigonometric functions are valid only for radian x. Thus, we have, on applying Taylor's Theorem with $x_0 = 0$, $f(x) = \mathrm{P}_n(x) + \mathrm{R}_n(x)$, that is

$$f(x) = f(0) + f'(0)x + \frac{f''(0)}{2!}x^2 + \cdots + \frac{f^n(0)}{n!}x^n + \mathrm{R}_n(x)$$

$$\sin 3^\circ = \sin\frac{\pi}{60} = \frac{\pi}{60} - \frac{(\pi/60)^3}{3!} + \frac{(\pi/60)^5}{5!} - \frac{(\pi/60)^7}{7!} + \cdots + \mathrm{R}_n(\pi/60)$$

We must now determine how many terms in the series are required to achieve five decimal place accuracy. That is, we need to choose n, satisfying

$$\left|\mathrm{R}_n\left(\frac{\pi}{60}\right)\right| < .000005.$$

Now,

$$\mathrm{R}_n(x) = f^{n+1}(c)\,\frac{x^{n+1}}{(n+1)!} \qquad \text{(Lagrange's form of remainder)}$$

for some c between 0 and x, so that

$$\left|\mathrm{R}_n\left(\frac{\pi}{60}\right)\right| \le \frac{(\pi/60)^{n+1}}{(n+1)!}$$

Therefore, we choose n satisfying $\frac{(\pi/60)^{n+1}}{(n+1)!} \le .000005.$

With the help of a calculating device, it can be verified that the smallest value of n that meets this criterion is $n = 3$. Thus, for the values of $\sin 3^\circ$ upto five places of accuracy is

$$\sin 3^\circ \approx \frac{\pi}{60} - \frac{(\pi/60)^3}{3!} \approx 0.05234$$

REMARK: In the above example, we need not use the concept of $\mathrm{R}_n(x) \to 0$ as $n \to \infty$. Thus, it was in fact a problem of Section 9.2, except for the word 'Maclaurin series' used in the statement.

EXAMPLE 9.18 Show that the Macluarin series for e^x converges to e^x for all x; that is

$$e^x = \sum_{k=0}^{\infty}\frac{x^k}{k!} = 1 + x + \frac{x^2}{2!} + \frac{x^3}{3!} + \cdots + \frac{x^k}{k!} + \cdots \qquad (-\infty < x < \infty)$$

SOLUTION

Let $f(x) = e^x$, so that $f^{n+1}(x) = e^x$ for all x in $(-\infty, \infty)$.
We need to show that $\mathrm{R}_n(x) \to 0$ as $n \to \infty$ for all x in the interval $(-\infty, \infty)$, in the Taylor's formula

$$f(x) = P_n(x) + R_n(x),$$

where $a = 0$ and

$$R_n(x) = \frac{f^{n+1}(x)}{(n+1)!}x^{n+1} = e^c \cdot \frac{x^{n+1}}{(n+1)!}, \qquad \text{(Lagrange's form of remainder)}$$

for some c between 0 and x.

It follows, from the formula $\lim_{n\to\infty} \frac{x^n}{n!} = 0$ that $R_n(x) \to 0$ as $n \to \infty$. Thus, we have,

$$f(x) = f(0) + \frac{f'(0)}{1!}x + \frac{f''(0)}{2!}x^2 + \frac{f'''(0)}{3!} + \cdots$$

That is,

$$e^x = 1 + x + \frac{x^2}{2!} + \frac{x^3}{3!} + \cdots$$

EXAMPLE 9.19 Expand $\ln(1 + x)$ as a power series of x, $-1 \le x \le 1$.

SOLUTION

The function $f(x) = \ln(1 + x)$ has derivatives of every order for $x > -1$ and $f^{n+1}(x) = \frac{(-1)^n n!}{(1+x)^{n+1}}$.

Therefore, by Taylor's formula with $a = 0$, $f(x) = P_n(x) + R_n(x)$, for every $x > -1$. However, the remainder $R_n(x)$ tends to zero only for $-1 < x \le 1$. Thus, the Maclaurin series for $\ln(1 + x)$ converges only for $-1 < x \le 1$. We explain below how $R_n(x) \to 0$ as $n \to \infty$ for different values of x.

Case I: $0 \le x \le 1$. In this case, we consider the Lagrange's form of remainder, that is,

$$\begin{aligned} R_n(x) &= f^{n+1}(c) \cdot \frac{x^{n+1}}{(n+1)!}, \qquad \text{for some c between 0 and } x. \\ &= \frac{(-1)^n \cdot n!}{(1+c)^{n+1}} \cdot \frac{x^{n+1}}{(n+1)!} \\ &= (-1)^n \cdot \frac{1}{n+1} \cdot \left(\frac{x}{1+c}\right)^{n+1} \end{aligned}$$

Since, $0 \le \frac{x}{1+c} < 1$; $\left(\frac{x}{1+c}\right)^{n+1} \to 0$ as $n \to \infty$.

Thus, $R_n \to 0$ as $n \to \infty$.

Case II: $-1 < x < 0$: In this case, we consider Cauchy's form of remainder, that is,

$$\begin{aligned} R_n(x) &= \frac{f^{n+1}(\theta x)}{n!}(1-\theta)^n x^{n+1}, \quad \text{where } \theta \text{ is some number lying between 0 and 1.} \\ &= \frac{(-1)^n}{(1+\theta x)^{n+1}} \cdot (1-\theta)^n \cdot x^{n+1} \\ &= (-1)^n \cdot x^{n+1} \cdot \left(\frac{1-\theta}{1+\theta x}\right)^n \cdot \frac{1}{1+\theta x} \end{aligned}$$

Now, $0 < 1 - \theta < 1 + \theta x$ so that $0 < \frac{1-\theta}{1+\theta x} < 1$; hence $\left(\frac{1-\theta}{1+\theta x}\right)^n \to 0$ as $n \to \infty$. Also, $x^n \to 0$ as $n \to \infty$. Thus,

$$R_n(x) \to 0 \text{ as } n \to \infty.$$

Therefore, $R_n(x) \to 0$ as $n \to \infty$ for $-1 < x \leq 1$; so the Maclaurin series converges to $f(x)$ for $-1 < x \leq 1$. Hence,

$$f(x) = f(0) + f'(0)x + \frac{f''(0)}{2!}x^2 + \frac{f'''(0)}{3!}x^3 + \cdots$$

Now, $f(0) = 0$ and $f^n(0) = (-1)^{n-1}(n-1)!$. Thus, we have

$$\ln(1+x) = x - \frac{x^2}{2} + \frac{x^3}{3} - \frac{x^4}{4} + \cdots, \qquad -1 < x \leq 1$$

REMARK: (i) The Lagrange's form of remainders in the above example involve the terms $\left(\frac{x}{1+c}\right)^{n+1}$ and x^{n+1} respectively, which do not converge to finite limits as $n \to \infty$, if $x > 1$.
(ii) The Cauchy's form of remainder mentioned in the Taylor's Theorem is

$$\frac{f^{n+1}(c)}{n!}(x-c)^n(x-a).$$

For $a = 0$ and writing c as θx (where $0 < \theta < 1$, so that $0 < \theta x < x$) we get Cauchy's form of remainder also as

$$\frac{f^{n+1}(\theta x)}{n!}(1-\theta)^n \cdot x^{n+1}.$$

EXAMPLE 9.20 Find the Maclaurin series for $\cos 2x$.

SOLUTION

We have already shown that the Macluarin series for $\cos x$ converges to $\cos x$ for all values of x in $(-\infty, \infty)$. That is,

$$\cos x = 1 - \frac{x^2}{2!} + \frac{x^4}{4!} - \frac{x^6}{6!} + \cdots$$

Substituting $2x$ for x in the above series, we get Macluarin series for $\cos 2x$ as

$$\cos 2x = 1 - \frac{(2x)^2}{2!} + \frac{(2x)^4}{4!} - \frac{(2x)^6}{6!} + \cdots$$

which holds for $-\infty < 2x < \infty$, so the newly created series converges for all x.

EXAMPLE 9.21 Find the Macluarin series for $x \sin x$.

SOLUTION

We can find Maclaurin series for $x \sin x$ by multiplying the Maclaurin series for $\sin x$ by x:

$$x \sin x = x\left(x - \frac{x^3}{3!} + \frac{x^5}{5!} - \frac{x^7}{7!} + \cdots\right)$$
$$= x^2 - \frac{x^4}{3!} + \frac{x^6}{5!} - \frac{x^8}{7!} + \cdots$$

The new series converges for $-\infty < x < \infty$ because the series for $\sin x$ converges for all x.

EXAMPLE 9.22 Find the Maclaurin series for $e^{-x/2}$.

SOLUTION

The Maclaurin series for e^x converges to e^x. That is,

$$e^x = 1 + x + \frac{x^2}{2!} + \frac{x^3}{3!} + \cdots$$

Substituting $-\frac{x}{2}$ for x in the above series, we get

$$e^{-x/2} = 1 - \frac{x}{2} + \left(-\frac{x}{2}\right)^2 \frac{1}{2!} + \left(-\frac{x}{2}\right)^3 \frac{1}{3!} + \cdots$$

$$= 1 - \frac{x}{2} + \frac{x^2}{2^2 \cdot 2!} - \frac{x^3}{2^3 \cdot 3!} + \cdots + (-1)^n \frac{x^n}{2^n \cdot n!} + \cdots$$

EXAMPLE 9.23 Find the Maclaurin series for $(1 + x)^m$.

SOLUTION

The function $f(x) = (1 + x)^m$ has derivatives of every order. In fact, we have

$$f'(x) = m(1 + x)^{m-1}, f''(x) = m(m - 1)(1 + x)^{m-2}, \ldots,$$
$$f^{k}(x) = m(m - 1) \ldots (m - k + 1)(1 + x)^{m-k}$$

In case m is a non-negative integer, then $f^n(x) = 0$ for all $n > m$. Thus $R_n(x) = 0$ whenever $n \geq m$. Therefore, by taking $f^n(0) = 0$ for $n > m$ in Taylor's Theorem $f(x) = P_n(x) + R_n(x)$ with Lagrange's form of remainder, and letting n tend to ∞, we get

$$f(x) = f(0) + \frac{f'(0)}{1!}x + \frac{f''(0)}{2!}x^2 + \cdots + \frac{f^m(0)}{m!}x^m + 0.$$

That is,

$$(1 + x)^m = 1 + m + \frac{m(m - 1)}{2!}x^2 + \cdots + \frac{m(m - 1)\cdots 2.1}{m!}x^m,$$

whenever m is a non-negative integer, and $-\infty < x < \infty$.

We now discuss the case where m is not a non-negative integer. It can be proved that if m is not a non-negative integer, then the Macluarin series of $(1 + x)^m$, which is also called the binomial series, converges for $|x| < 1$. The Cauchy's form of remainder is

$$R_n(x) = \frac{f^{n+1}(\theta x)}{n!}(1 - \theta)^n x^{n+1}, \quad \text{where } \theta \text{ is some number lying between 0 and 1,}$$

$$= \frac{m(m - 1)(m - 2)\cdots(m - n)}{n!}(1 + \theta x)^{m-n-1} \cdot (1 - \theta)^n \cdot x^{n+1}$$

$$= \frac{m(m - 1)(m - 2)\cdots(m - n)}{n!}\left(\frac{1 - \theta}{1 + \theta x}\right)^n \cdot (1 + \theta x)^{m-1} \cdot x^{n+1}$$

We observe that

(i) $\lim\limits_{n\to\infty} \frac{m(m - 1)\cdots(m - n)}{n!}x^{n+1} = 0$

In fact, by taking $u_n = \frac{m(m - 1)\cdot\cdots\cdot(m - n)}{n!}x^{n+1}$, we get

$$\frac{u_{n+1}}{u_n} = \frac{m-n-1}{n+1}x = \left(\frac{m}{n+1} - 1\right)x$$

so that $$\frac{u_{n+1}}{u_n} \to -x.$$

Now, from the result "If $\lim_{n\to\infty} \frac{u_{n+1}}{u_n} = l$ and $|l| < 1$, then $u_n \to 0$", it follows that for

$$|x| < 1, \lim_{n\to\infty} \frac{m(m-1)\cdots(m-n)}{n!}x^{n+1} = 0$$

(ii) $\lim_{n\to\infty}\left(\frac{1-\theta}{1+\theta x}\right)^n = 0$

This follows from the facts that $0 < \theta < 1$ and $-1 < x < 1$ so that $0 < \frac{1-\theta}{1+\theta x} < 1$.

(iii) $(1+\theta x)^{m-1}$ remains unchanged as n tends to ∞.

Thus, it follows that for $|x| < 1$,

$$\lim_{n\to\infty} R_n(x) = 0.$$

Hence, we have,

$$f(x) = f(0) + \frac{f'(0)}{1!}x + \frac{f''(0)}{2!}x^2 + \frac{f'''(x)}{3!}x^3 + \cdots$$

That is,

$$(1+x)^m = 1 + \frac{m}{1!}x + \frac{m(m-1)}{2!}x^2 + \frac{m(m-1)(m-2)}{3!}x^3 + \cdots, \qquad \text{for } |x| < 1.$$

EXAMPLE 9.24 Find the binomial series for

(i) $\frac{1}{(1+x)^2}$ (ii). $\frac{1}{\sqrt{1+x}}$ (iii) $\sqrt{1+x}$

SOLUTION

From the binomial series

$$(1+x)^m = 1 + \frac{m(m-1)}{2!}x^2 + \frac{m(m-1)(m-2)}{3!}x^3 + \frac{m(m-1)(m-2)(m-3)}{4!}x^4 + \cdots$$

it follows that

(i) $$\frac{1}{(1+x)^2} = (1+x)^{-2}$$

$$= 1 - 2x + \frac{(-2)(-3)}{2!}x^2 + \frac{(-2)(-3)(-4)}{3!}x^3 + \frac{(-2)(-3)(-4)(-5)}{4!}x^4 + \cdots$$

$$= 1 - 2x + 3x^2 - 4x^3 + 5x^4 - \cdots$$

(ii) $$\frac{1}{\sqrt{1+x}} = (1+x)^{-1/2} = 1 + \left(-\frac{1}{2}\right)x + \frac{\left(-\frac{1}{2}\right)\left(-\frac{1}{2}-1\right)}{2!}x^2$$

$$+ \frac{\left(-\frac{1}{2}\right)\left(-\frac{1}{2}-1\right)\left(-\frac{1}{2}-1\right)}{3!}x^3 + \cdots$$

$$= 1 - \frac{1}{2}x + \frac{1 \cdot 3}{2^2 \cdot 2!}x^2 - \frac{1 \cdot 3 \cdot 5}{2^3 \cdot 3!}x^3 + \cdots$$

(iii) $$\sqrt{1+x} = (1+x)^{1/2} = 1 + \frac{1}{2} \cdot x + \frac{\frac{1}{2}\left(\frac{1}{2} - 1\right)}{2^2}x^2 + \frac{\frac{1}{2}\left(\frac{1}{2} - 1\right)\left(\frac{1}{2} - 2\right)}{3!}x^3 + \cdots$$

$$= 1 + \frac{1}{2}x - \frac{1}{8}x^2 + \frac{1}{16}x^3 + \cdots$$

EXERCISE 9.4

Find the Maclaurin Series for the functions given in Questions 1–8.

1. $f(x) = x\cos \pi x$

2. $f(x) = \sin^2 x$ [Hint: $\sin^2 x = \frac{1}{2}(1 - \cos 2x)$]

3. $f(x) = \dfrac{x^2}{1 - 2x}$

4. $f(x) = \sin\left(\dfrac{\pi}{2}x\right)$

5. $f(x) = \cos^2 x$

6. $f(x) = x\ln(1 + 3x)$

7. $f(x) = \dfrac{x^2}{2} - 1 + \cos x$

8. $f(x) = \sin x - x + \dfrac{x^3}{3!}$

In Questions 9–13, Maclaurin series of a function at some point is provided. Identify the function and the point.

9. $(0.1) - \dfrac{(0.1)^3}{3!} + \dfrac{(0.1)^5}{5!} + \cdots + \dfrac{(-1)^{n-1}(0.1)^{2n-1}}{(2n-1)!} + \cdots$

10. $1 - \dfrac{(.01)^2}{2!} + \dfrac{(.01)^4}{4!} + \cdots + \dfrac{(-1)^{n-1}(.01)^{2n-2}}{(2n-2)!} + \cdots$

11. $1 - \dfrac{\pi^2}{2^2 \cdot 2!} + \dfrac{\pi^4}{2^4 \cdot 4!} + \cdots + \dfrac{(-1)^{n-1}(\pi)^{2n-2}}{2^{2n-2} \cdot (2n-2)!} + \cdots$

12. $.2 - \dfrac{(.2)^2}{2} + \dfrac{(.2)^3}{3} - \dfrac{(.2)^4}{4} + \cdots + \dfrac{(-1)^{n-1}(.2)^n}{n} + \cdots$

13. $1 + (.2) + \dfrac{(.2)^2}{2!} + \dfrac{(.2)^3}{3!} + \cdots + \dfrac{(.2)^n}{n!} + \cdots$

14. $\dfrac{\pi}{3} - \dfrac{\pi^3}{3^3 \cdot 3!} + \dfrac{\pi^5}{3^5 \cdot 5!} - \cdots + \dfrac{(-1)^n \pi^{2n+1}}{3^{2n+1} \cdot (2n+1)!} + \cdots$

CHAPTER TEN

Functions of Two Variables

10.1 FUNCTIONS OF TWO OR MORE VARIABLES

There are many familiar formulas in which a given variable depends on two or more other variables. For example, the area A of a triangle depends on the base length b and height h by the formula $A = \frac{1}{2}bh$; the volume of a rectangular box depends on the length l, the width w and the height h by the formula $V = lwh$; and the arithmetic average $\bar{x}$ of n real numbers, $x_1, x_2, \ldots, x_n$ depends on those numbers by the formula $\bar{x} = (x_1 + x_2 + \cdots + x_n)/n$. Thus, we may say that

A is a function of the two variables, b and h;

V is a function of the three variables l, w and h;

$\bar{x}$ is a function of the n variables $x_1, x_2, \ldots, x_n$.

The terminology and notation for function of two or more variables is similar to that of functions of one variable. For example, the expression

$$z = f(x, y)$$

means that z is a function of x and y in the sense that a unique value of the dependent variable z is determined by specifying the values for the independent variables x and y. Similarly, $w = f(x, y, z)$ expresses w as a function of x, y and z and $u = f(x_1, x_2, \ldots, x_n)$ expresses u as a function of $x_1, x_2, \ldots, x_n$.

As with functions of one variable, the independent variables of a function of two or more variables may be restricted to lie in some set D, which we call the 'domain' of f. Sometimes the domain will be determined by the physical restrictions on the variables. If the function is defined by a formula and if there are no physical restrictions or other restrictions stated explicitly, then it is understood that the domain consists of all points for which the formula yields a real value for the dependent variable. We call this the 'natural domain' of the function. The following definition summarizes this discussion.

> **? DEFINITION:** A function of two variables x and y is a rule that assigns a unique real number $f(x, y)$ to each point (x, y) in some set D in the xy-plane.

EXAMPLE 10.1 Let $f(x, y) = x^2\sqrt{y} + 2x + 1$. Find $f(0, 0), f(-1, 1), f(t^2, t), f(2a, b)$. Also find the natural domain of f.

SOLUTION

By substitution

$$
\begin{aligned}
f(0, 0) &= 0^2\sqrt{0} + 2 \cdot 0 + 1 = 1 \\
f(-1, 1) &= (-1)^2\sqrt{1} + 2(-1) + 1 = 0 \\
f(t^2, t) &= (t^2)^2\sqrt{t} + 2t^2 + 1 = t^{9/2} + 2t^2 + 1 \\
f(2a, b) &= (2a)^2\sqrt{b} + 2(2a) + 1 = 4a^2\sqrt{b} + 4a + 1
\end{aligned}
$$

The natural domain of f consists of all those points in the xy-plane for which $f(x, y)$ is real. This restricts the choice of y, that we must have $y \geq 0$. Thus the natural domain of f consists of all points in the xy-plane that are on or above the x-axis.

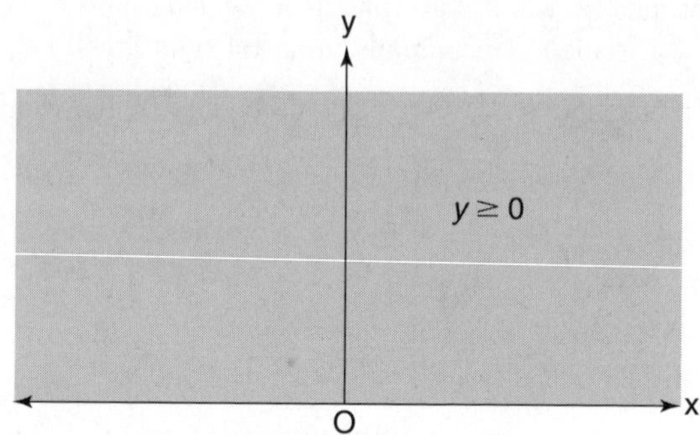

Fig. 10.1

EXAMPLE 10.2 Sketch the natural domain of the function $f(x, y) = \log(y^2 - x)$.

SOLUTION

$\log(y^2 - x)$ is defined only when $y^2 - x > 0$, that is, for $y^2 > x$. We first sketch the parabola $y^2 = x$. The region $y^2 > x$ consists of all the points above and below this curve.

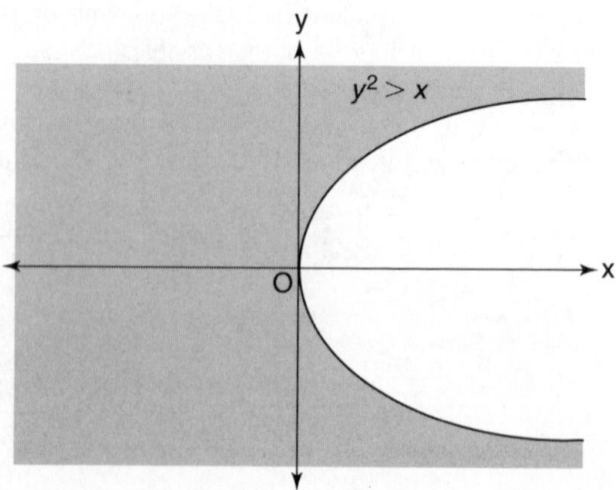

Fig. 10.2

10.1.1 Graphs and Level Curves of Functions of Two Variables

Recall that for a function f of one variable, the graph of $f(x)$ in the xy-plane was defined to be the graph of the equation $y = f(x)$. Similarly, if f is a function of two variables, we define the graph of $f(x, y)$ in the 3-dimensional xyz-space to be the 'graph' of the equation $z = f(x, y)$. In general, such a graph will be a surface in space.

EXAMPLE 10.3 Describe the graph of the following functions:

(a) $f(x, y) = 1 - x - 2y$ (b) $f(x, y) = \sqrt{4 - x^2 - y^2}$ (c) $f(x, y) = -\sqrt{x^2 + y^2}$

SOLUTION

(a) By definition, the graph of the given function is the graph of the equation

$$z = 1 - x - 2y$$

which, being a first degree equation, represents a plane. This plane will intersect the three axes at $A(1, 0, 0), B(0, 1/2, 0)$ and $C(0, 0, 1)$ respectively. The triangular portion ABC of the plane is shown in Fig. 10.3.

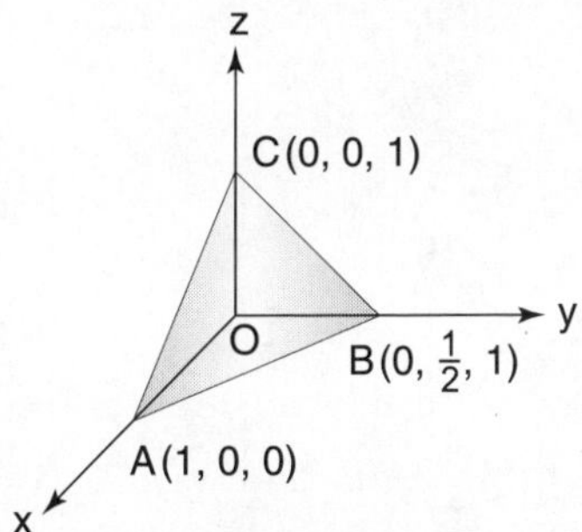

Fig. 10.3

(b) By definition, the graph of the given function is the graph of the equation

$$z = \sqrt{4 - x^2 - y^2}.$$

After squaring both sides this may be written as

$$x^2 + y^2 + z^2 = 4$$

which represents a sphere of radius 2, with origin as the centre. Since z can take only non-negative values in the original equation, therefore graph is the upper hemisphere of this sphere.

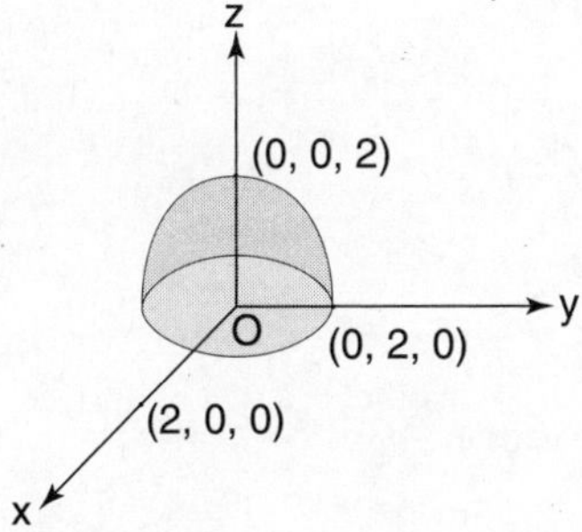

Fig. 10.4

(c) The required graph is the graph of the equation

$$z = -\sqrt{x^2 + y^2}$$

After squaring, we get

$$z^2 = x^2 + y^2$$

which is the equation of a circular cone. Since $z \leq 0$ in the original equation, the required graph is the lower half of this cone.

Before going for further discussion on function of two variables, first let us be familiar with the standard equation of a few common 3-dimensional surfaces.

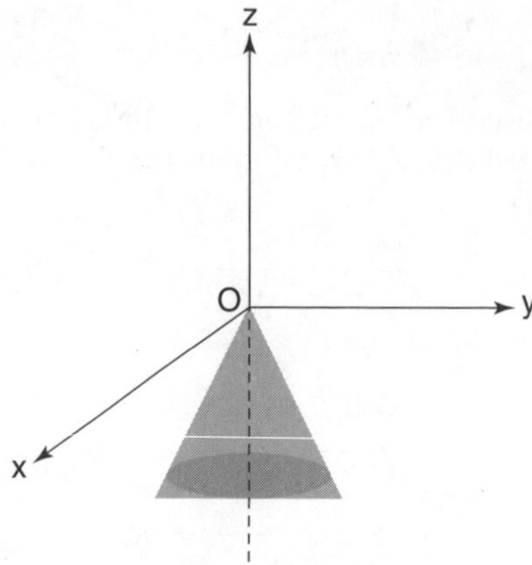

Fig. 10.5

Ellipsoid: Its equation is

$$\frac{x^2}{a^2} + \frac{y^2}{b^2} + \frac{z^2}{c^2} = 1 \qquad (1)$$

It is symmetric with respect to all the three co-ordinate planes, meets the axes at $(\pm a, 0, 0)$, $(0, \pm b, 0)$ and $(0, 0, \pm c)$; is bounded by the parallelopiped with surfaces $x = \pm a$, $y = \pm b$, $z = \pm c$.

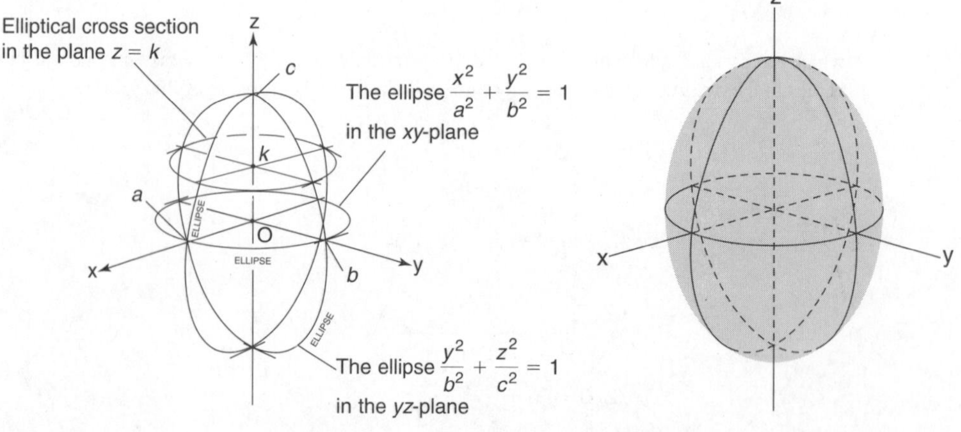

Fig. 10.6

The curves obtained by the intersection of the surface with the co-ordinate planes are

$$x = 0 \ : \ \text{the ellipse } \frac{y^2}{b^2} + \frac{z^2}{c^2} = 1;$$

$$y = 0 \ : \ \text{the ellipse } \frac{x^2}{a^2} + \frac{z^2}{c^2} = 1;$$

$$z = 0 \ : \ \text{the ellipse } \frac{x^2}{a^2} + \frac{y^2}{b^2} = 1;$$

Any plane $z = k(|k| < c)$ parallel to the xy-plane intersects the surface in an ellipse

$$\frac{x^2}{a^2\left(1 - \frac{k^2}{c^2}\right)} + \frac{y^2}{b^2\left(1 - \frac{k^2}{c^2}\right)} = 1.$$

If any two of the semi-axes a, b, c are equal, the surface is called an 'ellipsoid of revolution'. It may be noticed that if $b = c$, the surface may be generated by revolving the ellipse $\frac{x^2}{a^2} + \frac{y^2}{b^2} = 1$ about the x-axis. If all three are equal, it is a sphere.

Elliptic paraboloid: Its equation is

$$\frac{x^2}{a^2} + \frac{y^2}{b^2} = \frac{z}{c} \tag{2}$$

It is symmetric with respect to the co-ordinate planes $x = 0$ and $y = 0$ and meets the axes only at the origin. The surface lies entirely above or below the xy-plane, depending on the sign of c.

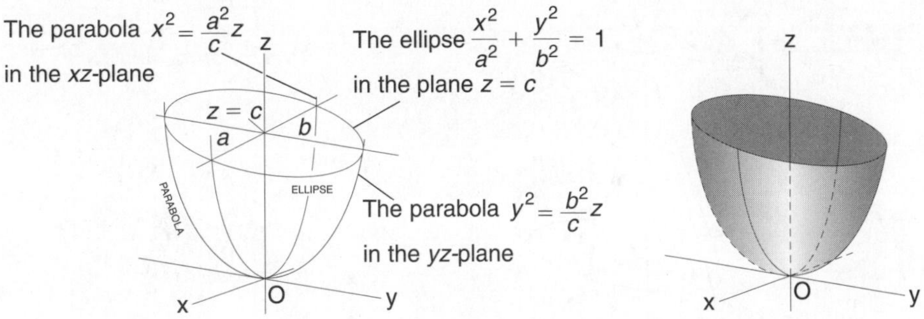

Fig. 10.7

The curves obtained by the intersection of the surface with the co-ordinate planes are

$$x = 0 \ : \ \text{the parabola } y^2 = \frac{b^2}{c}z;$$

$$y = 0 \ : \ \text{the parabola } x^2 = \frac{a^2}{c}z;$$

$$z = 0 \ : \ \text{the origin } (0, 0, 0).$$

Any plane $z = k$ parallel to xy-plane, intersects the surface in an ellipse

$$\frac{x^2}{ka^2/c} + \frac{y^2}{kb^2/c} = 1$$

If $a = b$ in (2) then we get the surface of an 'circular paraboloid', also called 'paraboloid of revolution'. Shapes cut out from a circular paraboloid are used for antennas in microwave transmissions, radio-telescopes, radars and so on.

The elliptic cone: Its equation is

$$\frac{x^2}{a^2} + \frac{y^2}{b^2} = \frac{z^2}{c^2} \tag{3}$$

It is symmetric with respect to all the three co-ordinate planes and meets the axes only at the origin. The curves obtained by its intersection with the co-ordinate planes are:

$$x = 0 \; : \; \text{the lines } z = \pm\frac{c}{b}y$$

$$y = 0 \; : \; \text{the lines } z = \pm\frac{c}{a}x$$

$$z = 0 \; : \; \text{the origin } (0, 0, 0).$$

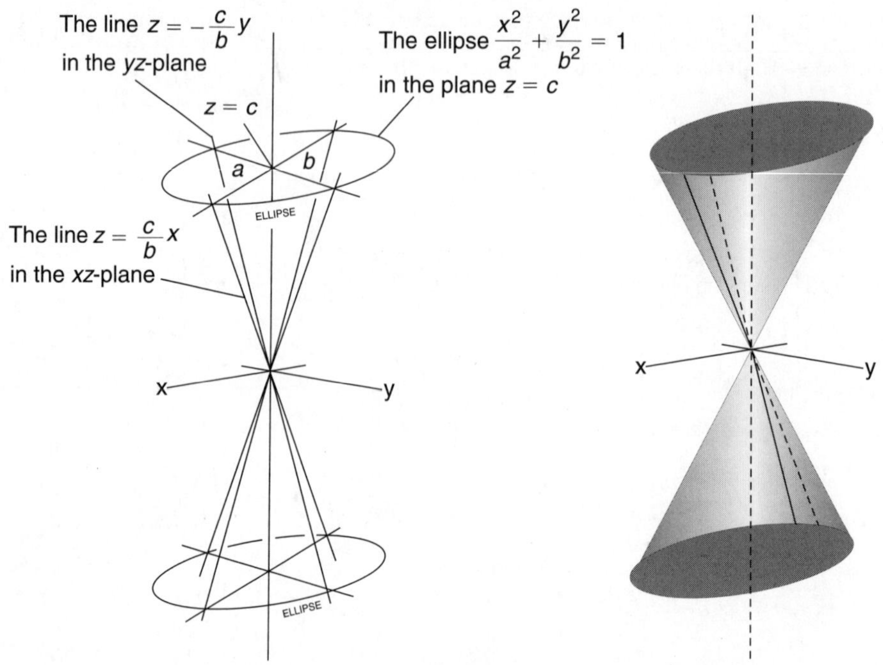

Fig. 10.8

Any plane $z = k$, parallel to the xy-plane intersects the surface in an ellipse

$$\frac{x^2}{a^2k^2/c^2} + \frac{y^2}{b^2k^2/c^2} = 1.$$

If $a = b$, the cone is a 'right circular cone'.

The hyperboloid of one sheet: Its equation is

$$\frac{x^2}{a^2} + \frac{y^2}{b^2} - \frac{z^2}{c^2} = 1 \tag{4}$$

It is symmetric with respect to all the three co-ordinate planes, meets the x-axis at $(\pm a, 0, 0)$ and the y-axis at $(0, \pm b, 0)$.

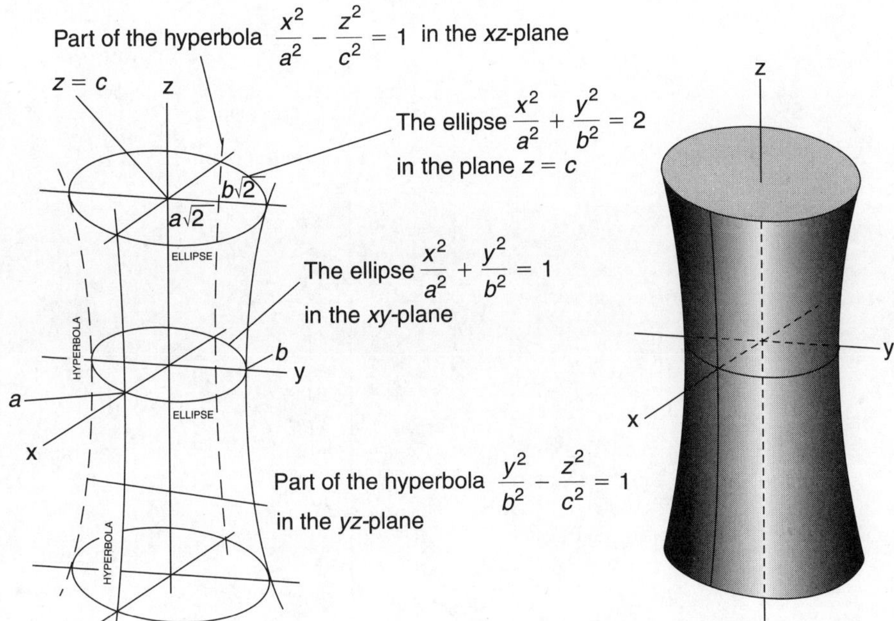

Fig. 10.9

The curves obtained by its intersection with the co-ordinate planes are

$$x = 0 : \text{ the hyperbola } \frac{y^2}{b^2} - \frac{z^2}{c^2} = 1;$$

$$y = 0 : \text{ the hyperbola } \frac{x^2}{a^2} - \frac{z^2}{c^2} = 1;$$

$$z = 0 : \text{ the ellipse } \frac{x^2}{a^2} + \frac{y^2}{b^2} = 1.$$

Any plane $z = k$ parallel to the xy-plane, intersects the surface in an ellipse

$$\frac{x^2}{a^2\left(1 + \frac{k^2}{c^2}\right)} + \frac{y^2}{b^2\left(1 + \frac{k^2}{c^2}\right)} = 1.$$

If $a = b$, the hyperboloid is a surface of revolution.

Hyperboloid of two sheets: Its equation is

$$\frac{z^2}{c^2} - \frac{x^2}{a^2} - \frac{y^2}{b^2} = 1 \qquad (5)$$

It is symmetric with respect to all the three co-ordinate planes; intersects only the z-axis at, $(\pm c, 0, 0)$. The curves obtained by its intersection with the co-ordinate planes are

$$x = 0 : \text{ the hyperbola } \frac{z^2}{c^2} - \frac{y^2}{b^2} = 1;$$

$$y = 0 \ : \ \text{the hyperbola } \frac{z^2}{c^2} - \frac{x^2}{a^2} = 1;$$

$$z = 0 \ : \ \text{no intersection.}$$

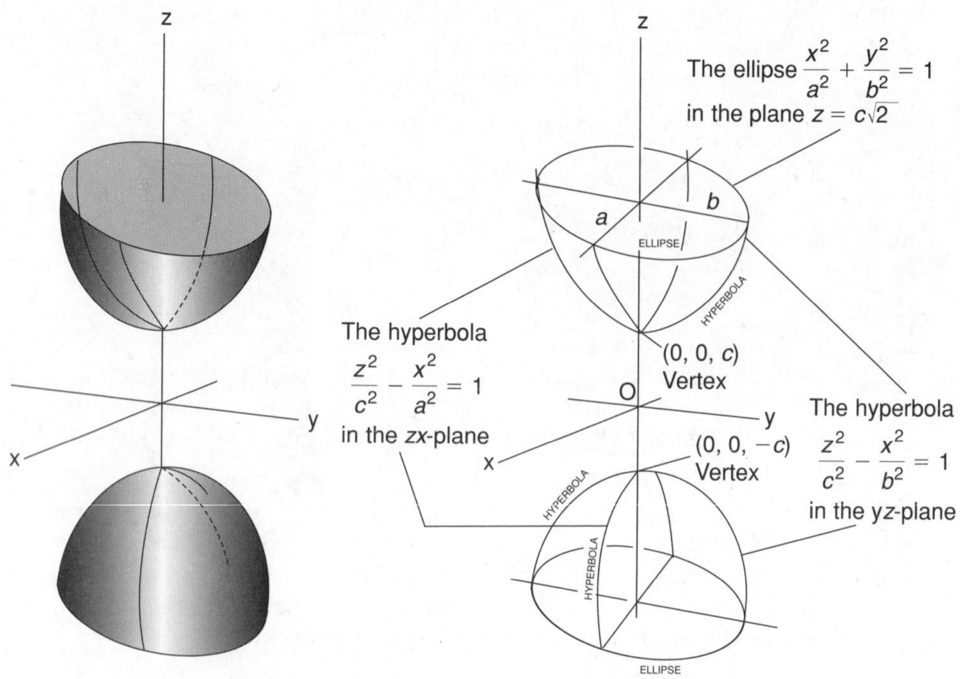

Fig. 10.10

Any plane $z = k\,(|k| > c)$ parallel to the xy-plane, intersects the surface in an ellipse

$$\frac{x^2}{a^2\left(\frac{k^2}{c^2} - 1\right)} + \frac{y^2}{b^2\left(\frac{k^2}{c^2} - 1\right)} = 1$$

Hyperbolic paraboloid: Its equation is

$$\frac{y^2}{b^2} - \frac{x^2}{a^2} = \frac{z}{c}, \quad c > 0 \tag{6}$$

It is symmetric with respect to $x = 0$ and $y = 0$ and meet the co-ordinate axes only at the origin. The curves obtained by the intersection of the surface with the co-ordinate planes are

$$x = 0 \ : \ \text{the parabola } y^2 = \frac{b^2}{c} z;$$

$$y = 0 \ : \ \text{the parabola } x^2 = -\frac{a^2}{c} z;$$

$$z = 0 \ : \ \text{the lines } y = \pm\frac{b}{a} x.$$

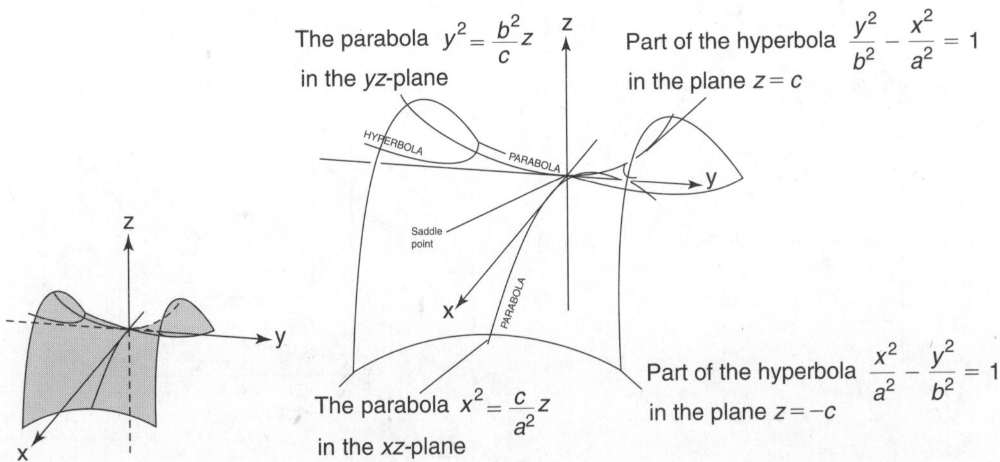

Fig. 10.11

Any plane $z = k\ (k > 0)$ parallel to the xy-plane, intersects the surface in a hyperbola

$$\frac{y^2}{b^2k/c} - \frac{x^2}{a^2k/c} = 1$$

If we move towards origin along the surface in the yz-plane, the origin appears to be minimum. On the other hand, if we move along the zx-plane, it appears to be a maximum. Such a point is called a 'saddle point', as the surface is shaped like saddle with such a point.

REMARK: Except the equations (2) and (6), none of the above equations can represent a function (why?). In fact, all the remaining equations represent pairs of functions. For example, equation (1) provides the functions

$$f(x, y) = c\sqrt{1 - \frac{x^2}{a^2} - \frac{y^2}{b^2}}$$

and

$$g(x, y) = -c\sqrt{1 - \frac{x^2}{a^2} - \frac{y^2}{b^2}}$$

respectively which represent the upper half (above xy-plane) and lower half (below xy-plane) of the surface of the ellipsoid.

Level Curves: When a plane intersects a 3-dimensional surface, we get a curve in space. If the plane is parallel to the xy-plane, that is, if the equation of the plane is of the type $z = k$, then the corresponding curve has a special name. It is called the contour line of the surface of height k. The projection of a contour line on the xy-plane is called the level curve of the surface of height k.

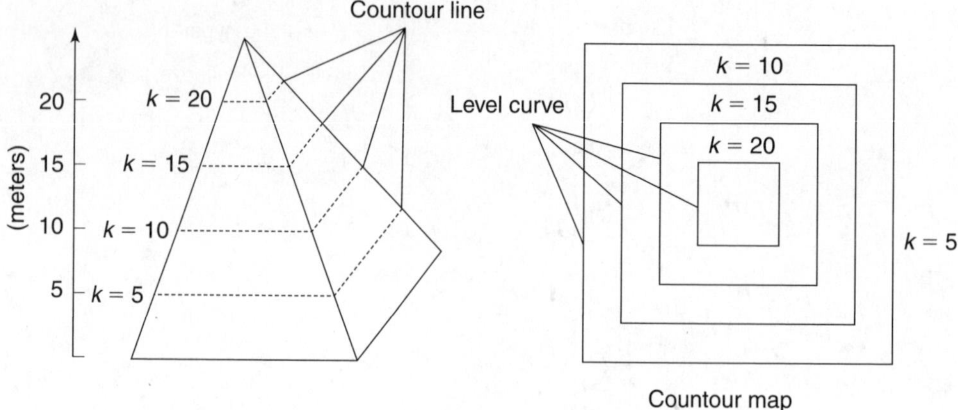

Fig. 10.12

? DEFINITION: (i) For a function $f(x, y)$ of two variables x and y, the set of all points $(x, y, f(x, y))$ in space, for (x, y) in the domain of f, is called the 'graph' of f. The graph of f is also called the 'surface' $z = f(x, y)$.

(ii) If the surface $z = f(x, y)$ is cut by the horizontal plane $z = k$, then the set of all points of intersection is called the 'contour line' of the surface of height k.

(iii) The projection of a contour line on the xy-plane is called the 'level curve' of f of height k. A collection of level curves of $z = f(x, y)$ is called a 'contour map' or 'contour plot' of f.

Contour lines are used to locate the positions at the same height on an elevated surface such as a hill, mountain, buildings and so on. They help in determining the shape of the surface at different heights. A contour map gives an idea about the steepness of the surface. See Fig. 10.12.

EXAMPLE 10.4 Draw the level curves of $f(x, y) = y^2 - x^2$ of height $k = 0, 1, 2, \ldots$

[DU, B.Sc., (Ph.Sc.), 2006]

SOLUTION

The level curves of the surface $f(x, y) = y^2 - x^2$ of height $k = 0, 1, 2$ are respectively,

$$y^2 - x^2 = 0, \quad y^2 - x^2 = 1, \quad y^2 - x^2 = 2$$

While $y^2 - x^2 = 0$ represents a pair of straight lines $y = x$ and $y = -x$, $y^2 - x^2 = 1$ and $y^2 - x^2 = 2$ represent hyperbolas.

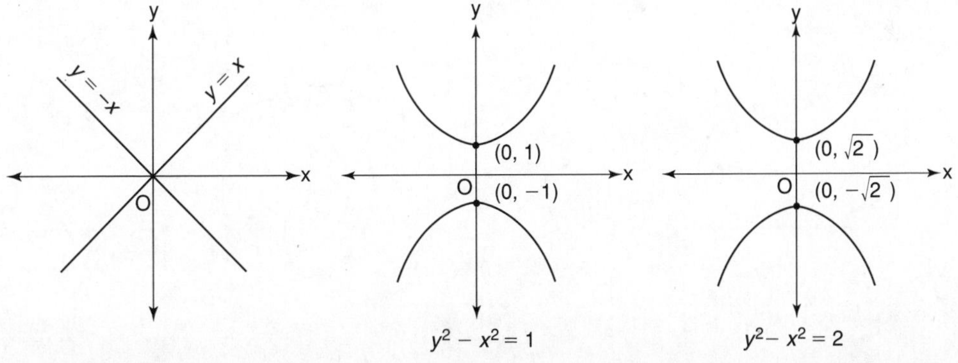

Fig. 10.13

EXAMPLE 10.5 Describe the surface represented by the following equations:

(i) $z = x^2 + y^2$

(ii) $z = 9x^2 + 25y^2$

(iii) $z = 16x^2 - 9y^2$

(iv) $z = \sqrt{16x^2 + 25y^2}$

(v) $z = \sqrt{4 - x^2 - y^2}$

(vi) $z = 5\sqrt{\frac{x^2}{16} + \frac{y^2}{9} - 1}$

(vii) $z = -4\sqrt{1 - \frac{x^2}{25} - \frac{y^2}{9}}$

(viii) $z = -2x - 3y + 5$

Draw the level curves of the above surfaces of height 1, 2, 5 and 7.

SOLUTION

(i) The intersection of the surface $z = x^2 + y^2$ and the yz-plane (that is, $x = 0$) is the parabola $y^2 = z$ with vertex at the origin, z-axis being the axis. Any plane parallel to the yz-plane, say, $x = c$, will intersect the surface in a parabola $y^2 = z - c^2$.

Similarly, intersection of the surface with zx-plane (that is $y = 0$) is the parabola $x^2 = z$, with vertex at the origin and z-axis being the axis.

Intersection of the plane $z = k$ is a circle $x^2 + y^2 = k$.

The surface is defined for $z \geq 0$ only.

Thus the surface is a paraboloid, z-axis as its axis, existing in the upper half space only.

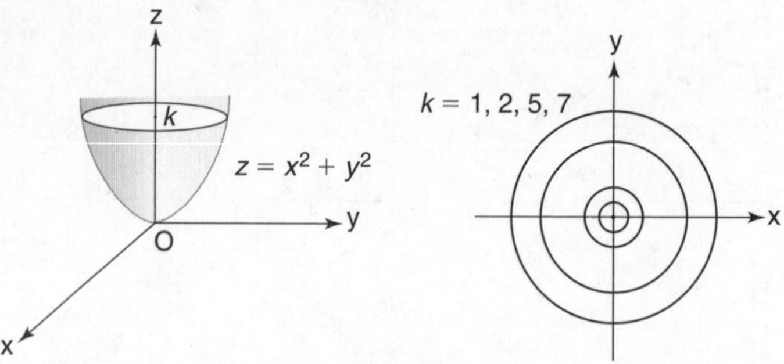

Fig. 10.14

The level curves of height 1, 2, 5, 7 are concentric circles with radii 1, $\sqrt{2}$, $\sqrt{5}$, $\sqrt{7}$ respectively given by

$$x^2 + y^2 = 1, \quad x^2 + y^2 = 2, \quad x^2 + y^2 = 5, \quad x^2 + y^2 = 7$$

(ii) $z = 9x^2 + 25y^2$

The plane $x = 0$ intersects the surface in the parabola $y^2 = \dfrac{1}{25}z$. Any plane $x = c$ parallel to $x = 0$ intersects the surface in a parabola.

Similarly the plane $y = 0$ intersects the surface in a parabola $x^2 = \dfrac{1}{9}z$. Any plane $y = c$ parallel to $y = 0$ will intersect the surface in a parabola.

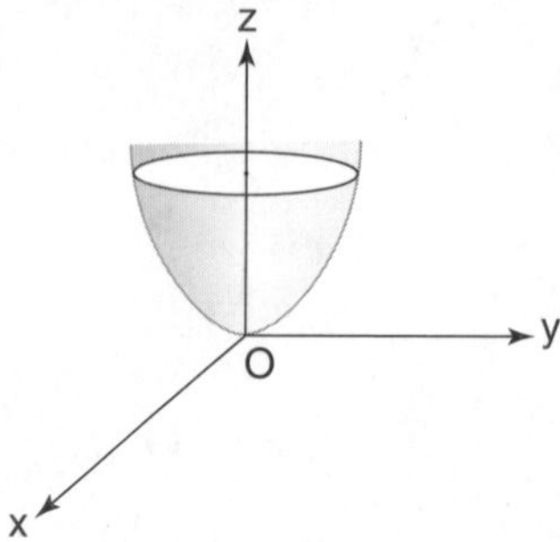

Fig. 10.15

The plane $z = k$ intersects the surface in an ellipse

$$\frac{x^2}{(\sqrt{k}/3)^2} + \frac{y^2}{(\sqrt{k}/5)^2} = 1.$$

The surface exists for $z \geq 0$ only.
Hence the surface is an elliptic paraboloid, with z-axis as its axis, existing in the upper half space only.
The level curves of height 1, 2, 5, 7 are ellipses with centre at the origin, given by

$$9x^2 + 25y^2 = 1, \qquad 9x^2 + 25y^2 = 2$$
$$9x^2 + 25y^2 = 5, \qquad 9x^2 + 25y^2 = 7$$

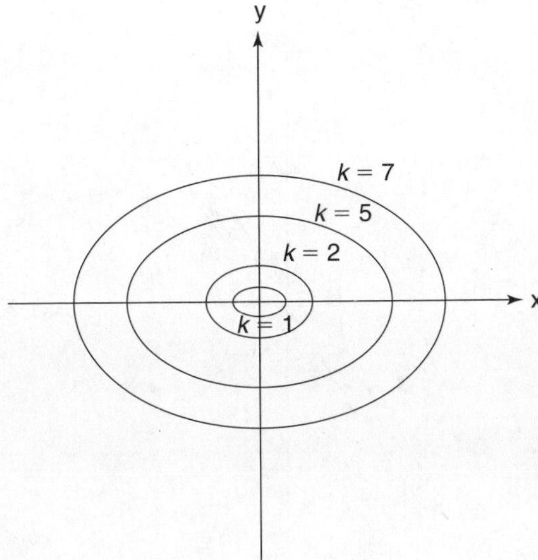

Fig. 10.16

(iii) $z = 16x^2 - 9y^2$

The intersection of the surface and the plane $x = 0$ is the parabola $y^2 = \left(-\frac{1}{9}\right)z$. Intersection of the surface and the planes parallel to $x = 0$ are also parabolas. Similarly, intersection of the surface and planes $y = 0$ and planes parallel to $y = 0$ are also parabolas. However, intersection of the surface and the planes of the form $z = k$, parallel to $z = 0$, are hyperbolas $\frac{x^2}{k/16} - \frac{y^2}{k/9} = 1$ (In case of $z = 0$, the hyperbola degenerates into a pair of straight lines).
The graph of the surface is a hyperbolic paraboloid.

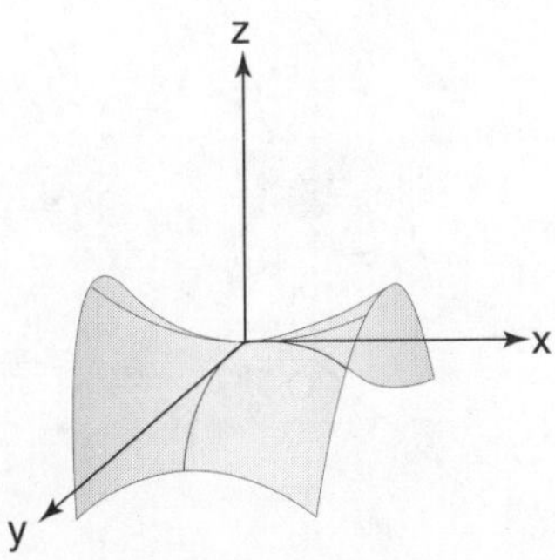

Fig. 10.17

The level curves of this paraboloid of height $k = 1, 2, 5, 7$ are respectively

$$16x^2 - 9y^2 = 1, \qquad 16x^2 - 9y^2 = 2$$
$$16x^2 - 9y^2 = 5 \quad \text{and} \quad 16x^2 - 9y^2 = 7$$

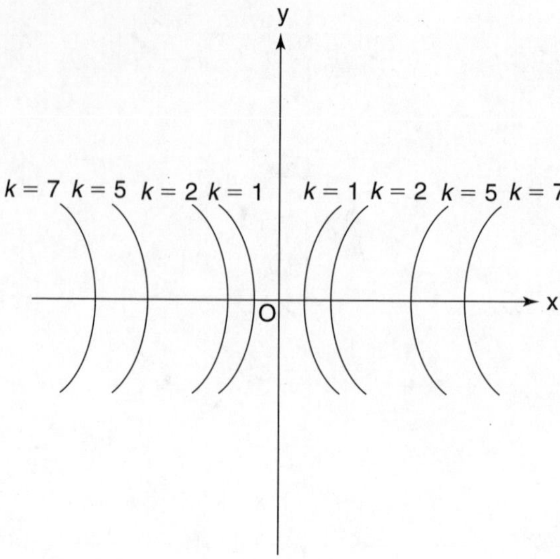

Fig. 10.18

(iv) $z = \sqrt{16x^2 + 25y^2}$

The graph of this surface is the portion of the surface

$$z^2 = 16x^2 + 25y^2$$

on and above the xy-plane (that is, $z = 0$)
The intersection of the surface with the plane $x = 0$ is the pair of straight lines $z = \pm 5y$.
The intersection of the plane $z = k$ (k constant) with the surface is an ellipse $16x^2 + 25y^2 = k^2$.
The graph of the surface is the upper half of the elliptic cone

$$16x^2 + 25y^2 = z^2$$

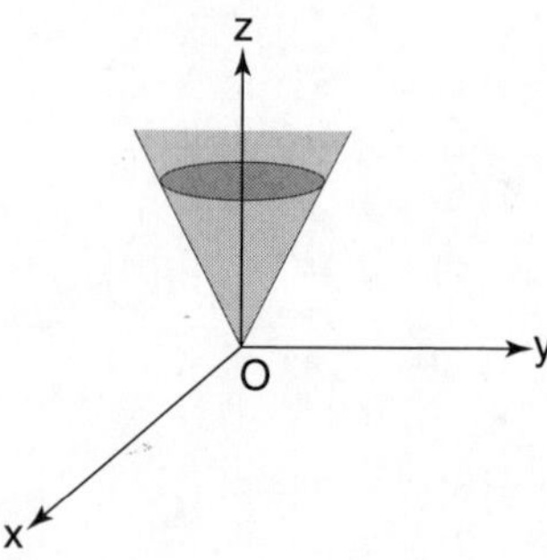

Fig. 10.19

The level curves of height $k = 1, 2, 5, 7$ are respectively,

$$16x^2 + 25y^2 = 1, \qquad 16x^2 + 25y^2 = 4$$
$$16x^2 + 25y^2 = 25, \qquad 16x^2 + 25y^2 = 49$$

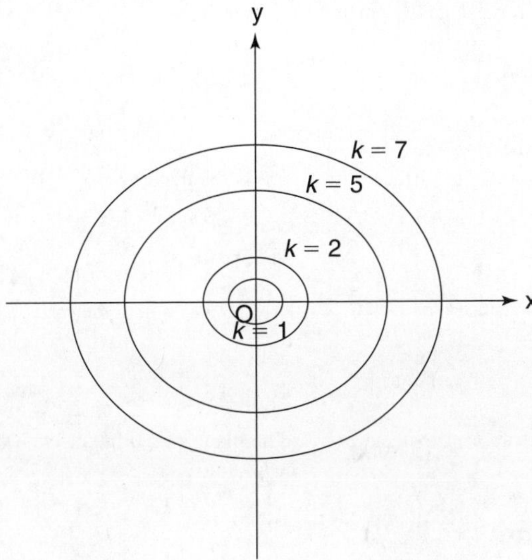

Fig. 10.20

(v) $z = \sqrt{4 - x^2 - y^2}$

The graph of this surface is the upper half of the sphere

$$x^2 + y^2 + z^2 = 4$$

on and above the xy-plane. The centre and radius of this sphere are (0, 0, 0) and 2 respectively. The level curves are

$$x^2 + y^2 = 3, \qquad x^2 + y^2 = 0,$$

for $k = 1$ and $k = 2$ respectively. For $k = 5, 7$, the level curves donot exist.

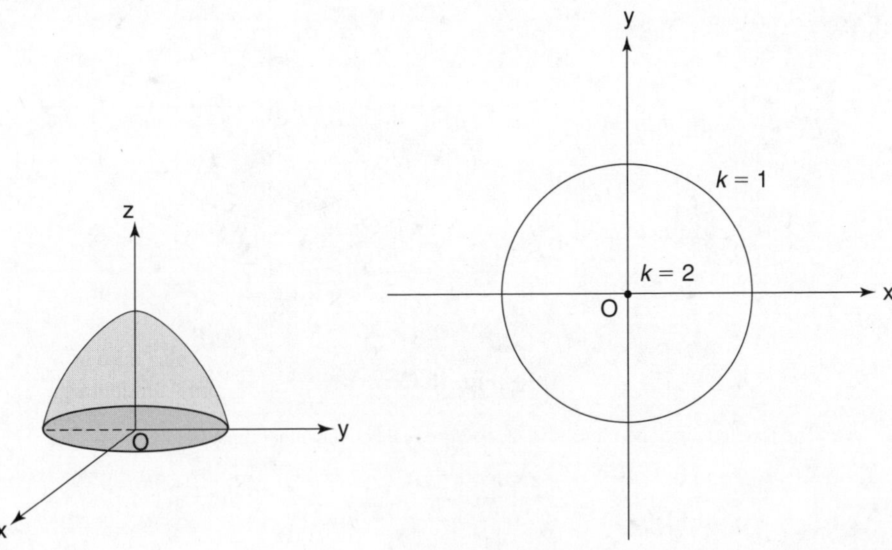

Fig. 10.21

(vi) Given surface represents the portion of the surface

$$\frac{x^2}{16} + \frac{y^2}{9} - \frac{z^2}{25} = 1$$

on and above the xy-plane $z = 0$
The intersection of the surface with $x = 0$ is the hyperbola

$$\frac{y^2}{9} - \frac{z^2}{25} = 1$$

on yz-plane above the plane $z = 0$. Similarly, intersection of the surface with $y = 0$ is the hyperbola

$$\frac{x^2}{16} - \frac{y^2}{25} = 1$$

on the zx-plane above the plane $z = 0$. The plane $z = 0$ intersects it on the ellipse

$$\frac{x^2}{16} + \frac{y^2}{9} = 1.$$

The plane $z = k$, parallel to xy-plane, intersects it in the ellipse

$$\frac{x^2}{16} + \frac{y^2}{9} = 1 + \frac{k^2}{25}$$

Thus the graph is the hyperboloid on and above the plane $z = 0$.

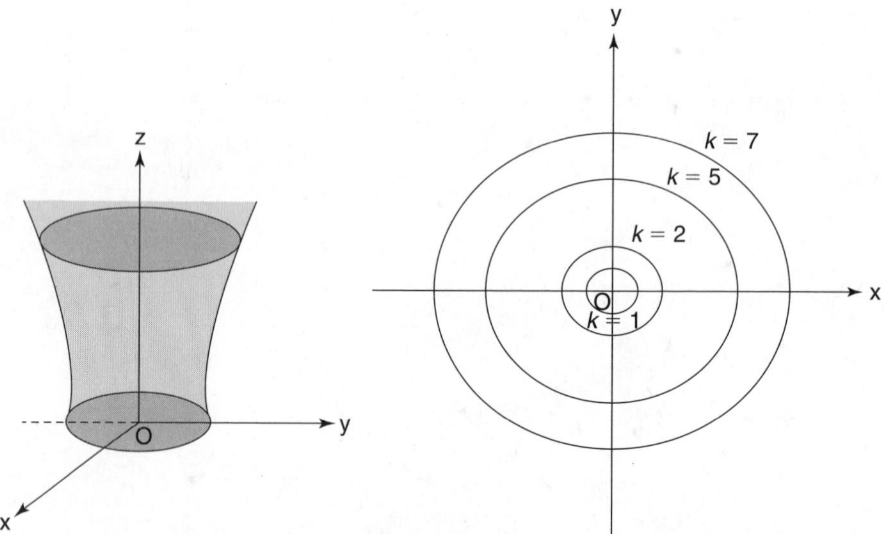

Fig. 10.22

The level curve of height $c = 1, 2, 5, 7$ are respectively, the ellipses

$$\frac{x^2}{16} + \frac{y^2}{9} = \frac{26}{25}, \qquad \frac{x^2}{16} + \frac{y^2}{9} = \frac{29}{25}$$

$$\frac{x^2}{16} + \frac{y^2}{9} = 2, \qquad \frac{x^2}{16} + \frac{y^2}{9} = \frac{74}{25}.$$

(vii) $z = -4\sqrt{1 - \frac{x^2}{25} - \frac{y^2}{9}}$

The given surface is the portion of

$$\frac{x^2}{25} + \frac{y^2}{9} + \frac{z^2}{16} = 1.$$

on and below the xy-plane.
The intersection of the surface and the plane $x = 0$ is the semi-ellipse

$\frac{y^2}{9} + \frac{z^2}{16} = 1, z \le 0$ on the yz-plane (that is, $x = 0$). Similarly, the plane $y = 0$ intersects it in the semi-ellipse,

$$\frac{x^2}{25} + \frac{z^2}{16} = 1, \qquad z \le 0$$

on the zx-plane (that is, $y = 0$)

The surface that intersects the plane $z = 0$ is the ellipse $\frac{x^2}{25} + \frac{y^2}{9} = 1.$

The surface is finite as $-5 \le x \le 5,\ -3 \le y < 3,\ -4 \le z \le 0.$

The graph is the lower half of the ellipsoid with centre at the origin and semi-axes along x, y and z-axes of length 5, 3 and 4 unit respectively.
There are no level curves of height $k = 1, 2, 5, 7$ as the surface exists below the xy-plane.

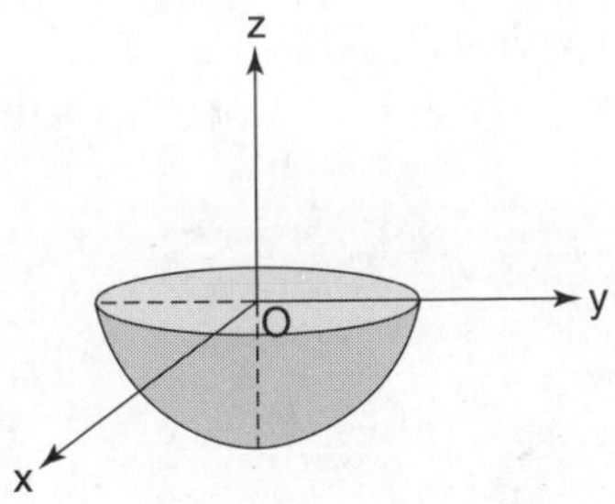

Fig. 10.23

(viii) $z = -2x - 3y + 5$

The surface is plane through A (0, 0, 5), B (0, 5/3, 0) and C (5/2, 0,0). The triangular portion ABC is shown in Fig. 10.24. The surface is extended infinitely. Its level curves of height $k = 1, 2, 5, 7$ are straight lines,

$$2x + 3y = 4, \qquad 2x + 3y = 3$$

$$2x + 3y = 0, \qquad 2x + 3y = -2$$

Fig. 10.23

EXERCISE 10.1

Describe the graph of the functions in Questions 1–10. Also draw the level curves at heights 0, 1, 2, −2, 3, wherever they exist.

1. $z = \sqrt{9 - x^2 - y^2}$
2. $z = 2 - x - y$
3. $z = -\frac{1}{2}(x^2 + y^2)$
4. $z = \sqrt{x^2 + \frac{y^2}{4} - 1}$
5. $z = +2\sqrt{1 - \frac{x^2}{9} - \frac{y^2}{4}}$
6. $z = -2x - 3y + 5$
7. $z = \sqrt{\frac{x^2}{4} + \frac{y^2}{9}}$
8. $z = -\sqrt{4x^2 + 2y^2}$
9. $z = 2\sqrt{1 - \frac{x^2}{16} - \frac{y^2}{9}}$
10. $z = 2\sqrt{\frac{x^2}{16} + \frac{y^2}{9} - 1}$

10.2 PARTIAL DERIVATIVES

If $z = f(x, y)$, then the effect in z with the change in only one of the two independent variables may be of interest to us. For example the ideal gas law in physics states that under appropriate conditions the pressure exerted by a gas is a function of the volume of the gas and its temperature. So a physicist may be interested in the rate of change of the pressure if the volume is held fixed and the temperature is allowed to vary or if the temperature is held fixed and the volume is allowed to vary. The mathematical tools for studying this type of situations is known as partial derivatives, which we define below.

? DEFINITION: If $z = f(x, y)$ and (x_0, y_0) is a point in the domain of f, then the 'partial derivative' of *f* with respect to *x* at (x_0, y_0) (also called the partial derivative of *z* with respect to *x* at (x_0, y_0)) is the derivative at x_0 of the function when $y = y_0$ is held fixed and *x* is allowed to vary. The partial derivative is denoted by $f_x(x_0, y_0)$ or $\left.\frac{\partial f}{\partial x}\right|_{(x_0,y_0)}$ or $\left.\frac{\partial z}{\partial x}\right|_{(x_0,y_0)}$ and is given by

$$f_x(x_0, y_0) = \left.\frac{d}{dx}[f(x, y_0)]\right|_{x=x_0} = \lim_{h\to 0}\frac{f(x_0 + h, y_0) - f(x_0, y_0)}{h}.$$

Similarly,

$$f_y(x_0, y_0) = \left.\frac{d}{dy}[f(x_0, y)]\right|_{y=y_0} = \lim_{h\to 0}\frac{f(x_0, y_0 + h) - f(x_0, y_0)}{h}$$

The slope of the tangent by $z = f(x, y_0)$ at the point P $f(x_0, y_0, f(x_0, y_0))$ in the plane $y = y_0$ is the value of the partial derivative of f with respect to x at (x_0, y_0). The tangent line to the curve at P is the line in the plane $y = y_0$ through P with this slope. The partial derivative $\frac{\partial f}{\partial x}$ at (x_0, y_0) gives the rate of change of f with respect to x when y is held fixed at the value y_0. This is the rate of change of f in the direction parallel to x-axis at (x_0, y_0).

Similarly, the slope of the curve $z = f(x_0, y)$ at the point $P(x_0, y_0, f(x_0, y_0))$ in the vertical plane $x = x_0$ is the partial derivative of f with respect to y at (x_0, y_0). The tangent line to the curve at P is the line in the plane $x = x_0$ that passes through P with this slope. This partial derivative gives the rate of change of f with respect to y at (x_0, y_0) when x is held fixed at the value x_0. This is the rate of change of f in the direction parallel to y-axis at (x_0, y_0).

EXAMPLE 10.6 Find the partial derivatives $\frac{\partial f}{\partial x}$ and $\frac{\partial f}{\partial y}$ at $(4, -5)$ if

$$f(x, y) = x^2 + 3xy + y - 1.$$

SOLUTION

$$\frac{\partial f}{\partial x} = \frac{\partial}{\partial x}(x^2 + 3xy + y - 1) = 2x + 3y$$

$$\therefore \quad \left.\frac{\partial f}{\partial x}\right|_{(4,-5)} = 2 \times 4 + 3(-5) = -7$$

$$\frac{\partial f}{\partial y} = \frac{\partial}{\partial y}(x^2 + 3xy + y - 1) = 3x + 1$$

$$\therefore \quad \left.\frac{\partial f}{\partial y}\right|_{(4,-5)} = 3(4) + 1 = 13$$

EXAMPLE 10.7 The plane $x = 1$ intersects the paraboloid $z = x^2 + y^2$ in a parabola. Find the slope of the tangent to the parabola at (1, 2, 5).

SOLUTION

The slope is the value of the partial derivative $\frac{\partial z}{\partial y}$ at (1, 2, 5).

$$\frac{\partial z}{\partial y} = \frac{\partial}{\partial y}(x^2 + y^2) = 2y$$

$$\therefore \quad \left.\frac{\partial z}{\partial y}\right|_{(1,2,5)} = 2 \times 2 = 4$$

EXAMPLE 10.8 Find $\frac{\partial z}{\partial x}$ if the equation

$$yz - \ln z = x + e^y$$

defines z as a function of two independent variables x and y and the partial derivative exists.

SOLUTION

We differentiate both sides of the equation with respect to x, holding y constant and treating z as a differentiable function of x:

$$\frac{\partial}{\partial x}(yz) - \frac{\partial}{\partial x}(\ln z) = \frac{\partial x}{\partial x} + \frac{\partial e^y}{\partial x}$$

$$y\frac{\partial z}{\partial x} - \frac{1}{z}\frac{\partial z}{\partial x} = 1 + 0 \qquad \left(\text{since } y \text{ is constant } \frac{\partial}{\partial x}(yz) = y\frac{\partial z}{\partial x}\right)$$

$$\left(y - \frac{1}{z}\right)\frac{\partial z}{\partial x} = 1$$

$$\frac{\partial z}{\partial x} = \frac{z}{yz - 1}$$

EXAMPLE 10.9 Find the slope of the sphere $x^2 + y^2 + z^2 = 1$ in the y-direction at the points $\left(\frac{2}{3}, \frac{1}{3}, \frac{2}{3}\right)$ and $\left(\frac{2}{3}, \frac{1}{3}, -\frac{2}{3}\right)$.

SOLUTION

The point $\left(\frac{2}{3}, \frac{1}{3}, \frac{2}{3}\right)$ lies on the upper hemisphere $z = \sqrt{1 - x^2 - y^2}$, and the point $\left(\frac{2}{3}, \frac{1}{3}, -\frac{2}{3}\right)$ lies on the lower hemisphere $z = -\sqrt{1 - x^2 - y^2}$. We could find the slopes by differentiating each expression for z separately with respect to y and then evaluating the derivative at $x = \frac{2}{3}$ and $y = \frac{1}{3}$. However, it is more convenient to differentiate the given equation

$$x^2 + y^2 + z^2 = 1$$

implicitly with respect y, since this will give us both the slopes with one differentiation. To perform the implicit differentiation, we view z as a function of x and y and differentiate both sides with respect to y, taking x to be fixed. The computations are as follows:

$$\frac{\partial}{\partial y}(x^2 + y^2 + z^2) = \frac{\partial}{\partial y}(1)$$

$$0 + 2y + 2z\frac{\partial z}{\partial y} = 0$$

$$\frac{\partial z}{\partial y} = -\frac{y}{z},$$

$$\therefore \qquad \left.\frac{\partial z}{\partial y}\right|_{(2/3,1/3,2/3)} = -\frac{1}{3}\Big/\frac{2}{3} = -\frac{1}{2}, \quad \left.\frac{\partial z}{\partial y}\right|_{(2/3,1/3,-2/3)} = -\frac{1}{3}\Big/\left(-\frac{2}{3}\right) = \frac{1}{2}.$$

Suppose that f is a function of two variables x and y. Since the partial derivatives $\frac{\partial f}{\partial x}$ and $\frac{\partial f}{\partial y}$ are also functions of x and y, these functions may themselves have partial derivatives. This gives rise to four possible second-order partial derivatives of f, which are

$$\frac{\partial^2 f}{\partial x^2} = \frac{\partial}{\partial x}\left(\frac{\partial f}{\partial x}\right) = f_{xx} \qquad \frac{\partial^2 f}{\partial y^2} = \frac{\partial}{\partial y}\left(\frac{\partial f}{\partial y}\right) = f_{yy}$$

$$\frac{\partial^2 f}{\partial y \partial x} = \frac{\partial}{\partial y}\left(\frac{\partial f}{\partial x}\right) = f_{xy} \qquad \frac{\partial^2 f}{\partial x \partial y} = \frac{\partial}{\partial x}\left(\frac{\partial f}{\partial y}\right) = f_{yx}$$

EXAMPLE 10.10 Find the second-order partial derivatives of $f(x, y) = x^2y^3 + x^4y$.

SOLUTION.

We have

$$\frac{\partial f}{\partial x} = 2xy^3 + 4x^3y \quad \text{and} \quad \frac{\partial f}{\partial y} = 3x^2y^2 + x^4$$

so that

$$\frac{\partial^2 f}{\partial x^2} = \frac{\partial}{\partial x}\left(\frac{\partial f}{\partial x}\right) = \frac{\partial}{\partial x}(2xy^3 + 4x^3y) = 2y^3 + 12x^2y$$

$$\frac{\partial^2 f}{\partial y^2} = \frac{\partial}{\partial y}\left(\frac{\partial f}{\partial y}\right) = \frac{\partial}{\partial y}(3x^2y^2 + x^4) = 6x^2y$$

$$\frac{\partial^2 f}{\partial x\partial y} = \frac{\partial}{\partial x}\left(\frac{\partial f}{\partial y}\right) = \frac{\partial}{\partial x}(3x^2y^2 + x^4) = 6xy^2 + 4x^3$$

$$\frac{\partial^2 f}{\partial y\partial x} = \frac{\partial}{\partial y}\left(\frac{\partial f}{\partial x}\right) = \frac{\partial}{\partial y}(2xy^3 + 4x^3y) = 6xy^2 + 4x^3$$

Third-order, fourth-order, and higher-order partial derivatives can be obtained by successive differentiation. Some possibilities are

$$\frac{\partial^3 f}{\partial x^3} = \frac{\partial}{\partial x}\left(\frac{\partial^2 f}{\partial x^2}\right) = f_{xxx} \qquad \frac{\partial^4 f}{\partial y^4} = \frac{\partial}{\partial y}\left(\frac{\partial^3 f}{\partial y^3}\right) = f_{yyyy}.$$

EXAMPLE 10.11 Let $f(x, y) = y^2e^x + y$. Find f_{xyy}.

SOLUTION

$$f_{xyy} = \frac{\partial^3 f}{\partial y^2\partial x} = \frac{\partial^2}{\partial y^2}\left(\frac{\partial f}{\partial x}\right) = \frac{\partial^2}{\partial y^2}(y^2e^x) = \frac{\partial}{\partial y}(2ye^x) = 2e^x.$$

EXERCISE 10.2

1. Find all the second order partial derivatives of

(i) e^{x-y} **(ii)** e^{xy} **(iii)** $\tan(\tan^{-1}x + \tan^{-1}y)$

2. Verify that $\dfrac{\partial^2 z}{\partial x\,\partial y} = \dfrac{\partial^2 z}{\partial y\,\partial x}$, where z is

(i) $\ln(y\sin x + x\sin y)$ **(ii)** $\dfrac{xy}{\sqrt{1 + x^2 + y^2}}$ **(iii)** $\sin^{-1}\left(\dfrac{x}{y}\right)$

3. Find the value of $\dfrac{1}{a^2}\cdot\dfrac{\partial^2 z}{\partial x^2} + \dfrac{1}{b^2}\cdot\dfrac{\partial^2 z}{\partial y^2}$, where

$$a^2x^2 + b^2y^2 - c^2z^2 = 0$$

4. Verify that $x\dfrac{\partial z}{\partial x} + y\dfrac{\partial z}{\partial y} = nz$, where

(i) $z = ax^2 + 2hxy + by^2, \quad n = 2$

(ii) $z = (x^2 + xy + y^2)^{-1}, \quad n = -2$

(iii) $z = \sin^{-1}\dfrac{x}{y} + \tan^{-1}\dfrac{y}{x}, \quad n = 0$

(iv) $z = x^k \ln\dfrac{y}{x}, \quad n = k$

(v) $z = \dfrac{x^{1/4} + y^{1/4}}{x^{1/5} + y^{1/5}}, \quad n = \dfrac{1}{20}$

5. If $z = 3xy - y^3 + (y^2 - 2x)^{3/2}$, verify that

$$\frac{\partial^2 z}{\partial x\, \partial y} = \frac{\partial^2 z}{\partial y\, \partial x} \text{ and } \frac{\partial^2 z}{\partial x^2} \cdot \frac{\partial^2 z}{\partial y^2} = \left(\frac{\partial^2 z}{\partial x\, \partial y}\right)^2.$$

6. If $z = f(x + ay) + \phi(x - ay)$, prove that $\frac{\partial^2 z}{\partial y^2} = a^2 \frac{\partial^2 z}{\partial x^2}$.

7. If $u = f(ax^2 + 2hxy + by^2)$, $v = \phi(ax^2 + 2hxy + by^2)$, prove that

$$\frac{\partial}{\partial y}\left(u\frac{\partial v}{\partial x}\right) = \frac{\partial}{\partial x}\left(u\frac{\partial v}{\partial y}\right)$$

8. If $\theta = t^n e^{-r^2/4t}$, find the value of n which makes

$$\frac{1}{r^2} \cdot \frac{\partial}{\partial r}\left(r^2 \frac{\partial \theta}{\partial r}\right) = \frac{\partial \theta}{\partial t}$$

9. If $u = f(r)$, where $r = \sqrt{x^2 + y^2}$ prove that

$$\frac{\partial^2 u}{\partial x^2} + \frac{\partial^2 u}{\partial y^2} = f''(r) + \frac{1}{r} f'(r).$$

10. If $v = \log(x^2 + y^2 + z^2)$, prove that

$$x\frac{\partial^2 v}{\partial y\, \partial z} = y\frac{\partial^2 v}{\partial z\, \partial x} = z\frac{\partial^2 v}{\partial x\, \partial y}$$

11. If $v = r^m$, where $r^2 = x^2 + y^2 + z^2$. Show that

$$\frac{\partial^2 v}{\partial x^2} + \frac{\partial^2 v}{\partial y^2} + \frac{\partial^2 v}{\partial z^2} = m(m + 1)r^{m-2}.$$

12. If $u = \frac{1}{\sqrt{(x^2 + y^2 + z^2)}}$; $x^2 + y^2 + z^2 \neq 0$, show that $\frac{\partial^2 u}{\partial x^2} + \frac{\partial^2 u}{\partial y^2} + \frac{\partial^2 u}{\partial z^2} = 0.$

13. Find the slope of the following surfaces

(i) $z = \sqrt{x^2 + 2y^2}$ along x-axis at $(1, 2, 3)$

(ii) $z = -\sqrt{4 - x^2 - y^2}$ along y-axis at $(1, 1, -\sqrt{2})$

(iii) $z = \sqrt{1 - (4x^2 + 9y^2)}$ along x-axis at $\left(\frac{1}{4}, \frac{1}{9}, \frac{\sqrt{23}}{6}\right)$

14. A point moves along the intersection of the elliptic paraboloid $z = x^2 + 3y^2$ and the plane $y = 1$. At what rate is z changing with x when the point is at $(2, 1, 7)$?

15. Find the slope of the tangent line at $(-1, 1, 5)$ to the curve of intersection of the surface $z = x^2 + 4y^2$ and

(a) the plane $x = -1$ **(b)** the plane $y = 1$

16. The volume V of a right circular cylinder is given by the formula $V = \pi r^2 h$, where r is the radius and h is the height.

(a) Find a formula for instantaneous rate of change of V

(i) with respect to r if r changes and h remains constant

(ii) with respect to h if h changes and r remains constant

(b) Suppose that r has a constant value of 8 cm, but h varies. Find the instantaneous change of V with respect to h at the point where $h = 6\,\text{cm}$.

(c) Suppose that h has a constant value of 4 ft, but r varies. Find the rate of change of V with respect to r at the point where $r = 6\,\text{ft}$.

17. According to the ideal gas law, the pressure, temperature and volume of a gas is related by $P = \frac{kT}{V}$, where k is a constant of proportionality. Suppose that V is measured in cubic inches $(\text{inch})^3$, T is measured in kelvins (K) and that for a certain gas the constant of proportionality is $k = 10$ in lb/K.

(a) Find the instantaneous rate of change of pressure with respect to temperature if the temperature is 80 K and volume remains fixed at $50\,(\text{inch})^3$.

(b) Find the instantaneous rate of change of volume with respect to pressure if the volume is $50\,(\text{inch})^3$ and the temperature remains fixed at 80 K.

18. The temperature at a point on a metal plate in the xy-plane is $T(x, y) = 4x^2 + 2xy + y^3$ degrees Celsius. Find the rate at which the temperature changes with respect to distance if we start at $(2, 3)$ and move

(i) to the right and parallel to the x-axis ;

(ii) upward and parallel to the y-axis .

■ 10.3 APPLICATIONS OF PARTIAL DERIVATIVES

10.3.1 The Wave Equation

Consider a string of length L that is stretched taut between $x = 0$ and $x = L$ on the x-axis, and suppose that the string is set into vibratory motion by "plucking" it at time $t = 0$. The displacement of a point on the string depends both on its coordinate x and the elapsed time t, and hence is described by a function $u(x, t)$ of two variables. For a fixed value t, the function $u(x, t)$ depends on x alone, and the graph of u versus x describes the shape of the string – think of it as a "snapshot" of the string at time t. It follows that at a fixed time t, the partial derivative $\frac{\partial u}{\partial x}$ represents the slope of the string at x, and the sign of the second partial derivative $\frac{\partial^2 u}{\partial x^2}$ tells us whether the string is concave up or concave down at x.

For a fixed value of x, the function $u(x, t)$ depends on t alone, and the graph of u versus t is the position versus time curve of the point on the string with coordinate x. Thus, for a fixed value of x, the partial derivative $\frac{\partial u}{\partial t}$ is the velocity of the point with coordinate x, and $\frac{\partial^2 u}{\partial t^2}$ is the acceleration of that point.

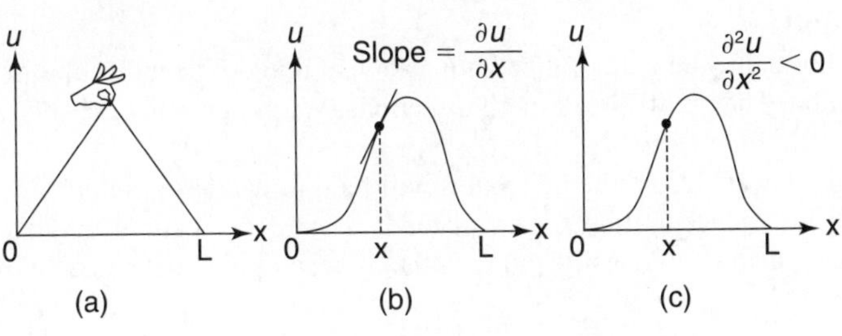

Fig. 10.25

It can be proved that under appropriate conditions the function $u(x, t)$ satisfies an equation of the form

$$\frac{\partial^2 u}{\partial t^2} = c^2 \frac{\partial^2 u}{\partial x^2}$$

where c is a positive constant that depends on the physical characteristics of the string. This equation, which is called the 'one-dimensional wave equation', involves partial derivatives of the unknown function $u(x, t)$ and hence is classified as a partial differential equation.

EXAMPLE 10.12 Show that the function $u(x, t) = \sin(x - ct)$ is a solution of the wave equation.

SOLUTION

We have

$$\frac{\partial u}{\partial x} = \cos(x - ct), \qquad \frac{\partial^2 u}{\partial x^2} = -\sin(x - ct)$$

$$\frac{\partial u}{\partial t} = -c\cos(x - ct), \qquad \frac{\partial^2 u}{\partial t^2} = -c^2\sin(x - ct)$$

Thus, $u(x, t)$ satisfies the one dimensional wave equation.

10.3.2. Laplace's Equation

For $z = f(x, y)$, the equation of the form $\frac{\partial^2 z}{\partial x^2} + \frac{\partial^2 z}{\partial y^2} = 0$ is known as 'Laplace's equation'.

EXAMPLE 10.13 The following functions satisfy Laplace's equation:

(i) $z = x^2 - y^2 + 2xy$

(ii) $z = e^x \sin y + e^y \cos x$

SOLUTION

(i) $\frac{\partial z}{\partial x} = 2x + 2y, \quad \frac{\partial^2 z}{\partial x^2} = 2$

$\frac{\partial z}{\partial y} = -2y + 2x, \quad \frac{\partial^2 z}{\partial y^2} = -2$

Thus, $\frac{\partial^2 z}{\partial x^2} + \frac{\partial^2 z}{\partial y^2} = 0$

(ii) Can be done similarly.

10.3.3 Diffusion Equation

Assume that we inject a substance into a tube filled with a solvent liquid. The molecules of the substance are in random motion. They will spread in all directions. We assume that the solvent liquid is not in motion.

At locations with high concentration, the molecules will tend to decrease in number and conversely, at locations with low concentration they will tend to increase in number. This random migration process, called diffusion, will finally end with the molecules at equal density throughout the tube.

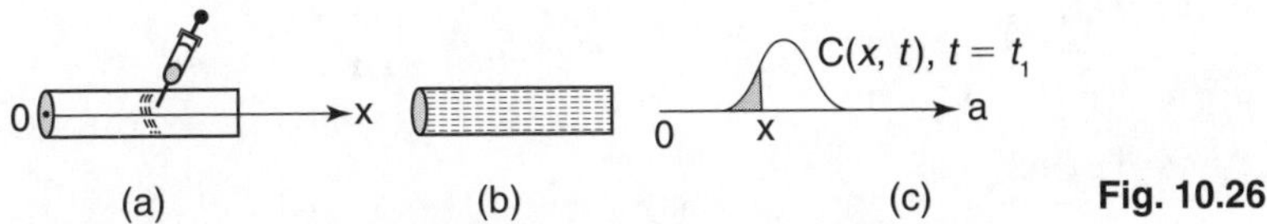

Fig. 10.26

We place the x-axis parallel to the axis of the tube. For simplicity we assume that concentration of the substance varies only in the x-direction. At each x, the concentration also depends on the time instant t. Thus we my denote the concentration by

$$\mathrm{C} = \mathrm{C}(x, t)$$

Its unit here is gm/cm^3. The function $\mathrm{C}(x,t)$ is plotted in Fig. 10.26 (c) for some time instant t_1. It can be shown that

$$\frac{\partial \mathrm{C}}{\partial t} = \mathrm{D}\frac{\partial^2 \mathrm{C}}{\partial x^2}, \tag{i}$$

where D is some positive constant, known as diffusion constant. The equation (i) is called the 'diffusion equation in one dimension'.

EXAMPLE 10.14 Find coefficients a and b such that

$$\mathrm{C}(x, t) = \exp(ax + bt)$$

satisfies the diffusion equation.

SOLUTION

$$\mathrm{C}(x, t) = \exp(ax + bt)$$

$$\frac{\partial \mathrm{C}}{\partial x} = ae^{ax+bt}, \qquad \frac{\partial \mathrm{C}}{\partial t} = b\,e^{ax+bt}$$

$$\frac{\partial^2 \mathrm{C}}{\partial x^2} = a^2\,e^{ax+bt}$$

Substituting these values in the diffusion equation $\frac{\partial \mathrm{C}}{\partial t} = \mathrm{D}\frac{\partial^2 \mathrm{C}}{\partial x^2}$, we get

$$b\,e^{ax+bt} = \mathrm{D}a^2\,e^{ax+bt}$$

This equation is satisfied for all values of x if we choose the coefficient arbitrarily and if we let $b = \mathrm{D}a^2$.

EXAMPLE 10.15 Verify that

$$\mathrm{C}(x, t) = t^{-1/2}\exp(-x^2/4\mathrm{D}t)$$

is a particular solution of the diffusion equation.

SOLUTION

Here $\mathrm{C}(x, t) = t^{-1/2} \cdot e^{-\frac{x^2}{4\mathrm{D}t}}$, so that

$$\frac{\partial \mathrm{C}}{\partial x} = t^{-1/2} \cdot e^{-x^2/4\mathrm{D}t} \cdot \left(-\frac{x}{2\mathrm{D}t}\right) = -\frac{1}{2\mathrm{D}}t^{-3/2} \cdot x \cdot e^{-x^2/4\mathrm{D}t}$$

$$\frac{\partial^2 \mathrm{C}}{\partial x^2} = -\frac{1}{2\mathrm{D}}t^{-3/2}\left(e^{-x^2/4\mathrm{D}t} + x \cdot e^{-x^2/4\mathrm{D}t}\left(-\frac{x}{2\mathrm{D}t}\right)\right) = \left(-\frac{1}{2\mathrm{D}}t^{-3/2} + \frac{x^2}{4\mathrm{D}^2}t^{-5/2}\right)e^{-x^2/4\mathrm{D}t}$$

$$\frac{\partial C}{\partial t} = -\frac{1}{2}t^{-3/2} \cdot e^{-x^2/4Dt} + t^{-1/2} \cdot e^{-x^2/4Dt}\left(\frac{x^2}{4Dt^2}\right)$$

$$= \left(-\frac{1}{2}t^{-3/2} + \frac{x^2}{4D}t^{-5/2}\right) e^{-x^2/4Dt}$$

$$= D\left(-\frac{1}{2D}t^{-3/2} + \frac{x^2}{4D^2}t^{-5/2}\right) e^{-x^2/4Dt}$$

$$= D\frac{\partial^2 C}{\partial x^2}$$

10.3.4. Heat Equation

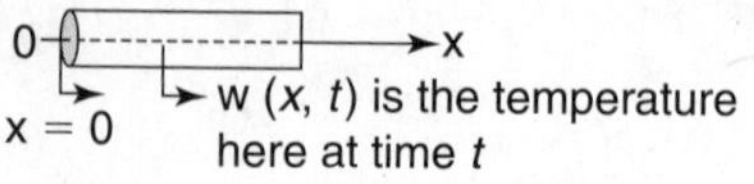

Fig. 10.27

If $w(x, t)$ represents the temperature at position x at time t in a uniform conducting rod with perfectly insulated sides, then the partial derivatives w_{xx} and w_t satisfy the following equation:

$$w_{xx} = \frac{1}{c^2} w_t$$

This equation is called 'one-dimensional heat equation'. The value of the positive constant c^2 is determined by the material from which the rod is made. It has been determined experimentally for a broad range of materials. For example, for dry soil $c^2 = .19 ft^2/day$.

You might have observed the similarity between the heat equation and the diffusion equation. In electrical engineering, the heat equation appears in the forms

$$v_{xx} = RCv_t \qquad \text{and} \qquad i_{xx} = RCi_t$$

which are known as 'telegraph equations'. These equations describe the voltage v and the flow of current i in a coaxial cable or in any other cable in which leakage and inductance are negligible. The function and constants in these equations are

$v(x, t)$ = voltage at point x at time t.

R = resistance per unit length

C = capacitance to ground per unit of cable length.

$i(x, t)$ = current at point x at time t.

EXAMPLE 10.16 Verify that

$$w(x, t) = e^{-c^2\pi^2 t} \sin \pi x$$

is a solution of the heat equation.

[DU, B.Sc., (Ph.Sc.), 2006]

SOLUTION

$$w(x, t) = e^{-c^2\pi^2 t} \sin \pi x$$

$$\frac{\partial w}{\partial x} = \pi e^{-c^2\pi^2 t} \cos \pi x, \quad \frac{\partial^2 w}{\partial x^2} = -\pi^2 e^{-c^2\pi^2 t} \sin \pi x$$

$$\frac{\partial w}{\partial t} = -c^2\pi^2 e^{-c^2\pi^2 t} \sin \pi x$$

$$\therefore \quad \frac{\partial^2 w}{\partial x^2} = \frac{1}{c^2} \cdot \frac{\partial w}{\partial t}$$

EXERCISE 10.3

1. Show that the function $u(x, t) = \sin \omega t \cdot \sin \omega x$, satisfies the wave equation for all real values of ω.
2. Show that a function of the form $u(x, t) = f(x + ct) + g(x - ct)$ satisfies the wave equation.

Verify that the function in Questions 3–9 are all solutions of wave equation.

3. $w = \sin(x + ct)$
4. $w = 4\cos(2x + 2ct) + e^{x+ct}$
5. $w = \ln(3x + 3ct)$
6. $w = \tan(2x - 2ct)$
7. $w = \cos(5x + 5ct)$
8. $w = \sin(x + ct) + \cos(3x + 3ct)$
9. $w = f(u)$, where f is differentiable function of u and $u = k(x + ct)$ where k is a constant.

Verify that the functions in Questions 10–16 satisfy Laplace's equation.

10. $f(x, y) = e^{-y}\cos x$
11. $f(x, y) = \ln\sqrt{(x^2 + y^2)}$
12. $f(x, y) = x^2 - y^2$
13. $f(x, y) = x^2 - y^2$
14. $f(x, y) = \tan^{-1}\left(\dfrac{y}{x}\right)$
15. $f(x, y) = \tan^{-1}\dfrac{2xy}{x^2 - y^2}$
16. $z = \ln(x^2 + y^2) + 2\tan^{-1}\left(\dfrac{y}{x}\right)$

Verify that the functions in Questions 17–18 are solutions to the heat equation.

17. $u(x, t) = e^{c^2\pi^2 t}(Ae^{\pi x} + Be^{-\pi x})$; A, B are constants.
18. $u(x, t) = Px + Q$; P, Q are constants.

Verify that the functions in Questions 19–21 are solutions of diffusion equation.

19. $c(x, t) = e^{-c^2k^2t}(a\cos kx + b\sin kx)$; a, b are constants.
20. $c(x, t) = e^{c^2k^2t}(ae^{kx} + be^{-kx})$; a, b, k are constants.
21. $c(x, t) = ax + b$; a, b are constants.

PART III

CHAPTER ELEVEN

Geometry of Complex Numbers

11.1 INTRODUCTION

The inadequacy of real numbers in solving quadratic equations of the type $x^2 + 1 = 0$ gave rise to a new type of numbers. Euler first defined the number i, called 'iota' or 'imaginary unit', as the number whose square is –1. With the help of i, square roots of a negative number can be interpreted in the following way:

$$\sqrt{-5} = \sqrt{5(-1)} = \sqrt{5}\,i, \quad \sqrt{-9} = \sqrt{9(-1)} = 3i \text{ etc.}$$

Also, $i^2 = -1$, $i^3 = -i$, $i^4 = 1$ and so on.

The numbers $\sqrt{-5}$, $\sqrt{-9}$ etc. are called 'imaginary numbers'.

A complex number is a number of the form $a + ib$ where a, b are real numbers and $i = \sqrt{-1}$. A complex number is usually denoted by z. If $z = a + ib$, then a and b are called 'real part' and 'imaginary part' of z and denoted by Re(z) and Im(z), respectively. Since any real number a can be expressed as $a + io$, every real number is a complex number. Thus, $\mathbb{R} \subseteq \mathbb{C}$ where $\mathbb{R}$ and $\mathbb{C}$ represent the set of real numbers and complex numbers, respectively.

We know that members of $\mathbb{R}$ are uniquely represented by the points of a number line. In the same way, members of $\mathbb{C}$ are uniquely represented by the points of a co-ordinate plane. The complex number $x + iy$ is represented by the point whose co-ordinates are (x, y). A co-ordinate plane used for representing complex numbers is called an 'Argand plane' or 'Gaussian plane'. The x-axis and y-axis are, respectively, called the 'real axis' and the 'imaginary axis'. The point P(x, y) is called the 'image' of the complex number $z = x + iy$, and the complex number z is called the 'affix' of the point P. The point P is also referred to as P(z) or simply as z.

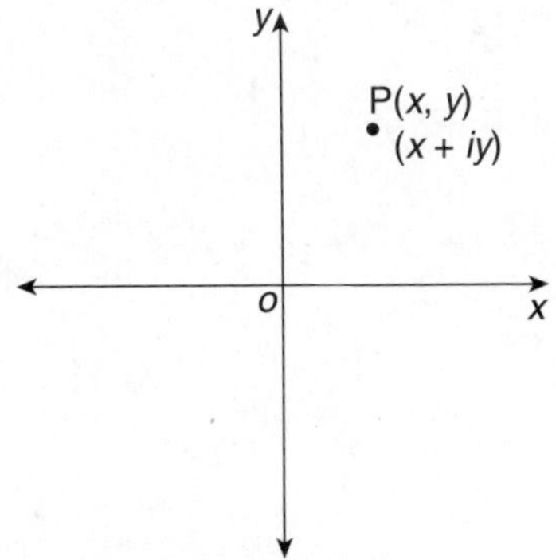

Fig. 11.1

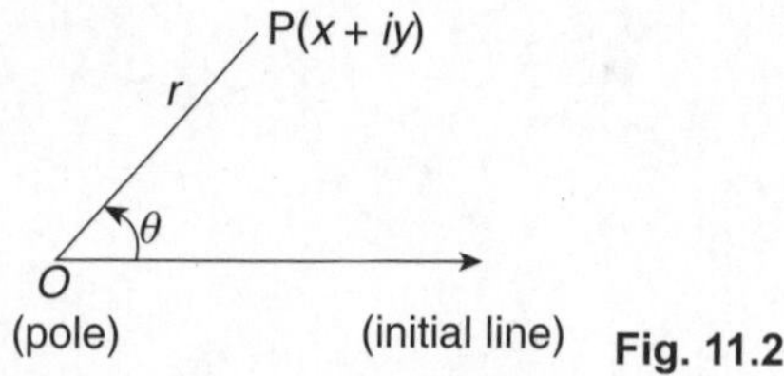

Fig. 11.2

Another way of representing a complex number is in its polar form, that is, $z = r(\cos\theta + i\sin\theta)$, where $r \in [0, \alpha)$ and $\theta \in [0, 2\pi)$. Here r is the distance of z from a fixed point O called the pole and θ is the angle made by OP in the positive direction, with a fixed line through O, called the initial line. The number r is called the 'modulus' and θ is called the 'amplitude or argument' of z. The modulus of a complex number z is usually denoted by $|z|$ and its argument is denoted by arg(z) or amp(z). For $z \neq 0$, the modulus and argument of z are uniquely defined. However, for $z = r(\cos\theta + i\sin\theta)$, $\theta \in [0, 2\pi)$, we have $z = r(\cos(\theta + 2k\pi) + i\sin(\theta + 2k\pi))$, $k \in Z$, where Z is the set of integers. The set

$$\text{Arg } z = \{\alpha : \alpha = \theta + 2k\pi, k \in Z, \theta \in (0, 2\pi)\}$$

is called the 'extended arguments' of the complex number z. Two non-zero complex numbers z_1 and z_2, represented as $z_1 = r_1(\cos\theta_1 + i\sin\theta_1)$, $z_2 = r_2(\cos\theta_2 + i\sin\theta_2)$ are equal if $r_1 = r_2$ and $\theta_1 - \theta_2 = 2k\pi$, where $k \in Z$.

REMARK: For defining argument we have taken the limit of θ from 0 to 2π. That is, $0 \leq \theta < 2\pi$. In some books, it is taken from $-\pi$ to π. Geometrically, they mean the same thing as $\cos(-\alpha) + i\sin(-\alpha) = \cos(2\pi-\alpha) + i\sin(2\pi-\alpha)$.

there is yet another approach to represent a complex number. Let P(x, y) denote the complex number $z = x + iy$ in a rectangular co-ordinate system. Then the vector $\vec{OP}$ has magnitude $= \sqrt{x^2 + y^2}$ and its direction is given by $\tan\theta = \frac{y}{x}$. Due to this fact, we can represent the complex number z by the vector $\vec{OP}$.

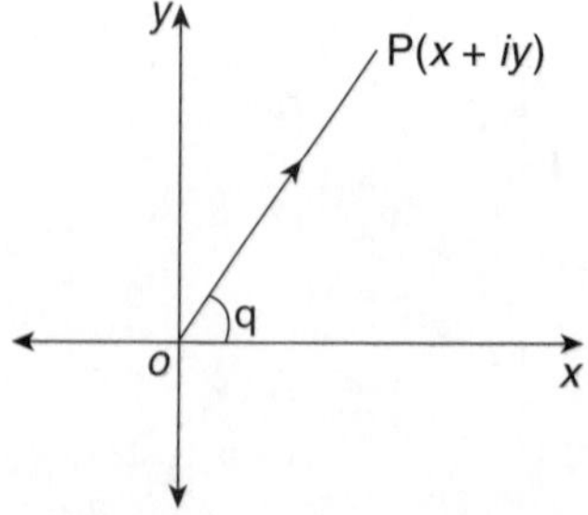

Fig. 11.3

REMARK: (i) The complex number $x - iy$ is called the 'conjugate' of the complex number $z = x + iy$ and is denoted by $\bar{z}$. It follows that

(i) $z\bar{z} = x^2 + y^2 = |z|^2$ (ii) $(\bar{\bar{z}}) = z$ (iii) $z + \bar{z} = 2\,\text{Re}\,(z)$.

EXAMPLE 11.1 Represent the complex numbers $2 - 3i, -4i, 2, -3 + 2i$ on the Argand plane.

SOLUTION

First draw two mutually perpendicular straight lines X'OX and Y'OY to represent the real axis and the imaginary axis. With suitable scale, plot the points (–2, –3), (0, –4), (2, 0) and (–3, 2), respectively, and name them as P, Q, R and S. The given numbers $-2, -3i, -4i, 2, -3 + 2i$ are represented by P, Q, R and S, respectively.

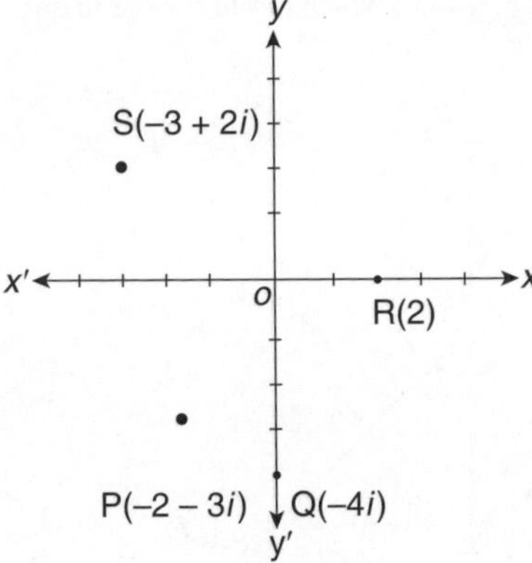

Fig. 11.4

EXAMPLE 11.2 Find the polar representation of the numbers

(a) $z_1 = -1 - i$ (b) $z_2 = 1 - i\sqrt{3}$

also determine their extended argument.

SOLUTION

(a) The image of z_1 is (–1, –1) which lies in the third quadrant.

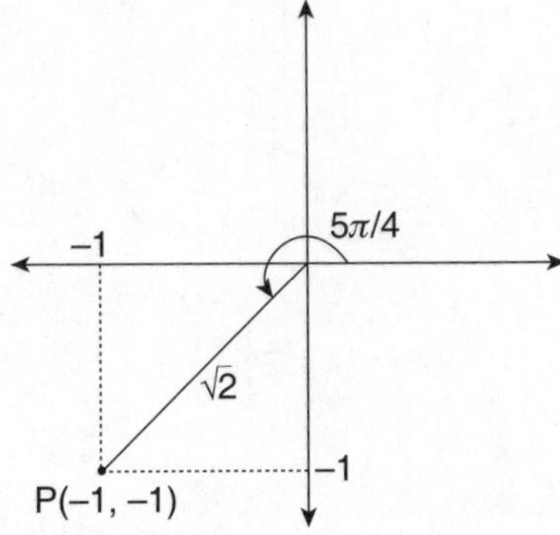

Fig. 11.5

Taking $r(\cos\theta + i\sin\theta) = -1 - i$,
we get, $r\cos\theta = -1$, $r\sin\theta = -1$

$$r = \sqrt{(-1)^2 + (-1)^2} = \sqrt{2}$$

$$\cos\theta = -\frac{1}{\sqrt{2}}, \sin\theta = -\frac{1}{\sqrt{2}}$$

$$\therefore\ \theta = \pi + \frac{\pi}{4} = \frac{5\pi}{4}$$

Hence, $z_1 = \sqrt{2}\left(\cos\frac{5\pi}{4} + i\sin\frac{5\pi}{4}\right)$

$$\text{Arg } z_1 = \left\{\frac{5\pi}{4} + 2k\pi \mid k \in Z\right\}$$

(b) The image of $z_2 = 1 - i\sqrt{3}$ is $(1, -\sqrt{3})$ which lies in the fourth quadrant.

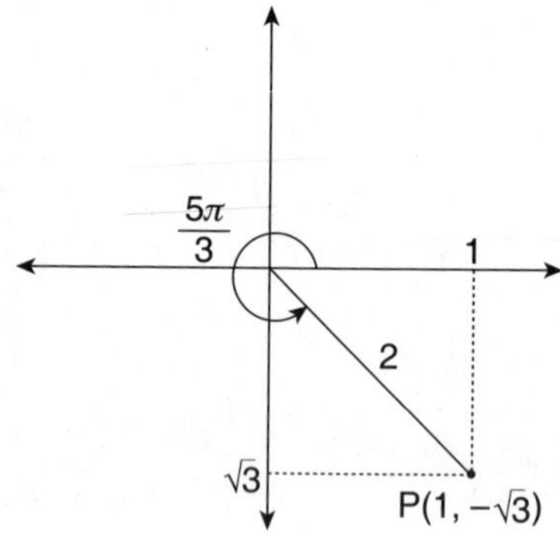

Fig. 11.6

Taking $r(\cos\theta + i\sin\theta) = 1 - i\sqrt{3}$,

we get, $r\cos\theta = 1$, $r\sin\theta = -\sqrt{3}$

squaring and adding $r^2 = 4$

$$\therefore\ r = 2, \cos\theta = \frac{1}{2}, \sin\theta = -\frac{\sqrt{3}}{2}$$

$$\therefore\ \theta = 2\pi - \frac{\pi}{3} = \frac{5\pi}{3}$$

Hence $$z_2 = 2\left(\cos\frac{5\pi}{3} + i\sin\frac{5\pi}{3}\right)$$

$$\text{Arg } z_2 = \left\{\frac{5\pi}{3} + 2k\pi \mid k \in Z\right\}$$

EXAMPLE 11.3 Find all complex numbers z such that $|z| = 1$ and $\left|\dfrac{z}{\bar{z}} + \dfrac{\bar{z}}{z}\right| = 1.$

SOLUTION

Let $z = \cos\theta + i\sin\theta$ $\quad (\because |z| = 1, r = 1)$
$\theta \in [0, 2\pi)$

$$\text{Then, } 1 = \left|\frac{z}{\bar{z}} + \frac{\bar{z}}{z}\right| = \frac{|z^2 + \bar{z}^2|}{|z\bar{z}|} = \frac{|z^2 + \bar{z}^2|}{|z|^2}$$

Now, $|z| = 1$, $z^2 = (\cos\theta + i\sin\theta)^2$

$$= \cos 2\theta + i\sin 2\theta$$

$$\bar{z}^2 = (\cos\theta - i\sin\theta)^2$$

$$= \cos 2\theta - i\sin 2\theta$$

$$\Rightarrow \quad 2|\cos 2\theta| = 1$$

$$\Rightarrow \quad \cos 2\theta = \frac{1}{2} \text{ or } \cos 2\theta = -\frac{1}{2}$$

If $\quad \cos 2\theta = \dfrac{1}{2}$, then $\theta = \dfrac{\pi}{6}, \dfrac{5\pi}{6}, \dfrac{7\pi}{6}, \dfrac{11\pi}{6}$

If $\quad \cos 2\theta = -\dfrac{1}{2}, \theta = \dfrac{\pi}{3}, \dfrac{2\pi}{3}, \dfrac{4\pi}{3}, \dfrac{5\pi}{3}.$

Hence, there are eight solutions
$z_k = \cos\theta_k + i\sin\theta_k$, $\quad k = 1, 2, \ldots, 8$

and $\quad \theta_k$'s are $\dfrac{\pi}{6}, \dfrac{\pi}{3}, \dfrac{2\pi}{3}, \dfrac{5\pi}{6}, \dfrac{7\pi}{6}, \dfrac{4\pi}{3}, \dfrac{5\pi}{3}, \dfrac{11\pi}{6}$, respectively.

EXAMPLE 11.4 Let z_1, z_2, z_3 be complex numbers such that $z_1 + z_2 + z_3 = 0$ and $|z_1| = |z_2| = |z_3| = 1$.

Prove that: $\quad z_1^2 + z_2^2 + z_3^2 = 0$

SOLUTION

$$z_1^2 + z_2^2 + z_3^2 = (z_1 + z_2 + z_3)^2 - 2(z_1 z_2 + z_2 z_3 + z_3 z_1)$$

$$= -2(z_1 z_2 + z_2 z_3 + z_3 z_1)$$

$$= -2 z_1 z_2 z_3 \left(\frac{1}{z_3} + \frac{1}{z_1} + \frac{1}{z_2}\right)$$

$$= -2 z_1 z_2 z_3 \left(\frac{\bar{z}_3}{z_3\bar{z}_3} + \frac{\bar{z}_1}{z_1\bar{z}_1} + \frac{\bar{z}_2}{z_2\bar{z}_2}\right)$$

$$= -2\,z_1 z_2 z_3\,(\bar{z}_3 + \bar{z}_1 + \bar{z}_2) \qquad (\because\ z_i\,\bar{z}_i = |z_i|^2 = 1, i = 1, 2, 3)$$

$$= -2\,z_1 z_2 z_3\left(\overline{z_3 + z_1 + z_2}\right)$$

$$= 0 \qquad (\because\ z_1 + z_2 + z_3 = 0)$$

11.1.1 Geometrical Representation of Sum, Difference, Product and Quotient of Complex Numbers

Sum of two complex numbers

Let $z_1 = x_1 + iy_1$ and $z_2 = x_2 + iy_2$ be two complex numbers. The sum of z_1 and z_2 is defined by

$$z_1 + z_2 = (x_1 + x_2) + i(y_1 + y_2).$$

Let P and Q be the images of z_1 and z_2 in the Argand plane. Then the vectors $\overrightarrow{OP}$ and $\overrightarrow{OQ}$ represent the complex numbers z_1 and z_2, respectively. We complete the parallelogram OPRQ with OP and OQ as two adjacent sides. By the law of parallelogram of vectors, $\overrightarrow{OR}$ is the resultant of $\overrightarrow{OP}$ and $\overrightarrow{OQ}$. Hence, $\overrightarrow{OR}$ represent the complex number $z_1 + z_2$. Thus, image of $z_1 + z_2$ is the point R, where OR is the diagonal of the parallelogram OPRQ. It may be verified that the co-ordinates of the point R are $(x_1 + x_2, y_1 + y_2)$.

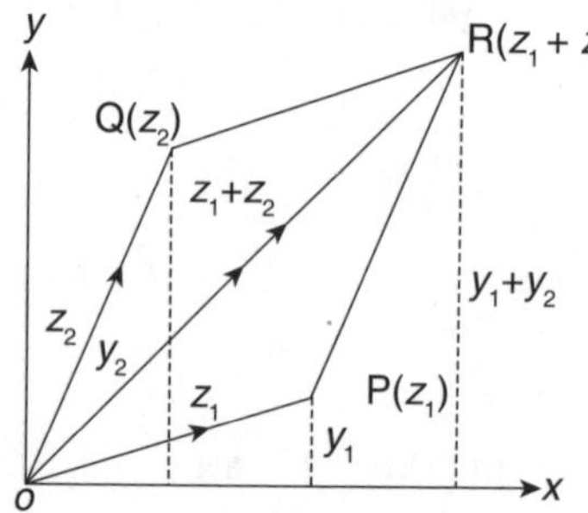

Fig. 11.7

Difference of two complex numbers

The difference of two complex numbers $z_1 = x_1 + iy_1$ and $z_2 = x_2 + iy_2$ is defined by

$$z_1 - z_2 = (x_1 - x_2) + i(y_1 - y_2)$$

As in the previous case, let $\overrightarrow{OP}$ and $\overrightarrow{OQ}$ represent the complex numbers z_1 and z_2, respectively. Then $\overrightarrow{OQ'}$ represent the complex number $-z_2$, where $\overrightarrow{OQ'}$ is equal in magnitude but opposite in direction to the vector $\overrightarrow{OQ}$. The diagonal vector $\overrightarrow{OR}$ represents the resultant of $\overrightarrow{OP}$ and $\overrightarrow{OQ'}$. Hence, $\overrightarrow{OR}$ represents the complex number $z_1 - z_2$. Thus, image of $z_1 - z_2$ is the point R, where OR is the diagonal of the parallelogram OPRQ′. It may be easily verified that the co-ordinates of the point R are $(x_1 - x_2, y_1 - y_2)$.

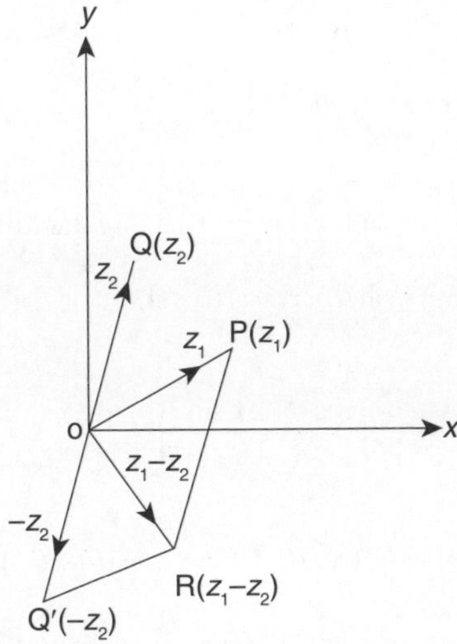

Fig. 11.8

Product of two complex numbers

Let the polar forms of $z_1 = x_1 + iy_1$ and $z_2 = x_2 + iy_2$ be $z_1 = r_1$ $(\cos \theta_1 + i \sin \theta_1)$ and $z_2 = r_2$ $(\cos \theta_2 + i \sin \theta_2)$.

Then
$$\begin{aligned} z_1 z_2 &= (x_1 + iy_1)(x_2 + iy_2) \\ &= r_1 r_2 (\cos \theta_1 + i \sin \theta_1)(\cos \theta_2 + i \sin \theta_2) \\ &= r_1 r_2 [\cos (\theta_1 + \theta_2) + i \sin (\theta_1 + \theta_2)] \end{aligned}$$

Let P and Q be the images of z_1 and z_2, respectively. Then, $OP = r_1$, $OQ = r_2$, $\angle POX = \theta_1$, $\angle QOX = \theta_2$.

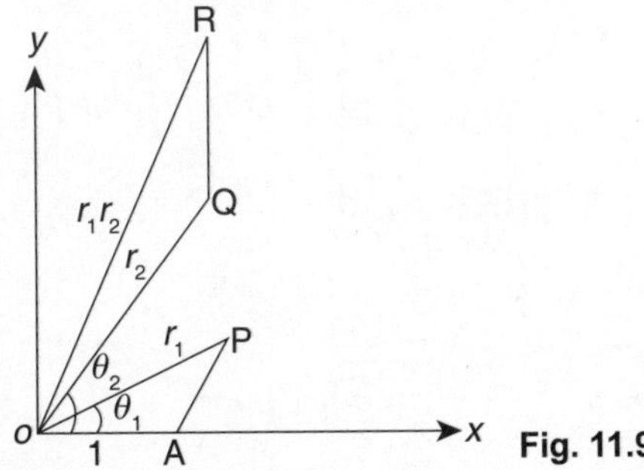

Fig. 11.9

Let A be a point on the real axis such that OA = 1. Join PA. Now draw triangle OQR similar to triangle OAP. From these two triangles, we get

$$\frac{OR}{OQ} = \frac{OP}{OA}$$

$$\Rightarrow \quad \frac{OR}{r_2} = \frac{r_1}{1}$$

$$\Rightarrow \quad OR = r_1 r_2$$

$$\angle ROX = \angle QOX + \angle ROQ$$
$$= \angle QOX + \angle POA$$
$$= \theta_2 + \theta_1$$

Hence R represents the complex number $z_1 z_2$.

Quotient of two complex numbers

Let $z_1 = x_1 + iy_1$ and $z_2 = x_2 + iy_2$ be two complex numbers having the following polar forms

$$z_1 = r_1 (\cos \theta_1 + i \sin \theta_1),\ z_2 = r_2 (\cos \theta_2 + i \sin \theta_2)$$

Then

$$\frac{z_1}{z_2} = \frac{r_1 (\cos \theta_1 + i \sin \theta_1)}{r_2 (\cos \theta_2 + i \sin \theta_2)}$$

$$= \frac{r_1}{r_2} [\cos (\theta_1 - \theta_2) + i \sin (\theta_1 - \theta_2)]$$

Let P and Q represent the complex numbers z_1 and z_2, respectively. Let A be a point on the real axis such that OA = 1. Now construct the triangle OPR similar to ΔOQA.

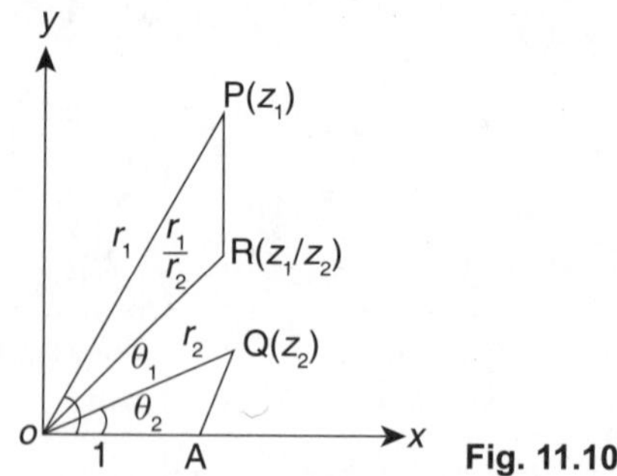

Fig. 11.10

We have

$$\frac{OP}{OR} = \frac{OQ}{OA}$$

$$\Rightarrow \frac{r_1}{OR} = \frac{r_2}{1}$$

$$\Rightarrow OR = \frac{r_1}{r_2}$$

Also,

$$\angle ROX = \angle POX - \angle POR$$
$$= \angle POX - \angle QOX$$
$$= \theta_1 - \theta_2$$

Thus, R has modulus $\frac{r_1}{r_2}$ and argument $\theta_1 - \theta_2$. Therefore, R represent the complex number $\frac{z_1}{z_2}$.

REMARK: If P and Q represent the complex numbers $z_1 = x_1 + iy_1$ and $z_2 = x_2 + iy_2$ respectively in the Argand plane, then co-ordinates of P and Q are (x_1, y_1) and (x_2, y_2), respectively. Therefore, we get,

$$PQ = \sqrt{(x_2 - x_1)^2 + (y_2 - y_1)^2}$$
$$= |(x_2 - x_1) + i(y_2 - y_1)|$$
$$= |z_2 - z_1|$$

EXAMPLE 11.5 Interpret geometrically: $(3 - 2i) - (-2 - 4i) = 5 + 2i$

SOLUTION

With a suitable scale on the Argand plane, P(3, –2) and Q(–2, –4) are plotted. Point Q′ is plotted on the line QO at a distance equal to OQ but in the opposite direction. With OP and OQ′ as adjacent sides, parallelogram OPRQ' is completed. Then, R represents the complex number $(3 - 2i) - (-2 - 4i)$. Co-ordinates of R are (5, 2).

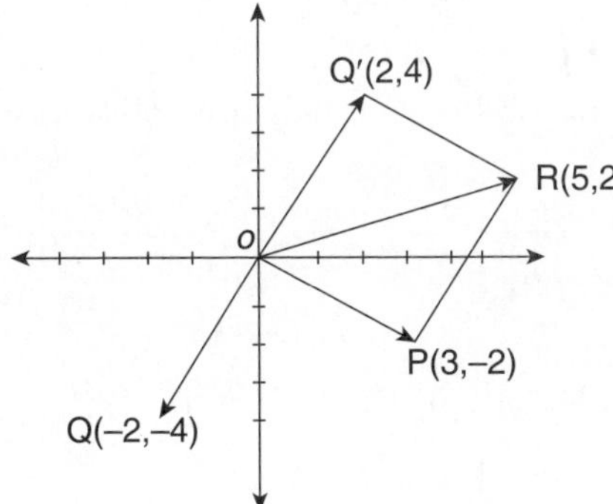

Fig. 11.11

EXAMPLE 11.6 Find the product of $z_1 = 1 - i$ and $z_2 = \sqrt{3} + i$ using polar representation of complex numbers.

SOLUTION

$$z_1 = 1 - i$$
$$= \sqrt{2}\left(\cos\frac{7\pi}{4} + i\sin\frac{7\pi}{4}\right) \quad \text{(verify)}$$
$$z_2 = \sqrt{3} + i$$
$$= 2\left(\cos\frac{\pi}{6} + i\sin\frac{\pi}{6}\right)$$

Hence

$$z_1 z_2 = 2\sqrt{2}\left[\cos\left(\frac{7\pi}{4} + \frac{\pi}{6}\right) + i\sin\left(\frac{7\pi}{4} + \frac{\pi}{6}\right)\right]$$
$$= 2\sqrt{2}\left(\cos\frac{23\pi}{12} + i\sin\frac{23\pi}{12}\right)$$

EXAMPLE 11.7 Show that the modulus of the sum of two complex numbers is always less than or equal to the sum of their moduli, that is,

$$|z_1 + z_2| \le |z_1| + |z_2|.$$

SOLUTION

Let the points P and Q represent the complex numbers z_1 and z_2, respectively, in the Argand plane. Complete the parallelogram OPRQ. Then R represents the complex number $z_1 + z_2$. Thus, we have OP = $|z_1|$, OQ = $|z_2|$ and OR = $|z_1 + z_2|$.

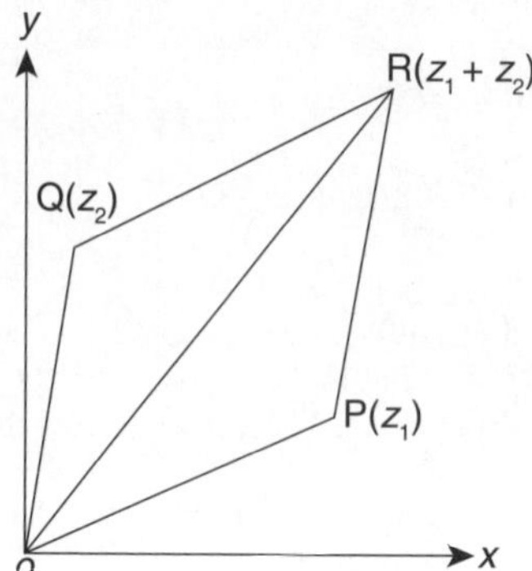

Fig. 11.12

Using the property that sum of two sides is greater than the third side in a triangle, we get

$$\text{OR} \le \text{OP} + \text{PR}$$

$$\Rightarrow \quad \text{OR} \le \text{OP} + \text{OQ} \qquad (\because\ \text{OQ} = \text{PR})$$

$$\Rightarrow \quad |z_1 + z_2| \le |z_1| + |z_2|$$

Alternatively, we have

$$|z_1 + z_2|^2 = (z_1 + z_2)\left(\overline{z_1 + z_2}\right)$$

$$= (z_1 + z_2)(\bar{z}_1 + \bar{z}_2)$$

$$= z_1\,\bar{z}_1 + z_2\,\bar{z}_2 + z_1\,\bar{z}_2 + z_2\,\bar{z}_1$$

$$= |z_1|^2 + |z_2|^2 + z_1\,\bar{z}_2 + \left(\overline{z_1 \bar{z}_2}\right)$$

$$= |z_1|^2 + |z_2|^2 + 2\,\text{Re}\,(z_1\,\bar{z}_2)$$

$$\le |z_1|^2 + |z_2|^2 + 2\,|z_1\, z_2|$$

$$= (|z_1| + |z_2|)^2$$

$$\therefore\ |z_1 + z_2| \le |z_1| + |z_2|$$

If O, P, Q are collinnear, then OR = OP + PR = OP + OQ; thus

$|z_1 + z_2| = |z_1| + |z_2|$.

EXAMPLE 11.8 Prove that modulus of the difference of two complex numbers is always greater than or equal to the difference of their moduli, that is,

$$|z_1 - z_2| \geq ||z_1| - |z_2||$$

SOLUTION

Let the points P and Q represent the complex numbers z_1 and z_2, respectively, in the Argand plane.

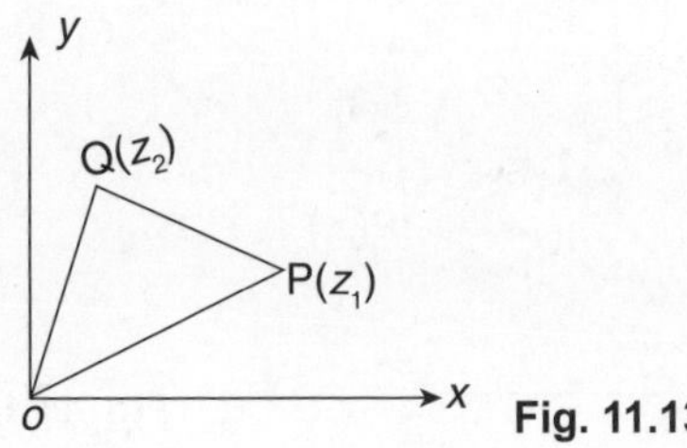

Fig. 11.13

Then,

$$OP = |z_1|, OQ = |z_2| \text{ and } PQ = |z_2 - z_1| = |z_1 - z_2|$$

In the triangle OPQ,

$$OP + PQ \geq OQ$$

$$OQ + PQ \geq OP$$

$$\Rightarrow |z_1| + |z_1 - z_2| \geq |z_2|$$

$$|z_2| + |z_1 - z_2| \geq |z_1|$$

$$\Rightarrow |z_1 - z_2| \geq |z_2| - |z_1|$$

and $$|z_1 - z_2| \geq |z_1| - |z_2|$$

$$\Rightarrow |z_1 - z_2| \geq ||z_1| - |z_2||$$

EXAMPLE 11.9 Prove that $|z_1 + z_2|^2 + |z_1 - z_2|^2 = 2(|z_1|^2 + |z_2|^2)$ for any two complex numbers z_1 and z_2. Also interpret the result geometrically.

SOLUTION

We have

$$|z_1 + z_2|^2 = (z_1 + z_2)\left(\overline{z_1 + z_2}\right) = (z_1 + z_2)\left(\bar{z}_1 + \bar{z}_2\right)$$

$$\text{or, } |z_1 + z_2|^2 = z_1\bar{z}_1 + z_2\bar{z}_2 + z_1\bar{z}_2 + z_2\bar{z}_1 = |z_1|^2 + |z_2|^2 + z_1\bar{z}_2 + z_2\bar{z}_1 \quad (1)$$

Similarly,

$$|z_1 - z_2|^2 = |z_1|^2 + |z_2|^2 - z_1\bar{z}_2 - \bar{z}_1 z_2 \quad (2)$$

Adding (1) and (2), we get

$$|z_1 + z_2|^2 + |z_1 - z_2|^2 = 2(|z_1|^2 + |z_2|^2) \quad (3)$$

Geometrical Interpretation

Let P and Q represent the complex numbers z_1 and z_2, respectively, in the Argand plane. We complete the parallelogram OPRQ with OP and OQ as adjacent sides. Then, we have

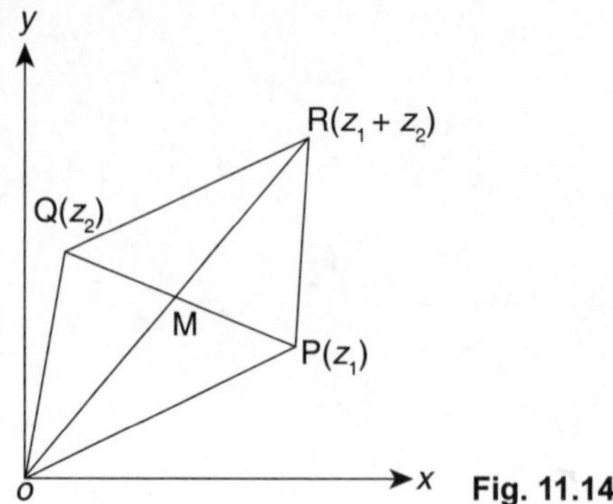

Fig. 11.14

$$OP = |z_1|, OQ = |z_2|, OR = |z_1 + z_2|, QP = |z_1 - z_2|$$

Hence (3) implies that

$$\begin{aligned} OR^2 + QP^2 &= 2\,(OP^2 + OQ^2) \\ &= 2\,OP^2 + 2\,OQ^2 \\ &= OP^2 + QR^2 + OQ^2 + PR^2 \qquad \text{as } OP = QR,\ OQ = PR \end{aligned}$$

This shows that the sum of squares of diagonal is equal to the sum of squares of four sides in a parallelogram.

EXAMPLE 11.10 Let z_1, z_2, z_3 be the affixes of the points P, Q and R, respectively. If $|z_1| = |z_2| = |z_3|$ and $z_1 + z_2 + z_3 = 0$, prove that the triangle PQR is equilateral.

SOLUTION

We have $z_1 + z_2 = -z_3$, so that $|z_1 + z_2| = |z_3|$.

$\therefore$ From the identity,

$$|z_1 + z_2|^2 + |z_1 - z_2|^2 = 2\,(|z_1|^2 + |z_2|^2)$$

We get $\qquad |z_1 - z_2|^2 = 3\,|z_3|^2 \qquad (\because |z_1| = |z_2| = |z_3|)$

Similarly, $\qquad |z_2 - z_3|^2 = 3\,|z_3|^2, |z_3 - z_1| = 3\,|z_3|^2$

$$\therefore |z_1 - z_2| = |z_2 - z_3| = |z_3 - z_1|$$

$$\Rightarrow \quad PQ = QR = RQ$$

Hence, triangle PQR is equilateral.

11.1.2 Point Dividing a Line Segment in a Given Ratio

Let $z_1 = x_1 + iy_1$ and $z_2 = x_2 + iy_2$ be two complex numbers and let P, Q be the points on the Argand plane representing z_1 and z_2, respectively. Then P, Q have co-ordinates (x_1, y_1) and (x_2, y_2), respectively.

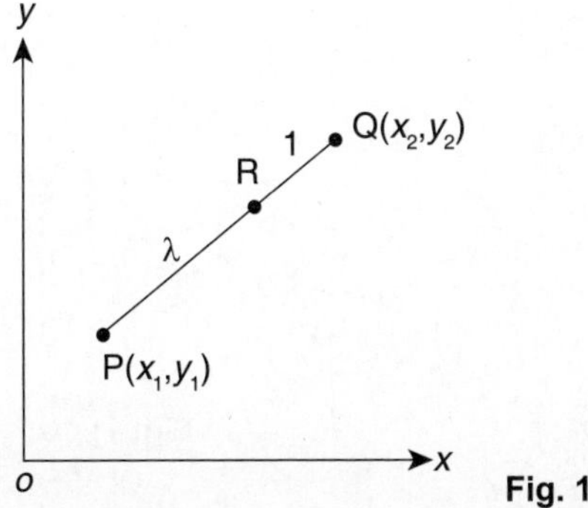

Fig. 11.15

Suppose R divides the line segment PQ in the ratio $\lambda : 1, \lambda \neq -1$. Then PR : RQ = $\lambda : 1$

Therefore, the co-ordinates of R are

$$\left(\frac{x_1 + \lambda\, x_2}{1+\lambda}, \frac{y_1 + \lambda\, y_2}{1+\lambda}\right)$$

The affix of R is, therefore,

$$\left(\frac{x_1 + \lambda\, x_2}{1+\lambda}\right) + i\left(\frac{y_1 + \lambda\, y_2}{1+\lambda}\right)$$

$$= \frac{x_1 + i\, y_1}{1+\lambda} + \lambda \frac{x_2 + i\, y_2}{1+\lambda}$$

$$= \frac{z_1 + \lambda\, z_2}{1+\lambda}.$$

REMARK: The affix of the midpoint of PQ is $\frac{1}{2}(z_1 + z_2)$.

EXAMPLE 11.11 Show that the centroid of a triangle whose vertices are z_1, z_2, z_3 is $\frac{1}{3}(z_1 + z_2 + z_3)$.

SOLUTION

Let P, Q, R be the points on the Argand plane for the affixes z_1, z_2 and z_3 respectively. Let $z_1 = x_1 + iy_1$, $z_2 = x_2 + iy_2$, $z_3 = x_3 + iy_3$, so that the co-ordinates of P, Q, R are (x_1, y_1), (x_2, y_2) and (x_3, y_3), respectively.

Let G be the centroid of Δ PQR so that its co-ordinates are $\left(\frac{x_1 + x_2 + x_3}{3}, \frac{y_1 + y_2 + y_3}{3}\right)$. Then affix of the point G is

$$\left(\frac{x_1 + x_2 + x_3}{3}\right) + i\left(\frac{y_1 + y_2 + y_3}{3}\right)$$

$$= \frac{1}{3}\left[(x_1 + iy_1) + (x_2 + iy_2) + (x_3 + iy_3)\right]$$

$$= \frac{1}{3}(z_1 + z_2 + z_3).$$

Alternatively, let P, Q and R be the points on the Argand plane representing z_1, z_2 and z_3, respectively. Let G be the centroid of Δ PQR.

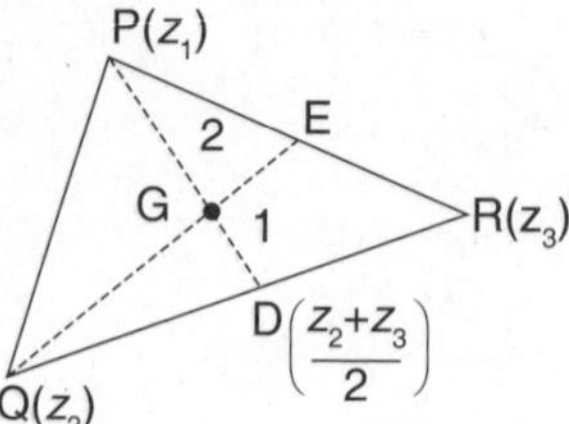

Fig. 11.16

If D is the mid-point of Q and R, then affix of D is $\frac{1}{2}(z_2 + z_3)$. Now, by property of centroid, G divides PD in the ratio 2 : 1. Hence the affix of G is

$$\frac{2\frac{1}{2}(z_2 + z_3) + 1\,z_1}{2+1}$$

$$= \frac{1}{3}(z_1 + z_2 + z_3).$$

11.1.3 Regarding Angles in the Argand Plane

The following result, regarding angles formed in the Argand plane, is a direct application of the geometrical representation of division of complex numbers.

THEOREM

If z_1, z_2, z_3 are vertices of a triangle ABC described in the anticlockwise direction, then

(i) $\arg\left(\frac{z_3 - z_1}{z_2 - z_1}\right) = \angle ABC = \alpha$ (say); and (ii) $\frac{z_3 - z_1}{z_2 - z_1} = \frac{AC}{AB}(\cos\alpha + i\sin\alpha)$.

Proof: Let P and Q be two points in the Argand plane representing $z_2 - z_1$ and $z_3 - z_1$, respectively.

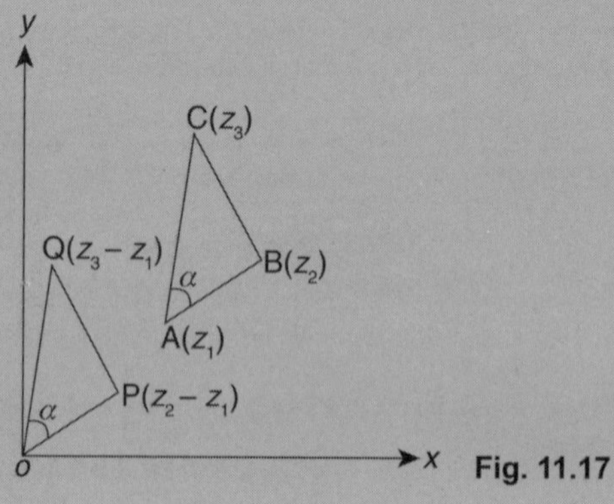

Fig. 11.17

Then,

$$AB = | z_2 - z_1 | = OP,$$

$$AC = | z_3 - z_1 | = OQ,$$

$$BC = | z_3 - z_2 | = PQ$$

Hence $\Delta ABC \cong \Delta OPQ$

Therefore $\angle POQ = \angle BAC = \alpha$

By the geometrical interpretation of quotient,

$\dfrac{z_3 - z_1}{z_2 - z_1}$ has modulus $\dfrac{OQ}{OP}$ and argument α.

Hence,
$$\frac{z_3 - z_1}{z_2 - z_1} = \frac{OQ}{OP}(\cos\alpha + i\sin\alpha)$$

$$= \frac{AC}{AB}(\cos\alpha + i\sin\alpha)$$

REMARK: The points z_1, z_2, z_3 will be collinear if and only if $\alpha = 0$ or π, in which case $\dfrac{z_3 - z_1}{z_2 - z_1}$ is purely real.

EXAMPLE 11.12 Find $\angle POQ$ and $\angle QOP$ where P and Q are two points on the Argand plane with affixes $z_1 = 1 + i$ and $z_2 = -1 + i$, respectively.

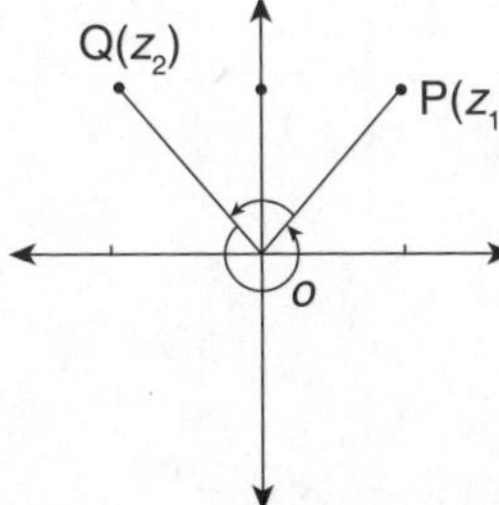

Fig. 11.18

SOLUTION

Here
$$z_1 = 1 + i$$
$$z_2 = -1 + i$$

$$\therefore \angle POQ = \arg\frac{z_2}{z_1} = \arg\frac{-1+i}{1+i}$$

$$= \arg\frac{(-1+i)(-1-i)}{2}$$

$$= \arg i$$

$$= \frac{\pi}{2}$$

$$\angle QOP = \arg \frac{z_1}{z_2} = \arg \frac{1}{i} = \arg(-i) = \frac{3\pi}{2}$$

EXAMPLE 11.13 For $P(z_1)$, $Q(z_2)$ and $R(z_3)$ on the Argand plane, find $\angle QPR$ and $\angle RPQ$ if $z_1 = 4 + 3i$, $z_2 = 4 + 7i$, $z_3 = 8 + 7i$.

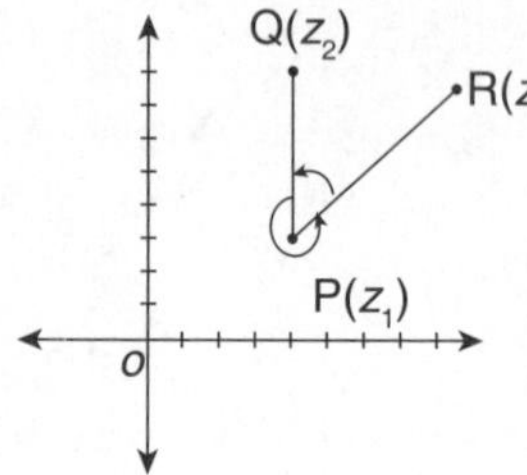

Fig. 11.19

SOLUTION

Here $\qquad z_1 = 4 + 3i,\ z_2 = 4 + 7i,\ z_3 = 8 + 7i$

$$\therefore\ \frac{z_2 - z_1}{z_3 - z_1} = \frac{4i}{4 + 4i} = \frac{i(1-i)}{(1+i)(1-i)} = \frac{1+i}{2}$$

$$\therefore\ \angle RPQ = \arg \frac{z_2 - z_1}{z_3 - z_1} = \arg\left(\frac{1+i}{2}\right) = \frac{\pi}{4}$$

and
$$\angle QPR = \arg \frac{z_3 - z_1}{z_2 - z_1} = \arg \frac{2}{1+i} = \arg(1-i) = \frac{7\pi}{4}$$

The following result can be proved by using the previous result.

THEOREM

Let z_1, z_2, z_3 and z_4 be four complex numbers. Then the line joining z_4 and z_3 is inclined to the line joining z_2 and z_1 at the following angle.

$$\arg\left(\frac{z_1 - z_2}{z_3 - z_4}\right) \text{ or } \arg\left(\frac{z_3 - z_4}{z_1 - z_2}\right)$$

Corollary: The line joining the complex numbers z_1 and z_2 is parallel (resp. perpendicular) to the line joining z_3 and z_4 if and only if $\frac{z_1 - z_2}{z_3 - z_4}$ is purely real (resp. purely imaginary).

P.T.O.

EXAMPLE 11.14 If z_1, z_2, z_3 are vertices of an isosceles triangle, right angled at the vertex z_2, prove that

$$z_1^2 + z_2^2 + z_3^2 = 2z_2(z_1 + z_2)$$ (D.U., B.Sc. (Ph.Sc.) 2009)

SOLUTION

Let A, B and C be the points in the Argand plane whose affixes are z_1, z_2 and z_3, respectively. Now

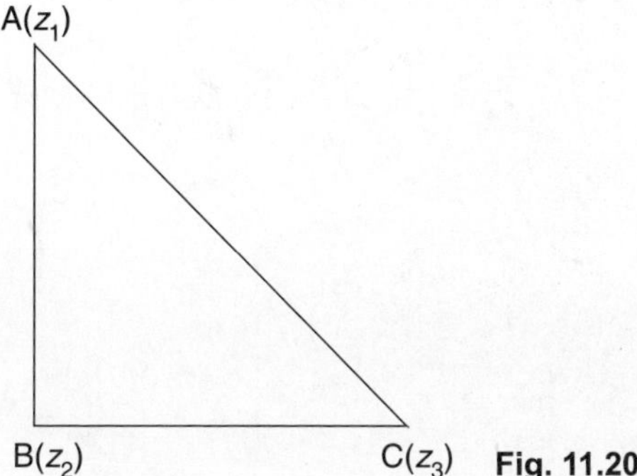

Fig. 11.20

$$\frac{z_1 - z_2}{z_3 - z_2} = \frac{AB}{BC}\left(\cos\frac{\pi}{2} + i\sin\frac{\pi}{2}\right)$$

$$= i$$

$$\Rightarrow \quad z_1 - z_2 = i\,(z_3 - z_2)$$

Squaring both the sides

$$(z_1 - z_2)^2 = -(z_3 - z_2)^2$$

$$\Rightarrow \quad z_1^2 + z_2^2 - 2z_1z_2 = -(z_3^2 + z_2^2 - 2z_2z_3)$$

$$\Rightarrow \quad z_1^2 + 2z_2^2 + z_3^2 = 2(z_1z_2 + z_2z_3)$$

$$= 2z_2(z_1 + z_3)$$

EXAMPLE 11.15 If A (a, b), C (c, d) be two opposite vertices of a square, find the other two vertices.

SOLUTION

Let z_1, z_2, z_3 and z_4 be the affixes of the points A, B, C and D, respectively, in the Argand plane.

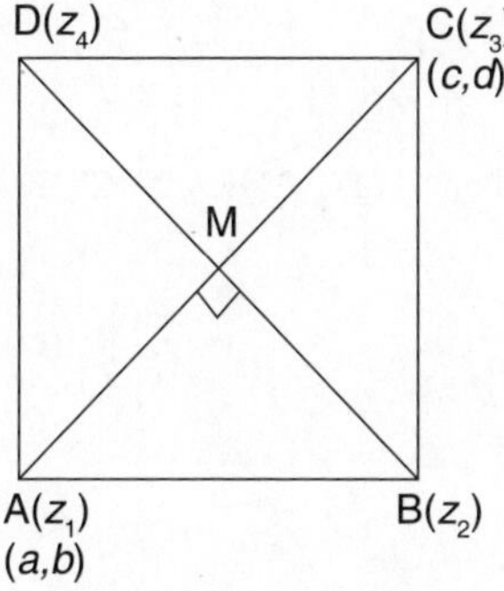

Fig. 11.21

In a square, the diagonals bisect each other perpendicularly. Therefore, the middle point M of AC is also the middle point of BD; also the angle between AC and BD is $\frac{\pi}{2}$. Accordingly, we have

$$\frac{z_1 + z_3}{2} = \frac{z_2 + z_4}{2} \tag{1}$$

and,
$$\frac{z_4 - \frac{z_2 + z_4}{2}}{z_3 - \frac{z_1 + z_3}{2}} = \frac{DM}{CM}\left(\cos\frac{\pi}{2} + i\sin\frac{\pi}{2}\right)$$

$$\Rightarrow \quad \frac{z_4 - z_2}{z_3 - z_1} = i \qquad (\because DM = CM)$$

$$\Rightarrow \quad z_4 - z_2 = i\,(z_3 - z_1) \tag{2}$$

Now, from (1) and (2), we get

$$\begin{aligned} z_4 + z_2 &= z_3 + z_1 \\ &= (c + id) + (a + ib) \\ &= (c + a) + i\,(d + b) \\ z_4 - z_2 &= i\,[(c + id) - (a + ib)] \\ &= (b - d) + i\,(c - a) \end{aligned}$$

$$\Rightarrow z_4 = \frac{1}{2}\,[(c + a + b - d) + i\,(d + b + c - a)]$$

$$= \frac{1}{2}\,[(a + b + c - d) + i\,(-a + b + c + d)]$$

and
$$z_2 = \frac{1}{2}\,[(c + a - b + d) + i\,(d + b - c + a)]$$

$$= \frac{1}{2}\,[(a - b + c + d) + i\,(a + b - c + d)]$$

EXAMPLE 11.16 Show that a necessary as well as sufficient condition for the points $P(z_1)$, $Q(z_2)$ and $R(z_3)$ to represent the vertices of an equilateral triangle is that

$$z_1^2 + z_2^2 + z_3^2 = z_1 z_2 + z_2 z_3 + z_3 z_1$$

SOLUTION

The triangle PQR is equilateral if and only if PQ = QR and $\angle PQR = \frac{\pi}{3}$ or $2\pi - \frac{\pi}{3}$, and this holds if and only if

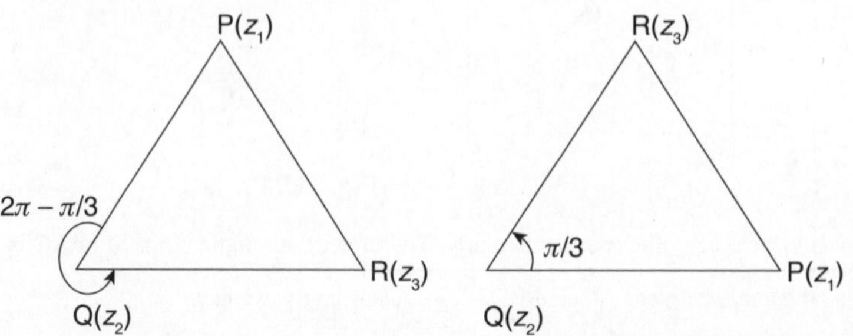

Fig. 11.22

$$\frac{z_3 - z_2}{z_1 - z_2} = \cos\frac{\pi}{3} \pm i\sin\frac{\pi}{3}$$

We write $\quad z_1 - z_2 = \alpha,\ z_2 - z_3 = \beta$

and $\quad z_3 - z_1 = \gamma$. Then,

$$\frac{-\beta}{\alpha} = \frac{1}{2} \pm i\frac{\sqrt{3}}{2}$$

$$\Rightarrow \left(\frac{\alpha}{2} + \beta\right)^2 = -\frac{3}{4}\alpha^2$$

$$\Rightarrow \alpha^2 + \beta^2 + \alpha\beta = 0$$

Similarly, we have $\beta^2 + \gamma^2 + \beta\gamma = 0$, $\gamma^2 + \alpha^2 + \gamma\alpha = 0$, so that on adding, we get

$$2(\alpha^2 + \beta^2 + \gamma^2) + (\alpha\beta + \beta\gamma + \gamma\alpha) = 0 \tag{1}$$

Again $\quad \alpha + \beta + \gamma = z_1 - z_2 + z_2 - z_3 + z_3 - z_1 = 0$

$$\Rightarrow \quad \alpha^2 + \beta^2 + \gamma^2 + 2(\alpha\beta + \beta\gamma + \gamma\alpha) = 0 \tag{2}$$

From (1) and (2), $\alpha^2 + \beta^2 + \gamma^2 = 0$

$$\Rightarrow (z_1 - z_2)^2 + (z_2 - z_3)^2 + (z_3 - z_1)^2 = 0$$

$$\Rightarrow z_1^2 + z_2^2 + z_3^2 = z_1 z_2 + z_2 z_3 + z_3 z_1$$

REMARK: In the above example, from (1) and (2), we get

$$\alpha^2 + \beta^2 + \gamma^2 = 0,$$

as well as $\quad \alpha\beta + \beta\gamma + \gamma\alpha = 0$

While $\quad \alpha^2 + \beta^2 + \gamma^2 = 0$ implies $(z_1 - z_2)^2 + (z_2 - z_3)^2 + (z_3 - z_1)^2 = 0 \quad$ (1)

as well as $\quad z_1^2 + z_2^2 + z_3^2 = z_1 z_2 + z_2 z_3 + z_3 z_1 \quad$ (2)

and

$$\begin{vmatrix} z_1 & z_2 & z_3 \\ z_2 & z_3 & z_1 \\ 1 & 1 & 1 \end{vmatrix} = 0, \tag{3}$$

from $\quad \alpha\beta + \beta\gamma + \gamma\alpha = 0$, we get $\dfrac{1}{\alpha} + \dfrac{1}{\beta} + \dfrac{1}{\gamma} = 0$,

that is,

$$\frac{1}{z_1 - z_2} + \frac{1}{z_2 - z_3} + \frac{1}{z_3 - z_1} = 0 \tag{4}$$

All these expressions provide necessary and sufficient conditions for the triangle PQR to be equilateral.

EXERCISE 11.1

1. Represent the following complex numbers on the Argand plane
   $z_1 = 3 + i$, $z_2 = -4 + 2i$, $z_3 = -5 - 4i$, $z_4 = 2i$, $z_5 = -4$
2. Find the polar representations of the numbers
   (a) $2 + 2i$ **(b)** $-1 + i\sqrt{3}$ **(c)** $1 + \cos\alpha + i\sin\alpha$ **(d)** 2 **(e)** –3i
3. Find all the complex numbers $z \neq 0$ such that $z + \dfrac{1}{z} \in \mathbb{R}$.
4. Find all complex numbers z such that $|z| = \left|\dfrac{1}{z}\right|$.
5. Let z_1, z_2 be two complex numbers such that $|z_1 + z_2| = \sqrt{3}$ and $|z_1| = |z_2| = 1$. Compute $|z_1 - z_2|$.
6. Find $|z|$, arg z, Arg z, arg $\bar{z}$, arg $(-z)$ for
   (a) $z = (1 - i)(6 + 6i)$ **(b)** $z = (7 - 7\sqrt{3}i)(-1 - i)$
7. Compute the following products using polar representation of complex numbers:
   (a) $(1 + i)(-2 - 2i)(i)$ **(b)** $-2i(-4 + 4\sqrt{3}i)(3 + 3i)$
   (c) $(1 - i)(-5 + 5i)$ **(d)** $\left(\dfrac{1}{2} - i\dfrac{\sqrt{3}}{2}\right)(-3 + 3i)(2\sqrt{3} + 2i)$
   Verify your results using algebraic form.
8. If P, Q, R are the points of affix z_1, z_2, $z_1 + z_2$, respectively, show that OPRQ is a parallelogram.
9. Find the modulus and argument of the centroid of the triangle whose vertices are given by $8 + 5i$, $-3 + i$, and $-2 - 3i$, respectively.
10. The points A, B, C represent the complex numbers z_1, z_2 and z_3, respectively, and G is the centroid of the triangle ABC. If $4z_1 + z_2 + z_3 = 0$ show that the origin is the midpoint of AG.
11. Show that a necessary and sufficient condition that the points P, Q, R representing the numbers z_1, z_2 and z_3, respectively, on the Argand plane are vertices of an equilateral triangle is that $\dfrac{1}{z_1 - z_2} + \dfrac{1}{z_2 - z_3} + \dfrac{1}{z_3 - z_1} = 0$.
12. Show that the modulus of the sum of two complex numbers is always greater than the difference of their moduli.
13. If z_1, z_2, z_3 be the affixes of the vertices of an equilateral triangle and z_0 be the affix of the centroid of the triangle, prove that
$$z_1^2 + z_2^2 + z_3^2 = 3z_0^2.$$
14. Prove that the complex numbers z_1, z_2, z_3 represent the vertices of an equilateral triangle if and only if $\begin{vmatrix} z_1 & z_2 & z_3 \\ z_2 & z_3 & z_1 \\ 1 & 1 & 1 \end{vmatrix} = 0$.
15. The numbers $2 + 3i$, $8 + 11i$, $17i$ are represented on the Argand diagram by the points A, B and C, respectively. Show that A, B and C are three vertices of a square. Find the fourth vertex.

16. OPQ is an equilateral triangle, named in counter-clockwise direction. If P and Q represent the complex numbers z_1 and z_2, show that $z_1 = z_2\left(\frac{1}{2} - i\frac{\sqrt{3}}{2}\right)$.

17. The vertices of a square taken in order in anti-clockwise direction represent the numbers z_1, z_2, z_3 and z_4. Prove that

$$z_2 = \frac{1}{2}(1+i)\,z_1 + \frac{1}{2}(1-i)\,z_3, \qquad z_4 = \frac{1}{2}(1-i)\,z_1 + \frac{1}{2}(1+i)\,z_3$$

■ 11.2 STRAIGHT LINES IN ARGAND PLANE

11.2.1 General Equation of a Straight Line

The general equation of a straight line in the Argand plane is

$$\bar{q}\,z + q\,\bar{z} + r = 0, \text{ where } r \text{ is real.}$$

Let $z = x + iy$ be any complex number. Then $\bar{z} = x - iy$, so that $x = \dfrac{z+\bar{z}}{2}$ and $y = \dfrac{z-\bar{z}}{2i}$.

We know that the general equation of a straight line is

$$ax + by + c = 0 \tag{1}$$

Substituting the values of x and y as obtained above, in (1), we get

$$a\left(\frac{z+\bar{z}}{2}\right) + b\left(\frac{z-\bar{z}}{2i}\right) + c = 0$$

$$\Rightarrow \quad ia\,(z + \bar{z}) + b\,(z - \bar{z}) + 2ic = 0$$

$$\Rightarrow \quad (ia + b)\,z + (ia - b)\,\bar{z} + 2ic = 0$$

Multiplying by i, we get

$$(-a + bi)\,z + (-a - bi)\,\bar{z} - 2c = 0$$

We write $q = -a - bi$, $\bar{q} = -a + bi$ and $r = -2c$.

The above equation then reduces to

$$\bar{q}z + q\bar{z} + r = 0 \tag{2}$$

We can make the following observation:

(A) If the straight line passes through two points with affixes z_1 and z_2, respectively, then

$$\bar{q}\,z_1 + q\,\bar{z}_1 + r = 0 \tag{3}$$

$$\bar{q}\,z_2 + q\,\bar{z}_2 + r = 0 \tag{4}$$

From (2), (3) and (4), we get

$$\begin{vmatrix} z & \bar{z} & 1 \\ z_1 & \bar{z}_1 & 1 \\ z_2 & \bar{z}_2 & 1 \end{vmatrix} = 0 \tag{5}$$

Also, from (2), (3) and (4), we get

$$\bar{q}\,(z - z_1) + q\,(\bar{z} - \bar{z}_1) = 0$$

$$\bar{q}\,(z_2 - z_1) + q\,(\bar{z}_2 - \bar{z}_1) = 0$$

$$\Rightarrow \frac{z - z_1}{z_2 - z_1} = \frac{\bar{z} - \bar{z}_1}{\bar{z}_2 - \bar{z}_1} \tag{6}$$

The equations (5) and (6) both represent a straight line passing through the two points whose affixes are z_1 and z_2.

(B) From (5), it follows that three complex numbers z_1, z_2 and z_3 are collinear if

$$\begin{vmatrix} z_1 & \bar{z}_1 & 1 \\ z_2 & \bar{z}_2 & 1 \\ z_3 & \bar{z}_3 & 1 \end{vmatrix} = 0$$

REMARKS:

(1) We may rewrite equation (2) in terms of x, y as

$$\bar{q}\,z + q\,\bar{z} + r = 0 \tag{2}$$

$$\Rightarrow \bar{q}\,(x + iy) + q(x - iy) + r = 0$$

$$\Rightarrow (\bar{q} + q)x + i\,(\bar{q} - q)\,y + r = 0$$

Comparing this equation with equation (1), that is,

$$ax + by + c = 0, \tag{1}$$

We get

$$m = \text{slope of the straight line (1)}$$

$$= -\frac{a}{b}$$

$$= -\frac{\bar{q} + q}{i\,(\bar{q} - q)}$$

$$= -i\left(\frac{q + \bar{q}}{q - \bar{q}}\right)$$

For $q \neq \bar{q}$, m is called the *angular coefficient* of the line (2).

(Remember: (1) and (2) in the above discussion represent the same straight line, only difference being that in (1), we have used Cartesian co-ordinates, whereas in (2) we have used complex co-ordinates).

(2) Let $\bar{q}_1 z + q_1 \bar{z} + r_1 = 0$

and $\bar{q}_2 z + q_2 \bar{z} + r_2 = 0$

be two straight lines in the Argand plane. Then proceeding as in above, we can conclude that the two lines are

parallel if and only if $-i\left(\frac{q_1 + \bar{q}_1}{q_1 - \bar{q}_1}\right) = -i\left(\frac{q_2 + \bar{q}_2}{q_2 - \bar{q}_2}\right)$, that is, $q_2 \bar{q}_1 = q_1 \bar{q}_2$,

whence, $\frac{q_1}{\bar{q}_1} = \frac{q_2}{\bar{q}_2}$

perpendicular if and only if $-i\left(\frac{q_1 + \bar{q}_1}{q_1 - \bar{q}_1}\right) \times -i\left(\frac{q_2 + \bar{q}_2}{q_2 - \bar{q}_2}\right) = -1$, that is, $\frac{q_1}{\bar{q}_1} + \frac{q_2}{\bar{q}_2} = 0$

concurrent if and only if $-i\left(\frac{q_1 + \bar{q}_1}{q_1 - \bar{q}_1}\right) \neq -i\left(\frac{q_2 + \bar{q}_2}{q_2 - \bar{q}_2}\right)$, that is, $\frac{q_1}{\bar{q}_1} \neq \frac{q_2}{\bar{q}_2}$

11.2.2 Parametric Equation of a Straight Line

The parametric equation of a straight line which passes through a point $Q(z_1)$ and is parallel to the straight line joining the origin to the point $R(z_0)$ is

$$z = z_1 + t\, z_0,$$

where t is a real parameter.

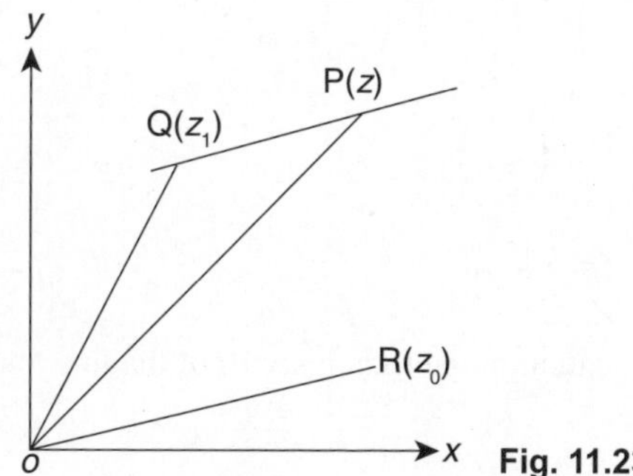

Fig. 11.23

Let P(z) be any point on the straight line. Following vectorial approach,

$$\overrightarrow{OP} = \overrightarrow{OQ} + \overrightarrow{QP}$$

Now, $\vec{QP} = t\,\vec{OR}$ for some real number t. Hence we have,

$$\vec{OP} = \vec{OQ} + t\,\vec{OR}$$

$$\Rightarrow z = z_1 + t\,z_0$$

which is the required equation.

REMARK: If the above straight line also passes through a point with affix z_2, then we get

$$z_2 = z_1 + t_0\,z_0 \text{ for some real number } t_0.$$

Hence, we get $z = z_1 + \frac{t}{t_0}(z_2 - z_1)$

$$\Rightarrow z = \lambda\,z_1 + (1 - \lambda)\,z_2, \text{ where } \lambda = 1 - \frac{t}{t_0},$$

which is the parametric equation of a straight line passing through two points whose affixes are z_1 and z_2, respectively.

EXAMPLE 11.17 Find the equation of the straight line joining the points whose affixes are $1 - i$ and $2 - 5i$.

SOLUTION

The equation of a straight line passing through two points with affixes z_1 and z_2 is

$$\frac{z - z_1}{z_2 - z_1} = \frac{\bar{z} - \bar{z}_1}{\bar{z}_2 - \bar{z}_1}$$

Taking $z_1 = 1 - i, z_2 = 2 - 5i$, we get

$$\frac{z - (1 - i)}{(2 - 5i) - (1 - i)} = \frac{\bar{z} - (\overline{1 - i})}{(\overline{2 - 5i}) - (\overline{1 - i})}$$

$$\Rightarrow \quad \frac{z - 1 + i}{1 - 4i} = \frac{\bar{z} - 1 - i}{1 + 4i}$$

$$\Rightarrow \quad (1 + 4i)z - (1 - 4i)\bar{z} - 6i = 0$$

$$\Rightarrow \quad (4 - i)z + (4 + i)\bar{z} + 6 = 0$$

EXAMPLE 11.18 Find the equation of the right bisector of the line joining the points z_1 and z_2.

(D.U., B.Sc. (Ph.Sc.) 2009)

SOLUTION

Let $P(z)$ be a point on the Argand plane. Then P lies on the right bisector of the line joining the points $Q(z_1)$ and $R(z_2)$ if P is equidistant from Q and R. Therefore, $PQ^2 = PR^2$

$$\Rightarrow |z - z_1|^2 = |z - z_2|^2$$

$$\Rightarrow (z - z_1)\left(\overline{z - z_1}\right) = (z - z_2)\left(\overline{z - z_2}\right)$$

$$\Rightarrow (z - z_1)(\bar{z} - \bar{z}_1) = (z - z_2)(\bar{z} - \bar{z}_2),$$

$$\Rightarrow z(\bar{z}_2 - \bar{z}_1) + \bar{z}(z_2 - z_1) + (z_1\bar{z}_1 - z_2\bar{z}_2) = 0,$$

which is the required equation.

EXAMPLE 11.19 Prove that the points with affixes z_1, z_2 and z_3 are collinear if and only if

$$z_1(\bar{z}_2 - \bar{z}_3) + z_2(\bar{z}_3 - \bar{z}_1) + z_3(\bar{z}_1 - \bar{z}_2) = 0$$

SOLUTION

The equation of the straight line passing through the points z_1 and z_2 is

$$\frac{z - z_1}{z_2 - z_1} = \frac{\bar{z} - \bar{z}_1}{\bar{z}_2 - \bar{z}_1} \qquad (1)$$

The points z_1, z_2 and z_3 are collinear iff z_3 lies on (1).

Hence the three points are collinear iff

$$\frac{z_3 - z_1}{z_2 - z_1} = \frac{\bar{z}_3 - \bar{z}_1}{\bar{z}_2 - \bar{z}_1}$$

$$\Rightarrow \quad z_1(\bar{z}_2 - \bar{z}_3) + z_2(\bar{z}_3 - \bar{z}_1) + z_3(\bar{z}_1 - \bar{z}_2) = 0$$

EXAMPLE 11.20 Show that point z_2 is the reflection of point z_1, with respect to the line $\bar{q}z + q\bar{z} + r = 0$ if $\bar{q}z_2 + q\bar{z}_1 = r$.

SOLUTION

(A point Q is called *reflection* or *inverse* of the point P with respect to a line L if L is the perpendicular bisector of PQ).

Necessity: Let P(z_1) and Q(z_2) be two points in the Argand plane such that Q is the reflection of P with respect to the line L. Let R(z) be any point on L. As per the given condition,

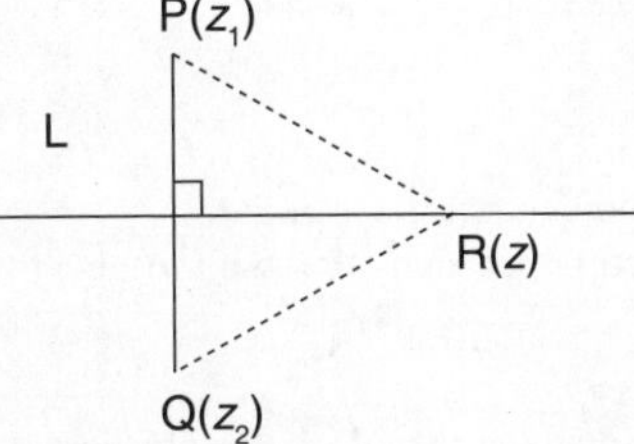

Fig. 11.24

PR = QR

$$\Rightarrow \quad |z - z_1| = |z - z_2|$$

$$\Rightarrow \quad |z - z_1|^2 = |z - z_2|^2$$

$$\Rightarrow \quad (z - z_1)(\bar{z} - \bar{z}_1) = (z - z_2)(\bar{z} - \bar{z}_2)$$

$$\Rightarrow \quad z(\bar{z}_2 - \bar{z}_1) + \bar{z}(z_2 - z_1) + (z_1\bar{z}_1 - z_2\bar{z}_2) = 0 \qquad (1)$$

Since R is any arbitrary point on the line, (1) can be treated as the equation of the given line. Hence (1) and

$$\bar{q}\,z + q\,\bar{z} + r = 0 \qquad (2)$$

represent the same straight line. Comparing (1) and (2), we get

$$\frac{\bar{z}_2 - \bar{z}_1}{\bar{q}} = \frac{z_2 - z_1}{q} = \frac{z_1\bar{z}_1 - z_2\bar{z}_2}{r} = k \text{ (say)}$$

$$\therefore\ \bar{q}\,z_2 + q\,\bar{z}_1 = \frac{1}{k}\,[(\bar{z}_2 - \bar{z}_1)\,z_2 + (z_2 - z_1)\,\bar{z}_1\,]$$

$$= \frac{1}{k}\,[z_1\,\bar{z}_1 - z_2\,\bar{z}_2\,]$$

$$= r$$

$\therefore\ \bar{q}\,z_2 + q\,\bar{z}_1 = r$, which is the required condition.

Sufficiency: Conversely, let the condition hold. That is,

$$\bar{q}\,z_2 + q\,\bar{z}_1 = r \qquad (3)$$

Subtracting (3) from (2), we get

$$\bar{q}\,(z - z_2) + q\,(\bar{z} - \bar{z}_1) = 0 \qquad (4)$$

Taking conjugates on both the sides, we get

$$q\,(\bar{z} - \bar{z}_2) + \bar{q}\,(z - z_1) = 0 \qquad (5)$$

From (4) and (5),

$$(z - z_1)\,(\bar{z} - \bar{z}_1) = (z - z_2)\,(\bar{z} - \bar{z}_2)$$

$$\Rightarrow\ (z - z_1)\left(\overline{z - z_1}\right) = (z - z_2)\left(\overline{z - z_2}\right)$$

$$\Rightarrow\ |z - z_1|^2 = |z - z_2|^2$$

$$\Rightarrow\ |z - z_1| = |z - z_2|$$

$\therefore$ The given line is perpendicular bisector of PQ.
Hence z_1 and z_2 are the reflections of each other with respect to the given line.

11.2.3 Half-Planes

Any straight line drawn on a plane divides the plane (excluding the line) into two halves; each half is called a half-plane. For example, the x-axis creates the half-planes $\{(x, y) : y > 0\}$ and $\{(x, y) : y < 0\}$, respectively. Treating the plane as an Argand plane, these half-planes are $\{z : \text{Im}\,(z) > 0\}$ and $\{z : \text{Im}\,(z) < 0\}$, respectively.

Given a straight line and two points, how to know that the two points lie on the same half-plane or not? The rule is same for both the Cartesian plane and the Argand plane. If the affixes, when put in the equation of the straight line, one by one, result in answers having the same sign, the points lie on the same half-plane, otherwise not.

EXAMPLE 11.21 Check whether the points $1 + 2i$, $-2 + i$ and the origin lie on the same side of the line

$$(3 - 5i)\, z + (3 + 5i)\, \bar{z} + 4 = 0$$

SOLUTION

Let $f(z) = (3 - 5i)\, z + (3 + 5i)\, \bar{z} + 4$

Then, $f(1 + 2i) = (3 - 5i)(1 + 2i) + (3 + 5i)(1 - 2i) + 4$

$= 30 > 0$

$f(-2 + i) = (3 - 5i)(-2 + i) + (3 + 5i)(-2 - i) + 4$

$= 2 > 0$

$f(0 + i0) = 4 > 0$

Since all of $f(1 + 2i)$, $f(-2 + i)$ and $f(0 + i0)$ bear the same sign, all the three points lie on the same side of the line.

EXERCISE 11.2

1. Show that the points $1 + 6i$, $3 + 10i$ and $4 + 12i$ are collinear.
2. Find the equation of the straight line joining the points whose affixes are $1 - i$ and $2 - 5i$.
3. Find the point of intersection of the two lines given by
   $\bar{a}_1 z + a_1 \bar{z} + b_1 = 0, \qquad \bar{a}_2 z + a_2 \bar{z} + b_2 = 0,$
   b_1, b_2 being real.
4. Prove that the complex numbers $a(\cos \alpha + i \sin \alpha)$, $b(\cos \beta + i \sin \beta)$ and $c(\cos \gamma + i \sin \gamma)$ will be represented by collinear points on the Argand plane if $ab \sin(\alpha - \beta) + bc \sin(\beta - \gamma) + ca \sin(\gamma - \alpha) = 0$.
5. Find the equation of the right bisector of the line joining the points z_1 and z_2.
6. If a, b, c be the affixes of the vertices of a triangle ABC, show that the equation of the perpendicular bisector of the side BC is
   $$z(\bar{b} - \bar{c}) + \bar{z}(b - c) = b\bar{b} - c\bar{c}.$$
7. Show that the points representing the numbers $1 + 2i$ and $-1 + 4i$ on the Argand plane are reflection points for the straight line
   $$z(1 + i) + \bar{z}(1 - i) + 6 = 0$$
8. Show that the straight lines
   $$\bar{q}_1 z + q_1 \bar{z} + r_1 = 0$$
   and
   $$\bar{q}_2 z + q_2 \bar{z} + r_2 = 0$$
   are parallel if $\bar{q}_1 q_2 = q_1 \bar{q}_2$ and are perpendicular if $\bar{q}_1 q_2 + q_1 \bar{q}_2 = 0$
9. Show that the points $1 + 3i$ and $3 - 5i$ lie on the same side of the line $(2 + 3i)z + (2 - 3i)\bar{z} + 8 = 0$. Does the origin also lie on the same side as these points?

■ 11.3 CIRCLES IN THE ARGAND PLANE

11.3.1 Equation of a Circle with Given Centre and Radius

The equation of a circle with centre a and radius r is

$$z\bar{z} - a\bar{z} - \bar{a}z + a\bar{a} - r^2 = 0$$

Let P(z) be any point on the circle whose centre is C(a) and radius r.

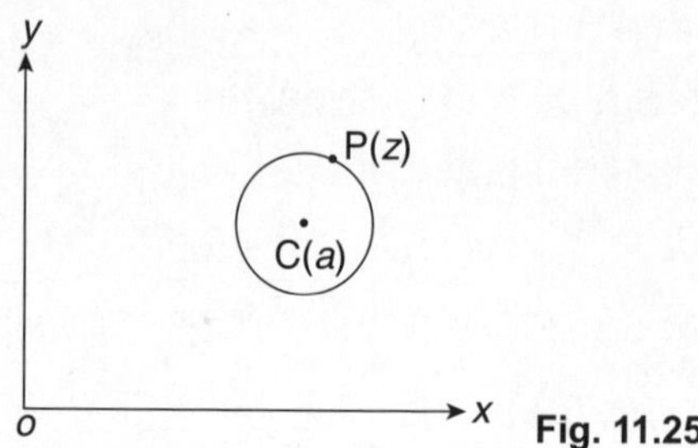

Fig. 11.25

Then

$$\text{CP} = r$$

$$\Rightarrow \quad |z - a| = r$$

$$\Rightarrow \quad |z - a|^2 = r^2$$

$$\Rightarrow \quad (z - a)(\bar{z} - \bar{a}) = r^2$$

$$\Rightarrow \quad z\bar{z} - \bar{a}z - a\bar{z} + a\bar{a} - r^2 = 0,$$

which is the required equation of the circle.

EXAMPLE 11.22 Find the equation of the circle whose radius is 3 and whose centre has affix $1 - i$.

SOLUTION

Equation of the circle whose centre is a and radius is r is given by

$$z\bar{z} - \bar{a}z - a\bar{z} + a\bar{a} - r^2 = 0$$

Here, $a = 1 - i, r = 3, \bar{a} = 1 + i$

$$\therefore a\bar{a} = (1 - i)(1 + i) = 1 + 1 = 2, r^2 = 9$$

Therefore, the equation of the circle is

$$z\bar{z} - (1 + i)z - (1 - i)\bar{z} - 7 = 0$$

EXAMPLE 11.23 Find the centre and radius of the circle whose equation is

$$|z - i| = 3|z + i|$$ (D.U., B.Sc. (Ph.Sc.) 2008)

SOLUTION

We have

$$|z-i| = 3\,|z+i|$$
$$\Rightarrow |z-i|^2 = 9\,|z+i|^2$$
$$\Rightarrow (z-i)\left(\overline{z-i}\right) = 9\,(z+i)\left(\overline{z+i}\right)$$
$$\Rightarrow (z-i)\left(\bar{z}-\bar{i}\right) = 9\,(z+i)\left(\bar{z}+\bar{i}\right)$$
$$\Rightarrow 8\,z\bar{z} - 10iz + 10i\bar{z} + 8 = 0$$
$$\Rightarrow z\bar{z} - \frac{5}{4}iz + \frac{5}{4}i\bar{z} + 1 = 0$$

Comparing this equation with

$$z\bar{z} - \bar{a}z - a\bar{z} + a\bar{a} - r^2 = 0,$$

We get

$$a = -\frac{5}{4}i, \quad \bar{a} = \frac{5}{4}i, \quad a\bar{a} - r^2 = 1$$

$$\Rightarrow \qquad a = -\frac{5}{4}i, \quad r = \sqrt{\frac{25}{16} - 1} = \frac{3}{4}$$

Hence, the centre is $-\frac{5}{4}i$ and radius is $\frac{3}{4}$.

EXAMPLE 11.24 If $\arg\left(\dfrac{z_2 - z_3}{z_1 - z_3}\right) = \arg\left(\dfrac{z_2 - z_4}{z_1 - z_4}\right)$ show that

(i) the points z_3, z_4 are on the same side of the line joining z_1 and z_2; and
(ii) z_1, z_2, z_3, z_4 are concyclic.

SOLUTION

Let A, B, C and D be the points on the Argand plane representing the complex numbers z_1, z_2, z_3 and z_4, respectively.

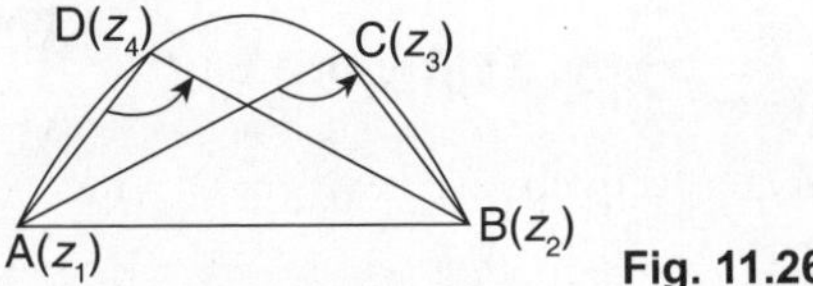

Fig. 11.26

Then,

$\arg\left(\dfrac{z_2 - z_3}{z_1 - z_3}\right) = \angle ACB$, the angle through which AC is to revolve to coincide with BC and

$\arg\left(\dfrac{z_2 - z_4}{z_1 - z_4}\right) = \angle ADB$, the angle through which AD is to revolve to coincide with BD.

$\therefore\ \angle ACB = \angle ADB$, both in magnitude and sense.

Hence both C and D must be on the same side of the side AB. (otherwise one of $\angle ACB$ and $\angle ADB$ will be greater than 180° while the other one will be less than 180°.)

Also, D must lie on the circle passing through A, B and C. Hence A, B, C and D are concyclic.

■ 11.3.2 The General Equation of a Circle

The general equation of a circle in the Argand plane is of the form

$$z\bar{z} + \bar{a}z + a\bar{z} + b = 0$$

Let $z = x + iy$ be a complex number so that $\bar{z} = x - iy$ and then

$$x = \frac{z+\bar{z}}{2}, \quad y = \frac{z-\bar{z}}{2i}$$

Now the general equation of a circle in the Cartesian plane is

$$x^2 + y^2 + 2gx + 2fy + c = 0 \qquad (1)$$

Putting $x = \dfrac{z+\bar{z}}{2}$, $y = \dfrac{z-\bar{z}}{2i}$ in (1), we get

$$\left(\frac{z+\bar{z}}{2}\right)^2 + \left(\frac{z-\bar{z}}{2i}\right)^2 + 2g\left(\frac{z+\bar{z}}{2}\right) + 2f\left(\frac{z-\bar{z}}{2i}\right) + c = 0$$

$$\Rightarrow \frac{1}{4}(z^2 + \bar{z}^2 + 2z\bar{z}) - \frac{1}{4}(z^2 + \bar{z}^2 - 2z\bar{z}) + g(z+\bar{z}) + f\left(\frac{z-\bar{z}}{i}\right) + c = 0$$

$$\Rightarrow z\bar{z} + g(z+\bar{z}) - if(z-\bar{z}) + c = 0 \qquad \left(\because \frac{1}{i} = -i\right)$$

$$\Rightarrow z\bar{z} + (g - if)z + (g + if)\bar{z} + c = 0$$

Writing $g - if = \bar{a}$, $g + if = a$, $c = b$, we get

$$z\bar{z} + \bar{a}z + a\bar{z} + b = 0 \qquad (2)$$

as the general equation of a circle in the Argand plane.

Centre and radius: The centre and radius of the circle

$$x^2 + y^2 + 2gx + 2fy + c = 0$$

are $(-g, -f)$ and $\sqrt{g^2 + f^2 - c}$, respectively. We have $g + if = a$, $g - if = \bar{a}$, $c = b$

so that $-a = -g + i(-f)$, $a\bar{a} = g^2 + f^2$

$\therefore$ Affix of $(-g, -f)$ is $-a$ and $g^2 + f^2 - c = a\bar{a} - b$.

Hence affix of the centre of the circle (1) is $-a$ and radius is $\sqrt{a\bar{a} - b}$.

EXAMPLE 11.25 Find the radius and centre of the circle $z\bar{z} - (2+3i)z - (2-3i)\bar{z} + 9 = 0$

SOLUTION

Comparing the given equation

$$z\bar{z} - (2+3i)z - (2-3i)\bar{z} + 9 = 0$$

with the general equation of a circle

$$z\bar{z} + \bar{a}z + a\bar{z} + b = 0,$$

whose centre is $-a$ and radius is $\sqrt{a\bar{a}-b}$, we get

$$\bar{a} = -(2+3i),\ a = -(2-3i),\ b = 9$$

$\therefore$ Centre is $2-3i$

$$\text{Radius} = \sqrt{a\bar{a}-b}$$

$$= \sqrt{(2+3i)(2-3i)-9}$$

$$= 2$$

11.3.3 Equation of a Circle with Given Endpoints of a Diameter

The equation of a circle having z_1 and z_2 as the affixes of the endpoints of a diameter is

$$(z-z_1)(\bar{z}-\bar{z}_2)+(z-z_2)(\bar{z}-\bar{z}_1)=0$$

Let A(z_1) and B(z_2) be the endpoints of a diameter of a circle. Let P(z) be any point on the circle.

Then $\angle APB = \dfrac{\pi}{2}$. Hence,

$$\frac{z_2-z}{z_1-z} = \frac{BP}{AP}\left(\cos\frac{\pi}{2}+i\sin\frac{\pi}{2}\right) = i\,k, \text{ where } k = \frac{BP}{AP}$$

Taking conjugate on both the sides,

$$\overline{\left(\frac{z_2-z}{z_1-z}\right)} = -i\,k$$

$$\Rightarrow \frac{\bar{z}_2-\bar{z}}{\bar{z}_1-\bar{z}} = -i\,k$$

Hence,

$$\frac{z_2-z}{z_1-z} = -\frac{\bar{z}_2-\bar{z}}{\bar{z}_1-\bar{z}}$$

$$\Rightarrow (z-z_1)(\bar{z}-\bar{z}_2)+(z-z_2)(\bar{z}-\bar{z}_1)=0,$$

which is the required equation.

Alternatively, since $\angle APB = \dfrac{\pi}{2}$, we have

$$AP^2+PB^2=AB^2$$

$$\Rightarrow |z-z_1|^2+|z-z_2|^2=|z_1-z_2|^2$$

$$\Rightarrow (z-z_1)(\bar{z}-\bar{z}_1)+(z-z_2)(\bar{z}-\bar{z}_2)=(z_1-z_2)(\bar{z}_1-\bar{z}_2)$$

$$\Rightarrow (z-z_1)\,\bar{z}+(z-z_2)\bar{z}-(z-z_1)\,\bar{z}_1-(z-z_2)\bar{z}_2-(z_1-z_2)(\bar{z}_1-\bar{z}_2)=0$$

$$\Rightarrow (z-z_1)\bar{z}+(z-z_2)\bar{z}-z\bar{z}_1-z\bar{z}_2+z_1\bar{z}_2+z_2\bar{z}_1=0$$

$$\Rightarrow (z-z_1)(\bar{z}-\bar{z}_2)+(z-z_2)(\bar{z}-\bar{z}_1)=0,$$

which is the required equation.

EXAMPLE 11.26 Find the equation of the circle described on the join of the points $(1 + i)$ and $(2 - i)$ as extremities of one of its diameters.

SOLUTION

If z_1 and z_2 are the extremities of a diameter, then equation of the circle

$$(z - z_1)(\bar{z} - \bar{z}_2) + (z - z_2)(\bar{z} - \bar{z}_1) = 0$$

Putting $z_1 = 1 + i, z_2 = 2 - i$, we get

$$(z - (1 + i))(\bar{z} - (2 + i)) + (z - (2 - i))(\bar{z} - (1 - i)) = 0$$

$$\Rightarrow z\bar{z} - (1 + i)\bar{z} - (2 + i)z + (1 + i)(2 + i) + z\bar{z} - (1 - i)z - (2 - i)\bar{z} + (2 - \text{i})(1 - i) = 0$$

$$\Rightarrow z\bar{z} - \frac{3}{2}z - \frac{3}{2}\bar{z} + 1 = 0,$$

which is the required equation.

Alternatively, if z_1 and z_2 are the extremities of a diameter, then centre of the circle is $\frac{1}{2}(z_1 + z_2)$ and radius is $\frac{1}{2}|z_2 - z_1|$.

Here $z_1 = 1 + i, \quad z_2 = 2 - i$

$\therefore$ Centre is $\frac{1}{2}(1 + i + 2 - i) = \frac{3}{2}$

Radius $= \frac{1}{2}|(2 - i) - (1 + i)|$

$= \frac{1}{2}\sqrt{1 + 4}$

$= \frac{\sqrt{5}}{2}$

Now equation of the circle is

$$z\bar{z} - \bar{a}z - a\bar{z} + a\bar{a} - r^2 = 0$$

Here $\quad a = \frac{3}{2}, r = \frac{\sqrt{5}}{2}$

$\therefore$ The equation of the given circle is

$$z\bar{z} - \frac{3}{2}z - \frac{3}{2}\bar{z} + \frac{9}{4} - \frac{5}{4} = 0$$

$$\Rightarrow \quad z\bar{z} - \frac{3}{2}z - \frac{3}{2}\bar{z} + 1 = 0$$

11.3.4 Discs in a Plane

Let z_0 be a fixed point in an Argand plane. Then the set $\{z : |z - z_0| < r\}$ is the collection of all points of the plane whose distance from z_0 is less than r. They are precisely the collection of all those points which lie strictly within the circle whose centre is z_0 and radius is r. Such a collection is called an *open disc* of radius r centred at z_0. On the other hand, the set $\{z : |z - z_0| \leq r\}$ is called a *closed disc* of radius r centred at z_0.

EXAMPLE 11.27 Sketch the set $\{z : |z + 1| \leq 2\}$

SOLUTION

This is a closed disc of radius 2 centred at $-1 + i\,0$

The shaded portion including the boundary is the required part of the sketch.

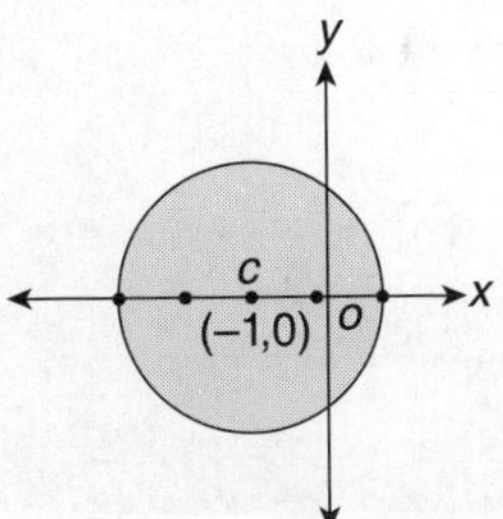

Fig. 11.27

EXERCISE 11.3

1. Find the radius and centre of the circle
   $z\bar{z} + (2 - 3i)z + (2 + 3i)\bar{z} + 4 = 0$
2. Show that the points represented by $-4 + 2i$, $1 - 3i$, $5 + 5i$ and $-2 + 6i$ are concyclic.
3. Find the equation of the circle described on the line joining the points given by $-1 -3i$ and $5 + 7i$ as diameter.
4. Find the equation of the circle whose radius is 3 and whose centre has the affix $1 - i$.
5. Represent graphically the set $S = \{z : |z| > |z - 1|\}$.
6. If $|(z - 1)/(z + 1)| = 2$, prove that the locus of z is a circle whose centre has affix $\left(-\frac{5}{3}, 0\right)$ and whose radius is $\frac{4}{3}$. Show that the point representing the number $(-5 + 4i)/3$ lies on this circle.
7. If $\frac{z - i}{z - 1}$ is purely imaginary, then prove that the image of z in the Argand plane lies on a circle of radius $\frac{1}{\sqrt{2}}$. Find the complex number represented by the centre of the circle.
8. If a and b are two points and z is a variable in the Argand diagram, then prove that if $\left|\frac{z - a}{z - b}\right| = 2$, the locus of z is a circle.
9. Find the centre and radius of the circle given by $|z - z_1|^2 + |z - z_2|^2 = k$, where k is a real positive constant not less than $\frac{1}{2}|z_1 - z_2|^2$.
10. If $z = x + iy$ and z moves on a circle $|z| = 1$, find the locus of $\frac{2}{z}$.

11.4 FUNDAMENTAL THEOREM OF ALGEBRA

Solving algebraic equation has been historically one of the favourite topics of the mathematicians. While linear equations are always solvable in real numbers, not all quadratic equations have this property. We mentioned in the beginning of this chapter that solving quadratic equations of the type $x^2 + 1 = 0$ gave rise to the idea of a complex number. Although Leonhard Euler introduced the notion $i = \sqrt{-1}$, it was Karl Friedrich Gauss, who fully exploited the idea of a complex number for solving an algebraic equation. He is credited with the following result which is known as 'Fundamental Theorem of Algebra'.

THEOREM

Every polynomial of degree $n \geq 1$ has at least one zero.

The zero in the above theorem belongs to the set of complex numbers. Therefore, it need not be a purely real number.

From the above result, it immediately follows that

Every polynomial equation of degree n has exactly n zeros.

Here the zeros need not be all distinct. This result may be stated in the following form:

THEOREM

Every polynomial equation of degree n has exactly n roots.

The above results tell us about the existence of roots of the polynomial equations. They do not hint at the methodology of solving such equations. In fact, there is no fixed method for solving polynomial equations. Below we mention few more results which will help us in solving a polynomial equation.

THEOREM

Let

$$a_0 x^n + a_1 x^{n-1} + \dots + a_n = 0 \qquad (1)$$

be a polynomial equation of degree n. Then we have the following:

(a) *(Real coefficients with complex roots)* If $a_0, a_1, \dots, a_n$ are all real numbers and $p + iq$ is a complex root of (1), then $p - iq$ is also a root of (1).

(b) *(Rational coefficients with irrational roots)* If $a_0, a_1, \dots, a_n$ are all rational numbers and $p + \sqrt{q}$ is an irrational root of (1), then $p - \sqrt{q}$ is also a root of (1).

(c) *(Integral coefficients with rational roots)* If $a_0, a_1, \dots, a_n$ are all integers and p/q is a rational (in lowest form) root of (1), then p is a divisor of a_n and q is a divisor of a_0. In particular, if $a_0 = 1$, then every rational root of (1) is an integer which divides a_n.

EXAMPLE 11.28 Form an equation in lowest degree with real coefficients which has $2 - 3i$, $3 + 2i$ as two of its roots.

SOLUTION

Since in a polynomial equation with real coefficients, complex roots occur in conjugate pairs, the required equation will have at least four roots, namely, $2 \pm 3i$, $3 \pm 2i$.

$\therefore$ A lowest degree equation with the given roots will have the form

$$(x - (2 + 3i))\,(x - (2 - 3i))\,(x - (3 + 2i))\,(x - (3 - 2i)) = 0$$

$$\Rightarrow ((x - 2) - 3i)\,((x - 2) + 3i)\,((x - 3) - 2i)\,((x - 3) + 2i) = 0$$

$$\Rightarrow ((x - 2)^2 + 9)\,((x - 3)^2 + 4) = 0$$

$$\Rightarrow (x^2 - 4x + 13)\,(x^2 - 6x + 13) = 0$$

EXAMPLE 11.29 Form an equation in lowest degree with rational coefficients which has $1 + \sqrt{2}$, and $2 - \sqrt{3}$ as two of its roots.

SOLUTION

Since in a polynomial equation with rational coefficients, irrational roots occur in pairs, the required equations will have at least four roots, namely $1 \pm \sqrt{2}$, $2 \pm \sqrt{3}$.

$\therefore$ A lowest degree equation with the given roots will have the form

$$(x - (1 + \sqrt{2}))\,(x - (1 - \sqrt{2}))\,(x - (2 + \sqrt{3}))\,(x - (2 - \sqrt{3})) = 0$$

$$\Rightarrow ((x - 1) - \sqrt{2})\,((x - 1) + \sqrt{2})\,((x - 2) - \sqrt{3})\,((x - 2) + \sqrt{3}) = 0$$

$$\Rightarrow ((x - 1)^2 - 2)\,((x - 2)^2 - 3) = 0$$

$$\Rightarrow (x^2 - 2x - 1)\,(x^2 - 4x + 1) = 0$$

EXAMPLE 11.30 Find the rational roots of the equation $x^3 + x^2 - 5x + 3 = 0$

SOLUTION

Since the given equation is a polynomial equation with integral coefficients with $a_0 = 1$, any rational root of this equation must be an integer. Moreover, the root must divide $a_n = 3$.

Hence the possible roots of the given equation are 1, –1, 3, –3.

By applying factor theorem, we find that $x = 1$ is a root of the given equation. We divide the polynomial expression by $x - 1$ to obtain

$$x^3 + x^2 - 5x + 3 = (x - 1)\,(x^2 + 2x - 3).$$

Now $x^2 + 2x - 3$ can be factored into $(x - 1)\,(x + 3)$.

$\therefore$ The roots of the given equation 1, 1, –3.

EXERCISE 11.4

1. Form an equation of lowest degree with real coefficients having $2 + \sqrt{(-3)}$ and $3 + \sqrt{(-5)}$ as two of its roots.
2. Form an equation whose roots are $1, -1, 2 \pm \sqrt{3}$.
3. Form an equation of lowest degree with rational coefficients having $2 + \sqrt{3}$ and $\sqrt{5} - 2$ as two of its roots.
4. One root of the equation $3x^3 - 23x^2 + 72x - 70 = 0$ is $3 + i\sqrt{5}$. Find the other roots.
5. Find the rational roots of the following equations:

   (i) $x^4 - 4x^3 + 6x^2 - 4x + 1 = 0$

   (ii) $2x^3 - 3x^2 - 11x + 6 = 0$

CHAPTER TWELVE

De Moivre's Theorem

12.1. DE MOIVRE'S THEOREM FOR A RATIONAL INDEX

Consider two complex numbers

$$z_1 = r_1 (\cos \theta_1 + i \sin \theta_1)$$

and

$$z_2 = r_2 (\cos \theta_2 + i \sin \theta_2)$$

By definition of multiplication, we have

$$z_1 z_2 = r_1 r_2 [\cos (\theta_1 + \theta_2) + i \sin (\theta_1 + \theta_2)] \tag{1}$$

Similarly, if $z_1, z_2, \ldots, z_n$ are n complex numbers, then by repeated use of (1), we get

$$z_1 z_2 \ldots z_n = r_1 r_2 \ldots r_n [\cos (\theta_1 + \theta_2 + \ldots + \theta_n) + i \sin (\theta_1 + \theta_2 + \ldots + \theta_n)] \tag{2}$$

If, in (2), we put $z_1 = z_2 = \ldots = z_n = z = r (\cos \theta + i \sin \theta)$, then

$$z^n = r^n (\cos n \theta + i \sin n \theta)$$

or

$$[r (\cos \theta + i \sin \theta)]^n = r^n (\cos n \theta + i \sin n \theta) \tag{3}$$

If now we take $r = 1$, we can write (3) as

$$(\cos \theta + i \sin \theta)^n = \cos n \theta + i \sin n \theta$$

This is called De Moivre's Formula for a positive integer n. In general, the formula holds for any integer (positive or negative). This is the statement of De Moivre's Theorem for an integral index. The general form of this theorem for a rational index is proved below.

THEOREM

When n is an integer (positive or negative),

$$(\cos \theta + i \sin \theta)^n = \cos n\theta + i \sin n\theta$$

When n is a fraction (positive or negative), $\cos n\theta + i \sin n\theta$ is one of the values of $(\cos \theta + i \sin \theta)^n$.

Proof:

Case I:

Let n be a positive integer. We shall prove the theorem by principle of mathematical induction.

For $n = 1$, the theorem is obviously true.

We assume that the theorem is true for $n = k$. Then

$$(\cos\theta + i\sin\theta)^k = \cos k\theta + i\sin k\theta \tag{1}$$

Now,

$$\begin{aligned}(\cos\theta + i\sin\theta)^{k+1} &= (\cos\theta + 1\sin\theta)^k(\cos\theta + i\sin\theta)\\ &= (\cos k\theta + i\sin k\theta)(\cos\theta + i\sin\theta)\end{aligned}$$

(Using (1))

$$\begin{aligned}&= (\cos k\theta\cos\theta - \sin k\theta\sin\theta)\\ &\quad + i(\cos k\theta\sin\theta + \cos\theta\sin k\theta)\\ &= \cos(k+1)\theta + i\sin(k+1)\theta.\end{aligned}$$

This shows that the theorem holds for $n = k + 1$. Hence by the principle of mathematical induction, the theorem is true for all positive integers n.

Case II:

Let n be a negative integer $= -m$ (say), where m is a positive integer. Therefore,

$$\begin{aligned}(\cos\theta + i\sin\theta)^n &= (\cos\theta + i\sin\theta)^{-m}\\ &= \frac{1}{(\cos\theta + i\sin\theta)^m}\\ &= \frac{1}{\cos m\theta + i\sin m\theta} \qquad \text{(By Case I)}\\ &= \frac{\cos m\theta - i\sin m\theta}{(\cos m\theta + i\sin m\theta)(\cos m\theta - i\sin m\theta)}\\ &= \cos m\theta - i\sin m\theta\\ &= \cos n\theta + i\sin n\theta\end{aligned}$$

Case III:

Let n be a fraction (positive or negative).

Let $n = \frac{p}{q}$, where p and q are integers and $q > 0$. Using Case I of the proof, we have

$$\begin{aligned}\left(\cos\frac{p}{q}\theta + i\sin\frac{p}{q}\theta\right)^q &= \cos q\left(\frac{p}{q}\theta\right) + i\sin q\left(\frac{p}{q}\theta\right)\\ &= \cos p\theta + i\sin p\theta.\end{aligned}$$

So, $\cos\frac{p}{q}\theta + i\sin\frac{p}{q}\theta$ is one of the values of $(\cos p\theta + i\sin p\theta)^{1/q}$.

Also, $\cos p\,\theta + i\sin p\,\theta = (\cos\theta + i\sin\theta)^p$, so

$$(\cos p\,\theta + i\sin p\,\theta)^{1/q} = (\cos\theta + i\sin\theta)^{p/q}.$$

Hence $\cos\frac{p}{q}\theta + i\sin\frac{p}{q}\theta$ is one of the values of $(\cos\theta + i\sin\theta)^{p/q}$.

REMARKS: 1. If n is an integer, then

$$\begin{aligned}(\cos\theta - i\sin\theta)^n &= \left(\frac{1}{\cos\theta + i\sin\theta}\right)^n \\ &= \frac{1}{(\cos\theta + i\sin\theta)^n} \\ &= \frac{1}{\cos n\theta + i\sin n\theta} \\ &= \cos n\theta - i\sin n\theta\end{aligned}$$

2. Many authors use cis θ for $\cos\theta + i\sin\theta$. In this notation, the De Moivre's theorem for integral index is

$$(\text{cis}\,\theta)^n = \text{cis}\,n\theta$$

3. If we put $r_1 = r_2 = \ldots = r_n = 1$ in equation (2) of the discussion before the theorem, we obtain

$$(\cos\theta_1 + i\sin\theta_1)(\cos\theta_2 + i\sin\theta_2)\ldots(\cos\theta_n + i\sin\theta_n)$$
$$= \cos(\theta_1 + \theta_2 + \ldots + \theta_n) + i\sin(\theta_1 + \theta_2 + \ldots + \theta_n).$$

EXAMPLE 12.1 Simplify $\dfrac{(\cos\alpha + i\sin\alpha)^4}{(\sin\beta + i\cos\beta)^5}$

SOLUTION

$$\frac{(\cos\alpha + i\sin\alpha)^4}{(\sin\beta + i\cos\beta)^5} = \frac{(\cos\alpha + i\sin\alpha)^4}{\left[\cos\left(\frac{\pi}{2} - \beta\right) + i\sin\left(\frac{\pi}{2} - \beta\right)\right]^5}$$

$$= (\cos\alpha + i\sin\alpha)^4\left[\cos\left(\frac{\pi}{2} - \beta\right) + i\sin\left(\frac{\pi}{2} - \beta\right)\right]^{-5}$$

$$= (\cos 4\alpha + i\sin 4\alpha)\left[\cos(-5)\left(\frac{\pi}{2} - \beta\right) + i\sin(-5)\left(\frac{\pi}{2} - \beta\right)\right]$$

$$= (\cos 4\alpha + i\sin 4\alpha)\left[\cos\left(5\beta - \frac{5\pi}{2}\right) + i\sin\left(5\beta - \frac{5\pi}{2}\right)\right]$$

$$= \cos\left(4\alpha + 5\beta - \frac{5\pi}{2}\right) + i\sin\left(4\alpha + 5\beta - \frac{5\pi}{2}\right)$$

$$= \cos\left(\frac{5\pi}{2} - (4\alpha + 5\beta)\right) - i\sin\left(\frac{5\pi}{2} - (4\alpha + 5\beta)\right)$$

$$= \sin(4\alpha + 5\beta) - i\cos(4\alpha + 5\beta)$$

EXAMPLE 12.2 Prove that

$$\left(\frac{1+\sin\phi + i\cos\phi}{1+\sin\phi - i\cos\phi}\right)^n = \cos\left(\frac{n\pi}{2} - n\phi\right) + i\sin\left(\frac{n\pi}{2} - n\phi\right)$$

SOLUTION

$$\left(\frac{1+\sin\phi + i\cos\phi}{1+\sin\phi - i\cos\phi}\right)^n = \left[\frac{1+\cos\left(\frac{\pi}{2}-\phi\right) + i\sin\left(\frac{\pi}{2}-\phi\right)}{1+\cos\left(\frac{\pi}{2}-\phi\right) - i\sin\left(\frac{\pi}{2}-\phi\right)}\right]^n$$

$$= \left[\frac{2\cos^2\left(\frac{\pi}{4}-\frac{\phi}{2}\right) + 2i\sin\left(\frac{\pi}{4}-\frac{\phi}{2}\right)\cos\left(\frac{\pi}{4}-\frac{\phi}{2}\right)}{2\cos^2\left(\frac{\pi}{4}-\frac{\phi}{2}\right) - 2i\sin\left(\frac{\pi}{4}-\frac{\phi}{2}\right)\cos\left(\frac{\pi}{4}-\frac{\phi}{2}\right)}\right]^n$$

$$= \frac{2^n\cos^n\left(\frac{\pi}{4}-\frac{\phi}{2}\right)\left[\cos\left(\frac{\pi}{4}-\frac{\phi}{2}\right) + i\sin\left(\frac{\pi}{4}-\frac{\phi}{2}\right)\right]^n}{2^n\cos^n\left(\frac{\pi}{4}-\frac{\phi}{2}\right)\left[\cos\left(\frac{\pi}{4}-\frac{\phi}{2}\right) - i\sin\left(\frac{\pi}{4}-\frac{\phi}{2}\right)\right]^n}$$

$$= \left[\cos\left(\frac{\pi}{4}-\frac{\phi}{2}\right) + i\sin\left(\frac{\pi}{4}-\frac{\phi}{2}\right)\right]^n \left[\cos\left(\frac{\pi}{4}-\frac{\phi}{2}\right) - i\sin\left(\frac{\pi}{4}-\frac{\phi}{2}\right)\right]^{-n}$$

$$= \left[\cos\left(\frac{\pi}{4}-\frac{\phi}{2}\right) + i\sin\left(\frac{\pi}{4}-\frac{\phi}{2}\right)\right]^n \left[\cos\left(\frac{\pi}{4}-\frac{\phi}{2}\right) + i\sin\left(\frac{\pi}{4}-\frac{\phi}{2}\right)\right]^n$$

$$= \left[\cos\left(\frac{\pi}{4}-\frac{\phi}{2}\right) + i\sin\left(\frac{\pi}{4}-\frac{\phi}{2}\right)\right]^{2n}$$

$$= \cos\left(\frac{n\pi}{2} - n\phi\right) + i\sin\left(\frac{n\pi}{2} - n\phi\right)$$

EXAMPLE 12.3 If $x = \cos\theta + i\sin\theta$ and $y = \cos\phi + i\sin\phi$, prove that $\dfrac{x-y}{x+y} = i\tan\dfrac{\theta-\phi}{2}$.

SOLUTION

$$\frac{x-y}{x+y} = \frac{(\cos\theta + i\sin\theta) - (\cos\phi + i\sin\phi)}{(\cos\theta + i\sin\theta) + (\cos\phi + i\sin\phi)}$$

$$= \frac{(\cos\theta - \cos\phi) + i(\sin\theta - \sin\phi)}{(\cos\theta + \cos\phi) + i(\sin\theta + \sin\phi)}$$

$$= \frac{-2\sin\frac{\theta+\phi}{2}\sin\frac{\theta-\phi}{2} + 2i\cos\frac{\theta+\phi}{2}\sin\frac{\theta-\phi}{2}}{2\cos\frac{\theta+\phi}{2}\cos\frac{\theta-\phi}{2} + 2i\sin\frac{\theta+\phi}{2}\cos\frac{\theta-\phi}{2}}$$

$$= \frac{2i\sin\frac{\theta-\phi}{2}\left[i\sin\frac{\theta+\phi}{2} + \cos\frac{\theta+\phi}{2}\right]}{2\cos\frac{\theta-\phi}{2}\left[\cos\frac{\theta+\phi}{2} + i\sin\frac{\theta+\phi}{2}\right]}$$

$$= i\tan\frac{\theta-\phi}{2}$$

EXAMPLE 12.4 Find an equation whose roots are the nth powers of the roots of the equation $x^2 - 2x\cos\theta + 1 = 0$.

SOLUTION

The roots of the equation $x^2 - 2x\cos\theta + 1 = 0$ are given by

$$x = \frac{2\cos\theta \pm \sqrt{4\cos^2\theta - 4}}{2}$$

$$= \cos\theta \pm \sqrt{-\sin^2\theta}$$

$$= \cos\theta \pm i\sin\theta$$

Let $\alpha = \cos\theta + i\sin\theta$ and $\beta = \cos\theta - i\sin\theta$.

We have to find an equation whose roots are α^n and β^n.

Now $\quad \alpha^n + \beta^n = (\cos\theta + i\sin\theta)^n + (\cos\theta - i\sin\theta)^n$

$$= \cos n\theta + i\sin n\theta + \cos n\theta - i\sin n\theta$$

$$= 2\cos n\theta,$$

and $\quad \alpha^n\beta^n = (\cos\theta + i\sin\theta)^n(\cos\theta - i\sin\theta)^n$

$$= (\cos n\theta + i\sin n\theta)(\cos n\theta - i\sin n\theta)$$

$$= \cos^2 n\theta + \sin^2 n\theta$$

$$= 1$$

Hence the required equation is

$$x^2 - (\alpha^n + \beta^n)x + \alpha^n\beta^n = 0$$

or $\quad x^n - 2x\cos n\theta + 1 = 0$

EXAMPLE 12.5 If $\cos\alpha + \cos\beta + \cos\gamma = \sin\alpha + \sin\beta + \sin\gamma = 0$, prove that

$$\cos 3\alpha + \cos 3\beta + \cos 3\gamma = 3\cos(\alpha + \beta + \gamma)$$

and $\quad \sin 3\alpha + \sin 3\beta + \sin 3\gamma = 3\sin(\alpha + \beta + \gamma)$

SOLUTION

Let $a = \cos\alpha + i\sin\alpha$, $b = \cos\beta + i\sin\beta$, $c = \cos\gamma + i\sin\gamma$.

Then $\quad a + b + c = (\cos\alpha + \cos\beta + \cos\gamma) + i(\sin\alpha + \sin\beta + \sin\gamma) = 0.$

Therefore, $\quad a^3 + b^3 + c^3 = 3abc$

or
$$(\cos\alpha + i\sin\alpha)^3 + (\cos\beta + i\sin\beta)^3 + (\cos\gamma + i\sin\gamma)^3$$
$$= 3(\cos\alpha + i\sin\alpha)(\cos\beta + i\sin\beta)(\cos\gamma + i\sin\gamma)$$
$$\Rightarrow (\cos 3\alpha + i\sin 3\alpha) + (\cos 3\beta + i\sin 3\beta) + (\cos 3\gamma + i\sin 3\gamma)$$
$$= 3[\cos(\alpha+\beta+\gamma) + i\sin(\alpha+\beta+\gamma)]$$

or
$$(\cos 3\alpha + \cos 3\beta + \cos 3\gamma) + i(\sin 3\alpha + \sin 3\beta + \sin 3\gamma)$$
$$= 3[\cos(\alpha+\beta+\gamma) + i\sin(\alpha+\beta+\gamma)]$$

Equating real and imaginary parts on both sides, we get
$$\cos 3\alpha + \cos 3\beta + \cos 3\gamma = 3\cos(\alpha+\beta+\gamma)$$
and
$$\sin 3\alpha + \sin 3\beta + \sin 3\gamma = 3\sin(\alpha+\beta+\gamma)$$

EXAMPLE 12.6 If $(1+x)^n = p_0 + p_1 x + p_2 x^2 + \ldots + p_n x^n$, then prove that
$$p_0 - p_2 + p_4 - \ldots = 2^{n/2}\cos\left(\frac{n\pi}{4}\right)$$
and
$$p_1 - p_3 + p_5 - \ldots = 2^{n/2}\sin\left(\frac{n\pi}{4}\right)$$

SOLUTION

We have
$$(1+x)^n = p_0 + p_1 x + p_2 x^2 + p_3 x^3 + p_4 x^4 + \ldots + p_n x^n, \tag{1}$$
Putting $x = i$ in (1), we get
$$(1+i)^n = p_0 + p_1 i + p_2 i^2 + p_3 i^3 + \ldots + p_n i^n$$
$$= (p_0 - p_2 + p_4 - \ldots) + i(p_1 - p_3 + p_5 - \ldots) \tag{2}$$
Now, let $1 + i = r(\cos\theta + i\sin\theta)$. Then
$$r\cos\theta = 1 \text{ and } r\sin\theta = 1$$
This implies $r = \sqrt{2}$ and $\theta = \frac{\pi}{4}$, so that
$$1 + i = \sqrt{2}\left(\cos\frac{\pi}{4} + i\sin\frac{\pi}{4}\right)$$
$$\therefore (1+i)^n = 2^{n/2}\left(\cos\frac{\pi}{4} + i\sin\frac{\pi}{4}\right)^n$$
$$\Rightarrow (1+i)^n = 2^{n/2}\left(\cos\frac{n\pi}{4} + i\sin\frac{n\pi}{4}\right) \tag{3}$$

Using (3) in (2), we get

$$2^{n/2}\left(\cos\frac{n\pi}{4}+i\sin\frac{n\pi}{4}\right)=(p_0-p_2+p_4-\ldots)+i\,(p_1-p_3+p_5-\ldots) \qquad (4)$$

Equating real and imaginary parts on both sides of (4), we get

$$p_0-p_2+p_4-\ldots=2^{n/2}\cos\left(\frac{n\pi}{4}\right)$$

and

$$p_1-p_3+p_5-\ldots=2^{n/2}\sin\left(\frac{n\pi}{4}\right)$$

EXAMPLE 12.7 By expanding $(1+z)^n$ by the binomial theorem and then putting $z=\cos 2\theta+i\sin 2\theta$, prove that

$$z^n\cos^n\theta\,\cos n\,\theta=\sum_{r=0}^{n}{}^{n}C_r\cos 2\,r\,\theta,$$

and

$$z^n\cos^n\theta\sin n\,\theta=\sum_{r=0}^{n}{}^{n}C_r\sin 2\,r\,\theta,$$

n being a positive integer.

SOLUTION

Expanding $(1+z)^n$ by the binomial theorem, we have

$$(1+z)^n={}^nC_0+{}^nC_1\,z+{}^nC_2\,z^2+\ldots+{}^nC_n\,z^n$$
$$=\sum_{r=0}^{n}{}^{n}C_r\,z^r.$$

Putting $z=\cos 2\theta+i\sin 2\theta$, we get

$$(1+\cos 2\theta+i\sin 2\theta)^n=\sum_{r=0}^{n}{}^{n}C_r\,(\cos 2\theta+i\sin 2\theta)^r$$

or

$$(2\cos^2\theta+2\,i\sin\theta\cos\theta)^n=\sum_{r=0}^{n}{}^{n}C_r\,(\cos 2\,r\,\theta+i\sin 2\,r\,\theta)$$

or

$$(2\cos\theta)^n\,(\cos\theta+i\sin\theta)^n=\sum_{r=0}^{n}{}^{n}C_r\,(\cos 2\,r\,\theta+i\sin 2\,r\,\theta)$$

or

$$2^n\cos^n\theta\,(\cos n\,\theta+i\sin n\,\theta)=\sum_{r=0}^{n}{}^{n}C_r\cos 2\,r\,\theta+i\sum_{r=0}^{n}{}^{n}C_r\sin 2\,r\,\theta$$

Equating real and imaginary parts of both sides, we get

$$2^n\cos^n\theta\cos n\,\theta=\sum_{r=0}^{n}{}^{n}C_r\,\cos 2\,r\,\theta$$

and

$$2^n\cos^n\theta\sin n\,\theta=\sum_{r=0}^{n}{}^{n}C_r\,\sin 2\,r\,\theta$$

EXERCISE 12.1

1. Simplify each of the following:

   (a) $\dfrac{(\cos 3\theta + i \sin 3\theta)^5 (\cos 2\theta - i \sin 2\theta)^4}{(\cos 2\theta + i \sin 2\theta)^5 (\cos \theta - i \sin \theta)^3}$

   (b) $\dfrac{\left(\cos \dfrac{\pi}{15} + i \sin \dfrac{\pi}{15}\right)^{10} + \left(\cos \dfrac{\pi}{15} - i \sin \dfrac{\pi}{15}\right)^{10}}{\left(\cos \dfrac{\pi}{3} + i \sin \dfrac{\pi}{3}\right)^6}$

   (c) $\dfrac{(\cos \theta + i \sin \theta)^4}{(\sin \theta + i \cos \theta)^5}$

   (d) $\left(\dfrac{1 + \cos \theta + i \sin \theta}{1 + \cos \theta - i \sin \theta}\right)^n$, n being an integer

   (e) $[(\cos \theta + \cos \phi) + i (\sin \theta + \sin \phi)]^n + [(\cos \theta + \cos \phi) - i (\sin \theta + \sin \phi)]^n$, n being an integer.

2. Show that $(1 + \cos \theta + i \sin \theta)^n + (1 + \cos \theta - i \sin \theta)^n$
   $= 2^{n+1} \cos^n \dfrac{\theta}{2} \cos n\theta$
   where n is an integer.
3. Find the modulus and argument of the complex number $\dfrac{(1 + \cos \theta + i \sin \theta)^5}{(\cos \theta + i \sin \theta)^3}$.
4. If $x_r = \cos \dfrac{\pi}{2^r} + i \sin \dfrac{\pi}{2^r}$, prove that $x_1 x_2 x_3 \ldots$ ad. inf. $= -1$
5. If $x = \cos \theta + i \sin \theta$ and $\sqrt{1 - c^2} = nc - 1$, show that $\dfrac{c}{2n}(1 + nx)\left(1 + \dfrac{n}{x}\right) = 1 + c \cos \theta$.
6. If n is a positive integer, show that

   (a) $\left(\sqrt{3} + i\right)^n + \left(\sqrt{3} - i\right)^n = 2^{n+1} \cos \dfrac{n\pi}{6}$.

   (b) $(1 + i)^n + (1 - i)^n = 2^{(n/2)+1} \cos \dfrac{n\pi}{4}$.
7. If α and β are the roots of the equation $x^2 - 2x + 4$, prove that $\alpha^n + \beta^n = 2^{n+1} \cos\left(\dfrac{n\pi}{3}\right)$. Hence find the value of $\alpha^6 + \beta^6$.
8. Prove that $(a + i b)^{m/n} + (a - i b)^{m/n} = 2\left(a^2 + b^2\right)^{m/2n} \cos\left[\left(\dfrac{m}{n}\right) \tan^{-1} \dfrac{b}{a}\right]$
9. If $z = \cos \theta + i \sin \theta$, show that $\dfrac{z^{2n-1}}{z^{2n+1}} = i \tan n\theta$
10. If $(a_1 + i b_1)(a_2 + i b_2) \ldots (a_n + i b_n) = A + i B$, prove that

    (a) $\tan^{-1} \dfrac{b_1}{a_1} + \tan^{-1} \dfrac{b_2}{a_2} + \ldots + \tan^{-1} \dfrac{b_n}{a_n} = \tan^{-1} \dfrac{B}{A}$

    (b) $(a_1^2 + b_1^2)(a_2^2 + b_2^2) \ldots (a_n^2 + b_n^2) = A^2 + B^2$

11. If $\sin\alpha + \sin\beta + \sin\gamma = \cos\alpha + \cos\beta + \cos\gamma = 0$, prove that
$\sin 2\alpha + \sin 2\beta + \sin 2\gamma = 0$
$\cos 2\alpha + \cos 2\beta + \cos 2\gamma = 0$
$\cos^2\alpha + \cos^2\beta + \cos^2\gamma = \dfrac{3}{2}$

12. If $\cos\alpha + 2\cos\beta + 3\cos\gamma = \sin\alpha + 2\sin\beta + 3\sin\gamma = 0$, prove that
$\cos 3\alpha + 8\cos 3\beta + 27\cos 3\gamma = 18\cos(\alpha+\beta+\gamma)$
$\sin 3\alpha + 8\sin 3\beta + 27\sin 3\gamma = 18\cos(\alpha+\beta+\gamma)$

13. If $a = \cos\alpha + i\sin\alpha$, $b = \cos\beta + i\sin\beta$, $c = \cos\gamma + i\sin\gamma$ and $\dfrac{a}{b} + \dfrac{b}{c} + \dfrac{c}{a} = -1$, prove that $\cos(\beta-\gamma) + \cos(\gamma-\alpha) + \cos(\alpha-\beta) + 1 = 0$

14. If $a = \cos\alpha + i\sin\alpha$, $b = \cos\beta + i\sin\beta$, $c = \cos\gamma + i\sin\gamma$, prove that $\dfrac{(a+b)(b+c)(c+a)}{abc}$ $= 8\cos\dfrac{\alpha-\beta}{2}\cos\dfrac{\beta-\gamma}{2}\cos\dfrac{\gamma-\alpha}{2}$

15. If $a = \cos\alpha + i\sin\alpha$, $b = \cos\beta + i\sin\beta$, $c = \cos\gamma + i\sin\gamma$, $d = \cos\delta + i\sin\delta$, prove that

(a) $(a+b)(c+d) = 4\cos\dfrac{\alpha-\beta}{2}\cos\dfrac{\gamma-\delta}{2}\left[\cos\dfrac{\alpha+\beta+\gamma+\delta}{2} + i\sin\dfrac{\alpha+\beta+\gamma+\delta}{2}\right]$

(b) $(a-b)(c-d) = -4\sin\dfrac{\alpha-\beta}{2}\sin\dfrac{\gamma-\delta}{2}\left[\cos\dfrac{\alpha+\beta+\gamma+\delta}{2} + i\sin\dfrac{\alpha+\beta+\gamma+\delta}{2}\right]$

(c) $(ab+cd) = 2\cos\dfrac{\alpha+\beta-\gamma-\delta}{2}\left[\cos\dfrac{\alpha+\beta+\gamma+\delta}{2} + i\sin\dfrac{\alpha+\beta+\gamma+\delta}{2}\right]$

16. Use the identity
$(a^2-b^2)(c^2-d^2) = (a^2-c^2)(b^2-d^2) - (a^2-d^2)(b^2-c^2)$,
to prove that
$\sin(\alpha-\beta)\sin(\gamma-\delta) = \sin(\alpha-\gamma)\sin(\beta-\delta) - \sin(\alpha-\delta)\sin(\beta-\gamma)$.

17. **(a)** From the De Moivre's Formula
$\cos n\theta + i\sin n\theta = (\cos\theta + i\sin\theta)^n$, where n is a positive integer, deduce that
$\cos n\theta = \cos^n\theta - {}^nC_2\cos^{n-2}\theta\sin^2\theta + {}^nC_4\cos^{n-4}\theta\sin^4\theta - \ldots$
$\sin n\theta = {}^nC_1\cos^{n-1}\theta\sin\theta - {}^nC_3\cos^{n-3}\theta\sin^3\theta + \ldots$
$$\tan n\theta = \frac{{}^nC_1\tan\theta - {}^nC_3\tan^3\theta + \ldots}{1 - {}^nC_2\tan^2\theta + {}^nC_4\tan^4\theta - \ldots}$$
[Hint: Expand $(\cos\theta + i\sin\theta)^n$ by binomial theorem]

(b) Use part (a) to express $\sin 6\theta$ and $\cos 6\theta$ as powers of $\sin\theta$ and $\cos\theta$.

18. **(a)** If $z = \cos\theta + i\sin\theta$ and n is a positive integer, prove that
$z^n + \dfrac{1}{z^n} = 2\cos n\theta,\ z^n - \dfrac{1}{z^n} = 2i\sin\theta.$
Hence show that
$2^{n-1}\cos^n\theta = \cos n\theta + {}^nC_1\cos(n-2)\theta + {}^nC_2\cos(n-4)\theta + \ldots$
[Hint: $(z+z^{-1})^n = z^n + z^{-n} + {}^nC_1(z^{n-2} + z^{-(n-2)}) + \ldots$]

(b) By the method of part (a), prove that
$16\cos^3\theta\sin^2\theta = -\cos 5\theta - \cos 3\theta + 2\cos\theta$

19. If $2\cos\alpha = x + \dfrac{1}{x}$ and $2\cos\beta = y + \dfrac{1}{y}$, prove that one of the values of $x^m y^n + \dfrac{1}{x^m y^n}$ is $2\cos(m\alpha + n\beta)$.

12.2 ROOTS OF COMPLEX NUMBERS AND SOLUTION OF EQUATIONS

The De Moivre's Theorem for fractional indices say that $\cos \frac{p}{q}\theta + i \sin \frac{p}{q}\theta$ is one of the values of $(\cos\theta + i\sin\theta)^{p/q}$. In particular, $\cos \frac{\theta}{q} + i \sin \frac{\theta}{q}$ is one of the values of $(\cos\theta + i\sin\theta)^{1/q}$, where q is a positive integer. We now show that $(\cos\theta + i\sin\theta)^{1/q}$ has exactly q distinct values given by

$$\cos\frac{\theta + 2k\pi}{q} + i\sin\frac{\theta + 2k\pi}{q}, \quad k = 0, 1, \ldots, q-1.$$

Let $z = \cos\theta + i\sin\theta$ and $u = \cos\frac{\theta + 2k\pi}{q} + i\sin\frac{\theta + 2k\pi}{q}$, where k is any integer. Then by De Moivre's Theorem for integral indices,

$$\begin{aligned} u^q &= \left(\cos\frac{\theta + 2k\pi}{q} + i\sin\frac{\theta + 2k\pi}{q}\right)^q \\ &= \cos q\left(\frac{\theta + 2k\pi}{q}\right) + i\sin q\left(\frac{\theta + 2k\pi}{q}\right) \\ &= \cos(\theta + 2k\pi) + i\sin(\theta + 2k\pi) \\ &= \cos\theta + i\sin\theta \\ &= z \end{aligned}$$

Therefore, u is a qth root of z. That is, for any integer k, $u = \cos\frac{\theta + 2k\pi}{q} + i\sin\frac{\theta + 2k\pi}{q}$ is a value of $(\cos\theta + i\sin\theta)^{1/q}$.

Now any two values of u are equal if and only if the arguments of the two values have the same sine and cosine. In other words, the two arguments must differ by a multiple of 2π. If this is the case, then the corresponding values of k must differ by a multiple of q. Therefore, u has only q distinct values, namely those given by

$$k = 0, 1, 2, \ldots, q-1$$

Thus the qth roots of $z = \cos\theta + i\sin\theta$ are given by

$$\cos\frac{\theta + 2k\pi}{q} + i\sin\frac{\theta + 2k\pi}{q}, \text{ where } k = 0, 1, \ldots, q-1.$$

If p and q are two integers prime to each other, then $(\cos\theta + i\sin\theta)^{p/q}$ has q distinct values given by

$$\cos\frac{p}{q}(\theta + 2k\pi) + i\sin\frac{p}{q}(\theta + 2k\pi),\ k = 0, 1, \ldots, q-1$$

Let us formally state it as a theorem.

THEOREM

If p and q are integers prime to each other ($q \neq 0$), then all the values of $(\cos\theta + i\sin\theta)^{p/q}$ are given by

$$\cos\left[(\theta + 2k\pi)\frac{p}{q}\right] + i\sin\left[(\theta + 2k\pi)\frac{p}{q}\right], \text{ where } k = 0, 1, 2, \ldots, q-1$$

REMARK: If $z = r(\cos\theta + i\sin\theta)$ is a complex number and n is a positive integer, then the nth roots of z are given by

$$\begin{aligned} z^{1/n} &= [r(\cos\theta + i\sin\theta)]^{1/n} \\ &= r^{1/n}(\cos\theta + i\sin\theta)^{1/n} \\ &= r^{1/n}\left[\cos\frac{\theta + 2k\pi}{n} + i\sin\frac{\theta + 2k\pi}{n}\right], k = 0, 1, \ldots, n-1 \end{aligned}$$

where $r^{1/n}$ is the positive arithmetic nth root of r.

EXAMPLE 12.8 Write down all the values of $\left(\sqrt{3} + i\right)^{1/3}$.

SOLUTION

Let $\sqrt{3} + i = r(\cos\theta + i\sin\theta)$

Then $\quad r\cos\theta = \sqrt{3}$ and $r\sin\theta = 1$

This gives $r = 2$ and $\theta = \dfrac{\pi}{6}$.

Therefore, $\quad \sqrt{3} + i = 2\left(\cos\dfrac{\pi}{6} + i\sin\dfrac{\pi}{6}\right)$

So

$$\begin{aligned} \left(\sqrt{3} + i\right)^{1/3} &= 2^{1/3}\left(\cos\frac{\pi}{6} + i\sin\frac{\pi}{6}\right)^{1/3} \\ &= 2^{1/3}\left[\cos\frac{1}{3}\left(\frac{\pi}{6} + 2k\pi\right) + i\sin\frac{1}{3}\left(\frac{\pi}{6} + 2k\pi\right)\right], k = 0, 1, 2 \\ &= 2^{1/3}\left[\cos(1 + 12k)\frac{\pi}{18} + i\sin(1 + 12k)\frac{\pi}{18}\right], k = 0, 1, 2 \\ &= 2^{1/3}\left[\cos\frac{p\pi}{18} + i\sin\frac{p\pi}{18}\right], \text{ where } p = 1, 13, 25 \end{aligned}$$

EXAMPLE 12.9 Find all the nth roots of unity, where n is a positive integer, and prove that they lie at the vertices of a regular n-gon inscribed in a unit circle centred at origin and with one of the vertices at 1.

SOLUTION

The nth roots of unity are the solutions of the equation $z^n - 1 = 0$. So the nth roots of unity are given by

$$\begin{aligned} z &= (1)^{1/n} \\ &= (\cos 0 + i\sin 0)^{1/n} \\ &= \cos\frac{2k\pi}{n} + i\sin\frac{2k\pi}{n}, k = 0, 1, \ldots, n-1 \end{aligned}$$

Writing the nth roots explicitly, we have

$$\begin{aligned} z_0 &= \cos 0 + i\sin 0 = 1 \\ z_1 &= \cos\frac{2\pi}{n} + i\sin\frac{2\pi}{n} \\ z_2 &= \cos\frac{4\pi}{n} + i\sin\frac{4\pi}{n} \end{aligned}$$

$$\begin{matrix} \dots & \dots & \dots \\ \dots & \dots & \dots \\ \dots & \dots & \dots \end{matrix}$$

$$z_n = \cos\frac{2(n-1)}{n}\pi + i\sin\frac{2(n-1)}{n}\pi$$

Suppose $A_0, A_1, \ldots, A_{n-1}$ are the images of the complex numbers $z_0, z_1, \ldots, z_{n-1}$ respectively on the Argand plane. Then the vector $\overrightarrow{OA_k}$ represent the complex number Z_k where $k \in \{0, 1, \ldots, n-1\}$. Since $OA_k = |Z_k| = 1$ for $k \in \{0, 1, \ldots, n-1\}$, therefore the points A_k lie on the unit circle with centre at origin. Also, the measure of the arc $\widehat{A_k A_{k+1}}$ is equal to

$$\arg Z_{k+1} - \arg Z_k = \frac{2(k+1)\pi}{n} - \frac{2k\pi}{n} = \frac{2\pi}{n}$$

for all $k \in \{0, 1, \ldots, n-2\}$, and that of the remaining arc $\widehat{A_{k+1} A_0}$ is

$$2\pi - \frac{2(n-1)\pi}{n} = \frac{2\pi}{n}.$$

Therefore all the arcs $\widehat{A_0A_1}, \widehat{A_1A_2}, \ldots, \widehat{A_{n-1}A_0}$ are equal and hence the polygon $A_0 A_1 \ldots A_{n-1}$ is a regular polygon. Also note that A_0 is the image of $Z_0 = 1$. Hence, the images of the nth roots of unity are vertices of a regular n-gon inscribed in the unit circle with one of the vertices at 1.

REMARK: As a particular case of the above example let us find the cube roots of unity. The cube roots of unity are roots of the equation $z^3 - 1 = 0$ and are given by

$$Z_k = \cos\frac{2k\pi}{3} + i\sin\frac{2k\pi}{3}, \; k = 0, 1, 2$$

Hence

$$Z_0 = \cos 0 + i\sin 0 = 1$$

$$Z_1 = \cos\frac{2\pi}{3} + i\sin\frac{2\pi}{3} = \frac{-1}{2} + i\frac{\sqrt{3}}{2} = \omega$$

$$Z_2 = \cos\frac{4\pi}{3} + i\sin\frac{4\pi}{3} = \frac{-1}{2} - i\frac{\sqrt{3}}{2} = \omega^2$$

1, w, w^2 are the vertices of an equilateral triangle inscribed in the unit circle with centre at origin, as shown in the figure below.

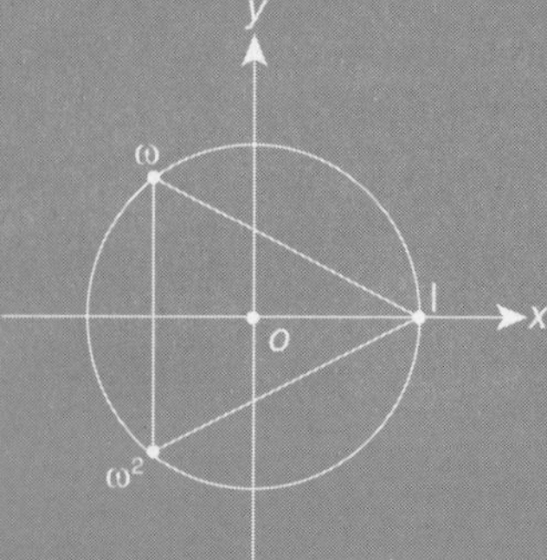

Fig. 12.1

EXAMPLE 12.10 Find the five fifth roots of unity and prove that the sum of their nth powers always vanishes unless n is a multiple of 5, in which case the sum is 5, n being an integer.

SOLUTION

We need to find roots of the equation $z^5 - 1 = 0$. So the fifth roots of unity are given by

$$z = 1^{1/5} = (\cos 0 + i \sin 0)^{1/5}$$

$$= \cos \frac{2k\pi}{5} + i \sin \frac{2k\pi}{5}, k = 0, 1, \ldots, 4$$

By De Moivre's Theorem, we can write

$$\cos \frac{2k\pi}{5} + i \sin \frac{2k\pi}{5} = \left(\cos \frac{2\pi}{5} + i \sin \frac{2\pi}{5} \right)^k, k = 0, 1, \ldots, 4$$

Let $w = \cos \frac{2\pi}{5} + i \sin \frac{2\pi}{5}$ so that the fifth roots of unity are given by w^k, $k = 0, 1, \ldots, 4$.

Now, the sum of the nth powers is

$$S = 1 + w^n + (w^2)^n + (w^3)^n + (w^4)^n$$

or
$$S = 1 + w^n + (w^n)^2 + (w^n)^3 + (w^n)^4 \quad (1)$$

Let n be a multiple of 5, say, $n = 5m$ for some integer m.

Then
$$w^n = w^{5m} = \left(\cos \frac{2\pi}{5} + i \sin \frac{2\pi}{5} \right)^{5m}$$

$$= \cos 2\pi m + i \sin 2\pi m = 1$$

So (1) becomes,

$$S = 1 + w^n + (w^n)^2 + (w^n)^3 + (w^n)^4$$

$$= 1 + 1 + 1 + 1 + 1$$

$$= 5$$

If n is not a multiple of 5, then

$$w^n = \left(\cos \frac{2\pi}{5} + i \sin \frac{2\pi}{5} \right)^n$$

$$= \cos \left[2\pi \left(\frac{n}{5} \right) \right] + i \sin \left[2\pi \left(\frac{n}{5} \right) \right]$$

$$\neq 1, \text{ since } \frac{n}{5} \text{ is not an integer.}$$

Now (1) is a geometric series with common ratio $w^n \neq 1$, therefore we have

$$S = \frac{1 - (w^n)^5}{1 - w^n} = \frac{1 - w^{5n}}{1 - w^n}$$

But $w^{5n} = 1$, therefore, $S = 0$.

EXAMPLE 12.11 Solve the equation $z^4 + 1 = 0$

SOLUTION

We have $z^4 = -1 = \cos\pi + i\sin\pi$

Therefore,

$$z = (\cos\pi + i\sin\pi)^{1/4}$$

$$= \cos(2k\pi + \pi)\frac{1}{4} + i\sin(2k\pi + \pi)\frac{1}{4}, \qquad k = 0, 1, 2, 3$$

$$= \cos(2k+1)\frac{\pi}{4} + i\sin(2k+1)\frac{\pi}{4}, \qquad k = 0, 1, 2, 3$$

On putting $k = 0, 1, 2, 3$, we get the roots as

$$\cos\frac{\pi}{4} + i\sin\frac{\pi}{4}, \cos\frac{3\pi}{4} + i\sin\frac{3\pi}{4}, \cos\frac{5\pi}{4} + i\sin\frac{5\pi}{4} \text{ and } \cos\frac{7\pi}{4} + i\sin\frac{7\pi}{4}$$

Hence the four roots are

$$\frac{1}{\sqrt{2}} + \frac{1}{\sqrt{2}}i, \quad \frac{-1}{\sqrt{2}} + \frac{1}{\sqrt{2}}i, \quad \frac{-1}{\sqrt{2}} - \frac{1}{\sqrt{2}}i \quad \text{and} \quad \frac{1}{\sqrt{2}} - \frac{1}{\sqrt{2}}i$$

EXAMPLE 12.12 Using De Moivre's theorem, solve the equation $z^7 + z = 0$

SOLUTION

The given equation is $z(z^6 + 1) = 0$

This gives $z = 0$ or $z^6 + 1 = 0$

Now the roots of equation $z^6 + 1 = 0$ are given by

$$z = (-1)^{1/6}$$

$$= (\cos\pi + i\sin\pi)^{1/6}$$

$$= \cos(2k+1)\frac{\pi}{6} + i\sin(2k+1)\frac{\pi}{6}, k = 0, 1, \ldots, 5$$

$$= \cos p\frac{\pi}{6} + i\sin p\frac{\pi}{6}, p = 1, 3, 5, 7, 9, 11$$

On putting values of p, the roots of $z^6 + 1 = 0$ are seen to be

$$\cos\frac{\pi}{2} \pm i\sin\frac{\pi}{2}, \cos\frac{\pi}{6} \pm i\sin\frac{\pi}{6}, -\cos\frac{\pi}{6} \pm i\sin\frac{\pi}{6}$$

Thus the roots of $z^6 + 1$ are

$$\pm i, \frac{\sqrt{3}}{2} \pm \frac{1}{2}i, \frac{-\sqrt{3}}{2} \pm i$$

Hence the seven roots of $z^7 + z = 0$ are

$$0, \pm i, \frac{\sqrt{3}}{2} \pm \frac{i}{2}, \frac{-\sqrt{3}}{2} \pm \frac{i}{2}$$

EXAMPLE 12.13 Use De Moivre's Theorem to solve the equation

$$z^7 - z^4 + z^3 - 1 = 0$$

SOLUTION

We have

$$z^7 - z^4 + z^3 - 1 = 0$$

or

$$(z^4 + 1)(z^3 - 1) = 0$$

This gives $z^4 + 1 = 0 \quad \text{or} \quad z^3 - 1 = 0.$

So the seven roots of the given equation are the roots of the equations $z^4 + 1 = 0$ and $z^3 - 1 = 0$.
Now, the roots of $z^4 + 1 = 0$ are given by

$$z = (-1)^{1/4} = (\cos \pi + i \sin \pi)^{1/4}$$
$$= \cos (2k+1)\frac{\pi}{4} + i \sin (2k+1)\frac{\pi}{4},\ k = 0, 1, 2, 3$$

So the four roots of $z^4 + 1 = 0$ are

$$\frac{1}{\sqrt{2}} \pm \frac{1}{\sqrt{2}} i,\ \frac{-1}{\sqrt{2}} \pm \frac{1}{\sqrt{2}} i$$

Again, $z^3 - 1 = 0$

$\Rightarrow$ $z = 1^{1/3}$

$$= \cos \frac{2p\pi}{3} + i \sin \frac{2p\pi}{3},\ p = 0, 1, 2$$

Thus, the three roots of $z^3 - 1 = 0$ are 0, $\frac{-1}{2} \pm \frac{\sqrt{3}}{2}$ i

Hence the seven roots of $z^7 - z^4 + z^3 - 1 = 0$ are

$$0,\ \frac{-1}{2} \pm \frac{\sqrt{3}}{2} i,\ \frac{1}{\sqrt{2}} \pm \frac{1}{\sqrt{2}} i,\ \frac{-1}{\sqrt{2}} \pm \frac{1}{\sqrt{2}} i$$

EXAMPLE 12.14 Solve the equation $z^6 + z^3 + 1 = 0$.

SOLUTION

The given equation $z^6 + z^3 + 1 = 0$
Solving the given equation as a quadratic in z^3, we get

$$z^3 = \frac{-1 \pm \sqrt{1-4}}{2} = \frac{-1}{2} \pm i \frac{\sqrt{3}}{2}$$

Let $\frac{-1}{2} \pm i \frac{\sqrt{3}}{2} = r(\cos \theta \pm i \sin \theta)$.

Then $r \cos \theta = \frac{-1}{2}$ and $r \sin \theta = \frac{\sqrt{3}}{2}$

$\Rightarrow$ $r = 1$ and $\theta = \frac{2\pi}{3}$

$\therefore$ $z^3 = 1\left(\cos \frac{2\pi}{3} \pm i \sin \frac{2\pi}{3}\right)$

or $z = \left(\cos \frac{2\pi}{3} \pm i \sin \frac{2\pi}{3}\right)^{1/3}$

$$= \cos \left[\left(2k\pi + \frac{2\pi}{3}\right)\frac{1}{3}\right] \pm i \sin \left[\left(2k\pi + \frac{2\pi}{3}\right)\frac{1}{3}\right],\ k = 0, 1, 2$$

$$= \cos \left[(6k+2)\frac{\pi}{9}\right] \pm i \sin \left[(6k+2)\frac{\pi}{9}\right],\ k = 0, 1, 2$$

So, the six roots of the given equation are

$$\cos\frac{2\pi}{9} \pm i\sin\frac{2\pi}{9}, \quad \cos\frac{8\pi}{9} \pm i\sin\frac{8\pi}{9}$$

and $\cos\frac{14\pi}{9} \pm i\sin\frac{14\pi}{9} = \cos\left(2\pi - \frac{4\pi}{9}\right) \pm i\sin\left(2\pi - \frac{4\pi}{9}\right) = \cos\frac{4\pi}{9} \pm i\sin\frac{4\pi}{9}$

Hence, the required roots are given by

$$z = \cos\frac{p\pi}{q} \pm i\sin\frac{p\pi}{q}, p = 2, 4, 8$$

EXAMPLE 12.15 Solve the equation $(2+z)^6 + (2-z)^6 = 0$.

SOLUTION

The given equation is

$$(2+z)^6 + (2-z)^6 = 0$$

or
$$\left(\frac{2+z}{2-z}\right)^6 + 1 = 0 \qquad (1)$$

Let $w = \frac{2+z}{2-z}$. Therefore (1) becomes

$$w^6 + 1 = 0$$

$$\Rightarrow \quad w = (-1)^{1/6}$$

$$= (\cos\pi + i\sin\pi)^{1/6}$$

$$= \cos\left[(2k\pi + \pi)\frac{1}{6}\right] + i\sin\left[(2k\pi + \pi)\frac{1}{6}\right], k = 0, 1, \ldots, 5$$

$$= \cos\left[(2k+1)\frac{\pi}{6}\right] + i\sin\left[(2k+1)\frac{\pi}{6}\right], k = 0, 1, \ldots, 5$$

Now, $w = \frac{2+z}{2-z}$ implies $z = \frac{2(w-1)}{w+1}$

So,
$$z = \frac{2\left[\cos\left\{(2k+1)\frac{\pi}{6}\right\} + i\sin\left\{(2k+1)\frac{\pi}{6}\right\} - 1\right]}{\cos\left\{(2k+1)\frac{\pi}{6}\right\} + i\sin\left\{(2k+1)\frac{\pi}{6}\right\} + 1}$$

$$= \frac{2\left[2i\sin\left\{(2k+1)\frac{\pi}{12}\right\}\cos\left\{(2k+1)\frac{\pi}{12}\right\} - 2\sin^2\left\{(2k+1)\frac{\pi}{12}\right\}\right]}{2\cos^2\left\{(2k+1)\frac{\pi}{12}\right\} + 2i\sin\left\{(2k+1)\frac{\pi}{12}\right\}\cos\left\{(2k+1)\frac{\pi}{12}\right\}}$$

$$= \frac{4i\sin\left\{(2k+1)\frac{\pi}{12}\right\}\left[\cos\left\{(2k+1)\frac{\pi}{12}\right\} + i\sin\left\{(2k+1)\frac{\pi}{12}\right\}\right]}{2\cos\left\{(2k+1)\frac{\pi}{12}\right\}\left[\cos\left\{(2k+1)\frac{\pi}{12}\right\} + i\sin\left\{(2k+1)\frac{\pi}{12}\right\}\right]}$$

$$= 2\,i \tan\left[(2\,k+1)\,\frac{\pi}{12}\right],\ k = 0, 1, \ldots, 5$$

On putting $k = 0, 1, 2, 3, 4, 5$, the roots are seen to be

$$\pm 2\,i \tan\frac{\pi}{12},\ \pm 2\,i \tan\frac{5\pi}{12},\ \pm 2i$$

EXAMPLE 12.16 Show that the roots of the equation $(z-1)^5 + z^5 = 0$ are given by

$$z = \frac{1}{2}\left[1 + i\cot\left(\frac{p\pi}{10}\right)\right],\ p = 1, 3, 5, 7, 9.$$

SOLUTION

The given equation is

$$(z-1)^5 + z^5 = 0$$

or $$\left(\frac{z-1}{z}\right)^5 + 1 = 0$$

or $$w^5 + 1 = 0, \text{ where } w = \frac{z-1}{z}$$

$$\Rightarrow \quad w = (-1)^{1/5}$$

$$= (\cos\pi + i\sin\pi)^{1/5}$$

$$= \cos\left[(2\,k+1)\,\frac{\pi}{5}\right] + i\sin\left[(2\,k+1)\,\frac{\pi}{5}\right],\ k = 0, 1, \ldots, 4$$

$$= \cos\frac{p\pi}{5} + i\sin\frac{p\pi}{5},\ p = 1, 3, 5, 7, 9$$

Now $w = \dfrac{z-1}{z}$ implies $z = \dfrac{1}{1-w}$

$$\therefore \quad z = \frac{1}{1 - \cos\left(\frac{p\pi}{5}\right) - i\sin\left(\frac{p\pi}{5}\right)}$$

$$= \frac{1}{2\sin^2\left(\frac{p\pi}{10}\right) - 2\,i\sin\left(\frac{p\pi}{10}\right)\cos\left(\frac{p\pi}{10}\right)}$$

$$= \frac{1}{2\sin\left(\frac{p\pi}{10}\right)\left[\sin\left(\frac{p\pi}{10}\right) - i\cos\left(\frac{p\pi}{10}\right)\right]}$$

$$= \frac{\sin\left(\frac{p\pi}{10}\right) + i\cos\left(\frac{p\pi}{10}\right)}{2\sin\left(\frac{p\pi}{10}\right)}$$

$$= \frac{1}{2}\left[1 + \cot\left(\frac{p\pi}{10}\right)\right],\ p = 1, 3, 5, 7, 9.$$

EXAMPLE 12.17 Solve the equation

$$z^4 + z^3 + z^2 + z + 1 = 0$$

SOLUTION

Multiplying both sides of the given equation by $(z - 1)$, we have

$$(z - 1)(z^4 + z^3 + z^2 + z + 1) = 0$$

or $$z^5 - 1 = 0$$

Thus, the roots of the given equation are the roots of $z^5 - 1 = 0$ except the root $z = 1$.

Now, $z^5 - 1 = 0 \Rightarrow \quad z = (1)^{1/5}$

or $$z = \cos\frac{2k\pi}{5} + i\sin\frac{2k\pi}{5},\ k = 0, 1, \ldots, 4.$$

On putting $k = 0, 1, \ldots, 4$ we get the five roots of $z^5 - 1 = 0$ as

$$1,\ \cos\left(\frac{2\pi}{5}\right) \pm i\sin\left(\frac{2\pi}{5}\right),\quad \cos\left(\frac{4\pi}{5}\right) \pm i\sin\left(\frac{4\pi}{5}\right)$$

Hence the roots of the equation $z^4 + z^3 + z^2 + z + 1 = 0$ are

$$\cos\left(\frac{2\pi}{5}\right) \pm i\sin\left(\frac{2\pi}{5}\right),\quad \cos\left(\frac{4\pi}{5}\right) \pm i\sin\left(\frac{4\pi}{5}\right)$$

EXERCISE 12.2

1. Find all the values of each of the following:
   (a) $(-1)^{1/4}$ **(b)** $(i)^{1/3}$ **(c)** $(1 + i)^{2/3}$
   (d) $\left(1 - i\sqrt{3}\right)^{1/3}$ **(e)** $\left(-1 + i\sqrt{3}\right)^{-1/5}$ **(f)** $\left(\sqrt{3} - i\right)^{2/5}$
2. Find all the fourth roots of unity and prove that they lie at the vertices of a square inscribed in the unit circle. Also draw a figure.
3. Prove that the product of the nth roots of unity is $(-1)^{n-1}$.
4. Find the seven seventh roots of unity and prove that for any positive integer n, the sum of their nth powers vanishes, unless n is a multiple of 7 in which case the sum is 7.
5. Use De Moivre's Theorem to solve the following equations:
   (a) $z^3 + 1 = 0$
   (b) $z^5 - 1 = 0$
   (c) $z^5 - z = 0$
   (d) $z^{10} - z^5 + 1 = 0$
   (e) $z^4 + 4z^2 + 16 = 0$
   (f) $z^9 + z^5 + z^4 + 1 = 0$
   (g) $z^8 - z^5 + z^3 - 1 = 0$
   (h) $(z + 1)^5 + (z - 1)^5 = 0$
6. Show that the roots of the equation $(1 - z)^7 + (1 + z)^7 = 0$

   are $z = \pm i\cot\left(\frac{p\pi}{7}\right), p = 1, 2, 3.$

7. Solve the equation:

$$(5 + x)^5 - (5 - x)^5 = 0$$

8. Using De Moivre's theorem solve the equation $(1 + x)^n = (1 - x)^n$, n being a positive integer.

9. Solve the equation $(z + 1)^{2n} + (z - 1)^{2n} = 0$
and show that the roots are purely imaginary.

10. Solve the equation $z^5 + z^4 + z^3 + z^2 + z + 1 = 0$

11. Solve the equation $z^{12} + z^{11} + z^{10} + \ldots + z + 1 = 0$

12. Use De Moivre's theorem to solve

$$z^4 - z^3 + z^2 - z + 1 = 0$$

13. Find the roots of the equation

$$x^8 - x^6 + x^4 - x^2 + 1 = 0$$

by using De Moivre's Theorem

[Hint: $(x^2 + 1)(x^8 - x^6 + x^4 - x^2 + 1) = x^{10} + 1$]

14. Solve the equation

$$(1 + i\,z)^{10} + (1 - i\,z)^{10} = 0$$

Answers to Exercises

CHAPTER 1

EXERCISE 1.1

1. (a) $A + B = \begin{bmatrix} 5 & 2 & 2 \\ -4 & 2 & 3 \end{bmatrix}$ **(b)** $A - B = \begin{bmatrix} -3 & 2 & 4 \\ 2 & -2 & 1 \end{bmatrix}$

(c) $2A = \begin{bmatrix} 2 & 4 & 6 \\ -2 & 0 & 4 \end{bmatrix}$ **(d)** $-B = \begin{bmatrix} -4 & 0 & 1 \\ 3 & -2 & -1 \end{bmatrix}$

2. (a) $A + B = \begin{bmatrix} 3 & 7 \\ 1 & 7 \end{bmatrix}$ **(b)** $A - B = \begin{bmatrix} 1 & 1 \\ 5 & -3 \end{bmatrix}$ **(c)** $3A - C = \begin{bmatrix} 8 & 7 \\ 6 & 2 \end{bmatrix}$

(d) $AB = \begin{bmatrix} -6 & 26 \\ -1 & 19 \end{bmatrix}$ **(e)** $BA = \begin{bmatrix} 11 & 10 \\ 11 & 2 \end{bmatrix}$ **(f)** $(BC)^T = \begin{bmatrix} 7 & 19 \\ 17 & 10 \end{bmatrix}$

3. (a) $BA = \begin{bmatrix} 3 & 14 & -19 \\ 4 & -8 & -2 \\ -5 & 14 & -1 \end{bmatrix}$ **(b)** $AC = \begin{bmatrix} 0 & 16 \\ 1 & 6 \end{bmatrix}$ **(c)** $CA = \begin{bmatrix} -2 & 12 & -6 \\ -5 & 14 & -1 \\ 5 & -6 & -6 \end{bmatrix}$

(d) $A(2B - 3C) = \begin{bmatrix} 2 & -78 \\ -21 & -32 \end{bmatrix}$ **(e)** $A(B + C) = \begin{bmatrix} 1 & 1 \\ -8 & -1 \end{bmatrix}$ **(f)** $AB + AC = \begin{bmatrix} 1 & 1 \\ -8 & -1 \end{bmatrix}$

15. (a) $\begin{bmatrix} a & b \\ 0 & 0 \end{bmatrix}$, a and b arbitrary **(b)** $\begin{bmatrix} -2a & a \\ -2b & b \end{bmatrix}$, a and b arbitrary

16. $\begin{bmatrix} a & b \\ c & -a \end{bmatrix}$, where b and c are arbitrary, and a is any solution of the equation $a^2 = -bc$.

17. $f(A) = \begin{bmatrix} 1 & -1 & -3 \\ -1 & -1 & -10 \\ -5 & 4 & 4 \end{bmatrix}$

18. $A^5 = \begin{bmatrix} 1 & 0 & -62 \\ 62 & 32 & 124 \\ 0 & 0 & 32 \end{bmatrix}$

20. $A^n = \begin{bmatrix} \cos n\theta & -\sin n\theta \\ \sin n\theta & \cos n\theta \end{bmatrix}$

21. $A^2 = \begin{bmatrix} 1 & 2 \\ 0 & 1 \end{bmatrix}$, $A^3 = \begin{bmatrix} 1 & 3 \\ 0 & 1 \end{bmatrix}$, $A^4 = \begin{bmatrix} 1 & 4 \\ 0 & 1 \end{bmatrix}$, $A^n = \begin{bmatrix} 1 & n \\ 0 & 1 \end{bmatrix}$

22. $A^2 = \begin{bmatrix} 2 & 2 \\ 2 & 2 \end{bmatrix}$, $A^3 = \begin{bmatrix} 4 & 4 \\ 4 & 4 \end{bmatrix}$, $A^n = \begin{bmatrix} 2^{n-1} & 2^{n-1} \\ 2^{n-1} & 2^{n-1} \end{bmatrix}$

23. $A^2 = \begin{bmatrix} 2 & 1 \\ 1 & 1 \end{bmatrix}$, $A^3 = \begin{bmatrix} 3 & 2 \\ 2 & 1 \end{bmatrix}$, $A^4 = \begin{bmatrix} 5 & 3 \\ 3 & 2 \end{bmatrix}$, $A^5 = \begin{bmatrix} 8 & 5 \\ 5 & 3 \end{bmatrix}$

$A^n = \begin{bmatrix} F_{n+1} & F_n \\ F_n & F_{n-1} \end{bmatrix}$ for $n > 2$, where F_n is the nth term of the Fibonacci sequence.

27. ***(a)*** and ***(d)***

29. AB gives the total cost of producing each kind of product in each city:

$$\begin{array}{cc} \text{Delhi} & \text{Chennai} \end{array}$$
$$\begin{pmatrix} 38 & 44 \\ 67 & 78 \end{pmatrix} \begin{array}{l} \text{Chair} \\ \text{Table} \end{array}$$

30. Units $\begin{array}{cccc} A & B & C & D \\ (6.3 & 3.3 & 3.6 & 5.0); \end{array}$ cost of one unit $\begin{pmatrix} 15 \\ 13 \\ 33 \end{pmatrix} \begin{array}{l} \text{Food} - \text{I} \\ \text{Food} - \text{II} \\ \text{Food} - \text{III} \end{array}$

Rs. 469.

EXERCISE 1.2

1. Different methods will give different answers.

3. (a) $\begin{bmatrix} 1 & 0 & 0 \\ 0 & 0 & 1 \\ 0 & 1 & 0 \end{bmatrix}$ **(b)** $\begin{bmatrix} 3 & 0 & 0 \\ 0 & 1 & 0 \\ 0 & 0 & 1 \end{bmatrix}$ **(c)** $\begin{bmatrix} 1 & 0 & -4 \\ 0 & 1 & 0 \\ 0 & 0 & 1 \end{bmatrix}$ **(d)** $\begin{bmatrix} 1 & 0 & 0 \\ 0 & 1 & 0 \\ 0 & -2 & 1 \end{bmatrix}$

EXERCISE 1.3

1. (a) Inverse does not exist. **(b)** $\begin{bmatrix} 1 & -\frac{1}{2} \\ -2 & \frac{3}{2} \end{bmatrix}$

2. (a) $\begin{bmatrix} 3 & -1 \\ -5 & 2 \end{bmatrix}$ **(b)** $\begin{bmatrix} 4 & -5 \\ -3 & 4 \end{bmatrix}$

3. (a) $A^{-1} = \begin{bmatrix} -\frac{1}{4} & \frac{3}{4} & -1 \\ \frac{3}{4} & -\frac{1}{4} & 0 \\ -\frac{1}{4} & -\frac{1}{4} & 1 \end{bmatrix}$ **(b)** $A^{-1} = \begin{bmatrix} 3 & -1 & 1 \\ -15 & 6 & -5 \\ 5 & -2 & 2 \end{bmatrix}$

(c) A^{-1} does not exist.

(d) $A^{-1} = \begin{bmatrix} 0 & 1 & -1 \\ 2 & -2 & -1 \\ -1 & 1 & 1 \end{bmatrix}$

(e) $A^{-1} = \begin{bmatrix} \frac{1}{2} & -\frac{1}{2} & \frac{1}{2} \\ -4 & 3 & -1 \\ \frac{5}{2} & -\frac{3}{2} & \frac{1}{2} \end{bmatrix}$

(f) $A^{-1} = \begin{bmatrix} 1 & 2 & -1 \\ -4 & -7 & 4 \\ -4 & -9 & 5 \end{bmatrix}$

(g) $A^{-1} = \begin{bmatrix} \frac{11}{3} & -3 & \frac{1}{3} \\ -\frac{7}{3} & 3 & -\frac{2}{3} \\ \frac{2}{3} & -1 & \frac{1}{3} \end{bmatrix}$

(h) $A^{-1} = \begin{bmatrix} \frac{2}{5} & -\frac{1}{5} & \frac{1}{5} \\ \frac{1}{5} & 0 & \frac{1}{5} \\ \frac{3}{5} & \frac{4}{5} & \frac{2}{5} \end{bmatrix}$

8. $ad - bc \neq 0$

9. $\begin{bmatrix} \frac{d}{ad - bc} & \frac{-b}{ad - bc} \\ \frac{-c}{ad - bc} & \frac{a}{ad - bc} \end{bmatrix}$

EXERCISE 1.4

1. (a) 2 (b) 3 (c) 3 (d) 2
2. (a) 3 (b) 3 (c) 3 (d) 3 (e) 3
   (f) 2 (g) 3 (h) 2 (i) 4 (j) 2

EXERCISE 1.5

1. $x = 1,\ y = 3,\ z = 3$
2. $x = \frac{35}{18},\ y = \frac{29}{18},\ z = \frac{5}{18}$
3. Inconsistent
4. $x = \frac{5}{7}k - \frac{8}{7},\ y = \frac{13}{7}k - \frac{25}{7},\ z = k$
5. $x = \frac{1}{2}(5 - 3k),\quad y = \frac{1}{2}(k - 3),\quad z = k$
6. $x = y = z = 1$
7. $x = -1, y = 4, z = 4$
8. $x = 3 + 2k, y = 1 - k, z = k$
9. Inconsistent
10. $x = 1, y = 2, z = 3$

11. $x = 3k - 2,\ y = \frac{10}{3}k - \frac{7}{3},\ z = k$

12. Inconsistent

13. $x = -\frac{11}{7}k,\ y = -\frac{1}{7}k,\ z = k$

14. $x = y = z = 0$

15. $x = y = z = 0$

16. $x = -\frac{10}{7}k,\ y = \frac{8}{7}k,\ z = k$

17. $\lambda = 1$ or 2 when $\lambda = 1, x = 1 + 2k, y = -3k, z = k$
when $\lambda = 2, x = 2k, y = 1 - 3k, z = k$

18. If $\lambda \neq 2$, then $x = \frac{\lambda + 6}{\lambda^2 + 2\lambda + 15},\ y = \frac{\lambda + \frac{3}{2}}{\lambda^2 + 2\lambda + 15},\ z = \frac{3\lambda - 6}{\lambda^2 + 2\lambda + 15}$

19. **(a)** unique solution for $\lambda \neq 3$, whatever μ be
(b) no solution for $\lambda = 3, \mu \neq 10$.
(c) infinitely many solutions for $\lambda = 3, \mu = 10$.

20. When $k \neq \pm 2, x = y = z = 0$ is the only solution
When $k = 2, x = y = z = 0$ is the only solution
When $k = -2, x = 4p, y = p, z = p$.

CHAPTER 2

EXERCISE 2.1

2. **(a)** $\sqrt{5}$ **(b)** 0 **(c)** $\sqrt{2}$

3. **(a)** 0 **(b)** -4 **(c)** 4

4. **(a)** $\cos\theta = -\frac{3}{\sqrt{10}}$ **(b)** $\theta = 45^\circ$

7. $c = \pm 3$

EXERCISE 2.2

1. $X + Y = \begin{bmatrix} 3 \\ 5 \end{bmatrix},\ 3X = \begin{bmatrix} 12 \\ 6 \end{bmatrix},\ 2X + 4Y = \begin{bmatrix} 4 \\ 16 \end{bmatrix},\ 3X - Y = \begin{bmatrix} 13 \\ 3 \end{bmatrix},\ Y + (-Y) = \begin{bmatrix} 0 \\ 0 \end{bmatrix}$

2. **(a)** $X + Y = (0, 8), X - Y = (4, -2), 2X = (4, 6), 3X - 2Y = (10, -1)$
(b) $X + Y = (1, 7), X - Y = (-3, -1), 2X = (-2, 6), 3X - 2Y = (-7, 1)$
(c) $X + Y = (1, 3, -5),\ X - Y = (1, 1, -1),\ 2X = (2, 4, -6),\ 3X - 2Y = (3, 4, -5)$
(d) $X + Y = (5, 2, -3),\ X - Y = (-1, -2, -5),\ 2X = (4, 0, -8),\ 3X - 2Y = (0, -4, -14)$

3. **(a)** $\sqrt{10} + \sqrt{97}$ **(b)** $4\sqrt{10}$ **(c)** $3\sqrt{17}$ **(d)** $\sqrt{61}$

9. 5 miles per hour

10. 13 pounds

11. The resultant force is (3, 4). To double the magnitude, replace (0, 3) by (3,7).

EXERCISE 2.3

4. **(a)** $(7, 5) = -3(1, -1) + 2(5, 1)$
   (b) Not a linear combination
   (c) Not a linear combination
   (d) $(0, 10, 8) = (2 - c)(-1, 2, 3) + (2 - 2c)(1, 3, 1) + c(1, 8, 5)$ where c is any real number.
5. **(a)** and **(b)**
6. **(a)** only
7. **(a)** $\begin{bmatrix} 1 \\ 3 \\ 2 \end{bmatrix} = \frac{14}{-5}\begin{bmatrix} 1 \\ 2 \\ 3 \end{bmatrix} + \frac{3}{-5}\begin{bmatrix} -1 \\ -1 \\ 0 \end{bmatrix} + \frac{8}{5}\begin{bmatrix} 2 \\ 5 \\ 4 \end{bmatrix}$
   (b) $\begin{bmatrix} 1 \\ 3 \\ -2 \end{bmatrix} = -5\begin{bmatrix} 5 \\ 1 \\ 3 \end{bmatrix} + \frac{31}{3}\begin{bmatrix} 2 \\ 0 \\ 1 \end{bmatrix} + \frac{8}{-3}\begin{bmatrix} -2 \\ -3 \\ -1 \end{bmatrix}$

EXERCISE 2.4

1. **(a)** Y is a scalar multiple of X
   (b) Any set of three vector in $\mathbf{R}^2$ is linearly dependent (See Theorem 1).
   (c) Y is a scalar multiple of X.
2. **(a)** linearly independent **(b)** linearly dependent
   (c) linearly dependent **(d)** linearly independent
   (e) linearly independent **(f)** linearly independent
3. $X = 3X_1 - 2X_2 - 5X_3$, S is linearly dependent
4. $\begin{bmatrix} 1 \\ 2 \\ 3 \end{bmatrix} = \begin{bmatrix} 1 \\ 0 \\ 0 \end{bmatrix} + 2\begin{bmatrix} 0 \\ 1 \\ 0 \end{bmatrix} + 3\begin{bmatrix} 0 \\ 0 \\ 1 \end{bmatrix} + 0\begin{bmatrix} 1 \\ 0 \\ 1 \end{bmatrix}$

   and $\begin{bmatrix} 1 \\ 2 \\ 3 \end{bmatrix} = \begin{bmatrix} 1 \\ 0 \\ 1 \end{bmatrix} + 2\begin{bmatrix} 0 \\ 1 \\ 0 \end{bmatrix} + 2\begin{bmatrix} 0 \\ 0 \\ 1 \end{bmatrix} + 0\begin{bmatrix} 1 \\ 0 \\ 0 \end{bmatrix}$
5. **(a)** $c = 2$ **(b)** $c = -7, c = 0$
6. **(a)** $\lambda = 1$ **(b)** $\lambda = -\frac{1}{2}, \lambda = 1$

EXERCISE 2.5

1. **(a)** A basis of $\mathbf{R}^2$ must contain two linearly independent vectors.
   (b) $X_2 = -2X_1$. $\therefore$ X_1 and X_2 are linearly dependent.
   (c) A basis for $\mathbf{R}^3$ must contain three linearly independent vectors.
2. **(a)**, **(d)** and **(e)**
3. **(a)**, **(c)** and **(d)**
8. All values of λ except 0 and -3 make the given set a basis for $\mathbf{R}^3$.

9. **(i)** False **(ii)** True **(iii)** False **(iv)** False **(v)** False
(vi) True **(vii)** True **(viii)** True **(ix)** True **(x)** True

EXERCISE 2.6

2. **(a)**, **(b)** and **(c)**

3. **(a)** subspace : line $y = 2x$
(b) subspace : y-axis
(c) Not a subspace
(d) Not a subspace
(e) Not a subspace
(f) Not a subspace
(g) Not a subspace

4. **(a)** Yes **(b)** Let $X = \begin{bmatrix} 1 \\ 1 \end{bmatrix}$ and $c = -1$. Then $cX \notin V$.

5. No

7. **(a)**, **(c)** and **(e)** are subspaces of $\mathbf{R}^3$

8. **(b)**, **(c)**, **(f)** and **(g)** are subspaces of $\mathbf{R}^3$

9. W is the plane $x + y + z = 0$ passing through origin · dim W = 2.

10. only **(c)** is a subspace of $\mathbf{R}^2$

12. No

14. Set of all vectors of the form $\begin{bmatrix} x \\ y \\ y \end{bmatrix}$; $x, y \in \mathbf{R}$

15. Span S = $\mathbf{R}^2$

CHAPTER 3

EXERCISE 3.1

1. $T\left(\begin{bmatrix} 1 \\ 2 \end{bmatrix}\right) = \begin{bmatrix} 2 \\ -1 \end{bmatrix}, T\left(\begin{bmatrix} -1 \\ 4 \end{bmatrix}\right) = \begin{bmatrix} -2 \\ -5 \end{bmatrix}$

2. **(a)** and (d)

3. **(a)**, **(b)**, **(c)** and **(d)**

4. **(a)**, **(b)** and **(d)**

6. No

7. (4, 11)

8. **(a)** (3, −3) **(b)** (1, 2) **(c)** (4, 1)

EXERCISE 3.2

1. $T\left(\begin{bmatrix} x \\ y \end{bmatrix}\right) = \begin{bmatrix} 0 \\ y \end{bmatrix}$

2. **(a)** $T\left(\begin{bmatrix} x \\ y \\ z \end{bmatrix}\right) = \begin{bmatrix} 0 \\ 0 \\ z \end{bmatrix}$ **(b)** $T\left(\begin{bmatrix} x \\ y \\ z \end{bmatrix}\right) = \begin{bmatrix} 0 \\ y \\ z \end{bmatrix}$

4. **(a)** reflection w.r.t. the line $y = x$
(b) reflection w.r.t. the line $y = -x$
(c) anticlockwise rotation through 90°
(d) reflection w.r.t. origin

5. **(a)** reflection in xz-plane
(b) reflection in xy-plane
(c) reflection in the x-axis
(d) reflection in the z-axis

6. **(a)** $(2\sqrt{3}, 4)$ **(b)** $(3, 2)$ **(c)** $\left(-\frac{1}{\sqrt{2}}, \frac{3}{\sqrt{2}}\right)$ **(d)** $(\sqrt{3}, -1)$

7. **(a)** $\begin{bmatrix} \frac{\sqrt{3}}{2} & -\frac{1}{2} \\ \frac{1}{2} & \frac{\sqrt{3}}{2} \end{bmatrix}$ **(b)** $\begin{bmatrix} \frac{1}{\sqrt{2}} & -\frac{1}{\sqrt{2}} \\ \frac{1}{\sqrt{2}} & \frac{1}{\sqrt{2}} \end{bmatrix}$ **(c)** $\begin{bmatrix} \frac{1}{2} & -\frac{\sqrt{3}}{2} \\ \frac{\sqrt{3}}{2} & \frac{1}{2} \end{bmatrix}$

(d) $\begin{bmatrix} 0 & -1 \\ 1 & 0 \end{bmatrix}$ **(e)** $\begin{bmatrix} -\frac{1}{2} & -\frac{\sqrt{3}}{2} \\ \frac{\sqrt{3}}{2} & -\frac{1}{2} \end{bmatrix}$ **(f)** $\begin{bmatrix} -1 & 0 \\ 0 & -1 \end{bmatrix}$

8. **(a)** $\theta = \frac{\pi}{2}$ **(b)** $\theta = \pi$

11. **(b)** **(i)** projection onto x-axis **(ii)** reflection w.r.t. y-axis
(iii) projection onto y-axis **(iv)** reflection w.r.t. line $y = x$
(v) reflection in origin **(vi)** rotation through 90°

14. $\frac{x^2}{9} + \frac{y^2}{4} = 1$

15. $x^2 + y^2 = 9$

16. **(a)** $y = -2x$ **(b)** $y = \frac{1}{2}x$ **(c)** $y = -\left(\frac{8 + 5\sqrt{3}}{11}\right)x$ **(d)** $y = 2x$

17. **(a)** $y = 3x$ **(b)** $y = 3(x + 1)$

19. **(a), (c), (d)**

20. Distance between two points is invariant under **(a)**, **(b)**, **(e)** and **(f)**.
Area of triangle is invariant under **(a)**, **(b)**, **(e)** and **(f)**.

22. $m = 7$ and $n = 5$

23. $T(X) = \begin{bmatrix} 1 \\ 4 \\ 1 \end{bmatrix}, T(Y) = \begin{bmatrix} 8 \\ 7 \\ 8 \end{bmatrix}, T(Z) = \begin{bmatrix} 9 \\ 1 \\ 9 \end{bmatrix}$

EXERCISE 3.3

1. $\begin{bmatrix} 2 & 3 \\ 1 & 2 \end{bmatrix}$

2. **(a)** $\begin{bmatrix} 1 & 1 \\ 1 & -1 \end{bmatrix}$ **(b)** $\begin{bmatrix} 1 & 0 \\ 0 & 1 \end{bmatrix}$ **(c)** $\begin{bmatrix} 2 & 3 \\ 3 & -2 \end{bmatrix}$

(d) $\begin{bmatrix} 0 & 0 \\ 0 & 0 \end{bmatrix}$ **(e)** $\begin{bmatrix} 2 & 0 \\ 0 & -1 \end{bmatrix}$ **(f)** $\begin{bmatrix} 1 & 2 \\ 0 & 1 \end{bmatrix}$

3. **(a)** $\begin{bmatrix} 0 & 0 \\ 0 & 1 \end{bmatrix}$ **(b)** $\begin{bmatrix} -1 & 0 \\ 0 & 1 \end{bmatrix}$ **(c)** $\begin{bmatrix} 0 & 1 \\ 1 & 0 \end{bmatrix}$ **(d)** $\begin{bmatrix} \frac{1}{2} & \frac{-\sqrt{3}}{2} \\ \frac{\sqrt{3}}{2} & \frac{1}{2} \end{bmatrix}$

(e) $\begin{bmatrix} \frac{1}{\sqrt{2}} & -\frac{1}{\sqrt{2}} \\ \frac{1}{\sqrt{2}} & \frac{1}{\sqrt{2}} \end{bmatrix}$ **(f)** $\begin{bmatrix} 3 & 0 \\ 0 & 3 \end{bmatrix}$ **(g)** $\begin{bmatrix} \frac{1}{3} & 0 \\ 0 & \frac{1}{3} \end{bmatrix}$ **(h)** $\begin{bmatrix} -1 & 0 \\ 0 & -1 \end{bmatrix}$

4. **(a)** $\begin{bmatrix} \frac{\sqrt{3}}{2} & -\frac{1}{2} \\ \frac{1}{2} & \frac{\sqrt{3}}{2} \end{bmatrix}$ **(b)** $\begin{bmatrix} 1 & 2 \\ 0 & 1 \end{bmatrix}$ **(c)** $\begin{bmatrix} 5 & 0 \\ 0 & 1 \end{bmatrix}$ **(d)** $\begin{bmatrix} 1 & 0 \\ 3 & 1 \end{bmatrix}$

5. **(a)** $\begin{bmatrix} 1 & 0 & 0 \\ 0 & -1 & 0 \\ 0 & 0 & 0 \end{bmatrix}$ **(b)** $\begin{bmatrix} 2 & 0 & 0 \\ 0 & 3 & 0 \\ 0 & 0 & 4 \end{bmatrix}$ **(c)** $\begin{bmatrix} 1 & -1 & 0 \\ 0 & 1 & 1 \\ -1 & 0 & 1 \end{bmatrix}$

(d) $\begin{bmatrix} 1 & -1 & 0 \\ 1 & 0 & 1 \\ 0 & 1 & -1 \end{bmatrix}$ **(e)** $\begin{bmatrix} -1 & 1 & 0 \\ 0 & 0 & 0 \\ -1 & 0 & 1 \end{bmatrix}$ **(f)** $\begin{bmatrix} 0 & 1 & 0 \\ 0 & 0 & 1 \\ 1 & 0 & 0 \end{bmatrix}$

6. **(a)** $\begin{bmatrix} 4 & 0 & 0 \\ 0 & 4 & 0 \\ 0 & 0 & 4 \end{bmatrix}$ **(b)** $\begin{bmatrix} 1 & 0 & 0 \\ 0 & 1 & 0 \\ 0 & 0 & 0 \end{bmatrix}$ **(c)** $\begin{bmatrix} 0 & 0 & 0 \\ 0 & 0 & 0 \\ 0 & 0 & 1 \end{bmatrix}$ **(d)** $\begin{bmatrix} 1 & 0 & 0 \\ 0 & 1 & 0 \\ 0 & 0 & -1 \end{bmatrix}$

(e) $\begin{bmatrix} 1 & 0 & 0 \\ 0 & -1 & 0 \\ 0 & 0 & 1 \end{bmatrix}$ **(f)** $\begin{bmatrix} 1 & 0 & 0 \\ 0 & -1 & 0 \\ 0 & 0 & -1 \end{bmatrix}$ **(g)** $\begin{bmatrix} -1 & 0 & 0 \\ 0 & -1 & 0 \\ 0 & 0 & -1 \end{bmatrix}$

7. **(a)** $\begin{bmatrix} 1 & 0 \\ 1 & 1 \\ 0 & 1 \end{bmatrix}$ **(b)** $\begin{bmatrix} 1 & 1 \\ 0 & 1 \\ 1 & -1 \end{bmatrix}$ **(c)** $\begin{bmatrix} 1 & 1 & 0 \\ -1 & 0 & 2 \end{bmatrix}$ **(d)** $\begin{bmatrix} 1 & -1 & 1 \\ 1 & 2 & 1 \end{bmatrix}$

(e) $\begin{bmatrix} 0 & 1 & 0 \end{bmatrix}$ **(f)** $\begin{bmatrix} 1 \\ -1 \\ 0 \end{bmatrix}$ **(g)** $\begin{bmatrix} a & b \\ c & d \end{bmatrix}$

8. **(a)** $n \times n$ identity matrix I_n **(b)** $n \times n$ zero matrix O

9. **(b)** $\begin{bmatrix} 1 & 0 \\ 0 & k \end{bmatrix}$

13. $\begin{bmatrix} 0 & -1 \\ 1 & 0 \end{bmatrix}$

14. **(a)** $y = 3x + 8$ **(b)** triangle with vertices (0, 0), (4, 8) and (12, 4)

15. Triangle with vertices (−10, 15), (3, 12), (5, −2)

16. $x^2 + y^2 = 9$

17. **(a)** $\begin{bmatrix} 0 & -2 \\ 2 & 0 \end{bmatrix}$ **(b)** $\begin{bmatrix} 3 & 0 \\ 0 & -3 \end{bmatrix}$ **(c)** $\begin{bmatrix} 1 & 0 \\ 0 & -1 \end{bmatrix}$

(d) $\begin{bmatrix} -1 & 0 \\ 0 & 0 \end{bmatrix}$ **(e)** $\begin{bmatrix} 0 & 0 \\ 0 & \frac{1}{2} \end{bmatrix}$

18. (a) $\begin{bmatrix} 0 & -2 \\ -2 & 0 \end{bmatrix}$ **(b)** $\begin{bmatrix} 3 & 6 \\ 0 & 3 \end{bmatrix}$ **(c)** $\begin{bmatrix} \frac{1}{2} & 0 \\ 0 & 5 \end{bmatrix}$

(d) $\begin{bmatrix} 1 & 0 \\ 2 & 5 \end{bmatrix}$ **(e)** $\begin{bmatrix} 0 & 1 \\ -5 & 0 \end{bmatrix}$

19. (a) $\begin{bmatrix} -1 & 0 & 0 \\ 0 & 0 & 0 \\ 0 & 0 & 1 \end{bmatrix}$ **(b)** $\begin{bmatrix} -1 & 0 & 0 \\ 0 & 1 & 0 \\ 0 & 0 & 0 \end{bmatrix}$ **(c)** $\begin{bmatrix} 1/5 & 0 & 0 \\ 0 & -1/5 & 0 \\ 0 & 0 & 1/5 \end{bmatrix}$

(d) $\begin{bmatrix} -1 & 0 & 0 \\ 0 & -1 & 0 \\ 0 & 0 & -1 \end{bmatrix}$

20. (a) Yes **(b)** No **(c)** Yes

(d) No **(e)** Yes

21. (a) $\begin{bmatrix} 1 \\ 2 \\ 0 \end{bmatrix}$ **(b)** $\begin{bmatrix} 3 \\ 6 \\ 9 \end{bmatrix}$ **(c)** $\begin{bmatrix} 1 \\ -2 \\ 3 \end{bmatrix}$ **(d)** $\begin{bmatrix} -1 \\ -2 \\ 3 \end{bmatrix}$

22. $\begin{bmatrix} 2 & 1 \\ -1 & 3 \end{bmatrix}$

23. (a) $(8, -2)$ **(b)** $(7, 1)$

CHAPTER 4

EXERCISE 4.1

1. $\lambda_1 = 1, \lambda_2 = 4$

2. (a) $\lambda^2 - 4\lambda - 5 = 0; \lambda_1 = -1$ and $\lambda_2 = 5$; $\begin{bmatrix} k \\ -k \end{bmatrix}$ and $\begin{bmatrix} 2k \\ k \end{bmatrix}$, $k \neq 0$

(b) $\lambda^2 - 2\lambda = 0; \lambda_1 = 0$ & $\lambda_2 = 2$; $\begin{bmatrix} 0 \\ k \end{bmatrix}$ and $\begin{bmatrix} k \\ 0 \end{bmatrix}$ where $k \neq 0$

(c) $\lambda^2 - 5\lambda + 4 = 0; \lambda_1 = 1$ & $\lambda_2 = 4$; $\begin{bmatrix} k \\ -2k \end{bmatrix}$ and $\begin{bmatrix} k \\ k \end{bmatrix}$ where $k \neq 0$

(d) $\lambda^2 - 2\lambda = 0; \lambda_1 = 0, \lambda_2 = 2$; $\begin{bmatrix} -k \\ k \end{bmatrix}$ and $\begin{bmatrix} k \\ k \end{bmatrix}$, $k \neq 0$

(e) $\lambda^2 = 0; \lambda = 0$; $k\begin{bmatrix} 1 \\ 0 \end{bmatrix}$ and $k\begin{bmatrix} 0 \\ 1 \end{bmatrix}$, $k \neq 0$

(f) $(3 - \lambda)^2 = 0; \lambda = 3$; $k\begin{bmatrix} 1 \\ 0 \end{bmatrix}$ and $k\begin{bmatrix} 0 \\ 1 \end{bmatrix}$, $k \neq 0$

(g) $\lambda^2 - 2\lambda = 0; \lambda_1 = 0, \lambda_2 = 2$; $\begin{bmatrix} k \\ -k \end{bmatrix}$ and $\begin{bmatrix} k \\ k \end{bmatrix}$, $k \neq 0$

(h) $\lambda^2 - 3\lambda = 0; \lambda_1 = 0, \lambda_2 = 3$; $k\begin{bmatrix} -1/\sqrt{2} \\ 1 \end{bmatrix}$ and $k = \begin{bmatrix} 1 \\ \sqrt{2} \end{bmatrix}$, $k \neq 0$

4. (a) $\lambda^3 - 7\lambda^2 + 11\lambda - 5 = 0$; $\lambda_1 = 1$, $\lambda_2 = 1$, $\lambda_3 = 5$; For $\lambda_1 = \lambda_2 = 1$, $k\begin{bmatrix}-2\\1\\0\end{bmatrix}$ and $k\begin{bmatrix}-1\\0\\1\end{bmatrix}$; For $\lambda = 5$, $k\begin{bmatrix}1\\1\\1\end{bmatrix}$, $k \neq 0$.

(b) $(\lambda - 2)^3 = 0$; $\lambda_1 = 2, \lambda_2 = 2, \lambda_3 = 2$; $\begin{bmatrix}k\\0\\0\end{bmatrix}$, $k \neq 0$.

(c) $(\lambda - 1)(\lambda - 3)(\lambda - 2) = 0$; $\lambda_1 = 1, \lambda_2 = 3, \lambda_3 = -2$; $k\begin{bmatrix}6\\3\\8\end{bmatrix}$, $k\begin{bmatrix}0\\5\\2\end{bmatrix}$, $k\begin{bmatrix}0\\0\\1\end{bmatrix}$, $k \neq 0$.

(d) $\lambda^3 - 18\lambda^2 + 45\lambda = 0$; $\lambda_1 = 0, \lambda_2 = 3, \lambda_3 = 15$; For $\lambda = 0$, $\begin{bmatrix}k\\2k\\2k\end{bmatrix}$, for $\lambda = 3$, $\begin{bmatrix}2k\\-k\\2k\end{bmatrix}$ and for $\lambda = 15$, $\begin{bmatrix}2k\\-2k\\k\end{bmatrix}$, where $k \neq 0$.

(e) $\lambda^3 - 5\lambda^2 + 2\lambda + 8 = 0$; $\lambda_1 = -1, \lambda_2 = 2, \lambda_3 = 4$; $k\begin{bmatrix}1\\0\\-1\end{bmatrix}$, $k\begin{bmatrix}-2\\-3\\2\end{bmatrix}$ and $k\begin{bmatrix}8\\5\\2\end{bmatrix}$ where $k \neq 0$.

(f) $(3 - \lambda)(\lambda^2 - 10\lambda + 21) = 0$; $\lambda_1 = 3, \lambda_2 = 3, \lambda_3 = 7$; For $\lambda = 3$, $\begin{bmatrix}1\\0\\0\end{bmatrix}$ and $\begin{bmatrix}0\\-\sqrt{3}\\1\end{bmatrix}$ are two linearly independent eigenvectors. For $\lambda = 9$, $\begin{bmatrix}0\\1\\\sqrt{3}\end{bmatrix}$ is an eigenvector.

5. (a) $\lambda = 0$, $\begin{bmatrix}k\\0\end{bmatrix}$, $k \neq 0$; $\lambda = 1$, $\begin{bmatrix}0\\k\end{bmatrix}$, $k \neq 0$

(b) $\lambda = 1$, $\begin{bmatrix}-k\\k\end{bmatrix}$, $k \neq 0$; $\lambda = -1$, $\begin{bmatrix}k\\k\end{bmatrix}$, $k \neq 0$.

(c) No eigenvalues

(d) $\lambda = 3$; every nonzero vector in $\mathbf{R}^2$ is an eigenvector

(e) $\lambda = -1$; every nonzero vector in $\mathbf{R}^2$ is an eigenvector

6. (a) $\lambda_1 = 1$; $\begin{bmatrix}0\\k\end{bmatrix}$, $k \neq 0$; $\lambda_2 = -1$; $\begin{bmatrix}k\\0\end{bmatrix}$, $k \neq 0$

(b) $\lambda = \frac{1}{2}$; all nonzero vectors in $\mathbf{R}^2$ are eigenvectors

(c) $\lambda = -1$; all nonzero vectors in $\mathbf{R}^2$ are eigenvectors

(d) $\lambda = 0$, $\begin{bmatrix}0\\0\\k\end{bmatrix}$; $\lambda = 1$, $\begin{bmatrix}p\\q\\0\end{bmatrix}$

(e) $\lambda = 1$, $\begin{bmatrix}0\\p\\q\end{bmatrix}$; $\lambda = -1$, $\begin{bmatrix}k\\0\\0\end{bmatrix}$, $k \neq 0$.

14. $(a - d)^2 + 4bc > 0$.

EXERCISE 4.2

2. (a) For $\lambda = 0$, eigenspace is $r\begin{bmatrix} 1 \\ -1 \end{bmatrix}$ and for $\lambda = -1$, eigenspace is $s\begin{bmatrix} -3 \\ 2 \end{bmatrix}$; $r, s \in \mathbf{R}$.

   (b) For $\lambda = 1$, eigenspace is $r\begin{bmatrix} 0 \\ 1 \end{bmatrix}$ and for $\lambda = 5$, eigenspace is $s\begin{bmatrix} 2 \\ 1 \end{bmatrix}$; $r, s \in \mathbf{R}$.

   (c) For $\lambda = 6$, eigenspace is $r\begin{bmatrix} 4 \\ 1 \end{bmatrix}$ and for $\lambda = 1$ eigenspace is $s\begin{bmatrix} -1 \\ 1 \end{bmatrix}$; $r, s \in \mathbf{R}$.

   (d) For $\lambda = 3$, eigenspace is $r\begin{bmatrix} 1 \\ 1 \end{bmatrix}$ and for $\lambda = -1$, eigenspace is $s\begin{bmatrix} 1 \\ -1 \end{bmatrix}$; $r, s \in \mathbf{R}$.

   (e) For $\lambda = 1$, $\left\{\begin{bmatrix} 1 \\ 0 \end{bmatrix}, \begin{bmatrix} 0 \\ 1 \end{bmatrix}\right\}$ is a basis for the eigenspace.

   (f) For $\lambda = 0$, $r\begin{bmatrix} -1 \\ 1 \end{bmatrix}$ and for $\lambda = 2$, $s\begin{bmatrix} 1 \\ 1 \end{bmatrix}$; $r, s \in \mathbf{R}$

3. (a) Set of vectors of the form $\begin{bmatrix} k \\ 0 \end{bmatrix}$ i.e. x-axis

   (b) Set of vectors of the form $\begin{bmatrix} 0 \\ k \end{bmatrix}$ i.e. y-axis

   (c) For $\lambda = 1$, set of vectors of the form $\begin{bmatrix} k \\ k \end{bmatrix}$ i.e. the line $y = x$ and for $\lambda = -1$, set of vectors of the form $\begin{bmatrix} -k \\ k \end{bmatrix}$ i.e. the line $y = -x$.

   (d) For $\lambda = 0$, set of vectors of the form $\begin{bmatrix} 0 \\ k \end{bmatrix}$ i.e. the y-axis and for $\lambda = 1$, set of vectors of the form $\begin{bmatrix} k \\ 0 \end{bmatrix}$ i.e. the x-axis

   (e) For $\lambda = 0$, set of vectors of the form $\begin{bmatrix} 0 \\ k \end{bmatrix}$ i.e. the y-axis and for $\lambda = 1$, set of vectors of the form $\begin{bmatrix} k \\ k \end{bmatrix}$ i.e. the line $y = x$.

   (f) $\mathbf{R}^2$ is an eigenspace.

5. (a) For $\lambda = 1$, basis $\left\{\begin{bmatrix} 1 \\ -1 \\ 0 \end{bmatrix}\right\}$, dim = 1;

   For $\lambda = -1$, basis $\left\{\begin{bmatrix} 0 \\ 1 \\ -1 \end{bmatrix}\right\}$, dim = 1;

   For $\lambda = 3$, basis $\left\{\begin{bmatrix} 2 \\ 3 \\ -1 \end{bmatrix}\right\}$, dim = 1.

   (b) For $\lambda = 10$, basis $\left\{\begin{bmatrix} 2 \\ 2 \\ 1 \end{bmatrix}\right\}$, dim = 1;

   For $\lambda = 1$, basis $\left\{\begin{bmatrix} -1 \\ 1 \\ 0 \end{bmatrix}, \begin{bmatrix} -1 \\ 0 \\ 2 \end{bmatrix}\right\}$, dim = 2.

(c) For $\lambda = 3$, basis $\left\{\begin{bmatrix}-1\\1\\0\end{bmatrix},\begin{bmatrix}0\\0\\1\end{bmatrix}\right\}$, dim = 2;

For $\lambda = 1$, basis $\left\{\begin{bmatrix}1\\1\\0\end{bmatrix}\right\}$, dim = 1

(d) For $\lambda = 2$, basis $\left\{\begin{bmatrix}-1\\0\\1\end{bmatrix},\begin{bmatrix}0\\1\\0\end{bmatrix}\right\}$, dim = 2;

For $\lambda = 1$, basis $\left\{\begin{bmatrix}-2\\1\\1\end{bmatrix}\right\}$, dim = 1

EXERCISE 4.3

1. $P = \begin{bmatrix}-1 & 2\\1 & 1\end{bmatrix}$, $D = \begin{bmatrix}3 & 0\\0 & 4\end{bmatrix}$

2. No

3. Yes

4. (a) $P = \begin{bmatrix}1 & 1\\1 & 2\end{bmatrix}$, $D = \begin{bmatrix}3 & 0\\0 & 2\end{bmatrix}$

(b) $P = \begin{bmatrix}1 & -1\\1 & 1\end{bmatrix}$, $D = \begin{bmatrix}0 & 0\\0 & 2\end{bmatrix}$

(c) $P = \begin{bmatrix}1 & 1\\0 & -1\end{bmatrix}$, $D = \begin{bmatrix}1 & 0\\0 & 3\end{bmatrix}$

(d) $P = \begin{bmatrix}1 & -(1+\sqrt{2})\\1+\sqrt{2} & 1\end{bmatrix}$, $D = \begin{bmatrix}1 & 0\\0 & -1\end{bmatrix}$

5. (a) Not diagonalisable

(b) Diagonalisable, $P = \begin{bmatrix}3 & 1\\2 & 1\end{bmatrix}$, $P^{-1}AP = \begin{bmatrix}2 & 0\\0 & 1\end{bmatrix}$

(c) No diagonalisable

(d) Not diagonalisable

6. (a) $P = \begin{bmatrix}-1 & 0 & -2\\0 & 1 & 1\\1 & 0 & 1\end{bmatrix}$, $P^{-1}AP = \begin{bmatrix}2 & 0 & 0\\0 & 2 & 0\\0 & 0 & 1\end{bmatrix}$

(b) $P = \begin{bmatrix}3 & 2 & 2\\-1 & 1 & 0\\3 & 0 & 1\end{bmatrix}$, $P^{-1}AP = \begin{bmatrix}1 & 0 & 0\\0 & 2 & 0\\0 & 0 & 2\end{bmatrix}$

(c) $P = \begin{bmatrix}-2 & 1 & 2\\-2 & 2 & 1\\1 & 2 & -2\end{bmatrix}$, $P^{-1}AP = \begin{bmatrix}0 & 0 & 0\\0 & 9 & 0\\0 & 0 & -9\end{bmatrix}$

(d) $P = \begin{bmatrix}1 & 0 & 0\\0 & 1 & 0\\-1 & 0 & 1\end{bmatrix}$, $D = \begin{bmatrix}0 & 0 & 0\\0 & 1 & 0\\0 & 0 & 1\end{bmatrix}$

7. (a) Not diagonalisable

(b) Diagonalisable, $P = \begin{bmatrix} 1 & -1 & -1 \\ -1 & 1 & 0 \\ 1 & 0 & 1 \end{bmatrix}$, $D = \begin{bmatrix} 1 & 0 & 0 \\ 0 & -2 & 0 \\ 0 & 0 & -2 \end{bmatrix}$

(c) Not diagonalisable

(d) Diagonalisable $P = \begin{bmatrix} 1 & 1 & 1 \\ 4 & 2+\sqrt{3} & 2-\sqrt{3} \\ 6 & 7+4\sqrt{3} & 7-4\sqrt{3} \end{bmatrix}$, $D = \begin{bmatrix} 4 & 0 & 0 \\ 0 & 2+\sqrt{3} & 0 \\ 0 & 0 & 2-\sqrt{3} \end{bmatrix}$

8. $\lambda = 0$: $k\begin{bmatrix} 0 \\ 1 \end{bmatrix}$, $k \neq 0$. A is not diagonalizable

9. (a) Diagonalisable

(b) Diagonalisable

(c) Not diagonalisable

10. Diagonalisable, $P = \begin{bmatrix} \cos\theta & -\sin\theta \\ \sin\theta & \cos\theta \end{bmatrix}$, $P^{-1}AP = \begin{bmatrix} 1 & 0 \\ 0 & -1 \end{bmatrix}$

12. $(a-d)^2 + 4bc > 0$

■ CHAPTER 5

EXERCISE 5.1

1. (a) $\frac{(-1)^{n-1}}{3^{n-1}}$ **(b)** $\frac{2n-1}{2n}$ **(c)** $\frac{n^2}{(n+1)\sqrt{e}}$ **(d)** $n^2 - 1$ **(e)** $(-1)^{n-1}(4n-2)$

(f) $\frac{1}{2} + \frac{1}{2}(-1)^{n-1}$ **(g)** $3 + (-1)^n$ **(h)** $1 + \frac{(-1)^n}{n}$ **(i)** $\frac{n^2-1}{\sqrt[n+1]{\pi}}$ **(j)** $n + \frac{1}{n}$

2. (a) $(-r)^{n-1}, (-r)^n$ **(b)** $(-1)^{n-1}r^n, (-1)^n r^{n+1}$

EXERCISE 5.2

2. 55, 233, 610.

EXERCISE 5.3

1. $\Delta x_n = .4x_n$, 38

2. $\Delta x_n = \frac{1}{332}x_n(500 - x_n)$, 31, 169

3. 39th month, 40950

4. $\Delta x_n = .1x_n$, 8 years

EXERCISE 5.4

1. Bounded **2.** Bounded **3.** Bounded below **4.** Unbounded

5. Bounded **6.** Bounded **7.** Bounded below **8.** Bounded above

9. Bounded **10.** Bounded below **11.** Strictly decreasing **12.** Strictly increasing

13. Strictly decreasing **14.** Strictly decreasing **15.** Strictly decreasing **16.** Strictly decreasing
17. Strictly decreasing **18.** Strictly increasing **19.** Strictly decreasing **20.** Strictly decreasing
21. Strictly increasing **22.** Strictly decreasing **23.** Strictly decreasing **24.** Strictly decreasing
25. Strictly increasing **26.** Strictly decreasing **27.** Strictly increasing

EXERCISE 5.5

5. **(a)** 0 **(b)** 0 **(c)** $\frac{3}{2}$ **(d)** $\frac{1}{18}$ **(e)** 0 **(f)** 1 **(g)** $\frac{4}{3}$ **(h)** $\frac{4}{3}$ **(i)** 1 **(j)** $\frac{3}{16}$

6. Converges, 1 **7.** Converges, 4 **8.** Converges, 2 **9.** Diverges **10.** Diverges

11. Converges **12.** Converges, 1 **13.** Diverges **14.** Diverges **15.** Converges, 0

16. Converges, 1 **17.** Diverges **18.** Converges, 0 **19.** Converges, e^{-2} **20.** Converges, 1

21. Converges, 1 **22.** Converges, 3 **23.** Converges, 0 **24.** Converges, 9 **25.** Diverges

26. Converges, e^{-1} **27.** Converges, 0 **28.** Converges, 1 **29.** Converges, $e^{1/2}$ **30.** Converges, 1

31. Converges, 0 **32.** Converges, $\pi/2$ **33.** Diverges **34.** Converges, 1 **35.** Converges, x

36. Converges, 0 **37.** Converges, 3 **38.** Converges, 6 **39.** Converges, $\frac{1}{p-1}$

CHAPTER 7

EXERCISE 7.1

1. 7070 **2.** 784 **3.** 13.93 hours **4.** 3.1 hours

5. **(a)** $\frac{dy}{dx} = .01y,\ y_0 = 10{,}000$ **(b)** $y = 10{,}000\, e^{t/100}$ **(c)** 69.31 hours **(d)** 150.41 hours

6. 196 days

7. **(a)** $y \approx 2e^{0.1386t}$ **(b)** $y = 5e^{0.015t}$ **(c)** $y \approx .5995\, e^{.5117t}$ **(d)** $y \approx 0.8706\, e^{0.1386t}$

8. 694

9. 9 times

EXERCISE 7.2

1. **(a)** $y = 10.2^{-1/6}t^{(-1/6)t}$ **(b)** $y = 10^{19/9} \cdot e^{(-1/9)t}$ **(c)** $y = 20e^{(-.02)t}$ **(d)** $y = 10.2^{(-1/5)t}$

2. 3.31 days

3. **(a)** $\frac{dy}{dt} = -ky,\ k \approx .1810$ **(b)** $y = 5 \times 10^7 e^{-.18t}$

(c) $\approx$ 219000 atoms **(d)** 12.72 days

4. 0.000121, 88.6 gms. **5.** 9954 years **6.** 13.3 years

7. 10.3 hours **8.** 34 min. **9.** $T = 100\,e^{-0.35t}$

10. 18.5 min **11.** **(a)** $\frac{dT}{dt} = -k(T - 21),\ T(0) = 95,\ T = 21 + 74e^{-kt}$

(b) 6.22 minutes

EXERCISE 7.3

1. **(a)** $y_0 = 5$ **(b)** $C = 12$, **(c)** $k = 1$, **(d)** $t = .3365$ units

(e) $\frac{dy}{dt} = \frac{1}{12}y(12 - y), y(0) = 5$

2. 50,000; 50; 25,000 **3.** 4.6 weeks **4.** 13 years

■ CHAPTER 8

EXERCISE 8.1

6. $\frac{d^2y}{d\theta^2} + y = 0$ **9.** $-\frac{3}{2}$ **10.** $\frac{b}{3a^2}\sec t \operatorname{cosec}^4 t$

EXERCISE 8.2

(i) $\frac{1}{4}\left[3(-1)^n n!(x - 2)^{-(n+1)} + (-1)^n n!(x + 2)^{-(n+1)}\right]$

(ii) $(-1)^n n!\left[\frac{1}{(x - 1)^{n+1}} + \frac{2(n + 1)}{(x - 1)^{n+2}} - \frac{1}{(x + 1)^{n+1}}\right]$

(iii) $(-1)^n n! \sin^{n+1}\theta \sin(n + 1)\theta$, where $\theta = \cot^y x$

(iv) $(-1)^n n!\left[\frac{1}{(x + 1)^{n+1}} - \frac{2^n}{(2x + 1)^{n+1}}\right]$

(v) $\frac{(-1)^n n!}{4a^3}\left[\frac{1}{(x - 1)^{n+1}} - \frac{1}{(x + a)^{n+1}} - \frac{2\sin\left\{(n + 1)\cot^{-1}\frac{x}{a}\right\}}{(x^2 + a^2)^{(n+1)/2}}\right]$

(vi) $(-1)^{n-1}(n - 1)!\sin^n\theta \sin n\theta$, where $\theta = \tan^{-1}\left(\frac{1}{x}\right)$

(vii) $(-1)^{n-1}(n - 1)!\sin^n\theta \sin n\theta$, where $\theta = \tan^{-1}\left(\frac{1}{x}\right)$

(viii) $2 \cdot (-1)^{n-1}(n - 1)! \sin^n\theta \sin n\theta$, where $\theta = \cot^{-1}x$

(ix) $(-1)^{n-1}(n - 1)!a^{-n}\sin^n\theta \sin n\theta$, where $\theta = \tan^{-1}\left(\frac{a}{x}\right)$

(x) $(-1)^n(n - 1)!\operatorname{cosec}^n\alpha \sin^n\theta \sin n\theta$, where $\theta = \cot^{-1}\left(\frac{x - \cos\alpha}{\sin\alpha}\right)$.

EXERCISE 8.3

6. $\dfrac{(-1)^{n-2}(n-1)!}{x^{n-1}}$

CHAPTER 9

EXERCISE 9.1

1. $P_0(x) = \ln 4;\quad P_1(x) = \ln 4 + \dfrac{1}{4}(x-4);\quad P_2(x) = \ln 4 + \dfrac{1}{4}(x-4) - \dfrac{1}{32}(x-4)^2;$

   $P_3(x) = \ln 4 + \dfrac{1}{4}(x-4) - \dfrac{1}{32}(x-4)^2 + \dfrac{1}{192}(x-4)^3.$

2. $P_0(x) = \ln 1 = 0;\quad P_1(x) = x;\quad P_2(x) = x - \dfrac{x^2}{2};\quad P_3(x) = x - \dfrac{x^2}{2} + \dfrac{x^3}{3}.$

3. $P_0(x) = 1;\quad P_1(x) = 2 - x;\quad P_2(x) = 2 - x + (x-1)^2;$

   $P_3(x) = 2 - x + (x-1)^2 - (x-1)^3.$

4. $P_0(x) = \dfrac{1}{2};\quad P_1(x) = \dfrac{1}{2} - \dfrac{1}{2^2}x;\quad P_2(x) = \dfrac{1}{2} - \dfrac{1}{2^2}x + \dfrac{1}{2^3}x^2;$

   $P_3(x) = \dfrac{1}{2} - \dfrac{1}{2^2}x + \dfrac{1}{2^3}x^2 + \dfrac{1}{2^4}x^3.$

5. $P_0(x) = \dfrac{1}{\sqrt{2}};\quad P_1(x) = \dfrac{1}{\sqrt{2}} + \dfrac{1}{\sqrt{2}}\left(x - \dfrac{\pi}{4}\right);$

   $P_2(x) = \dfrac{1}{\sqrt{2}} + \dfrac{1}{\sqrt{2}}\left(x - \dfrac{\pi}{4}\right) - \dfrac{1}{\sqrt{2}} \cdot \dfrac{(x-\pi/4)^2}{2!}$

   $P_3(x) = \dfrac{1}{\sqrt{2}} + \dfrac{1}{\sqrt{2}}\left(x - \dfrac{\pi}{4}\right) - \dfrac{1}{\sqrt{2}} \cdot \dfrac{(x-\pi/4)^2}{2!} - \dfrac{1}{\sqrt{2}} \cdot \dfrac{(x-\pi/4)^3}{3!}$

6. $P_0(x) = \dfrac{1}{\sqrt{2}};\quad P_1(x) = \dfrac{1}{\sqrt{2}} - \dfrac{1}{\sqrt{2}}\left(x - \dfrac{\pi}{4}\right);$

   $P_2(x) = \dfrac{1}{\sqrt{2}} - \dfrac{1}{\sqrt{2}}\left(x - \dfrac{\pi}{4}\right) - \dfrac{1}{\sqrt{2}} \cdot \dfrac{(x-\pi/4)^2}{2!};$

   $P_3(x) = \dfrac{1}{\sqrt{2}} - \dfrac{1}{\sqrt{2}}\left(x - \dfrac{\pi}{4}\right) - \dfrac{1}{\sqrt{2}} \cdot \dfrac{(x-\pi/4)^2}{2!} + \dfrac{1}{\sqrt{2}} \cdot \dfrac{(x-\pi/4)^3}{3!}$

7. $P_0(x) = \sqrt{3};\quad P_1(x) = \sqrt{3} + \dfrac{1}{2\sqrt{3}}(x-3);$

   $P_2(x) = \sqrt{3} + \dfrac{1}{2\sqrt{3}}(x-3) - \dfrac{1}{24\sqrt{3}}(x-3)^2;$

   $P_3(x) = \sqrt{3} + \dfrac{1}{2\sqrt{3}}(x-3) - \dfrac{1}{24\sqrt{3}}(x-3)^2 + \dfrac{1}{144\sqrt{3}}(x-3)^2.$

8. $\sqrt{6} + \dfrac{1}{2\sqrt{6}}(x-2) - \dfrac{1}{48\sqrt{6}}(x-2)^2 + \dfrac{1}{576\sqrt{6}}(x-2)^3$

9. $P_3(x) = 1 - x^2 - \dfrac{x^3}{3}$

10. $1-2x + \dfrac{(-2)^2x^2}{2!} + \dfrac{(-2)^3x^3}{3!} + \cdots + \dfrac{(-2)^n}{n!}x^n$

11. $P_{2n+1}(x) = 3x + (-1)\cdot 3^3 \cdot x^3 + \cdots + (-1)^n \cdot 3^{2n+1} \cdot x^{2n+1}$

12. $1- \dfrac{2^2x^2}{2!} + \dfrac{2^4x^4}{4!} + \cdots + (-1)^n\dfrac{2^{2n}x^{2n}}{(2n)!}$

13. $P_n(x) = \dfrac{x}{1!} + \dfrac{x^3}{3!} + \dfrac{x^5}{5!} + \cdots + \dfrac{x^n}{n!}$, if n is odd;

$= \dfrac{x}{1!} + \dfrac{x^3}{3!} + \dfrac{x^5}{5!} + \cdots + \dfrac{x^{n-1}}{(n-1)!}$, if n is even.

15. $P_6(x) = -\dfrac{x^2}{2!} - \dfrac{2x^4}{4!} - \dfrac{4x^5}{5!} - \dfrac{16x^6}{6!}$

16. $P_3(x) = x - \dfrac{x^3}{3}$

EXERCISE 9.2

2. $-.0673 < x < .0467$

3. $|x| < (0.06)^{1/5} < 0.5678$

4. $|\text{Error}| < .0003906$

5. $|\text{Error}| < \left(\dfrac{0.2}{24}\right)^4 < 6.7 \times 10^{-5}$, under estimation

6. $|\text{Error}| < \dfrac{(10^{-3})^3}{6} < 1.67 \times 10^{-10}, -10^{-3} < x < 0$

7. $|\text{Error}| < (3^{0.1})\dfrac{(0.1)^3}{6} < 1.87 \times 10^{-5}$

8. $0 < x < .02$

10. 1.693×10^{-4}

12. $|\text{Error}| < 1.26 \times 10^{-5}$

EXERCISE 9.3

1. $7 + 8(x-1) + 15(x-1)^2 + 9(x-1)^3$

2. $8 + 10(x-2) + 6(x-2)^2 + (x-2)^3$

3. $-17(x+1) + 24(x+1)^2 - 14(x+1)^3 + 3(x+1)^4$

4. $21 - 36(x+2) + 25(x+2)^2 - 8(x+2)^3 + (x+2)^4$

5. $\dfrac{1}{3}\left[\dfrac{\ln 3}{1!}(x+1) + \dfrac{(\ln 3)^2}{2!}(x+1)^2 + \dfrac{(\ln 3)^3}{3!}(x+1)^3 + \cdots\right]$

6. $\sum_{n=0}^{\infty}\dfrac{e}{n!}(x-1)^n,\ \sum_{n=0}^{\infty}\dfrac{e^2}{n!}(x-2)^n$

7. $\frac{1}{2}\left[1 - (x - \ln 2) + \frac{(x - \ln 2)^2}{2!} - \frac{(x - \ln 2)^3}{3!} + \cdots\right]$

8. $\sum_{n=0}^{\infty}(-1)(x + 1)^n$

9. $\sum_{n=0}^{\infty}\frac{(-1)^n}{(2n)!}\pi^{2n}\left(x - \frac{1}{2}\right)^{2n}$

10. $\sum_{n=1}^{\infty}\frac{(-1)^{n-1}}{n}(x - 1)^n$

11. $1 - \frac{2x}{1!} + \frac{(2x)^2}{2!} - \frac{(2x)^3}{3!} + \cdots$

12. $\sum_{n=0}^{\infty}\frac{(-1)^n\pi^{2n}x^{2n}}{(2n)!}$

13. $\sum_{n=0}^{\infty}(-1)^n\frac{3^{2n+1}x^{2n+1}}{(2n + 1)!}$

14. $\sum_{n=0}^{\infty}(-1)^n x^n$

15. $9\sum_{n=0}^{\infty}\frac{(-1)^n x^{2n}}{(2n)!}$

16. $\sum_{n=0}^{\infty}\frac{x^{2n}}{(2n)!}$

17. $\sum_{n=0}^{\infty}\frac{x^{2n+1}}{(2n + 1)!}$

18. $\sum_{n=0}^{\infty}\frac{(-1)^n}{(2n + 1)!}x^{2n+1}$

19. $x^5 - 7x^3 + 2x^2 - 4$

20. $(x - 1)^3$

EXERCISE 9.4

1. $x - \frac{\pi^2x^3}{2!} + \frac{\pi^4x^5}{4!} - \frac{\pi^6x^7}{6!} + \cdots = \sum_{n=0}^{\infty}\frac{(-1)^n\pi^{2n}x^{2n+1}}{(2n)!}$

2. $2x^2 - \frac{2}{3}x^4 + \cdots$

3. $2^2x^2 + 2^3x^3 + 2^4x^4 + \cdots = \sum_{n=0}^{\infty}(2x)^{n+2}$

4. $\frac{\pi}{2}x - \frac{1}{3!}\left(\frac{\pi}{2}x\right)^3 + \frac{1}{5!}\left(\frac{\pi}{2}x\right)^5 + \cdots$

5. $1 + \sum_{n=1}^{\infty}\frac{(-1)^n(2x)^{2n}}{2.(2n)!}$

6. $3x^2 - \frac{3^2}{2}x^3 + \frac{3^3}{3}x^4 - \frac{3^4}{4}x^5 + \cdots$

7. $\frac{x^4}{4!} - \frac{x^6}{6!} + \frac{x^8}{8!} - \frac{x^{10}}{10!} + \cdots = \sum_{n=2}^{\infty}\frac{(-1)^n x^{2n}}{(2n)!}$

8. $\frac{x^5}{5!} - \frac{x^7}{7!} + \frac{x^9}{9!} - \frac{x^{11}}{11!} + \cdots$

9. $\sin x, 0.1$

10. $\cos x, 0.01$

11. $\cos x, \frac{\pi}{2}$

12. $\ln(1 + x), 0.2$

13. $e^x, 0.2$

14. $\sin x, \frac{\pi}{3}$

CHAPTER 10

EXERCISE 10.2

1. (i) $e^{x-y}, -e^{x-y}, -e^{x-y}, e^{x-y}$

(ii) $l^{x^y}x^{y-2}y\,(x^y.y + y - 1), l^{x^y}.x^{y-1}[1 + (x^y + 1)ylnx],$
$l^{x^y}x^{y-1}[1 + (x^y + 1)ylnx], l^{x^y}.x^y(lnx)^2[x^y + 1]$

(ii) $\dfrac{2y(1+y^2)}{1-xy}, \dfrac{2(x+y)}{(1-xy)^3}, \dfrac{2x(1+x^2)}{(1-xy)^3}$

3. $\dfrac{1}{c^2z}$

8. $-\dfrac{3}{2}$

13. (i) $\dfrac{1}{3}$ **(ii)** $\dfrac{1}{\sqrt{2}}$ **(iii)** $\dfrac{-6}{\sqrt{23}}$

14. 4

15. (a) 8 **(b)** -2

16. (a) (i) $2\pi rh$ **(ii)** π **(b)** 64π **(c)** 48π

17. (a) $\dfrac{1}{5}\cdot\dfrac{\text{lb}}{in^2k}$ **(b)** $-\dfrac{25}{8}\cdot\dfrac{in^5}{\text{lb}}$

18. (i) $8x+6$, **(ii)** $3y^2+4$

■ CHAPTER 11

EXERCISE 11.1

2. (a) $2\sqrt{2}\left(\cos\dfrac{\pi}{4}+i\sin\dfrac{\pi}{4}\right)$ **(b)** $2\left(\cos\dfrac{2\pi}{3}+i\sin\dfrac{2\pi}{3}\right)$

(c) $2\cos\dfrac{\alpha}{2}\left(\cos\dfrac{\alpha}{2}+i\sin\dfrac{\alpha}{2}\right)$, if $\alpha\in(0,\pi)$, $-2\cos\dfrac{\alpha}{2}\left(\cos\left(\dfrac{\alpha}{2}+\pi\right)+i\sin\left(\dfrac{\alpha}{2}+\pi\right)\right)$ if $a\in(\pi, 2\pi)$; 0, if $a=\pi$

(d) $2(\cos 0+i\sin 0)$

(e) $2\left(\cos\dfrac{3\pi}{2}+i\sin\dfrac{3\pi}{2}\right)$

3. $z\in\mathbb{R}$ or $z=x+iy$ with $x^2+y^2=1$

4. $z=x+iy$ with $x^2+y^2=1$

5. 1

6. (a) $|z|=12$, $\arg z=0$, $\text{Arg } z=2k\pi$, $\arg\bar{z}=0$, $\arg(-z)=\pi$

(b) $|z|=14\sqrt{2}$, $\arg z=\dfrac{11\pi}{12}$, $\text{Arg } z=\dfrac{11\pi}{12}+2k\pi$, $\arg\bar{z}=\dfrac{13\pi}{12}$, $\arg(-z)=\dfrac{\pi}{12}$.

7. (a) $4(\cos 0+i\sin 0)$ **(b)** $48\sqrt{2}\left(\cos\dfrac{5\pi}{12}+i\sin\dfrac{5\pi}{12}\right)$

(c) $30\left(\cos\dfrac{\pi}{2}+i\sin\dfrac{\pi}{2}\right)$ **(d)** $12\sqrt{2}\left(\cos\dfrac{7\pi}{4}+i\sin\dfrac{7\pi}{4}\right)$

9. $\sqrt{2}, \dfrac{\pi}{4}$

15. $-6+9i$

EXERCISE 11.2

2. $(i-4)z-(i+4)\bar{z}+6=0$

3. $\dfrac{a_1 b_2 - a_2 b_1}{\bar{a} a_2 - a_1 \bar{a}_2}$

5. $(z-z_1)(\bar{z}-\bar{z}_1)=(z-z_2)(\bar{z}-\bar{z}_2)$

9. No

EXERCISE 11.3

1. $-2-3i, 3$

3. $z\bar{z}+(-2+2i)z+(-2-2i)\bar{z}-26=0$

4. $z\bar{z}-(1+i)z-(1-i)\bar{z}-7=0$

5. Right half-plane of the line $x=\dfrac{1}{2}$

7. $\dfrac{1}{2}(1+i)$

9. $\dfrac{1}{2}(z_1+z_2), r=\dfrac{1}{2}\sqrt{\left[2k-|z_1-z_2|^2\right]}$

10. Circle with centre at origin, radius = 2.

EXERCISE 11.4

1. $x^4-10x^3+45x^2-98x+98=0$

2. $x^4-4x^3+4x-1=0$

3. $x^4-16x^2+8x-1=0$

4. $3-i\sqrt{5}, \dfrac{5}{3}$

5. **(i)** 1, 1, 1, 1

(ii) $\dfrac{1}{2}, -2, 3$

CHAPTER 12

EXERCISE 12.1

1. **(a)** 1

(b) -1

(c) $\sin 9\theta + i\cos 9\theta$

(d) $\cos n\theta + i\sin n\theta$

(e) $2^{n+1} \cos^n \left(\dfrac{\theta - \phi}{2}\right) \cos \left[n\left(\dfrac{\theta + \phi}{2}\right)\right]$

3. Modulus is $32 \left|\cos^5\left(\dfrac{\theta}{2}\right)\right|$ and argument is $\left(-\dfrac{\theta}{2}\right)$ or $2\pi - \dfrac{\theta}{2}$.

17. (b) $\cos 6\theta = \cos^6\theta - 15\cos^4\theta\sin^2\theta + 15\cos^2\theta\sin^4\theta - \sin^6\theta$.

$\sin 6\theta = 6\cos^5\theta\sin\theta - 20\cos^3\theta\sin^3\theta + 6\cos\theta\sin^5\theta$.

EXERCISE 12.2

1. **(a)** $\dfrac{1}{\sqrt{2}} \pm \dfrac{i}{\sqrt{2}}, \dfrac{-1}{\sqrt{2}} \pm \dfrac{i}{\sqrt{2}}$

(b) $-i, \pm\dfrac{\sqrt{3}}{2} + \dfrac{i}{2}$

(c) $2^{1/3}\left(\pm\dfrac{\sqrt{3}}{2} + \dfrac{i}{2}\right), -2^{1/3}i$

(d) $2^{1/3}\left[\cos\left(\dfrac{p\pi}{18}\right) + i\sin\left(\dfrac{p\pi}{18}\right)\right]$, where $p = 1, 13, 25$.

(e) $2^{-1/5}\left[\cos\left(\dfrac{p\pi}{15}\right) - i\sin\left(\dfrac{p\pi}{15}\right)\right]$, where $p = 2, 8, 14, 20, 26$.

(f) $4^{1/5}\left[\cos\left(\dfrac{p\pi}{15}\right) + i\sin\left(\dfrac{p\pi}{15}\right)\right]$, where $p = 1, 7, 13, 19, 25$.

2. $\pm 1, \pm i$

4. Seventh roots of unity: $\cos\dfrac{2k\pi}{7} + i\sin\dfrac{2k\pi}{7}, k = 0, 1, \ldots, 6$.

5. **(a)** $-1, \dfrac{1}{2}, \pm i\dfrac{\sqrt{3}}{2}$

(b) $\cos\dfrac{2k\pi}{5} + i\sin\dfrac{2k\pi}{5}, k = 0, 1, 2, 3, 4$.

(c) $0, \pm 1, \pm i$

(d) $\cos(6k+1)\dfrac{\pi}{15} \pm i\sin(6k+1)\dfrac{\pi}{15}, k = 0, 1, 2, 3, 4$.

(e) $\pm 1 \pm i\sqrt{3}$

(f) $\cos\dfrac{(2k+1)}{5}\pi + i\sin\dfrac{(2k+1)}{5}\pi, k = 0, 1, 2, 3, 4$ and $\left(\pm\dfrac{1}{\sqrt{2}} \pm \dfrac{i}{\sqrt{2}}\right)$

(g) $-1, \dfrac{-1}{2} \pm i\dfrac{\sqrt{3}}{2}, \cos\dfrac{p\pi}{5} + i\sin\dfrac{p\pi}{5}, p = 1, 3, 5, 7, 9$.

(h) $0, \pm i\cot\dfrac{\pi}{10}, \pm i\cot\dfrac{3\pi}{10}$

7. $0, \pm i\tan\dfrac{\pi}{5}, \pm i\tan\dfrac{2\pi}{5}$

8. $i \tan \dfrac{k\pi}{n}$, $k = 0, 1, \ldots, n - 1$

9. $-i \cot \left[\dfrac{(2k+1)\pi}{4n}\right]$, $k = 0, 1, \ldots, 2n - 1$

10. $-1, \pm \dfrac{1}{2} \pm i \dfrac{\sqrt{3}}{2}$

11. $\cos \dfrac{2k\pi}{13} \pm i \sin \dfrac{2k\pi}{13}$, $k = 1, 2, \ldots, 6$

12. $\cos \dfrac{\pi}{5} \pm i \sin \dfrac{\pi}{5}$, $\cos \dfrac{3\pi}{5} \pm i \sin \dfrac{3\pi}{5}$

13. $\cos \dfrac{p\pi}{10} \pm i \sin \dfrac{p\pi}{10}$, $p = 1, 3, 7, 9$

14. $\pm \tan \left(\dfrac{k\pi}{20}\right)$, $k = 1, 3, 5, 7, 9$

APPENDIX

Table A: Natural exponential function of *x*

x	e^x	e^{-x}	x	e^x	e^{-x}
0.00	1.0000	1.0000	0.30	1.3499	0.7408
0.01	1.0101	0.9901	0.35	1.4191	0.7047
0.02	1.0202	0.9802	0.40	1.4918	0.6703
0.03	1.0305	0.9705	0.45	1.5683	0.6376
0.04	1.0408	0.9608	0.50	1.6487	0.6065
0.05	1.0513	0.9512	0.55	1.7333	0.5769
0.06	1.0618	0.9418	0.60	1.8221	6.5488
0.07	1.0725	0.9324	0.65	1.9155	0.5220
0.08	1.0833	0.9231	0.70	2.0138	0.4966
0.09	1.0942	0.9139	0.75	2.1170	0.4724
0.10	1.1052	0.9048	0.80	2.2255	0.4493
0.11	1.1163	0.8958	0.85	2.3396	0.4274
0.12	1.1275	0.8869	0.90	2.4596	0.4066
0.13	1.1388	0.8781	0.95	2.5857	0.3867
0.14	1.1503	0.8694	1.0	2.7183	0.3679
0.15	1.1618	0.8607	1.1	3.0042	0.3329
0.16	1.1735	0.8521	1.2	3.3201	0.3012
0.17	1.1853	0.8437	1.3	3. 6693	0.2725
0.18	1.1972	0.8353	1.4	4.0552	0.2466
0.19	1.2092	0.8270	1.5	4.4817	0.2231
0.20	1.2214	0.8187	1.6	4.9530	0.2019
0.21	1.2337	0.8106	1.7	5.4739	0.1827
0.22	1.2461	0.8025	1.8	6.0496	0.1653
0.23	1.2586	0.7945	1.9	6.6859	0.1496
0.24	1.2712	0.7866	2.0	7.3891	0.1353
0.25	1.2840	0.7788	2.1	8.1662	0.1225

(*Continued*)

Table A: (Contd.)

x	e^x	e^{-x}	x	e^x	e^{-x}
2.2	9.0250	0.1108	4.1	60.340	0.0166
2.3	9.9742	0.1003	4.2	66.686	0.0150
2.4	11.023	0.0907	4.3	73.700	0.0136
2.5	12.182	0.0821	4.4	81.451	0.0123
2.6	13.464	0.0743	4.5	90.017	0.0111
2.7	14.880	0.0672	4.6	99.484	0.0101
2.8	16.445	0.0608	4.7	109.95	0.0091
2.9	18.174	0.0550	4.8	121.51	0.0082
3.0	20.086	0.0498	4.9	134.29	0.0074
3.1	22.198	0.0450	5.0	148.41	0.0067
3.2	24.533	0.0408	5.5	244.69	0.0041
3.3	27.113	0.0369	6.0	403.43	0.0025
3.4	29.964	0.0334	6.5	665.14	0.0015
3.5	33.115	0.0302	7.0	1096.6	0.0009
3.6	36.598	0.0273	7.5	1808.0	0.0006
3.7	40.447	0.0247	8.0	2981.0	0.0003
3.8	44.701	0.0224	8.5	4914.8	0.0002
3.9	49.402	0.0202	9.0	8103.1	0.0001
4.0	54.598	0.0183	10.0	22026.0	0.00005

Table B: Natural logarithm of *x*

x	$\ln x$	x	$\ln x$	x	$\ln x$
		1.4	0.3365	2.8	1.0296
0.1	−2.3026	1.5	0.4055	2.9	1.0647
0.2	−1.6094	1.6	0.4700	3.0	1.0986
0.3	−1.2040	1.7	0.5306	3.1	1.1314
0.4	−0.9163	1.8	0.5878	3.2	1.1632
0.5	−0.6931	1.9	0.6419	3.3	1.1939
0.6	−0.5108	2.0	0.6931	3.4	1.2238
0.7	−0.3567	2.1	0.7419	3.5	1.2528
0.8	−0.2231	2.2	0.7885	3.6	1.2809
0.9	−0.1054	2.3	0.8329	3.7	1.3083
1.0	0.0000	2.4	0.8755	3.8	1.3350
1.1	0.0953	2.5	0.9163	3.9	1.3610
1.2	0.1823	2.6	0.9555	4.0	1.3863
1.3	0.2624	2.7	0.9933	4.1	1.4110

(Continued)

x	ln x	x	ln x	x	ln x
4.2	1.4351	7.3	1.9879	14	2.6391
4.3	1.4586	7.4	2.0015	15	2.7081
4.4	1.4816	7.5	2.0149	16	2.7726
4.5	1.5041	7.6	2.0281	17	2.8332
4.6	1.5261	7.7	2.0412	18	2.8904
4.7	1.5476	7.8	2.0541	19	2.9444
4.8	1.5686	7.9	2.0669	20	2.9957
4.9	1.5892	8.0	2.0794	25	3.2189
5.0	1.6094	8.1	2.0919	30	3.4012
5.1	1.6292	8.2	2.1041	35	3.5553
5.2	1.6487	8.3	2.1163	40	3.6889
5.3	1.6677	8.4	2.1282	45	3.8067
5.4	1.6864	8.5	2.1401	50	3.9120
5.5	1.7047	8.6	2.1518	55	4.0073
5.6	1.7228	8.7	2.1633	60	4.0943
5.7	1.7405	8.8	2.1748	65	4.1744
5.8	1.7579	8.9	2.1861	70	4.2485
5.9	1.7750	9.0	2.1972	75	4.3175
6.0	1.7918	9.1	2.2083	80	4.3820
6.1	1.8083	9.2	2.2192	85	4.4427
6.2	1.8245	9.3	2.2300	90	4.4998
6.3	1.8405	9.4	2.2407	100	4.6052
6.4	1.8563	9.5	2.2513	110	4.7005
6.5	1.8718	9.6	2.2618	120	4.7875
6.6	1.8871	9.7	2.2721	130	4.8675
6.7	1.9021	9.8	2.2824	140	4.9416
6.8	1.9169	9.9	2.2925	150	5.0106
6.9	1.9315	10	2.3026	160	5.0752
7.0	1.9459	11	2.3979	170	5.1358
7.1	1.9601	12	2.4849	180	5.1930
7.2	1.9741	13	2.5649	190	5.2470

Table C: Two thousand five hundred random digits

	00–04	05–09	10–14	15–19	20–24	25–29	30–34	35–39	40–44	45–49
00	22808	04391	45529	53968	57136	98228	85485	13801	68194	56382
01	49305	36965	44849	64987	59501	35141	50159	57369	76913	75739
02	81934	19920	73316	69243	69605	17022	53264	83417	55193	92929
03	10840	13508	48120	22467	54505	70536	91206	81038	22418	34800
04	99555	73289	59605	37105	24621	44100	72832	12268	97089	68112
05	32677	45709	62337	35132	45128	96761	08745	53388	98353	46724
06	09401	75407	27704	11569	52842	83543	44750	03177	50511	15301
07	73424	31711	65519	74869	56744	40864	75315	89866	96563	75142
08	37075	81378	59472	71858	86903	66860	03757	32723	54273	45477
09	02060	37158	55244	44812	45369	78939	08048	28036	40946	03898
10	94719	43565	40028	79866	43137	28063	52513	66405	71511	66135
11	70234	48272	59621	88778	16536	36505	41724	24776	63971	01685
12	07972	71752	92745	86465	01845	27416	50519	48458	68460	63113
13	58521	64882	26993	48104	61307	73933	17214	44827	88306	78177
14	32580	45202	21148	09684	39411	04892	02055	75276	51831	85686
15	88796	30829	35009	22695	23694	11220	71006	26720	39476	60538
16	31525	82746	78935	82980	61236	28940	96341	13790	66247	33839
17	02747	35989	70387	89571	34570	17002	79223	96817	31681	15207
18	46651	28987	20625	61347	63981	41085	67412	29053	00724	14841
19	43598	14436	33521	55637	39789	26560	66404	71802	18763	80560
20	30596	92319	11474	64546	60030	73795	60809	24016	29166	36059
21	56198	64370	85771	62633	78240	05766	32419	35769	14057	80674
22	68266	67544	06464	84956	18431	04015	89049	15098	12018	89338
23	31107	28597	65102	75599	17496	87590	68848	33021	69855	54015
24	37555	05069	38680	87274	55152	21792	77219	48732	03377	01160
25	90463	27249	43845	94391	12145	36882	48906	52336	00780	74407
26	99189	88731	93531	52638	54989	04237	32978	59902	05463	09245
27	37631	74016	89072	59598	55356	27346	80856	80875	52850	36548
28	73829	21651	50141	76142	72303	06694	61697	76662	23745	96282
29	15634	89428	47090	12094	42134	62381	87236	90118	53463	46969
30	00571	45172	78532	63863	98597	15742	41967	11821	91389	07476
31	83374	10184	56384	27050	77700	13875	96607	76479	80535	17454
32	78666	85645	13181	08700	08289	62956	64439	39150	95690	18555
33	47890	88197	21368	65254	35917	54035	83028	84636	38186	50581
34	56238	13559	79344	83198	94642	35165	40188	21456	67024	62771
35	36369	32234	38129	59963	99237	72648	66504	99065	61161	16186
36	42934	34578	28968	74028	42164	56647	76806	61023	33099	48293
37	09010	15226	43474	30174	26727	39317	48508	55438	85336	40762

(Continued)

	00–04	05–09	10–14	15–19	20–24	25–29	30–34	35–39	40–44	45–49
38	83897	90073	72941	85613	85569	24183	08247	15946	02957	68504
39	82206	01230	93252	89045	25141	91943	75531	87420	99012	80751
40	14175	32992	49046	41272	94040	44929	98531	27712	05106	35242
41	58968	88367	70927	74765	18635	85122	27722	95388	61523	91745
42	62601	04595	76926	11007	67631	64641	07994	04639	39314	83126
43	97030	71165	47032	85021	65554	66774	21560	04121	57297	85415
44	89074	31587	21360	41673	71192	85795	82757	52928	62586	02179
45	07806	81312	81215	99858	26762	28993	74951	64680	50934	32011
46	91540	86466	13229	76624	44092	96604	08590	89705	03424	48033
47	99279	27334	33804	77988	93592	90708	56780	70097	39907	51006
48	63224	05074	83941	25034	43516	22840	35230	66048	80754	46302
49	98361	97513	27529	66419	35328	19738	82366	38573	50967	72754

Table D: Area under the normal curve

The area is measured from the mean $z = 0$, to any ordinate z.

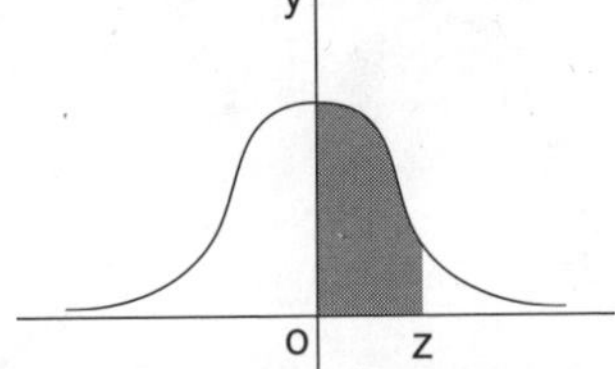

z	0.00	0.01	0.02	0.03	0.04	0.05	0.06	0.07	0.08	0.09
0.0	.0000	.0040	.0080	.0120	.0159	.0199	.0239	.0279	.0319	.0359
0.1	.0398	.0438	.0478	.0517	.0557	.0596	.0636	.0675	.0714	.0753
0.2	.0793	.0832	.0871	.0910	.0948	.0987	.1026	.1064	.1103	.1141
0.3	.1179	.1217	.1255	.1293	.1331	.1368	.1406	.1443	.1480	.1517
0.4	.1554	.1591	.1628	.1664	.1700	.1736	.1772	.1808	.1844	.1879
0.5	.1916	.1950	.1985	.2019	.2054	.2088	.2123	.2157	.2190	.2224
0.6	.2257	.2291	.2324	.2357	.2389	.2422	.2454	.2486	.2518	.2549
0.7	.2580	.2611	.2642	.2673	.2704	.2734	.2764	.2794	.2823	.2852
0.8	.2881	.2910	.2939	.2967	.2995	.3023	.3051	.3078	.3106	.3133
0.9	.3159	.3186	.3212	.3238	.3264	.3289	.3315	.3340	.3365	.3389
1.0	.3413	.3438	.3461	.3485	.3508	.3531	.3554	.3577	.3599	.3621
1.1	.3643	.3665	.3686	.3708	.3729	.3749	.3770	.3790	.3810	.3830
1.2	.3849	.3869	.3888	.3907	.3925	.3944	.3962	.3980	.3997	.4015
1.3	.4032	.4049	.4066	.4082	.4099	.4115	.4131	.4147	.4162	.4177
1.4	.4192	.4207	.4222	.4236	.4251	.4265	.4279	.4292	.4306	.4319
1.5	.4332	.4345	.4357	.4370	.4382	.4394	.4406	.4418	.4430	.4441
1.6	.4452	.4463	.4474	.4485	.4495	.4505	.4515	.4525	.4535	.4545

(*Continued*)

Table D: (Contd.)

z	0.00	0.01	0.02	0.03	0.04	0.05	0.06	0.07	0.08	0.09
1.7	.4554	.4564	.4573	.4582	.4591	.4599	.4608	.4616	.4625	.4633
1.8	.4641	.4649	.4656	.4664	.4671	.4678	.4686	.4693	.4699	.4706
1.9	.4713	.4719	.4726	.4732	.4738	.4744	.4750	.4756	.4762	.4767
2.0	.4772	.4778	.4783	.4788	.4793	.4798	.4803	.4808	.4812	.4817
2.1	.4821	.4826	.4830	.4834	.4838	.4842	.4846	.4850	.4854	.4857
2.2	.4861	.4865	.4868	.4871	.4875	.4878	.4881	.4884	.4887	.4890
2.3	.4893	.4896	.4898	.4901	.4904	.4906	.4909	.4911	.4913	.4916
2.4	.4918	.4920	.4922	.4925	.4927	.4929	.4931	.4932	.4934	.4936
2.5	.4938	.4940	.4941	.4943	.4945	.4946	.4948	.4949	.4951	.4952
2.6	.4953	.4955	.4958	.4957	.4959	.4960	.4961	.4962	.4963	.4964
2.7	.4965	.4966	.4967	.4968	.4969	.4970	.4971	.4972	.4973	.4974
2.8	.4974	.4975	.4976	.4977	.4977	.4978	.4979	.4980	.4980	.4981
2.9	.4981	.4982	.4983	.4983	.4984	.4984	.4985	.4985	.4986	.4986
3.0	.49865	.4987	.4987	.4988	.4988	.4989	.4989	.4989	.4990	.4990
3.1	.49903	.4991	.4991	.4991	.4992	.4992	.4992	.4992	.4993	.4993

Table E: 5% and 1% values of *t*

(*d.f.*)	$t_{0.05}$	$t_{0.01}$	(*d.f.*)	$t_{0.05}$	$t_{0.01}$
1	12.71	63.66	13	2.16	3.01
2	4.30	9.92	14	2.14	2.98
3	3.18	5.84	15	2.13	2.95
4	2.78	4.60	16	2.12	2.92
5	2.57	4.03	17	2.11	2.89
6	2.45	3.71	18	2.10	2.88
7	2.36	3.50	19	2.09	2.86
8	2.31	3.36	20	2.09	2.84
9	2.26	3.25	21	2.08	2.83
10	2.23	3.17	22	2.07	2.82
11	2.20	3.11	23	2.07	2.81
12	2.18	3.06	24	2.06	2.80
			25	2.06	2.79

Year 2008–09

B.Sc. Prog./I

MA-107(*a*) — MATHEMATICS–I

(For Physical Sciences)

(NC—Admission of 2008 onwards)

Time: 3 Hours *Maximum Marks: 75*

Attempt any *two* questions from each section.

Section I

1. (*a*) A force of 12 pounds is applied to an object along the negative x-axis and a force of 5 pounds is applied to the object along the positive y-axis. Find the magnitude of the resultant force, and also sketch its direction. 6

(*b*) Show that the set $\left\{\begin{bmatrix}1\\1\end{bmatrix}, \begin{bmatrix}1\\-1\end{bmatrix}\right\}$ is a basis for $\mathbf{R}^2$. 6

2. (*a*) Let T: $\mathbf{R}^2 \to \mathbf{R}^2$ be defined by

$$T\left(\begin{bmatrix}x\\b\end{bmatrix}\right) = \begin{bmatrix}x\\0\end{bmatrix}.$$

Show that T is a linear transformation. 6

(*b*) Find the eigenvalues and eigenvectors of the linear transformation denoting the reflection about x-axis in $\mathbf{R}^2$. 6

3. (*a*) Solve, by elementary row operation, the following system of equations: 6

$$x - 3y + z = -1$$
$$2x + y - 4z = -1$$
$$6x - 7y + 8z = 7$$

(*b*) Find the rank of the following matrix: 6

$$\begin{bmatrix}-2 & -1 & -3 & -1\\ 1 & 2 & 3 & -1\\ 1 & 0 & 1 & 1\\ 0 & 1 & 1 & -1\end{bmatrix}.$$

Section II

4. (*a*) Discuss the convergence of the sequence $\left\{\dfrac{(-1)^n}{n^2}\right\}$. 6

(*b*) Find the nth derivative of $y = \dfrac{x+1}{x^2-4}$. 6

(c) If $y = e^{m\cos^{-1}x}$, show that:

$$(1 - x^2)y_{n+2} - (2n + 1)xy_{n+1} - (n^2 + m^2)y_n = 0.$$ 6

5. (a) Sketch the graph of $y = \sqrt{x - 3}$. 6

(b) A cell of the bacterium *E. coli* divides into two cells every 20 minutes when placed in a nutrient culture. Let $x = x(t)$ be the number of cells that are present t minutes after a single cell is placed in the culture. Assume that the growth of bacteria is approximated by a continuous exponential growth model.

(i) Find an initial value problem whose solution is $x(t)$.

(ii) Find a formula for $x(t)$.

(iii) How many cells are present after 2 hours? 6

(c) Find the Maclaurin series expansion of $y = e^{-x}$, assuming $\lim_{n\to\infty} \text{Rn}(x) = 0$. 6

6. (a) Draw the level curves at heights $k = 0, 1, -2$ of the function:

$$z = 2 - x - y.$$ 6

(b) Find the slope of the surface $z = -\sqrt{4 - x^2 - y^2}$ along y-axis at $(1,1, -\sqrt{2})$. 6

(c) Verify that the function:

$$z = \text{In}(x^2 + y^2) + 2\tan^{-1}\left(\frac{y}{x}\right)$$

satisfies Laplace's equation $\frac{\partial^2 z}{\partial x^2} + \frac{\partial^2 z}{\partial y^2} = 0.$ 6

Section III

7. (a) A student is to appear for two tests. He is told that his chances of winning test 1 is 0.5, loosing test 2 is 0.3 and loosing both tests is 0.2. Find the probability that the student will win test 2 when he has already won test 1. 3 1/2

(b) Define Binomial distribution. If mean and variance of a binomial distribution are 12 and 4 respectively, what is the probability of success at a given trial? 4

8. (a) Derive Poisson Distribution as a limiting case of Binomial Distribution. 4 1/2

(b) A man wins Rs. 100 for heads and loses Rs. 100 for tail when a coin is tossed. Suppose he tosses once and quits if he wins, but tries once more if he loses on the first toss. What are his expected winnings? 3

9. (*a*) Calculate the correlation coefficient for the following heights in inches of fathers (x) and their sons (y). 4

x	y
65	67
66	68
67	65
68	72
69	72
70	69
72	71

(*b*) A process of making certain bearing is under control if the diameters of the bearing have a mean of 0.5 cm. What can we say about this process if a sample of 100 of these bearings has a mean diameter of 0.506 cm and a standard deviation of 0.004 cm? 3 ½

Year 2009–10

B.Sc. Prog./I

MA-107 (*a*) – MATHEMATICS – I

(For Physical Sciences)

(NC—Admission of 2008 onwards)

Time: 3 Hours *Maximum Marks: 75*

Attempt any *two* questions from each section.

Section I

1. (a) A boat is travelling towards east across a river at the rate of 4 miles per hour while the river's current is flowing at the rate of 3 miles per hour towards south. Find the resultant velocity and sketch its approximate direction. 6

   (b) Show that the set $\left\{\begin{bmatrix}1\\1\end{bmatrix}, \begin{bmatrix}1\\2\end{bmatrix}\right\}$ is a basis for $\mathbf{R}^2$. 6

2. (a) Let $T : \mathbf{R}^2 \rightarrow \mathbf{R}^2$ be defined by

$$T\left(\begin{bmatrix}x\\y\end{bmatrix}\right) = \begin{bmatrix}-y\\-x\end{bmatrix}$$

   Show that T is a linear transform. 6

   (b) Let $T : \mathbf{R}^2 \rightarrow \mathbf{R}^2$ be the linear transformation denoting reflection about y-axis. Find the eigenvalues and eigenvectors of T. 6

3. (a) Solve by elementary row operations, the following system of equations

$$\begin{aligned} 2x + 3y + z &= 9 \\ x + 2y + 3z &= 6 \\ 3x + y + 2z &= 8 \end{aligned}$$ 6

   (b) Find the rank of the following matrix:

$$\begin{bmatrix} 1 & 1 & 2 & 3 \\ 1 & 3 & 0 & 3 \\ 1 & -2 & -3 & 0 \\ 1 & 1 & 2 & 3 \end{bmatrix}$$ 6

Section II

4. (a) Examine the convergence of the sequence $\left\{\dfrac{\cos n}{n}\right\}$. 6

   (b) Find the nth derivative of $y = \dfrac{x}{1+3x+2x^2}$ 6

   (c) If $y = e^{m \sin^{-1} x}$, show that

$$(1-x^2)\, y_{n+2} - (2n+1)\, xy_{n+1} - (n^2+m^2)\, y_n = 0$$ 6

5. (a) Sketch the graph of $y = |x - 2| + 1$ 6

(b) If the population of a country doubles itself in 50 years, in how many years will it be three times under the assumption that the rate of increase is proportional to the number of inhabitants? 6

(c) Find the Maclaurin's series for l^{x+1}, assuming that $\lim_{n \to \infty} Rn(x) = 0$. 6

6. (a) Draw the level curves at heights $R = 0, 1, 2$ of the function $Z = \sqrt{9 - x^2 - y^2}$. 6

(b) Find the slope of the surface $Z = \sqrt{x^2 + 2y^2}$ along x-axis at (1, 2, 3) 6

(c) Verify that the function $u(x, t) = \sin(x + ct) + \cos(3x + 3ct)$ is a solution of the wave equation

$$\frac{\partial^2 u}{\partial t^2} = c^2 \frac{\partial^2 u}{\partial x^2}$$ 6

Section III

7. (a) The probability that a student of Mathematics will not pass M.B.A. examination is $\frac{3}{5}$ and that a student of History will not pass is $\frac{4}{5}$. Calculate the probability that at least one of them passes the examination. 4½

(b) In a throw of dice, a random variable X is defined in the following way:
X takes the value equal to 2 times
The outcome if it is odd and –1 times
The outcome if it is even, write down X.
Also write down the probability distribution of X. 3

8. (a) Find mean and variance of Binomial distribution. 3½

(b) A random variable X has the following probability distribution:

X	:	0	1	2	3	4	5	6	7
$P(X)$	:	0	k	$2k$	$2k$	$3k$	k^2	$2k^2$	$7k^2 + k$

Find k. Also evaluate $P(0 < X < 5)$. 4

9. (a) The following statistics were collected during a chemistry experiment to study a possible relationship between output (X) in milligrams and temperature setting (Y) in degrees Celsius. 3½

$N = 4, \bar{X} = 639, \bar{Y} = 188, \sigma_x^2 = 210, \sigma_y^2 = 407, \Sigma XY = 240700.$

Calculate Karl Pearson's coefficient of correlation between X and Y.

(b) A sample of 900 members is found to have a mean of 3.4 cm. Can it be reasonably regarded as a simple sample from a large population with mean 3.2 cm and s.d. 2.3 cm? 4

Year 2010–11

B.Sc. 1st Sem.

MAPT-101—MATHEMATICS—I (Calculus and Matrices)
(For Physical Sciences)
(Admission of 2010 and onwards)

Time : 3 Hours *Maximum Marks : 75*

Attempt any *two* questions from each section.

Section I

1. (*a*) Verify that the set:

$$\{(1, 1), (1, -1)\}$$

is a basis of $\mathbf{R}^2$. 6

(*b*) Examine which of the following is a subspace of $\mathbf{R}^2$. If it is a subspace, give its geometric interpretation: 6

$$V_1 = \{(a, b^2) \mid a, b \in \mathbf{R}\}$$
$$V_2 = \{(a, b) \mid a > 0; a, b \in \mathbf{R}\}.$$

2. (*a*) Which of the following transformations are linear? Also give the justification: 6

(i) $T : \mathbf{R}^2 \to \mathbf{R}^2$ defined as

$T(x, y) = (1, 2)$

(ii) $T : \mathbf{R}^3 \to \mathbf{R}^2$ as

$T(x, y, z) = (x, 4y)$.

(*b*) Sketch the image of unit square with vertices (0, 0), (0, 1), (1, 1), (1, 0) under the following transformations of the plane: 6

(*i*) reflection about origin;

(*ii*) dialation by a factor of 3.

3. (*a*) Reduce the matrix:

$$A = \begin{bmatrix} 1 & 1 & 1 & 1 \\ 1 & 3 & -2 & 1 \\ 2 & 0 & -3 & 2 \\ 3 & 3 & -3 & 3 \end{bmatrix}$$

to triangular form by elementary row operation and hence determine its rank. 6

(*b*) Let:

$$A = \begin{bmatrix} 2 & 1 & 0 \\ 0 & 2 & 1 \\ 0 & 0 & 2 \end{bmatrix}.$$

Find eigen values and eigen vector corresponding to one of them. 6

4. (*a*) Solve the system of equations: 6

$$x - 4y + 7z = 8$$

$$3x + 8y - 2z = 6$$

$$7x - 8y + 26z = 31$$

(*b*) Show that the only real value of λ for which the following equations have non-zero solution:

$$x + 2y + 3z = \lambda x$$

$$3x + y + 2z = \lambda y$$

$$2x + 3y + z = \lambda z$$

is **6**. 6

Section II

5. (*a*) Discuss the convergence of a sequence: 6

$$\left\langle \frac{(-1)^n}{n^3} \right\rangle.$$

(*b*) Find the *n*th derivative of: 6

$$y = \frac{x+2}{x^2 - 9}.$$

(*c*) If

$$y = (\sin^{-1} x)^2,$$

prove that:

$$(1-x^2)\frac{d^2y}{dx^2} - x\frac{dy}{dx} - 2 = 0$$

Hence show that:

$$(1 - x^2)y_{n+2} - (2n + 1)xy_{x+1} - n^2y_n = 0.$$ 6

6. (*a*) Sketch the graph of a function: 6

$$f(x) = \cos 2x; \qquad x \in \mathbf{R}$$

(*b*) Find the Maclaurin series expansion of:

$$y = \sin x$$

assuming that:

$$\lim_{n\to\infty} \mathbf{R}_n(x) = 0$$ 6

(*c*) A body with initial temperature of 100°C is allowed to cool in air which remains at a constant temperature of 20°C. It is given that after 10 minutes, the body has cooled to 40°C. 6

(i) Find the temperature of body after half an hour.

(ii) Draw a graph of temperature *vs.* time.

7. (*a*) Draw the level curves of height $k = 1, 2, 5$ for the surface:

$$f(x, y) = 9x^2 + 25y^2.$$ 6

(*b*) If

$$u = f(r)$$

where $r = \sqrt{x^2 + y^2}$,

prove that:

$$\frac{\partial^2 u}{\partial x^2} + \frac{\partial^2 u}{\partial y^2} = f''(r) + \frac{1}{r} f'(r).$$ 6

(*c*) Verify that:

$$f(x, y, z) = 2z^3 - 3(x^2 + y^2)z$$

is a solution of 3-dimensional Laplace equation:

$$\frac{\partial^2 f}{\partial x^2} + \frac{\partial^2 f}{\partial y^2} + \frac{\partial^2 f}{\partial z^2} = 0.$$ 6

8. (*a*) Show the following:

(i) $\lim_{n\to\infty} \sqrt{n+1} - \sqrt{n} = 0$

(ii) $\lim_{n\to\infty} \frac{\sin(n^2)}{n+1} = 0.$ 9

(*b*) Give *one* example each of the following: 9

(i) bounded sequence but not convergent

(ii) unbounded sequence

(iii) monotonically increasing sequence

(iv) monotonically decreasing sequence

Section III

9. (*a*) Find the equation of a circle whose radius is 3 and whose centre has affix $1 - i$. 3½

(*b*) State Fundamental Theorem of Algebra and hence form an equation in lowest degree with rational coefficients which has $1+\sqrt{2}$ and $2-\sqrt{3}$ as two of its roots. 4

10. (*a*) Give the geometrical representation of sum of two complex numbers. 3½

(*b*) Prove that for any *two* complex numbers z_1 and z_2:

$$|z_1 + z_2| \leq |z_1| + |z_2|.$$ 4

11. (*a*) Solve the equation:

$$z^4 + 1 = 0$$

(*b*) Prove that:

$$\left(\frac{1+\sin\phi+i\cos\phi}{1+\sin\phi-i\cos\phi}\right)^n = \cos\left(\frac{n\pi}{2}-n\phi\right)+i\sin\left(\frac{n\pi}{2}-n\phi\right).$$ 4

12. (*a*) Find the inverse of the matrix using elementary operations:

$$A = \begin{bmatrix} 1 & 2 & -3 \\ 1 & -2 & 1 \\ 5 & -2 & -3 \end{bmatrix}.$$ 4½

(*b*) Are the vectors:

$X_1 = (1, 2, 3)$, $X_2 = (0, 1, 2)$ and $(0, 0, 1)$ in $\mathbf{R}^3$ linearly independent or linearly dependent? 3

Solution to Examination Paper
MAPT – 101
B.Sc. (Physical Science) 1st Sem.
2010–11

Question No.	Solution/Reference to Solution
	Section I
1 (a)	Refer Example 2.27, page 62
1 (b)	Refer Q. 3(d), (e), Exercise 2.6, page 73
2 (a)	(i) Refer Q. 5(a), Exercise 3.1, page 83 (ii) Similar to Example 3.2, page 78
2 (b)	Refer Q. 10(h), (c) Exercise 3.2, page 95
3 (a)	Refer Example 1.30, page 28
3 (b)	Refer Q. 4(b), Exercise 4.1, page 128
4 (a)	Refer Example 1.35, page 36
4 (b)	Refer Example 1.37, page 37
	Section II
5 (a)	Similar to Q. 4(b), Exercise 5.5, page 181
5 (b)	Refer Q. 1(i), Exercise 8.2, page 252
5 (c)	Refer Example 8.18, page 256
6 (a)	For similar graph, refer Fig. 6.37, page 211
6 (b)	Refer Example 9.15, page 277
6 (c)	Refer Example 7.8, page 229
7 (a)	Refer Example 10.5 (ii), page 297
7 (b)	Refer Q. 9, Exercise 10.2, page 308

Question No.	Solution/Reference to Solution
7 (c)	Solution

$$f(x, y, z) = 2Z^3 - 3(x^2 + y^2)\,Z$$

$$\frac{\partial f}{\partial x} = -6z, \frac{\partial f}{\partial y} = -6yz, \frac{\partial f}{\partial z} = 6z^2 - 3(x^2 + y^2)$$

$$\frac{\partial^2 f}{\partial x^2} = -6z, \frac{\partial^2 f}{\partial y^2} = -6z, \frac{\partial^2 f}{\partial z^2} = 12z$$

$$\therefore \frac{\partial^2 f}{\partial x^2} + \frac{\partial^2 f}{\partial y^2} + \frac{\partial^2 f}{\partial z^2} = -6z - 6z + 12z = 0$$

8 (*a*) (i) Solution

$$0 \le \sqrt{n+1} - \sqrt{n} = \frac{\left(\sqrt{n+1} - \sqrt{n}\right)\left(\sqrt{n+1} + \sqrt{n}\right)}{\sqrt{n+1} + \sqrt{n}}$$

$$= \frac{1}{\sqrt{n+1} + \sqrt{n}} \le \frac{1}{2\sqrt{n}}$$

$$\therefore 0 \le \lim_{n\to\infty}(\sqrt{n+1} - \sqrt{n}) \le \lim_{n\to\infty} \frac{1}{2\sqrt{n}} = \frac{1}{2}\lim_{n\to\infty}\frac{1}{\sqrt{n}}$$

$$\therefore \lim_{n\to\infty}(\sqrt{n+1} - \sqrt{n}) = 0 \text{ (By Sandwich Theorem)}$$

(ii) Refer Q. 4(d), Exercise 5.5, page 181

8 (*b*) Hints:

(i) $a_n = (-1)^n$ (ii) $a_n = n$ (iii) $a_n = n^2$ (iv) $a_n = -n^2$

Section III

9 (a)	Refer Example 11.22, page 344
9 (b)	Refer Example 11.29, page 351
10 (a)	Refer Article 11.1.1, page 322
10 (b)	Refer Example 11.7, page 316
11 (a)	Refer Example 12.11, page 366
11 (b)	Refer Example 12.2, page 356
12 (a)	Refer Example 1.23, page 23
12 (b)	Refer Example 2.18, page 58.

Year 2011–12

B.Sc./Ist Sem.

MAPT-101—MATHEMATICS—I (Calculus and Matrices)

(For Physical Sciences)

(Admission of 2011 and onwards)

Time : 3 Hours *Maximum Marks :* 75

Attempt any *two* questions from each section.

Section I

1. (*a*) Verify that the set:

$$\left\{\begin{bmatrix}\pi\\0\end{bmatrix},\begin{bmatrix}1\\e\end{bmatrix}\right\}$$

of vectors is a basis of $\mathbf{R}^2$. 6

(*b*) Examine which of the following is a subspace of $\mathbf{R}^2$. If it is a subspace, give its geometric interpretation:

$$V_1=\{(a,2a):a\in\mathbf{R}\}$$

$$V_2=\{(a,b):a>0,a,b\in\mathbf{R}\}.$$ 6

2. (*a*) Let $T:\mathbf{R}^3\to\mathbf{R}^2$ be a linear transformation such that $T(1,0,1)=(2,-1)$, $T(0,1,1)=(1,1)$, and $T(1,1,0)=(-1,4)$. Find $T(1,1,1)$. 6

(*b*) Let **R** be the rectangle with vertices (1, 1), (1, 4), (3, 1) and (3, 4). Determine and sketch the image of **R** under:

(i) a reflection about the y-axis.

(ii) a translation by vector (1, 1). 6

3. (*a*) Reduce the matrix:

$$A=\begin{bmatrix}2&3&3\\3&6&12\\2&4&8\end{bmatrix}$$

to triangular form by elementary row operations and hence determine its rank. 6

(*b*) Find the characteristic equation, eigen values and eigen vector corresponding to one of them for the matrix:

$$A=\begin{bmatrix}1&2&-1\\1&0&1\\4&-4&5\end{bmatrix}.$$ 6

4. (*a*) Solve the system of equations:

$$x+2y-z=3$$
$$3x-y+2z=1$$
$$2x-2y+3z=2.$$ 6

(*b*) For what values of λ and μ do the following system of equations:

$$x+y+z=6$$
$$x+2y+3z=10$$
$$x+2y+\lambda z=\mu$$

have an infinite number of solutions. 6

Section II

5. (*a*) Discuss the convergence of a sequence:

$$\left\langle \frac{n}{3n+1} \right\rangle.$$ 6

(*b*) Find the *n*th derivative of:

$$y=\frac{x+1}{x^2-4}.$$ 6

(*c*) If

$$y=e^{m\sin^{-1}x},$$

show that:

$$(1-x^2)y_{n+2}-(2n+1)xy_{n+1}-(n^2+m^2)y_n=0.$$ 6

6. (*a*) Sketch the graph of a function:

$$f(x)=-(x+2)^2+3.$$ 6

(*b*) Find the Maclaurin series expansion of $y=\cos 2x$, assuming that:

$$\lim_{n\to\infty} R_n(x)=0.$$ 6

(*c*) In a school, there are 1000 students and all are likely to get infected with eye-flu virus. Initially, 20 students got infected and within 2 weeks, 100 students got infected with the disease. In how much time would the majority of students be infected by the eye-flu virus? It is given that the disease spreads with logistic growth model. 6

7. (*a*) Draw the level curves of height $k = 0, 1, 2$ for the surface:

$$z = f(x, y) = 5\sqrt{\frac{x^2}{16} + \frac{y^2}{9} - 1}. \qquad 6$$

(*b*) If $v = r^m$, where $r^2 = x^2 + y^2 + z^2$, show that:

$$\frac{\partial^2 v}{\partial x^2} + \frac{\partial^2 v}{\partial y^2} + \frac{\partial^2 v}{\partial z^2} = m(m+1)r^{m-2}. \qquad 6$$

(*c*) Verify that the function:

$$w = \cos(5x + 5ct)$$

is a solution of the wave equation. 6

8. (*a*) For what values of x can we replace $\sin x$ by $x - \frac{x^3}{6}$ with an error of magnitude no greater than 3×10^{-4}. 6

(*b*) Verify which of the following sequences are monotonic and bounded:

(i) $\left\langle \frac{2^n}{n!} \right\rangle$ 6

(ii) $\langle n - 2^n \rangle$. 6

Section III

9. (*a*) Find the radius and centre of the circle whose equation is:

$$z\bar{z} - (2+3i)z - (2-3i)\bar{z} + 9 = 0. \qquad 3\tfrac{1}{2}$$

(*b*) Form an equation in the lowest degree with real coefficients which has $2 - 3i$ and $3 + 2i$ as two of its roots. 4

10. (*a*) Find the modulus and argument of the centroid of the triangle whose vertices are given by $8 + 5i, -3 + i$ and $-2 - 3i$, respectively. 3½

(*b*) Let z_1, z_2, z_3 be complex numbers such that $z_1 + z_2 + z_3 = 0$ and $|z_1| = |z_2| = |z_3| = 1$. Prove that:

$$z_1^2 + z_2^2 + z_3^2 = 0. \qquad 4$$

11. (*a*) If $x = \cos\theta + i\sin\theta$ and $y = \cos\phi + i\sin\phi$, prove that:

$$\frac{x-y}{x+y} = i\tan\frac{\theta - \phi}{2}. \qquad 4$$

(*b*) Use De Moivre's theorem to solve the equation:

$$z^7 + z = 0.$$ 3½

12. (*a*) Find the inverse of the matrix:

$$A = \begin{bmatrix} 2 & 4 & 6 \\ 3 & 1 & 2 \\ 0 & 1 & -1 \end{bmatrix}$$

using elementary operations. 4

(*b*) Express vector $X = (3, 1, -4)$ as a linear combination of the vectors $X_1 = (1, 1, 1)$, $X_2 = (0, 1, 1)$ and $X_3 = (0, 0, 1)$. 3½

Solution to Examination Paper
MAPT – 101
B.Sc. (Physical Science) 1st Sem.
2011–12
Mathematics – I (Calculus and Matrices)

Question No.	Solution/Reference to Solution
	Section I
1 (a)	Refer Q. 2 (d) of Exercise 2.5, page 66
1 (b)	Refer Q. 3 (a) and (e) of Exercise 2.6, page 73
2 (a)	Refer Q. 8 (b) of Exercise 3.1, page 83
2 (b)	Similar to Q. 10 and Q. 18 of Exercise 3.2, pages 95–96
3 (a)	Refer Q. 1 (a) of Exercise 1.4, page 29. Solve similarly as Example 1.28, page 27
3 (b)	Refer Example 4.8, page 121
4 (a)	Refer Q. 7 of Exercise 1.5, page 38
4 (b)	Refer Q. 19 of Exercise 1.5, page 39
	Section II
5 (a)	Refer Example 5.17 (e), page 171
5 (b)	Refer Q. 1 (i) of Exercise 8.2, page 252
5 (c)	Refer Example 8.15, page 254
6 (a)	Refer Q. 3 (l) (iii) of Exercise 6.1, page 217
6 (b)	Refer Example 9.20, page 282
6 (c)	Refer Q. 3 of Exercise 7.3, page 239
7 (a)	Refer Example 10.5 (vi), page 297
7 (b)	Refer Q. 11 of Exercise 10.2, page 308
7 (c)	Refer Q. 7 of Exercise 10.3, page 313

Question No.	Solution/Reference to Solution
8 (a)	Refer Example 9.7, page 270
8 (b) (i)	Refer Example 5.14 (g), page 162
8 (b) (ii)	Refer Example 5.14 (b), page 162
	Section III
9 (a)	Refer Example 11.25, page 346
9 (b)	Refer Example 11.28, page 351
10 (a)	Refer Q. 9 of Exercise 11.1, page 336
10 (b)	Refer Example 11.4, page 321
11 (a)	Refer Example 12.3, page 356
11 (b)	Refer Example 12.12, page 366
12 (a)	Refer Example 1.22, page 22
12 (b)	Refer Q. 3 of Exercise 2.4, page 61

Year 2011–12

B.Sc. II Sem.

MAPT-101—MATHEMATICS—I (Calculus and Matrices)
(For Physical Sciences)
(Admission of 2010 and onwards)

Time : 3 Hours *Maximum Marks : 75*

Attempt any *two* questions from each section.

Section I

1. (*a*) Show that the set $S = \{(0, 1, 0), (1, 0, 1), (1, 1, 0)\}$ forms a basis for $\mathbf{R}^3$. 6

   (*b*) Examine which of the following is a subspace of $\mathbf{R}^2$. Also justify:

$$S_1 = \{(a, 0) : a \in \mathbf{R}\}$$

$$S_2 = \{(1, y) : y \in \mathbf{R}\}$$ 6

2. (*a*) Which of the following transformations are linear? Justify:

   (i) $T(x, y) = 2x - y$

   (ii) $T(x, y) = x + 1$. 6

   (*b*) Find the image of triangle having vertices (0, 0), (1, 2) and (2, 0) under translation by vector $\begin{pmatrix} 1 \\ 1 \end{pmatrix}$. 6

3. (*a*) Reduce the matrix

$$A = \begin{pmatrix} 5 & 3 & 14 & 4 \\ 5 & 4 & 16 & 5 \\ 1 & -1 & 2 & 0 \end{pmatrix}$$

   to triangular form by elementary row operations and hence determine its rank. 6

   (*b*) Find eigen values and eigen vector corresponding to one of them for the matrix

$$A = \begin{pmatrix} 2 & 2 & 1 \\ 1 & 3 & 1 \\ 1 & 2 & 2 \end{pmatrix}$$ 6

4. (*a*) Solve the system of equations:

$$2x + 3y + z = 9$$
$$x + 2y + 3z = 6$$
$$3x + y + 2z = 8$$ 6

(*b*) For what value of λ, does the following system of equations have a solution:

$$x+y+z=1$$
$$x+2y+4z=\lambda$$
$$x+4y+10z=\lambda^2.$$ 6

Section II

5. (*a*) Discuss the convergence of sequence:

$$\left\langle \frac{\sin n}{n} \right\rangle.$$ 6

(*b*) Find the *n*th derivative of:

$$y=\frac{x}{2x^2+3x+1}.$$ 6

(*c*) If

$$y=\sin\left(m\sin^{-1}x\right),$$

prove that

$$(1-x^2)y_{n+2}-(2n+1)x\,y_{n+1}-(n^2-m^2)y_n=0.$$ 6

6. (*a*) Sketch the graph of function

$$f(x)=\sin 2x,\ \ x\in \mathbf{R}.$$ 6

(*b*) Find the Maclaurin series expansion of $y=x\cos x$, assuming that $\lim_{n\to\infty} R_n(x)=0$. 6

(*c*) In a reserve forest, there is capacity to maintain 500 elephants. Initially there were 20 elephants. Within 2 years time, the number rose to 30. Assuming logistic growth model, find out how much time it will take for their population to grow to 200. 6

7. (*a*) Draw the level curves of height $k=0,1,2$ for the surface $z=\sqrt{4-x^2-y^2}$. 6

(*b*) If

$$u=\frac{1}{\sqrt{x^2+y^2+z^2}},$$

show that

$$\frac{\partial^2 u}{\partial x^2}+\frac{\partial^2 u}{\partial y^2}+\frac{\partial^2 u}{\partial z^2}=0.$$ 6

(*c*) Verify that the function $\omega=e^{-c^2k^2t}\left(a\cos kx+b\sin kx\right)$ is a solution of diffusion equation. 6

8. (*a*) For what value of *x*, can we approximate e^x by $1+x+\frac{x^2}{2}$, correct to four decimal places? 6

(*b*) Show that the sequence $<a_n>$ defined by

$$a_1 = 1 \quad a_{n+1} = \sqrt{2+a_n} \quad n \geq 1$$

converges. Also find its limit. 6

(*c*) Show that the sequence $\left\langle \frac{\underline{|n}}{n^n} \right\rangle$ converges. 6

Section III

9. (*a*) Prove that for any two complex numbers z_1 and z_2

$$\left| |z_1| - |z_2| \right| \leq |z_1 - z_2|.$$ 4

(*b*) Form an equation of the lowest degree with rational coefficients having $2+\sqrt{3}$ and $\sqrt{5}-2$ as two of its roots. 3½

10. (*a*) Find the centre and radius of the circle whose equation is

$$|z-i| = 3|z+i|.$$ 3½

(*b*) Let z_1, z_2, z_3 be the affixes of the points, P, Q, R respectively. If $|z_1| = |z_2| = |z_3|$ and $z_1 + z_2 + z_3 = 0$, prove that the triangle PQR is equilateral. 4

11. (*a*) Show that

$$(1+\cos\theta + i\sin\theta)^n + (1+\cos\theta - i\sin\theta)^n = 2^{n+1}\cos^n\frac{\theta}{2}\cos n\frac{\theta}{2}.$$ 3½

(*b*) Solve the equation:

$$z^4 + z^3 + z^2 + z + 1 = 0.$$ 4

12. (*a*) Using elementary row operations find the inverse of the matrix:

$$A = \begin{pmatrix} 1 & -1 & 1 \\ 4 & 1 & 0 \\ 8 & 1 & 1 \end{pmatrix}.$$ 4

(*b*) Express the vector $\begin{pmatrix} 8 \\ 7 \end{pmatrix}$ as a linear combination of vectors $\begin{pmatrix} 1 \\ 2 \end{pmatrix}$ and $\begin{pmatrix} 2 \\ 1 \end{pmatrix}$. 3½

Solution to Examination Paper
MAPT – 101
B.Sc. (Physical Science) 2nd Sem.
2011–12
Mathematics – I (Calculus and Matrices)

Question No.	Solution/Reference to Solution
	Section I
1 (a)	Refer Example 2.28, page 63
1 (b)	Refer Example 2.34, page 67 Example 2.40, page 71
2 (a) (i)	Refer Q. 4(d), page 83
2 (a) (ii)	$T(0, 0) = 0 + 1 \neq T(0)$ $\therefore$ T is not linear.
2 (b)	Refer discussion for Fig. 3.17, page 93
3 (a)	Refer Example 1.28, page 27
3 (b)	Refer Q. 4(a), Exercise 4.1, page 128
4 (a)	Refer Q. 2, Exercise 1.5, page 38
4 (b)	Refer Q. 17, Exercise 1.5, page 39
	Section II
5 (a)	Refer Example 5.17(viii), page 170
5 (b)	Refer Example 8.8, page 248
5 (c)	Refer Example 8.16, page 255
6 (a)	Refer Fig. 6.37, page 211
6 (b)	Similar to Example 9.21, page 282
6 (c)	Refer Example 7.3, page 239

Question No.	Solution/Reference to Solution
7 (a)	Refer Example 10.5(iv), page 297
7 (b)	Refer Q. 12, Exercise 10.2, page 308
7 (c)	Refer Q. 19, Exercise 10.3, page 313
8 (a)	Refer Q. 2, Exercise 9.2, page 273
8 (b)	Replace $2a_n$ by $2 + a_n$ in Example 5.21, page 176
8 (c)	Use the result: If $\lim_{n\to\infty}\left(\frac{a_{n+1}}{a_n}\right) = l,\ l < 1$, then $\lim_{n\to\infty} a_n = 0$. Here $a_n = \frac{\lfloor n}{n^n}$, $\frac{a_{n+1}}{a_n} = \frac{n^n}{(n+1)^n} = \frac{1}{\left(1+\frac{1}{n}\right)^n}$ $\lim_{n\to\infty}\left(\frac{a_{n+1}}{a_n}\right) = 1/e < 1$. Hence $\lim_{n\to\infty} a_n = 0$.
	Section III
9 (a)	Refer Example 11.8, page 327
9 (b)	Refer Q. 3, Exercise 11.4, page 351
10 (a)	Refer Example 11.23, page 344
10 (b)	Refer Example 11.10, page 328
11 (a)	Refer Q. 2, Exercise 12.1, page 360
11 (b)	Refer Example 12.17, page 370
12 (a)	Refer Q. 3(f), Exercise 1.3, page 24
12 (b)	Refer Example 2.10, page 53

INDEX

3-vector 44

additive inverse 7
affine transformation 93–94
 between two vectors 43
Archimedian property 168
arrangement of leaves 155
associative 6, 8–9
augmented matrix 31–32

basis 61–62, 64, 99, 110
 theorem 65
bounded 161, 174
 above 161, 174
 below 161, 174
branching habit of trees 155
breeding habit of rabbits 152, 157

carrying capacity 232–233
case 322
catenary 216
Cauchy's form 266, 282
Cayley-Hamilton theorem 129
cell-division 222
change of origin and scale 03
characteristic equation 119
 polynomial 119
 roots 119
coefficient matrix 30
columns 3
common logarithm 206–07
commutative 6, 8, 106
components 42, 44, 48
composition 105–06, 109, 115
compression 105, 194
constant growth 159
contour lines 296
contraction 84–85, 109, 115
convergent sequences 165, 168, 170
culture of bacteria 221

De Moivre's Theorem 248
decay constant 225, 232
decibel scale 207
degree of the differential equation 219
determinant 9, 10
diagonal elements 5, 139
 matrix 5, 109, 136, 138
diagonalizable 136–38
 problem 136–37
difference equation 158
differential equation 219
diffusion constant 311
dilation 84, 85, 109, 115
dimension 61, 65, 144
directed line segments 41
divergent 168–69
divergent sequences 168
domain 77, 186, 287
dot product of two vectors 43
drug 225, 226

earthquake 206
eigenspace 129, 130, 144
eigenvalue 115, 137–38, 144
eigenvector 115–16, 120, 122, 130, 136, 138
 problem 136
elementary column operation 17
 matrix 17–18
 row operation 13, 17
elements 3
ellipsoid 290–91, 295
elliptic cone 292
 paraboloid 291, 298
equality of two matrices 4
equivalent 13
 vectors 41
error estimation 265
exponential decay model 224–26
 function 202–03
 growth model 219–22

Fibonacci sequence 152
finite sequences 149
fission reaction 150
flexible cable 216
free fall 215
function of two variables 287, 289, 296, 306
fundamental theorem of algebra 119

general sine curves 210
geometric effect 107, 109
 growth 159
geometrical significance of an eigenvector 116
graph of a function 184
 of cot x, sec x, cosec x 209
 of f 184, 186, 296
 of the equation y = f(x) 184
 of the equation z = f(x, y) 289
 of the power function 194
growth of bacteria 222

half-life 225
heat equation 312
Hermitian matrix 6
homogeneous system 30
hydrogen potential 207
hyperbolic functions 214
 paraboloid 294
hyperboloid of one sheet 292
 of two sheets 293

identity matrix 5, 18, 138
 transformation 80, 84
infinite sequence 150
 series 274
inhibited growth model 233
initial point 41
 value problem 220–21
interval 207, 212
invariant 96, 131
 subsapces 131
inverse 18–20
 trigonometric functions 212–13
invertible 18, 20, 136

L' Hospital's Rule 177
Lagrange's form 266, 282
Laplace's equation 310
Leibnitz's Theorem 253–54
length of the vector 44
level curves 289, 295–96
linear approximation 261–62
 combination 52, 72
 equations 30
 non-homogeneous system 30
 operators 115
 systems 20
 transformation 77, 81, 91, 97
linearly dependent 56
 independent 56, 136, 138
logarithmic functions 205
 growth 159
 Richter Scale 206
logistic differential equation 233
 growth 159, 232
 model 233, 235

Maclaurin polynomial 261–62
 series 274, 277, 279
magnification 109, 194
magnitude of the vector 42
mathematical modelling 220
matrix 3
 addition 6
 operations, 6
 representation 98, 109
 transformation 91, 93
 transpose, 4–5
 types, 5–6
Mechanic's rule 151
minor 10
monotonic 161

natural basis 62
 domain 287
 logarithm 206
Newton's Law of Cooling 229
non-linear transformation 80, 81, 84, 92
non-singular 10, 15, 17, 20, 25
 matrix 5

observations 232
 of a differential equation 219
 of a matrix 3
orthogonal 43

parabolic functions 200
parallelogram of forces 46
partial derivatives 304, 309–10
 sum 274
period 207, 210
phyllotaxis 155
polynomial function 199–200
 growth 219–20, 232
power function 194–95, 199
powers of the matrix 9
principal diagonal 5, 14
probabilistic growth 159
projection 79, 84, 124

proportional growth 159
proportionality constant 201
Pythagoras theorem 42

quadratic approximation 261–62

radiant energy 212
radioactive disintegration 225
rank 25–27
rate constant of the equation 203
  of change 304–05
real vector space 49
rectangular coordinate system 41
recursion formula 151–52
reflection 86, 87, 105–06, 109, 115, 188, 191, 202
  about the y-axis 86
  in origin 87, 115
  with respect to the x-axis 86, 105
  with respect to yz-plane 87
resultant force 46
rotation 87–88
rows 3
r-rowed minors 25

Sandwich theorem 172
scalar matrix 5
  multiplication 7, 45–46, 48
second-order chemical reaction 201
sequence 149–51, 158, 161, 168
shear 91
similar matrices 136
simple harmonic motion 212
singular 10, 24
sinusoidal function 212
skew symmetric matrix 5
  Hermitian matrix 6
skydiving 215
slope of the tangent 304
solution of the difference equation 160
span 54, 72
  of S 72
square matrix 5–6, 9–10
squeeze property 173
standard basis 50–51, 98
  matrix 98
stochastic growth 159
strictly decreasing 161
  increasing 161
  monotonic 161
sub-matrices 25
  sequence 161
  spaces 67
  spaces of R2 68
  spaces of R3 68, 69
successive differentiation 241
sum 6, 45, 274, 320–22
symmetric matrix 5, 11

Taylor polynomial 261–62, 265–66, 275, 278
  series 275, 278
  theorem 266, 278, 282
telegraph equations 312
terminal point 41–42
Tower of Brahma 156–57
  Hanoi 156–57
transformation 77, 80, 82, 84, 98, 104
  linear, 77, 82, 97, 100
translation 92–93, 188–90
  matrix 114
transpose of a matrix 4
triangular matrix 6, 14–5
trigonometric functions 207, 212
trivial solution 31–32

unit matrix 5, 15
  vector 50

vector 41
  in a plane 41–42
  in space 44
  space over R 49

wave equation 309–10

zero matrix 5, 25
  transformation 81
  vector 42, 65, 67, 81